Modern Power System

IEEE Press
445 Hoes Lane
Piscataway, NJ 08854

Modern Power System

Arindam Ghosh
Curtin University
Australia

IEEE Press Series on Power and Energy Systems
Ganesh Kumar Venayagamoorthy, Series Editor

IEEE Press

WILEY

Published by John Wiley & Sons, Inc., Hoboken, New Jersey.

Library of Congress Cataloging-in-Publication Data

Names: Ghosh, Arindam author
Title: Modern power system / Arindam Ghosh.
Description: Hoboken, New Jersey: Wiley-IEEE Press [2026] | Series: IEEE Press series on power and energy systems | Includes index.
Identifiers: LCCN 2025031906 (print) | LCCN 2025031907 (ebook) | ISBN 9781394289912 cloth | ISBN 9781394289936 adobe pdf | ISBN 9781394289929 epub
Subjects: LCSH: Electric power systems
Classification: LCC TK1001.G47 2026 (print) |LCC TK1001 (ebook)
LC record available at https://lccn.loc.gov/2025031906
LC ebook record available at https://lccn.loc.gov/2025031907

Cover Design: Wiley
Cover Image: © xuanhuongho/Shutterstock

Set in 9.5/12.5pt STIXTwoText by Straive, Pondicherry, India

Dedicated to
my wife, Supriya, and son, Aviroop, and
to all my students

Contents

About the Author

Arindam Ghosh has held academic positions at the Indian Institute of Technology Kanpur for over two decades, at the Queensland University of Technology in Brisbane, and is currently a Research Professor at Curtin University in Perth, Australia. His work has focused on power systems, power electronics and controllers that enhance the reliability and sustainability of electricity networks.

He has been recognized internationally as a Fulbright Scholar at the University of Illinois at Urbana–Champaign, elected a Fellow of both the Indian National Academy of Engineering and the IEEE, and awarded the IEEE PES Nari Hingorani Custom Power Award in 2019. He earned his Ph.D. in Electrical Engineering from the University of Calgary, Canada, in 1983.

Passionate about bridging theory and practice, he continues to inspire students, researchers, and professionals working to shape the future of energy systems.

Preface

The electric power industry is a cornerstone of global infrastructure, vital for modern society and economic development. According to the *International Energy Agency* report on world energy employment in 2023, the power sector employed over 68 million people worldwide. Of these, over 36 million people were employed in the clean energy sector, while over 32 million people were employed in the fossil fuel-based power industry. According to the *US Bureau of Labor Statistics* report of 2023, 17,870 electrical engineers were employed in the power sector. Amongst all the technical societies of the *Institute of Electrical and Electronic Engineers (IEEE)*, the *Power and Energy Society (PES)* is the second largest, having around 42,000 members worldwide.

Given its vast impact, the power industry requires a multidisciplinary approach for its secured operation and continues to evolve, a journey that I have followed with fascination. My early studies of foundational texts like W. D. Stevenson's book [1] have deeply influenced the structure and focus of this book, blending traditional principles with modern advancements. Most of the topics covered in Stevenson's book are still valuable to gain knowledge in the area. However, the power sector has seen a sea of changes since the time the fourth edition of the book appeared in 1982. These days, power electronic technology plays a crucial role in both power transmission and distribution systems. Thyristor-based high voltage DC (HVDC) transmission systems started appearing in the 1970s. Subsequently, voltage source converter (VSC)-based HVDC systems were adopted on a large scale at the turn of this century. Currently, VSC-HVDC systems are used for offshore windfarms. Moreover, point-to-point HVDC systems have given way to multiterminal HVDC systems for offshore wind collection systems.

Also, thyristor-based static var compensators (SVCs) also started appearing in large scales during the 1970s to enhance voltage stability in long transmission lines, as well as, for power oscillation damping. There were hundreds of SVCs installed throughout the world. Fixed series compensation of transmission lines to enhance power flow was initially hindered by incidents at Mojave power station, where resonance issues caused turbine damage in the early 1970s. These were caused due to the resonance between the series capacitors and line reactors at frequencies that are below the synchronous frequency. However, the initial hesitation was overcome using thyristor-controlled series compensators (TCSCs), which can effectively change the series reactance to avoid the subsynchronous oscillations reaching the rotor shafts. Moreover, other thyristor-based devices have become common like voltage regulator, phase angle regulator, etc.

With the advancement in power electronic technology, voltage source converter-based flexible AC transmission (FACTS) devices have found their applications in both voltage regulation and power flow control in long transmission systems. Shunt compensation was achieved using static compensators (STATCOMs), which started replacing the SVCs. On the other hand, static synchronous series compensators were placed in series with the lines to replace TCSCs. Both shunt and series compensations can be achieved simultaneously using a unified power flow compensator.

Due to the rising concerns of climate change and the resultant global temperature rise, more and more renewable energy generators are getting integrated into both power transmission and distribution systems. This has caused disruptions in the traditional operations of power systems. Most of the renewable energy generators are connected to power systems through power converters, which cannot provide inertia to maintain stability margins required in bulk power transmission systems. These systems require smarter converter controls and storage devices. In distribution systems, for instance, rooftop photovoltaics introduce challenges like voltage imbalances, voltage rises, and reverse power flow. Furthermore, there is a concern that renewable generators are intermittent and thus they cannot supply the required baseload.

To modernize the power system, the concept of the smart grid has been introduced, through which the power system is integrated with information and communication technology (ICT) to facilitate a smooth two-way power flow and to provide near instantaneous balance between generation and consumption. Power transmission systems can have modern energy management systems integrating phasor measurement units, which can be used in load control centers for more accurate state estimation and power dispatch. Distribution systems will be equipped with smart meters, through which the load demand can be managed. Parts of distribution systems can form virtual power plants or can have several microgrids. Substations can be modernized using tailored computer programs that can communicate between different protective relays without the complicated layout of cables. Since the smart grid relies heavily on ICT, measures must be taken to ensure that the communication and computation devices are cybersecure.

Against the backdrop of all the changes that have occurred in the power systems over the last three decades, this book aims to combine the traditional power systems with the newer technologies that are increasingly appearing in power systems. The materials covered in the book have been taught over several years over different courses at four different universities. The book can be used for a basic course on power systems on the undergraduate level, as well as, for a higher-level undergraduate course or a first-level graduate course on power engineering. For example, Chapters 2–7 (excluding Sections 3.4, 6.6, 7.2, and 7.3) can be used for a first-level course, and the rest of the book can be used for a second-level course.

The book is organized into 12 chapters. Chapter 1 introduces the book. Most of us take the use of electricity for granted – for comfort and household appliances, for entertainment, for knowledge, for medical treatment, or for transportation. However, the history of how we came to this stage is fascinating. In Section 1.1, a brief history of electricity is presented. In the subsequent sections, the development stages leading to the modern power systems are discussed, including interconnections of electric grids, deregulations, blackouts, and smart grid.

Chapter 2 discusses the main components of power systems. It begins with discussions on transmission system parameters. It is easy to comprehend that transmission lines will have resistance. However, how they are represented by line inductance and capacitance is derived using the laws of magnetics. Following these, simplified models of synchronous generators and transformers are presented. A power system may contain different power equipment with different voltage and power levels connected together through various step-up or step-down transformers. The presence of different voltage levels makes power system calculations extremely difficult. To simplify this, a power system is represented in its per unit form where all quantities are normalized to a common base. The final section of this chapter discusses different ways of modeling a power transmission system depending on its length and how they can be simplified for power system calculations.

Chapter 3 discusses load flow techniques. A power system is a network of transmission lines, loads and generators. Even though such a system can be visualized as an RLC circuit, the network is so complicated that the node voltage and loop current analyses are impossible to perform. For a set of given loads and generations, the complex bus voltages and power flow through different lines are determined using load flow (or power flow) studies. The first step in this process is to combine all the elements of the power system in a bus admittance matrix. Then, step-by-step iterative procedures are executed for the accurate determination of the bus voltages. Three different load flow procedures – Gauss–Siedel, Newton–Raphson, and fast decouples – are presented. Furthermore, the DC load flow is also presented, through which rough estimates of bus voltage magnitudes and angle can be computed using a simplified noniterative procedure. However, power flow calculations may not be accurate due to erroneous measurements. Power system state estimation, on the other hand, is a mixture of load flow and statistical estimation theory that can provide a much more accurate snapshot of the power system status. This is discussed in Section 3.4.

A power system may contain several generators. How these generators must be scheduled to cater to load demands economically is discussed in Chapter 4. Economic operations depend on the most economically efficient generators catering for a higher portion of load demand. Furthermore, the method of committing a particular number of units to serve the load demand is also discussed in the chapter. The basic concepts of automatic generation control and load frequency control are also discussed in the chapter.

Power system fault studies are presented in Chapter 5. Power system faults can be balanced or unbalanced. For balanced faults, it is assumed that all the three phases have been short-circuited to the earth at the same location. These faults can be analyzed using the single-line diagram assuming the balanced operation of the faulted circuit. For unbalanced faults, however, the circuit becomes unbalanced, which is resolved into three balanced components that are called symmetrical components. Delta- and wye-connected loads and transformers have different characteristics for different symmetrical components. These are analyzed separately to form three balanced sequence networks under unbalanced conditions. Unbalanced faults can be single-line-to-ground, double-line-to-ground, or line-to-line faults. How the fault current can be calculated for these three different types of faults using the sequence networks is presented at the end of this chapter.

The purpose of the fault studies is to find the levels of fault current that must be interrupted by the protective devices. Chapter 6 discusses power system protection. Specifically, the protective elements like circuit breakers and fuses, instrument transformers that are used for reducing the fault currents and/or voltages to be used in protective relays, and different types of protective relays are discussed. It is possible that a particular protective relay or device may malfunction. Therefore, a backup must be provided for each protective relay so that they not only protect a certain segment of the network, but also provide a backup for its neighboring segments. The traditional way of protection was designed for power distribution systems assuming unidirectional flow of power – from substations down to the line feeding loads. However, with the advent of renewable energy generators that are connected throughout distribution networks, the unidirectional feature will be lost. Moreover, power converter-interfaced renewable generators are not capable of feeding the required amount of fault current to trip the relays. Therefore, alternative protection strategies must be developed for such situations. These are also discussed in the chapter.

Chapter 7 discusses power system stability. First, the transient stability problem for a single-machine, infinite bus system is discussed. Two important aspects, namely, critical clearing angle and critical clearing time, are introduced. The transient stability concept is then extended to finding the transient behavior in a multimachine power system using bus admittance matrices. Transient stability deals with the response of the power system with respect to a fault or a large disturbance in a power network. However, the so-called dynamic stability occurs in an interconnected power system due to excitation control. These pertain to low-frequency oscillations that can grow unless proper actions are taken. A power system stabilizer (PSS) is used for damping these oscillations. The design of PSS using small signal model is also discussed in the chapter.

Power system bus voltage can be progressively depressed due to the lack of reactive power resulting in voltage stability. Chapter 8 begins with a brief discussion on voltage stability. Reactive power can be compensated to avoid this problem. Ideal reactive power compensation is discussed in this chapter. These compensators can be connected in parallel (shunt) to a transmission line or in series with a line. In the ideal mode, the compensators can be represented by ideal current or voltage sources. Interestingly, these ideal compensators can not only improve the voltage profile, they also can improve the power transfer and can damp power systems oscillations, as have been explained in the chapter.

Chapter 9 is an extension of Chapter 8, where the physical power electric devices that can be used for shunt and series compensators are presented under the FACTS umbrella. Additionally, other FACTS devices that can enhance power system stability or increase power flow are also discussed. Generally, FACTS devices are of two types – either thyristor-based or power converter-based. The fundamental characteristics of the thyristor-based devices are discussed. Some of the high-power converter topologies are discussed that can be used for the realization of STATCOMs and SSSC. Also, subsynchronous oscillations that can result in a series-compensated power system are also discussed.

HVDC transmission is discussed in Chapter 10. In the first portion of the chapter, HVDC systems using line-commutated converters are discussed. This is followed by an introduction to VSC-based HVDC systems. Two different control structures of VSC-HVDC systems are discussed. Following this, multiterminal HVDC systems, their features, and controls are discussed. A brief discussion on DC system protection is presented at the end of the chapter.

A review of different renewable energy technologies is presented in Chapter 11. The topics covered are hydropower, solar power, wind power, hydrogen, and nuclear fusion. Different hydropower technologies such as pumped hydro, tidal, and wave energy are discussed briefly. Both solar photovoltaic systems and concentrated solar power are introduced, along with the maximum power point tracking that is used for getting the maximum benefit from solar irradiance. Regarding wind power, concepts such as tip speed ratio and pitch angle control, along with different wind power collector systems are discussed. Hydrogen production, storage, transmission, and utilization are discussed in Section 11.4. Some of the issues of integration of renewable sources in power transmission and distribution are discussed in Sections 11.6 and 11.7. Particularly, topics such as fault ride through, voltage rise, and line loss in distribution feeders and reverse power flow are also covered.

Fundamental concepts in smart grid are presented in Chapter 12. Diverse topics such as phasor measurement units, smart meters, demand response, cybersecurity, electric vehicles, smart grid communications, and standards are discussed, followed by the features of smart distribution grids. A smart grid is a very vast field of study, most of the topics are still in the research and development domains. This chapter presents a summary of different aspects of smart grid. However, several components of smart grid are discussed in the previous chapters of the book in terms of power systems monitoring, operation, and control.

November 2024 Arindam Ghosh

Reference

1 Stevenson, W.D. Jr. *Elements of Power Systems Analysis*. New York: McGraw-Hill, 1982.

Acknowledgments

Numerous students who attended my lectures at four different universities have posed various queries related to the subjects presented in the book. These questions have helped me to develop the general idea of the book. I thank all of them, especially all my students at the Indian Institute of Technology Kanpur whose queries were challenging and very thought-provoking.

I thank my wife, Supriya, for carefully proofreading the entire manuscript and my son, Aviroop, for making critical comments about several technical elements in the book. I am particularly grateful to Aviroop for his inputs on the smart grid communication section. I also thank both Supriya and Aviroop for their moral support and continuous encouragement during the preparation of the manuscript.

Finally, I thank the Wiley editorial team for their professional and helpful approach to the preparation of the manuscript.

About the Companion Website

This book is accompanied by a companion website:

www.wiley.com/go/ghoshpowersystem

The website includes Solution Manuals.

1

Introduction

The electric energy industry ranks among the largest global industries. As per [1], the electric energy generated in 2022 was close to 30,000 TWh. The main sources of electricity generation are coal, gas, oil, hydro, nuclear, wind, and solar. Table 1.1 shows the worldwide energy mix for the years 2014 and 2022 [1]. Fossil fuel-based greenhouse gas emitting generation sources have reduced their share from 67% in 2014 to 61.26% in 2022. The share of low-carbon-emitting sources such as hydro and nuclear has also declined from 26.95% to 24.14%. However, to offset these, the share of wind and solar power generation has increased from 3.88% to 11.89%. Wind and solar power generation are projected to experience rapid growth over the next 10–15 years. The percentage energy mix is shown in Figure 1.1.

A better picture of the generation mix is given in Table 1.2, where the total energy generation is shown for the years 2014 and 2022. It can be seen from this table that energy generation has increased by about 4,898 TWh between 2014 and 2022, where the energy output from all the sources has increased, except for oil, which has decreased. Another important statistic that has been reported in [1] is the share of electricity from low-carbon sources, including nuclear, hydro, wind, solar, biomass, geothermal, wave, and tidal. The total share of these sources has increased from 32.99% in 2014 to 38.73% in 2022.

1.1 A Brief History of Electricity

The use of electric energy is ubiquitous these days with applications in every facet of human endeavor. However, to reach this stage, it has taken several centuries and spanned many countries. The history of electricity is really interesting with many geniuses and talented individuals contributing to its development. In this section, a brief history of the modern discovery and origin of electricity is presented.

Modern Power System, First Edition. Arindam Ghosh.

Published 2026 by John Wiley & Sons, Inc.
Companion website: www.wiley.com/go/ghoshpowersystem

Table 1.1 Worldwide energy mix in percentages.

Energy sources	Percentage share	
	2014	2022
Coal	40.61	35.63
Gas	21.61	22.48
Hydropower	16.43	14.96
Nuclear	10.52	9.18
Wind	3.04	7.32
Solar	0.84	4.57
Oil	4.78	3.15
Bioenergy	1.84	2.36
Other renewables	0.33	0.34

Source: Adapter from Ritchie and Rosado [1].

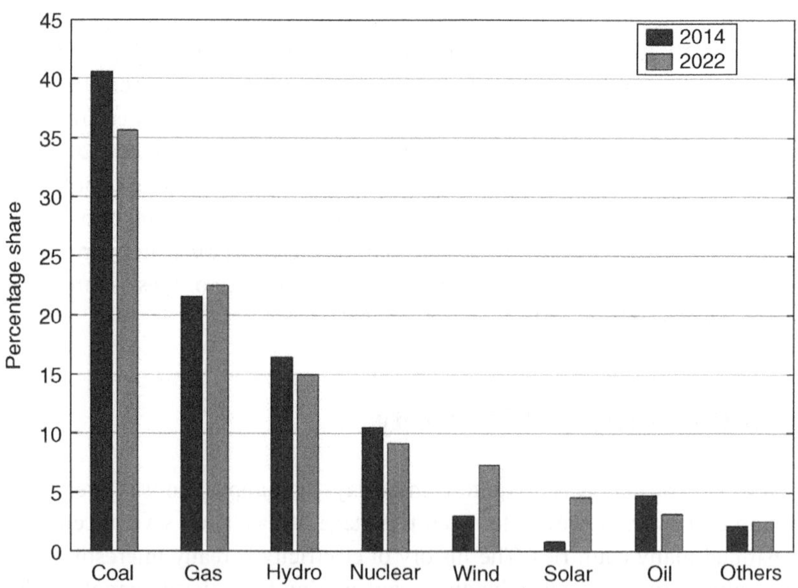

Figure 1.1 Comparison of energy mix percentage between the years 2014 and 2022. *Source:* Adapter from Ritchie and Rosado [1].

Table 1.2 Worldwide energy mix in TWh.

Energy sources	2014 (TWh)	2022 (TWh)
Coal	9,226.71	10,212.22
Gas	5,402.71	6,443.60
Hydropower	3,872.76	4,288.59
Nuclear	2,504.73	2,632.06
Wind	828.60	2,098.46
Solar	255.79	1,310.02
Oil	1,120.57	904.15
Bioenergy	470.66	675.11
Other renewables	80.30	96.80
Total	23,762.83	28,660.98

Source: Adapter from Ritchie and Rosado [1].

1.1.1 The Dawn of Electricity

In the beginning, the early pioneers carried out experiments to satisfy their intellectual curiosity. William Gilbert was a Doctor of Medicine who was the personal physician of Queen Elizabeth I of England. Despite being a physician, he is known for his work on the Earth's magnetic field because he speculated that the magnetic north pole attracts compasses. He was the first one to propose that the Earth is a giant magnet. However, about 1,000 years before Gilbert, ancient Indian philosopher Varhamihira, described the effects of the Earth's magnetic field, even though he could not predict the Earth's magnetic field like Gilbert. Interestingly, Gilbert studied static electricity using amber, which is called *elektron* in Greek. Thus, Gilbert called its effects *electric force*. The term *electricity* was coined in 1646 by English polymath Sir Thomas Browne using Gilbert's work.

Benjamin Franklin was an American statesman and one of its founding fathers. He is known for his role in several administrative capacities and with the Declaration of American Independence (his picture can be seen on $100 bills in the USA). He was a multitalented person who has been credited for several inventions such as the lightning rod, bifocal lens, and flexible urinary catheter. In 1752, he flew a kite with a metal key attached; possibly even a metallic rod to see sparks being generated from clouds. To temper scientific curiosity, he risked his life by flying a kite in cloudy conditions. Franklin is credited with the discovery of static electricity. Other notable individuals such as Charles-Augustin de Coulomb discovered electrostatic force attractions and repulsion (Coulomb's law) in 1785.

Meanwhile, Italian physician and scientist Luigi Galvani studied the effects of electricity on living organisms. He discovered that the muscles of dead frogs twitch when electricity is passed through them, therefore inspiring many high school experiments today. Galvanic isolation (which refers to the isolation between two electrical systems) and galvanometer (for measuring electric current) are named after him. Alessandro Volta invented a voltaic pile by putting alternate layers of zinc and copper with isolation between them. This inspired the development of electric batteries.

Hans Christian Ørsted and André-Marie Ampère made significant strides in recognizing that there is a relationship between electricity and magnetism. In 1820, Ørsted discovered that a compass needle deflects in the near vicinity of an electric current. In his honor, the unit of the magnetic field is named Oersted. Building on this, Ampère demonstrated that parallel wires carrying electric currents attract or repel each other depending on the direction of current flow. Today, the unit of electric current, Ampere or Amp bears his name, recognizing his contributions to electromagnetism.

An English scientist Michael Faraday started developing a keen interest in science while working as an apprentice in a bookshop and bookbinder, where he had access to many scientific books. Later he was able to gather work as an assistant of English chemist Humphry Davy. In 1821, Faraday started developing an interest in working on electromagnetism. He realized that an electric current could be produced by passing a magnet through a copper wire. This is the basic principle based on which all motors and generators are constructed today. His series of experiments resulted in the discovery of electromagnetic induction in 1931. Faraday subsequently used his induction principle to build a machine to generate voltage. Around the same time, American engineer Joseph Henry also worked independently on the induction principle and applied his work to electromagnets and telegraphs. However, Faraday was the first to publish his results. It is only befitting that the units of capacitance and inductance are named after Faraday and Henry, respectively.

In 1827, German physicist and mathematician Georg Ohm found a direct proportionality between voltage applied to and current through a conductor. This has resulted in the well-known Ohm's law. The unit of resistance, as well as impedance, is named after him. Scottish physicist James Clark Maxwell explained electromagnetism in his paper "On physical lines of force" in 1861. The timeline for the early development of electricity is listed in Table 1.3.

1.1.2 Development of Electrical Power Plant

Another pioneer introduced earlier, Sir Humphry Davy, is famous for the invention of Davy's safety lamp in 1809. This was used for underground coal mines in the early 19th century to prevent flammable gases from igniting. In Davy's lamp,

Table 1.3 Timeline for the early development of electricity.

Name	Year	Contributions	Acknowledgment
William Gilbert	1600	He conducted experiments to conclude that the Earth is magnetic, and the center of the Earth is iron. He also studied the production of static electricity by rubbing ambers.	Recognized for proposing that the Earth is a giant magnetic field.
Sir Thomas Browne	1646	From the Greek word "electron" for amber, Brown coined the word electricity.	Recognized for coining the term electricity
Benjamin Franklin	1752	Franklin discovered static electricity by observing sparks generated by clouds.	His face appears in $100 bill of the USA.
C. A. de Coulomb	1785	Coulomb's law describes the electrostatic force for attractions and repulsions.	The unit of electric charge is named after him.
Luigi Galvani	1791	Galvani demonstrated that electricity is a medium by which nerve cells pass signals to muscles.	Galvanic isolation and galvanic meters are named after him.
Alessandro Volta	1800	Volta invented the Voltaic pile, which is the precursor of modern batteries.	The potential difference between two electric conductors is called voltage and is measured in volts in his honor.
Hans Christian Ørsted and André-Marie Ampère	1819–1820	They recognized there is a relation between electricity and magnetism.	Magnetic field is measured in Orsted.
Michael Faraday	1821–1831	Faraday started working on electromagnetic induction.	The unit of capacitance is named after him.
Joseph Henrey	1832	Electromagnetic Induction.	The unit of inductance is named after him.
Georg Ohm	1827	Ohm mathematically electrical circuit, which is the famous Ohm's law.	The unit of impedance is named after him.
Joseph Henry	1832	Electromagnetic induction.	The unit of inductance is named after him.
James Clerk Maxwell	1861	Maxwell explained electromagnetism in his paper "On physical lines of force."	Best known for Maxwell's equation of electromagnetism.

an arc is produced when pieces of charcoal are connected to two wires of a battery. However, it took several decades to produce commercial arc lighting. The use of batteries for arc lighting was not commercially viable because the current requirement to generate arc would drain the batteries quickly. Moreover, the intense heat of the arc will also burn the charcoal. The generators started to become widely available following the development of dynamos. In 1870, Russian inventor Paul Jablochkoff devised a version of the arc lamp in which he used two carbon rods to increase the life span of the lamps [2]. Arc lamps were used in lighthouses and streets and rarely indoors due to the high intensity of these lights.

From the work of Davy and others, it was well-known that a current-carrying conductor could be heated to the point of incandescence. However, incandescent materials burned very quickly in the atmosphere to be of no use for lighting. In 1879, American inventor Thomas Alva Edison devised the first incandescent lamp using carbonized cotton thread in a vacuum space inside a glass bulb. The first bulb glowed for 44 hours before it burned out. Edison improved the design of the lamp so that it could last longer.

The first demonstration of electric motors was done by Michael Faraday. He not only proposed electromagnetic induction but also invented the Faraday Disk in which a disk is placed between two magnetic poles. An electric current is produced when the disk rotates. The current is drawn through two sliding contacts (armatures) that are placed at the center and edge of the disk. Therefore, Faraday can be called the first pioneer of DC machines. Many other pioneers further improved the design of DC generators such as Werner von Siemens and Thomas Edison, which led to the development of power plants.

Edison set up the first commercial power station at 255 Pearl Street in New York City on September 4, 1882 (see Figure 1.2). The station had a capacity of 110 kW, which was powered by 14 boilers and 4 direct current generators. Initially, the station supplied 82 customers in the lower Manhattan area with enough power to light 400 lamps. There were several advantages of a DC system at that time – it could be used for incandescent lamps and DC motors, which were the principal loads in the 1880s. DC generators can directly be used for battery charging and can be easily paralleled. These factors, along with the fact that AC machines were not available at that time, made DC systems the prime power supply candidates at the beginning of electricity supply.

Edison's DC system needed heavy distribution conductors. For the convenience of lamp manufacturers, the operating voltage was chosen as 110 V. To reduce the cost of copper conductors, a three-wire distribution system was used in which 0 V neutral was placed between +110 V and -110 V. However, even with this innovation, the voltage drop due to conductor resistance was so high that the generating plant could only supply customers that were located within 1–2 km from the plant.

Long-distance power transmission became a reality in 1895 when George Westinghouse opened the first major power plant in Niagara Falls using alternating

Figure 1.2 Pearl Street power station and Edison. Generated with AI using ChatGPT (version GPT-4).

current. This was able to transmit power to a distance of 200 miles (360 km) to the city of Buffalo, New York. It was the partnership between Westinghouse and Nikola Tesla (see Figure 1.3) that made this possible. Tesla was a Serbian-born, naturalized American citizen, whose genius made the alternating current feasible.

Figure 1.3 Nikola Tesla. Generated with AI using ChatGPT (version GPT-4).

He invented the induction motor that would run on AC power. It is still the workhorse of many industries. He also conceptualized the polyphase system, which was the game changer that made the AC system feasible. During his lifetime, Tesla was awarded 101 US patents. He also invented Tesla Coil, which can transmit and receive energy wirelessly when the coils are tuned at the same frequency. Even though Guglielmo Marconi is credited with the invention of wireless and received the Nobel Prize in 1909, both Tesla and the Indian physicist Jagadish Chandra Bose have made separate contributions to this technology. There are some very interesting anecdotes about Tesla. He had a deep friendship with Samuel Langhorn Clemens, better known as Mark Twain, the author of the adventures of Tom Swayer and Huckleberry Finn. Tesla passed away on January 7, 1943, in a hotel in New York City. Upon his death, the US government ordered the FBI to seize all his papers under the suspicion that they might contain advanced weapons technology. However, the papers were scrutinized by MIT professor John G. Trump (the uncle of 45th and 47th American President Donald J. Trump). Professor Trump concluded that there were no weapon technologies of interest in Tesla's papers.

The AC versus DC resulted in the famous *War of Currents*, with Edison proposing the continuation of DC, while Tesla and Westinghouse proposed AC (see Figure 1.4). This war resulted in disinformation campaigns, litigation, etc., and,

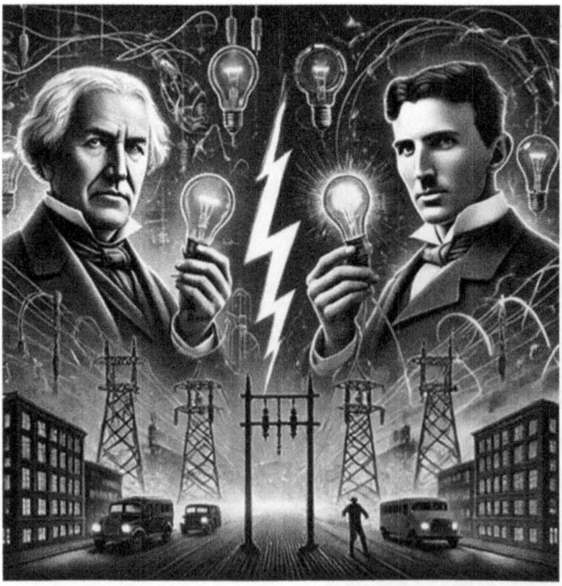

Figure 1.4 War of currents. Generated with AI using ChatGPT (version GPT-4).

in general, was quite contentious. In 1886, William Stanley developed a prototype system for power transformers to boost up or step down voltages. Westinghouse started pilot projects of using transformers in his laboratory in Pittsburg. The transformers were then used in Buffalo. This was the beginning of AC power generation and transmission.

When the dust settled after the War of Currents, AC won, and DC lost. The AC prevailed over DC due to the following factors:

- At that time, power electronics were unheard of. Therefore, DC voltages could not easily be converted to higher or lower voltages.
- Due to lower operating voltages, the DC power distribution had a limited range due to voltage drop.
- Transformers could boost AC voltage for transmission and could step it down for distribution. Therefore, power could be transmitted over long distances at higher voltages, reducing losses.
- AC systems use thinner copper wires, thereby reducing costs.
- The construction of AC generators was simpler.
- The construction of AC motors was also simpler. Moreover, they were more robust and cheaper than the DC motors even though they were not very sophisticated.

However, DC has made a strong comeback from the second half of the 20th century and is now considered one of the preferred modes of transmission in many instances, as will be discussed in Chapter 10.

1.2 Interconnection of Electricity Grids

Electricity supply was mandated by most countries by the public utility act. A public utility is an organization that maintains infrastructures for a public service, such as electricity and gas. Using these infrastructures, it provides a public service at the consumer level, be it residential, commercial, or industrial consumers. Electric utilities are subject to forms of public control and regulations. Public electric utilities usually were, and even now in many places are, statewide government monopolies. The utilities are regulated by public service commissions that fix their prices and services.

In the beginning, the electric utilities were responsible for generation, transmission, and distribution. However, starting in the 1920s, the utilities formed joint operations in the United States. The advantages of these were that the peak load could be covered, and backup power could be provided. As the transmission technologies improved, long-distance transmission became feasible and grids from different regions got interconnected to form a mega grid. Some of the advantages of forming interconnected grids are as follows:

- *Improving Reliability*: Electric power can be shared among the different utilities reliably, thereby reducing the requirement of spinning reserve.
- *Reducing Investment*: By sharing the generating resources, individual utilities do not need to add to their generating capacity.

From 1967, the East and West interconnections in the United States were directly connected together. However, the AC ties did not have high capacity and were subject to oscillations, so their connection proved unreliable. In 1975, the AC ties were disconnected because DC ties were found to work more reliably.

The US power system, including Canada and a portion of Mexico, is divided into three major grids: the Western interconnection, the Eastern interconnection, and the Texas interconnection. All these regions are coordinated by the *North American Electric Reliability Corporation* (NERC). European power grid is a massive synchronous grid that spans about 24 countries and is managed by the European Network of Transmission System Operators for Electricity (ENTSO-E). In addition to the countries in continental Europe, some of the African countries (e.g., Morocco, Algeria, and Tunisia) are also synchronized with this grid. In Australia, all the eastern seaboard states of Queensland, New South Wales, Victoria, and South Australia are connected through AC ties, while Tasmania is connected to Victoria through a subsea 370 km high-voltage DC (HVDC) link called Basslink. Similar interconnections can also be found on the continent of Africa and other countries. In India, the National Grid is a state-owned 50 Hz synchronous grid that is managed by the Power Grid Corporation of India. The state-owned State Grid Corporation of China is the largest utility company in the world.

Usually, AC connections are the more prevalent interconnection types. However, they can observe stability problems, some of which can be solved by power system stabilizers, which will be discussed in Chapter 7. The inclusion of flexible AC transmission system (FACTS) devices can alleviate stability problems. As mentioned earlier, the east–west connection in the NERC system resulted in system oscillations. To counteract these oscillations, the AC ties were disconnected in 1975 and were replaced by HVDC tie lines. The DC tie lines can isolate disturbances of one region from propagating to the other region, and therefore, are preferred in many interconnections.

1.3 Deregulation

The Energy Policy Act (EPACT) was established in 1992. Under the act, the power transmission companies were required to allow smaller electric generation companies to give access to their networks. This policy aimed to foster competition in power generation. The power generation, transmission, and distribution can now be split among various companies so that customers are benefited overall. The EPACT of 2005 was more stringent. It allowed incentives for alternate energy technologies to reduce greenhouse gas emissions. The key aspects of power system deregulation are as follows:

- *Unbundling*: The vertical integration of electricity utilities is broken down into separate entities for generation, transmission, and distribution. This allows for competition in electricity generation and sometimes in retail supply.
- *Competition*: Deregulation aims to introduce competition, especially in the generation and retail segments. Independent power producers can enter the market, and consumers can choose their electricity supplier.
- *Wholesale Markets*: Establishment of wholesale electricity markets where electricity is bought and sold. These markets are often managed by independent system operators (ISOs) or regional transmission organizations (RTOs).
- *Retail Choice*: Consumers, including residential, commercial, and industrial users, have the option to choose their electricity provider, similar to choosing a phone or internet service provider.
- *Regulation of Transmission and Distribution*: While generation and retail are open to competition, the transmission and distribution networks often remain regulated due to their natural monopoly characteristics. Access to these networks is typically provided on a nondiscriminatory basis.
- *Market Efficiency*: Deregulation is intended to improve efficiency by encouraging competition, which can lead to lower prices, innovation, and better customer service.
- *Challenges and Risks*: Deregulation can present challenges such as market manipulation, price volatility, and ensuring reliability of supply. Effective regulatory oversight and market design are crucial to mitigate these risks.

An ISO is an entity responsible for overseeing the operation of the electricity grid of a region. It ensures reliable electricity delivery and facilitates competitive wholesale electricity markets. ISOs are typically independent of electricity generators and transmission owners to prevent conflicts of interest and to ensure impartial management of the grid. They are responsible for the management of the electricity grid to ensure a balance between supply and demand in real time by monitoring grid conditions and coordinating the flow of electricity across transmission lines. Furthermore, they operate competitive wholesale electricity markets where electricity is bought and sold through auctions and manage market transactions to ensure fair pricing and market efficiency. The ISOs are also responsible for maintaining reserve margins, managing grid stability, and coordinating emergencies.

An RTO also performs the same task as an ISO. However, while an ISO typically operates within a single sector or smaller region, an RTO covers a larger geographic area covering multiple regions. ISOs often cater to local needs and market conditions, while RTOs must comply with the Federal Energy Regulatory Commission (FERC) order of 2000, which mandates that they meet specific characteristics and functions, such as ensuring open access to transmission and maintaining regional grid reliability.

In general, the potential benefits and drawbacks of deregulated markets are as follows:

- *Lower Prices*: Increased competition can drive down prices for consumers.
- *Innovation*: Competitive markets can foster innovation in technologies and services.
- *Customer Choice*: Consumers can choose their provider based on price, service quality, or other preferences.
- *Market Power*: Large players can dominate the market, reducing the intended benefits of competition.
- *Reliability Issues*: Deregulation can sometimes lead to underinvestment in infrastructure, affecting the reliability of supply.
- *Price Volatility*: Without proper market mechanisms, deregulated markets can experience significant price fluctuations.

Some of the examples of deregulated markets are as follows:

- *United States*: Various states have undertaken deregulation with different levels of success. Texas (ERCOT) is often cited as a successful example, while California faced significant challenges during its deregulation process in the early 2000s.
- *United Kingdom*: The United Kingdom was one of the first countries to deregulate its electricity market, beginning in the late 1980s and early 1990s.
- *European Union*: Many EU countries have moved toward deregulated electricity markets as part of broader energy market liberalization initiatives.
- *Australia*: The Australian Energy Market Operator (AEMO) was established in July 2009 to manage the National Electricity Market (NEM) in the eastern and southeastern states of the country and the Australian gas markets.

However, if a deregulated energy market does not follow a strict regulatory framework, it can lead to price gauging by the corporate bodies at the expense of the public. One example is the California energy crisis in 2000–2001, which resulted in widespread blackouts and soaring energy prices. This was caused primarily due to market manipulation by Enron and other minor players. Enron created artificial shortages by intentionally taking power plants offline or rerouting energy to other states and falsely reporting maintenance issues. After investigations, Enron was found guilty and the company collapsed in 2001.

In response to Enron's market manipulation, *Sarbanes-Oxley (SOX) Act* was passed by the US Congress in 2002 [3]. The threefold aims of the act were to combat fraud, improve financial reporting, and restore investor confidence. The act mandated the following:

- The company CEOs must personally certify all financial reports.
- The financial statements must be verified by external auditors.
- Strict rules must be set for record keeping and penalties will apply for falsifying or destroying records.
- Protections must be provided for whistleblowers who report corporate fraud.

This act restored public confidence not only in the energy deregulation but also the in US capital markets.

1.4 Renewable Energy

Water is the essential element of all lifeforms on this planet. It also has been used for the production of electric energy, which is termed hydropower or hydroelectricity. Two of the largest hydropower projects are the Three Gorges Project on the Yangtze River in Yichang in China and Itaipu Dam on the Parana River which is located on the border between Brazil and Paraguay. Even though hydropower is a form of renewable energy, its production is location dependent. It requires large catchment areas, continuous flow of water, and water height. The other forms of hydropower, which are still mostly in the experiment stages, are as follows:

- *Wave power plant*, which utilizes continuous wave movement in oceans to generate energy basically converting wave energy to potential energy.
- *Tidal power plant*, in which the rising and falling of sea levels due to the gravitational attractions of the Moon and the Sun alter the potential energy of water, which is converted into electricity.

The other form of location-dependent renewable power generation that is in existence is geothermal energy, in which water is injected into the Earth's mantle, which is always hot. The water is converted into steam, which is used to drive turbines to produce electricity. Basically, there are three types of geothermal plants. These are as follows:

- *Dry steam plant*, which uses steam extracted directly from geothermal reservoirs through production wells to turn generator turbines. After passing through turbines, the steam can be condensed back into water and reinjected back into the well.

- *Flash steam plant*, which uses high-pressure hot water from geothermal reservoirs. As the pressure is reduced, some hot water flashes into steam that is used to drive turbines. Flash steam plants can handle higher temperature resources compared to dry steam plants.
- *Binary cycle plant*, which requires moderate-temperature geothermal water to heat a secondary fluid with a lower boiling point than water. The secondary fluid vaporizes and drives the turbines. Binary cycle plants can utilize lower temperature resources and have the advantage of being able to operate without emitting any geothermal fluids or gases.

The first geothermal plant was established at Larderello, Italy in 1911. This is a dry steam-type plant. The world's largest geothermal field is located at Geysers in northern California. It contains 18 dry steam plants. Flash steam geothermal plants are located in regions with high-temperature geothermal resources, where water is present under pressure at temperatures typically above 182 °C. This is the most popular form of geothermal power that is located in several countries. Major flash steam plants are the Geysers in California, Larderello, and Wayang Windu in Indonesia. Major binary cycle plants are Blue Mountains in Nevada and Berlin Geothermal Field in Germany. Several other countries, including New Zealand, Iceland, the Philippines, Japan, and Kenya, have also geothermal power plants.

Alternate energy technologies that have enormous potential are hydrogen and nuclear fusion. While green hydrogen production or hydrogen production with carbon capture and storage is within the realm of possibilities, hydrogen storage, transportation, and utilization will require significant investment in research and development. However, effective utilization of hydrogen can alleviate greenhouse gas emissions through both electricity generation and automotive applications. Nuclear fusion can solve energy problems forever. However, so far, no experiment has been able to show net energy gain, that is, the energy output is more than the energy input. It is unclear when commercial-scale nuclear fusion will be feasible.

Currently, the most popular form of renewable energy sources is wind and solar. It can be seen from Tables 1.1 and 1.2 that their combined share is still less than hydropower. It is indeed true that wind or solar power potential varies from region to region. However, they are not totally location-sensitive. In most parts of the world, large-scale solar and wind farms are getting connected to the power grid. It is expected that the total share of these two energy sources will climb rapidly in the coming decade.

However, solar and wind energy are inherently intermittent, as the Sun does not always shine and the wind does not always blow. Consequently, these renewable sources cannot provide continuous energy generation all year around. Therefore, these sources are called *intermittent*. The dictionary definition of intermittent is that "which occurs at irregular intervals or that is not steady." Following this definition, defining these sources as intermittent may not be grammatically correct because they can produce steady power for hours. However, the fact remains that environmental conditions play a crucial role in the power generation capability of solar and wind power.

Consider, for instance, solar PVs. A passing cloud can partially or fully shade an array thereby causing fluctuations in the output of the power level. Different types of clouds and their heights are listed in [4]. Clouds that are below the height of 2,000 m have a major impact on solar PV power output. Cumulus clouds, due to their fluffy, white appearance, are one of the most recognizable types of clouds. These clouds are commonly associated with fair weather, though they can develop into more significant storm clouds under the right conditions. Stratocumulus clouds, on the other hand, typically appear as a blanket of gray or white patches, often with darker spots, covering a large portion of the sky. They are one of the most common types of clouds and can be seen in a variety of weather conditions. While both cumulus and stratocumulus clouds will affect the PV output, cumulus clouds can cause fluctuations in the PV power output when they pass overhead causing shadows in solar arrays. To smoothen such fluctuations, battery energy storage (BES) units may be required to be connected in parallel with PVs such that they can supply the sudden power shortfall.

When a fault occurs in some location in a transmission system, the voltage on the faulted phases will be zero. Due to the low impedance of transmission lines, a large voltage drop would be experienced across large areas of the transmission system until the fault is cleared by the opening of circuit breakers. Older types of wind turbines can trip even when voltages transiently drop to 70% from their nominal values. Wind turbines employing induction generators tend to increase their reactive power demand, thereby causing further voltage depression. This will slow down the voltage recovery process once the fault is cleared. This is why the fault ride-through regulations of wind turbines have been mandated by several authorities.

1.5 Blackouts

The primary causes of blackouts include the following:

- *Equipment Failure*: Aging or malfunctioning equipment can lead to failures and cascading outages.

- *Human Error*: Mistakes in grid management, maintenance, or operations can trigger blackouts.
- *Natural Disasters*: Severe weather events such as storms, hurricanes, earthquakes, and floods can damage infrastructure and cause widespread outages.
- *Cyber Attacks*: Increasing reliance on digital communication and control systems makes grids vulnerable to malicious attacks.
- *Demand Overload*: Excessive demand, especially during peak times, can overwhelm the grid if not properly managed.
- *Inadequate Infrastructure*: Poorly maintained or outdated infrastructure increases the risk of failures.

There have been several blackout events affecting several millions of people that have occurred in this century. Some of these are described below:

- *Northeast Blackout of August 2003*: Occurred in northeastern United States and Canada lasting up to two days and affecting 50 million people. The cause of the blackout was a software bug in the alarm system at an Ohio control room, which led to a cascade of failures after power lines brushed against trees.
- *Italy Blackout of September 2003*: A tree falling on a power line in Switzerland triggered a cascading failure across the interconnected European grid. Entire Italy was blacked out affecting 56 million people for 12 hours.
- *Java-Bali Blackout of August 2005*: A transmission line failure that led to a cascading failure in the grid, impacting 100 million people.
- *European Blackout of November 2006*: A routine disconnection of a high-voltage line over the Ems River in Germany triggered a cascading failure affecting 10 million people in Western Europe.
- *Brazil and Paraguay Blackout of November 2009*: This was caused by a failure at the Itaipu Dam impacting 87 million people.
- *Indian Blackout of July 2012*: Overdraw by states from the grid, combined with maintenance issues resulted in affecting 620 million people in northern, eastern, and northeastern India. This is the largest blackout in history so far.
- *South Australia Blackout of September 2016*: A severe storm with high winds and lightning caused multiple transmission lines to trip and a cascading failure. The entire state of South Australia lost power, affecting about 1.7 million people due to lack of fault ride-through capability of wind turbines.

1.5.1 Power System Oscillations

Most power system calculations are performed using phasor analysis, which assumes that the power system is in a sinusoidal steady state. However, in practice, oscillations resulting from small perturbations (such as load changes) are always present due to different mechanical components in the grid. Fortunately, these oscillations are well-damped and do not cause many problems to the systems operation and stability. The most concerning are the oscillations that result in high energy exchange that can result in system instability.

Power system oscillations are of two types: natural and forced. These are defined in the NERC report as follows [5]:

- *Natural*: These are low-frequency oscillations that can be of different types, such as
  - *Local*: Occurs when a plant or a generating unit oscillates with the rest of the system.
  - *Intraplant*: Occurs when two or more units of a plant oscillate against each other.
  - *Inter-area*: Occurs when several coherent units oscillate against other coherent group(s).
  - *Torsional*: Occurs when a series compensated system oscillates with the turbine shaft modes. This is usually called subsynchronous oscillations.
- *Forced*: These can result in sustained oscillations at any frequency resulting from events such as equipment failures and control interactions.

Oscillation analyses for different parts of the North American system are analyzed in [6].

The detection and countermeasures to mitigate forced oscillations are recommended in [5]. There are five steps involved in the detection process. The first step is to detect the presence of oscillations. It has been suggested that supervisory control and data acquisition (SCADA) or phasor measurement unit (PMU) data can be used for the detection of oscillations. The second step is to determine or estimate the magnitude and frequency of the oscillations. This might require advanced tools for accurate detection. The third step is to detect the extent of the oscillations, that is, are they localized or are they spread over a large area? The last two steps involve the determination of the general proximity of oscillations and the determination of the specific component that is causing the oscillations. The most predominant source of forced oscillations is the generating resources. Other components of power systems, such as renewable energy sources due to their converter control actions, FACTS devices, and HVDC lines, can also force oscillations in a power system. Mitigation of these oscillations is not an easy process. Obviously, control systems play a crucial role in damping oscillations. Therefore, careful tuning actions must be performed on various control elements such as generator excitation systems, FACTS and HVDC controllers, and converter control actions of renewable energy resources.

1.6 Smart Grid

The population of the world has doubled in the last 50 years. At the same time frame, the per capita income has risen by two and a half times worldwide. With the increasing prosperity, we now have bigger houses, more appliances and heating/cooling apparatus, and electric vehicles, which are increasing the demand for electricity. At the same time, the electricity infrastructure in many places is aging. The high volume of electricity trading is resulting in power flow and uncertainties in the systems, which were not designed to handle these. Moreover, due to the concerns of climate change, more and more renewable energy resources are getting connected to the power grids resulting in voltage rise, reverse power flow, and steep ramp rate in generation, causing further stress in the power infrastructure.

Against the backdrop of a deteriorating energy delivery system, it was decided that the power grid needs unprecedented innovations to improve the efficiency, reliability, and sustainability of electricity production and distribution. Smart grid was mandated under the Energy Independence and Security Act of 2007, which was approved by the US Congress and signed as law by the American President in the same year. The American and European definitions of smart grid are somewhat overlapping. However, the common theme among them is that the smart grid is the advancement of the power grid by integration of information and communication technology (ICT) with the power systems operation and infrastructure to create a dynamic and responsive system.

Some of the key components that a smart grid must include are as follows:

- *Data Communication*: Secure and fast communication networks that connect various components of the grid, including smart meters, sensors, and control centers.
- *Advanced Metering*: Two-way digital communication between the utility and customers for electricity usage and consumer participation.
- *Demand Response*: The consumers adjust the demand for electricity in response to supply conditions and real-time pricing signals.
- *Renewable Energy Integration*: Integration of solar, wind, and other renewable energy sources, along with battery storage systems to maximize greenhouse gas emission reductions.
- *Grid Automation and Control*: To operate a safe grid that can self-heal through the use of sensors, PMUs, FACTS devices, and HVDC systems.
- *Cybersecurity*: Protection against cyberattacks against grid operation and data confidentiality breaches.

Some of the benefits of the smart grid are as follows:

- *Reliability*: Enhancing the system reliability through early detection of potential hazards that can lead to outages and automatically reroute power and restore service after disruptions through self-healing processes.
- *Efficiency*: Improving system efficiency by reducing transmission losses through better infrastructure and better management of energy usage.
- *Greenhouse Gas Emission Reduction*: Through better utilization and integration of renewable energy sources in both power transmission and distribution systems.
- *Customer Awareness*: By providing customers with detailed information about their energy usage, real-time pricing, and reducing peak demand through demand response programs.

Overall, smart grid will play a crucial role in the power delivery systems in the 21st century and beyond. A truly smart grid will incorporate generators of all sizes and technologies. It will mitigate environmental impacts by facilitating the integration of renewable energy sources. It will improve the system reliability and improve the quality of service and will enable customer participation in energy usage and conservation. The comparison between smart grid and traditional grid is shown in Figure 1.5.

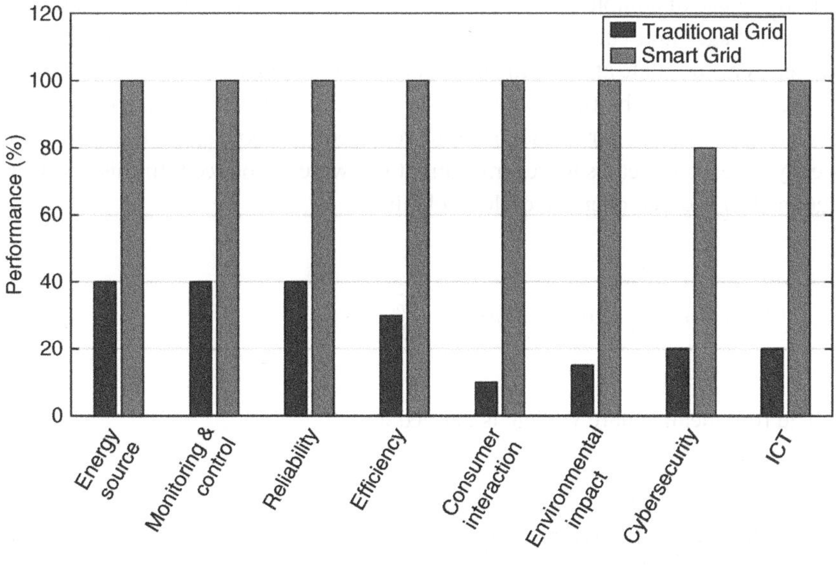

Figure 1.5 Performance comparison between smart and traditional grids.

1.7 Phasor Analysis

Consider an AC voltage source, given by

$$v(t) = V_m \sin(\omega t + \phi) \tag{1.1}$$

where ω is the angular frequency in rad/s and t is the time in seconds. Now assume that this voltage source is supplying a load containing a series combination of a resistor R, an inductor L, and a capacitor C. Let the current flowing through this load be denoted by i. Then, the loop equation is given by

$$v(t) = Ri(t) + L\frac{di(t)}{dt} + \frac{1}{C}\int i(t)dt \tag{1.2}$$

Combining (1.1) and (1.2), the following second-order differential equation is obtained.

$$\omega V_m \cos(\omega t + \phi) = \frac{1}{C}i(t) + R\frac{di(t)}{dt} + L\frac{d^2i(t)}{dt^2} \tag{1.3}$$

Even though it is possible to solve this second-order equation, a power system will involve multiple such equations. Solving several differential equations simultaneously will become a very complicated problem.

This problem was solved by another genius called Charles Steinmetz. He introduced the use of complex vectors for having both amplitudes and phases to represent voltages and currents in *sinusoidal steady state*. In this state, all voltages and currents oscillate at the same frequency with a stable amplitude and constant phase relative to each other. Note that the integration of the voltage given in (1.1) over a cycle will be zero. Therefore, to represent the voltage in the sinusoidal steady state, root mean square (rms) quantities were introduced. In this, the voltage in (1.1) squared over the cycle to obtain

$$v^2(t) = V_m^2 \int_0^{2\pi/\omega} \sin^2(\omega t + \phi)dt = \frac{V_m^2}{2} \tag{1.4}$$

The root mean square voltage is defined as

$$V_{RMS} = \sqrt{\int_0^{2\pi/\omega} v^2(t)dt} = \frac{V_m}{\sqrt{2}} \tag{1.5}$$

The phasor for the voltage in (1.1) is given by

$$V = \frac{V_m}{\sqrt{2}} \angle\phi \tag{1.6}$$

Note that the instantaneous quantities are denoted by lowercase letters, whereas the phasor quantities are denoted by uppercase letters.

In this book, complex power will be used extensively. For example, consider the following voltage and current:

$$V = |V|\angle 0^\circ \text{ and } I = |I|\angle -\theta$$

We know that real and reactive power are given respectively by

$$P = |V||I|\cos\theta \text{ and } Q = |V||I|\sin\theta$$

Now note that

$$VI^* = |V||I|\angle\theta = |V||I|(\cos\theta + j\sin\theta) = P + jQ \tag{1.7}$$

$$V^*I = |V||I|\angle -\theta = |V||I|(\cos\theta - j\sin\theta) = P - jQ \tag{1.8}$$

1.8 Concluding Remarks

Modern power systems are changing rapidly with the traditional way of power delivery system getting replaced by new technologies and power equipment. However, the power system analysis concepts and techniques still remain the same. They, however, are continuously getting upgraded as and when new technologies are incorporated into the systems. This book seeks to blend traditional power system analysis methods with emerging technologies.

References

1 H. Ritchie and P. Rosado, "*Energy Mix: Explore Data on Where Our Electricity Comes from, and this is Changing.*" Available. https://ourworldindata.org/electricity-mix

2 APS News, "This month in physics history," Vol. 19, No. 11, December 2010. Available. https://www.aps.org/publications/apsnews/201012/physicshistory.cfm#:~:text=British%20chemist%20Humphrey%20Davy%20is,component%20of%20his%20public%20lectures

3 S. Wagner and L. Dittmar, "The unexpected benefits of Sarbanes-Oxley," *Harvard Business Review*, Available. https://hbr.org/2006/04/the-unexpected-benefits-of-sarbanes-oxley

4 F. Barbieri, S. Rajakaruna and A. Ghosh, “Very short-term photovoltaic power forecasting with cloud modeling: a review,” *Renewable and Sustainable Energy Reviews*, Vol. 75, pp. 242–263, 2017.

5 NERC (2017). *Reliability Guidelines: Forced Oscillations Monitoring and Mitigation*. Atlanta.

6 NERC (2019). *Interconnection Oscillation Analysis: Reliability Assessment*. Atlanta.

2

Power System Components

The basic structure of a power system is shown in Figure 2.1. It contains generating plants, a transmission system, a subtransmission system, and a distribution system. These subsystems are interconnected through the transformers T_1, T_2, and T_3. A typical thermal plant generates electric power at a typical voltage level of 22 kV (voltage levels are usually specified line-to-line). This is boosted up to levels like 400 kV through the transformer T_1 for power transmission. Windfarms generate power at lower-voltage levels, which is then boosted by a transformer. A solar farm generates dc power, which is then boosted, converted into ac, and connected to the transmission system through a transformer. The transmission level voltage is stepped down by the transformer T_2 to the subtransmission level (66 or 132 kV) to supply industrial loads that require bulk power at a higher voltage. Most major industrial customers have their own transformers to step down the subtransmission level voltage supply to their desired levels. The voltage levels are stepped up or stepped down to reduce line losses and improve transmission efficiency. Distribution systems are designed to operate for much lower-power levels and are supplied with medium-level voltages by stepping down the transmission voltage through transformer T_3.

Power distribution network starts with transformer T_3, which steps down the voltage to 33 kV or 11 kV. The distribution system contains loads that are either commercial type (like office buildings, huge apartment complexes, data centers, shopping complexes, etc.) or residential (domestic) type. Usually, the commercial customers are supplied power at a voltage level of 11 kV, whereas the domestic consumers get power supply at 400–440 V. Since domestic customers get single-phase supplies, they usually receive 230–250 V at their inlet points. While a domestic customer with low-power consumption gets a single-phase supply, both industrial and commercial consumers get three-phase supplies not only because their consumption is high but also because many of them use three-phase motors.

Modern Power System, First Edition. Arindam Ghosh.

Published 2026 by John Wiley & Sons, Inc.
Companion website: www.wiley.com/go/ghoshpowersystem

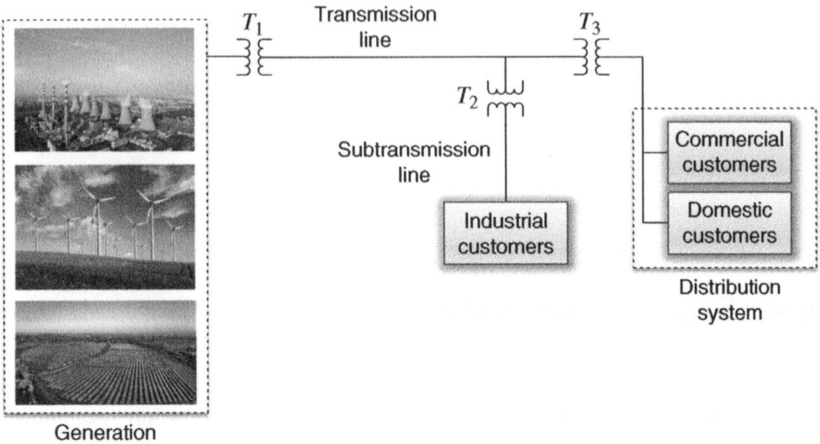

Figure 2.1 A typical power system. *Source:* chungking / Adobe Systems Incorporated; majeczka / Adobe Systems Incorporated; Stephen Davies / Adobe Stock Photos.

For example, the use of an induction motor is very common amongst industrial customers who run pumps, compressors, elevators, rolling mills, etc. It is to be noted that the voltage levels at the distribution level vary from country to country. For example, in Australia, the low distribution voltage level is 400 V (L-L), i.e., 230 V (L-N), while in the United States or Canada, it is 208 V (L-L), i.e., 120 V (L-N). Similarly, medium- and high-voltage levels also vary depending on the countries.

The main components of a power system are generators, transformers, and transmission lines. In this chapter, we shall discuss the models of these components that will be used subsequently in power system studies. Usually, these models are represented by

- voltage sources and reactances for generators;
- leakage reactances for transformers;
- resistance, inductance, and capacitance for transmission lines.

A power system may contain different power equipment with different voltages and power levels connected together through various step-up or step-down transformers. However, the presence of various voltage and power levels causes problems in finding out the currents (or voltages) at different points in the network. To avoid this, all the quantities are converted to a common per unit base. Once all the quantities are converted in this common base, a power system can be graphically represented by the impedances of the various parts of the network. This is called an impedance diagram (or reactance diagram if the resistances are neglected). It will be shown how such diagrams are formed. A high-power transmission line can be hundreds of kilometers long. The line parameters are usually dependent on the line length. Three different representations, namely, short, medium, and long transmission lines, are characterized depending on the line length. The line representation will be discussed in detail in Section 2.5. The so-called ABCD parameters using two-port network theory will be introduced. These can relate the sending end voltage and current to the receiving end voltage and current.

Power transmission lines are left to the vagaries of Mother Nature. A lightning strike on a line can cause voltage or current traveling along the line at a very high speed. These waves can reflect back along the line once they reach a terminating point or can be further transmitted in a complex pattern. Therefore, an understanding of this phenomenon is important from the protection point of view. The basic concept of traveling waves is introduced in Section 2.6.1.

2.1 Transmission Line Parameters

Overhead transmission lines and transmission towers are a common sight around the world. The transmission towers are usually made of steel and are solidly erected with a concrete base. The three-phase conductors are supported by the towers through insulators. The conductors are usually made of aluminum or its alloys. Aluminum is preferred over copper because an aluminum conductor is lighter in weight and cheaper in cost than copper conductor of the same resistance.

The conductors are not straight wires, but strands of wire twisted together to form a single conductor to give it higher tensile strength. One of the most common conductors is aluminum conductor, steel reinforced (ACSR). The cross-sectional view of such a conductor is shown in Figure 2.2. The central core is formed with strands of steel, while two layers of aluminum strands are put in the outer layer. The other types of conductors that are in use are all aluminum conductor (AAC), all aluminum alloy conductor (AAAC), aluminum conductor, alloy reinforced (ACAR), etc.

Power transmission lines are represented by series resistance and inductance and parallel capacitance. In the following sections, we shall briefly discuss these parameters. For a more detailed description, please refer to [1]. References [2, 3] also outline how these parameters are calculated.

2.1.1 Line Resistance

The dc resistance of a conductor is given by

$$R_{dc} = \frac{\rho l}{A} \Omega \tag{2.1}$$

where ρ is the resistivity of the wire in Ω-m; l is the length in m; and A is the cross-sectional area of the wire in m^2.

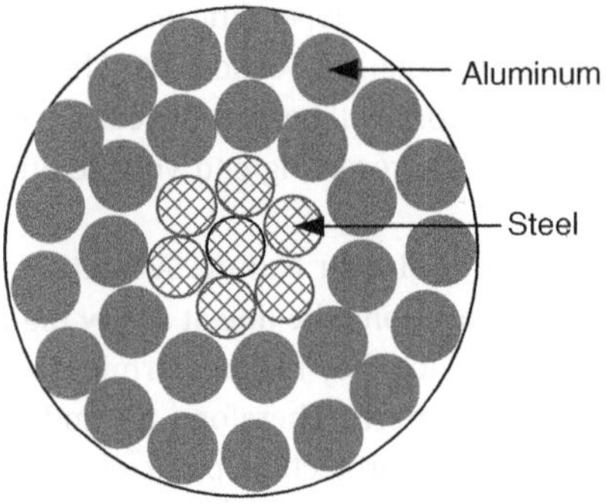

Figure 2.2 Cross-sectional view of ACSR conductor.

However, the resistance of an overhead conductor is not the same as that given by (2.1). In a DC line, the current density is uniform throughout the conductor. However, when an alternating current flows through a conductor, the current density is not uniform over the entire cross-section but gets pushed toward the surface of the conductor, especially at higher frequencies. This is called the *skin effect*. It results in the reduction of surface area available for current flow, thereby increasing the effective resistance of the conductors.

Skin depth usually defines how deep the current penetrates a conductor. It is inversely proportional to the square root of frequency. Therefore, the higher the frequency, the smaller the skin depth, which will push the current closer to the surface. In order to reduce the skin effect, the conductor surface area can be increased. This, however, will result in bulkier conductors. Alternatively, stranded conductors can be used, where several conductors will carry current individually, thereby reducing the skin effect.

In addition to skin effects, the temperature also affects the resistivity of the conductor. The temperature rise in the metallic conductors is almost linear in the practical range of operation, given by

$$\frac{R_2}{R_1} = \frac{T_0 + t_2}{T_0 + t_1} \tag{2.2}$$

where R_1 is the resistance at temperature t_1 in °C; R_2 is the resistance at temperature t_2 in °C; and T_0 is a constant that depends on the conductor materials.

Since the resistance of a conductor cannot be determined accurately, it is best to determine it from the data supplied by the manufacturer.

It is to be noted that resistance and temperature play a crucial role in the maximum power that can be transmitted over a transmission line. A conductor should be able to carry the maximum rated current over a long period of time. It must be able to withstand a temperature up to 75 °C. This is usually defined as the *thermal capacity of a line*, which is a function of the heat generated from the current through resistive (I^2R) drop, heat from the sun, and cooling from the wind as the transmission lines are exposed to nature.

2.1.2 Line Inductance

Series inductance appears in transmission lines due to the flow of alternating current in the conductors. The flow of current creates a magnetic field around the conductor as per Ampere's law. Now an alternating current changes with time. Therefore, the magnetic field also changes with time. The changing magnetic field induces a voltage that opposes the changes in current giving rise to inductance. The relationship between the inductance, flux linkage, and the phasor current is expressed as

$$L = \frac{\lambda}{I} \text{ H} \tag{2.3}$$

where L is the inductance in Henry, λ is the flux linkage in Weber-turns, and I is the phasor current in Ampere.

The inductance of a single-phase, two-wire line (see Figure 2.3 a) is given by

$$L = \frac{\mu_0}{2\pi} \ln\left(\frac{D}{r'}\right) \text{H/m} \tag{2.4}$$

where L is the inductance per unit length in Henry (H)/m; $\mu_0 = 4\pi \times 10^{-7}$ is the permeability of free space in H/m; D is the distance between conductor spacing

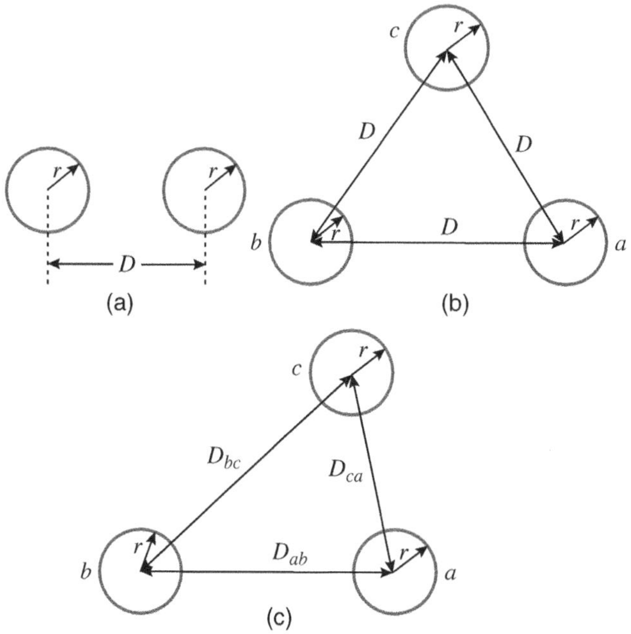

Figure 2.3 Placement of conductors for (a) single-phase, two-wire, (b) three-phase with equal spacing, and (c) three-phase with unequal spacing.

in m; r is the radius of the conductor in m; and $r' = re^{-1/4}$ in m is a fictitious conductor that has no internal flux but has the same inductance as that of a conductor with radius r. This can be termed as an equivalent radius accounting for internal flux and skin effect.

Substituting the values of μ_0 and $e^{-1/4}$, (2.4) is rewritten as

$$L = 2 \times 10^{-7} \ln\left(\frac{D}{0.7788r}\right) \text{H/m} \tag{2.5}$$

For example, let us assume that $D = 1$ m and $r = 0.01$ m. Then

$$L = 2 \times 10^{-7} \ln\left(\frac{1}{0.007788}\right) = 9.71 \times 10^{-7} \text{ H/m}$$

It is interesting to note that the inductance increases as the distance between the conductors increases, as shown in Figure 2.4.

A three-phase line when the conductors are placed at equal distance in an equilateral triangle is shown in Figure 2.3 (b). The inductance per unit length is the same as that given by (2.4) or (2.5), where D is the length between the conductors.

A three-phase line where the conductors are placed at unequal lengths is shown in Figure 2.3 (c). The inductances of the three phases will be unequal. These are given by

$$L_a = 2 \times 10^{-7} \ln\left(\frac{\sqrt{D_{ab}D_{ca}}}{r'}\right) \text{H/m} \tag{2.6}$$

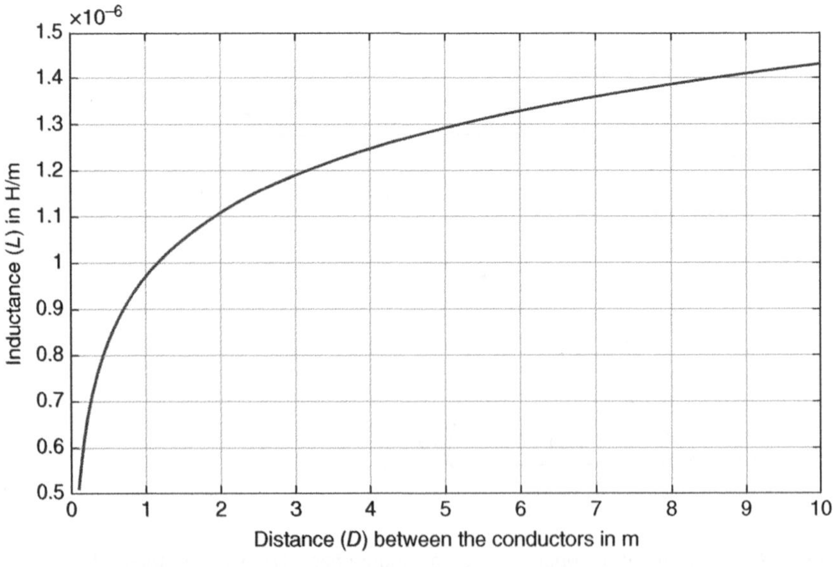

Figure 2.4 Variation of inductance versus the distance between the conductors.

$$L_b = 2 \times 10^{-7} \ln \left(\frac{\sqrt{D_{ab}D_{bc}}}{r'} \right) \text{H/m} \tag{2.7}$$

$$L_c = 2 \times 10^{-7} \ln \left(\frac{\sqrt{D_{bc}D_{ca}}}{r'} \right) \text{H/m} \tag{2.8}$$

For example, let $D_{ab} = 2$ m, $D_{bc} = 3$ m, $D_{ca} = 6$ m, and $r = 0.01$ m. Then

$$L_a = 2 \times 10^{-7} \ln \left(\frac{\sqrt{2 \times 6}}{0.007788} \right) = 1.2195 \times 10^{-6} \text{ H/m}$$

$$L_b = 2 \times 10^{-7} \ln \left(\frac{\sqrt{2 \times 3}}{0.007788} \right) = 1.1502 \times 10^{-6} \text{ H/m}$$

$$L_c = 2 \times 10^{-7} \ln \left(\frac{\sqrt{2 \times 6}}{0.007788} \right) = 1.2601 \times 10^{-6} \text{ H/m}$$

The asymmetrical line is undesirable. As it can give rise to unbalanced current flow, which will result in unbalanced voltages downstream. One of the ways of restoring the balanced nature of the circuit is to exchange the positions of the conductors at regular intervals. This is called transposition of line and is shown in Figure 2.5. In this, each segment of the line is divided into three equal subsegments. The conductors of each of the phases a, b, and c are exchanged after every subsegment such that each of them is placed in each of the three positions once in the entire segment. For example, the conductor of phase-a occupies positions in the sequence 1, 2, and 3 in the three subsegments while that of phase-b occupies 2, 3, and 1. The transmission line consists of several such segments.

The per phase inductance of a transposed line is given by

$$L = 2 \times 10^{-7} \ln \left(\frac{GMD}{r'} \right) \text{H/m} \tag{2.9}$$

where GMD is the geometric mean distance that is expressed as

$$GMD = \sqrt[3]{D_{ab}D_{bc}D_{ca}}$$

For example, let $D_{ab} = 2$ m, $D_{bc} = 4$ m, $D_{ca} = 3$ m, and $r = 0.01$ m. Then

$$GMD = \sqrt[3]{2 \times 4 \times 3} = \sqrt[3]{24} = 2.8845 \text{ m}$$

The inductance is then $L = 1.1829 \times 10^{-6}$ H/m.

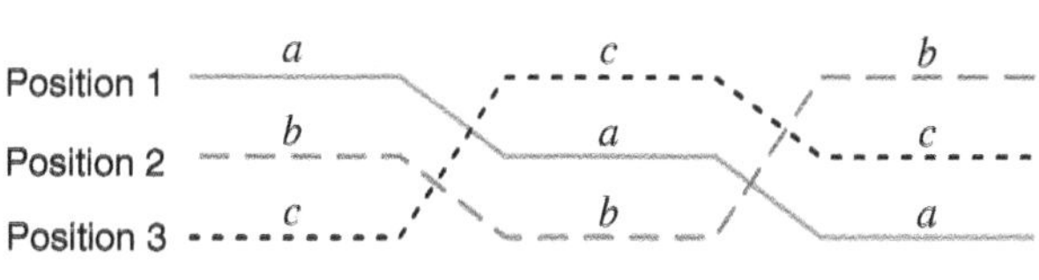

Figure 2.5 A segment of a transposed line.

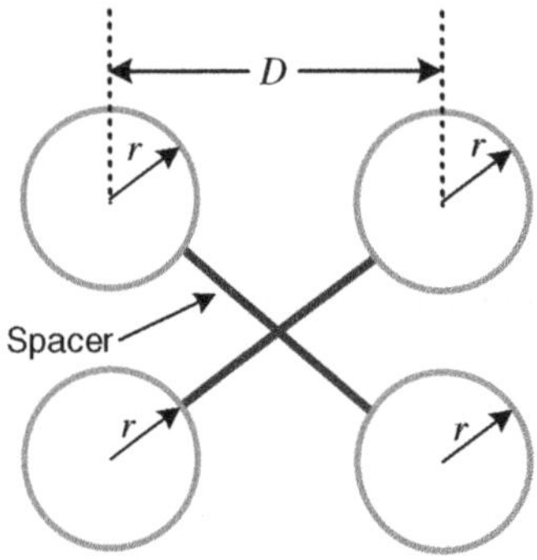

Figure 2.6 A four-conductor bundle.

To improve the electrical performance of transmission lines, sometimes the conductors are *bundled*, where multiple conductors are grouped together and spaced apart using spacers (see Figure 2.6). The spacers are used to maintain fixed spacing between the conductors in the bundle. Usually, two conductors are bundled when the transmission voltage is 220 kV, four conductors are used for a voltage of 400 kV, and for higher voltages, six or eight conductors are used in a bundle. There are several advantages of using bundled conductors. The conductors in a bundle divide the current flowing among themselves. Therefore, the magnitude of the current reduces, thereby reducing I^2R loss. It reduces line reactance as well. Moreover, bundling increases the effective surface area, which reduces the magnetic field intensity. This will reduce corona loss and radio interference.

The inductance of a bundled conductor is given by

$$L = 2 \times 10^{-7} \ln\left(\frac{GMD}{GMR}\right) \text{H/m} \tag{2.10}$$

where GMR is the effective radius of the bundle. For a bundle that has a spacing of D m between the bundles and has n number of sub-conductors in a bundle, the GMR is given by the formula

$$GMR = \sqrt[n]{r' D^{n(n-1)/2}}$$

For example, consider a bundled conductor with two sub-conductors, where $r = 0.01$ m, $D = 0.4$ m, and $GMD = 4$ m. Then

$$GMR = \sqrt{0.007788 \times 0.4} = 0.0558 \text{ m}$$

The inductance is then $L = 8.544 \times 10^{-7}$ H/m. If a non-bundled conductor is used, then for the same specifications, the inductance will be

$$L = 2 \times 10^{-7} \ln\left(\frac{GMD}{r'}\right) = 1.2438 \times 10^{-6} \text{H/m}$$

Obviously, bundling has reduced the inductance.

2.1.3 Line Charging Capacitance

Capacitance in a transmission line forms due to the potential difference between the conductors, which can store energy, just like capacitors. The conductors get charged in the same way as the parallel plates of a capacitor, where the air forms a dielectric medium between the conductors. This will cause transmission lines to store energy between conductors and even between conductors that are grounded. Therefore, the sending end must supply the charging current to charge the capacitance, even under no-load conditions. This is why they are sometimes called line charging capacitors. Capacitance between two parallel conductors depends on the size and the spacing between the conductors. Usually, the capacitance is neglected for the transmission lines that are less than 50 miles (80 km) long. However, the capacitance becomes significant for longer lines with higher voltages.

For a round conductor with a radius of r in m, the capacitance C is the ratio of charge q in coulombs of the conductor to the impressed voltage V, i.e.,

$$C = \frac{q}{V} \text{Farad(F)} \tag{2.11}$$

The charge on the conductor gives rise to an electric field with radial flux lines where the total electric flux is equal to the charge on the conductor. This is the basis for capacitor calculations.

Consider a single-phase, two-wire line with a distance D between the wires (see Figure 2.7 a). If the radius of each conductor is r, then the voltage between the lines a and b is given by

$$V_{ab} = \frac{q}{\pi\varepsilon_0} \ln\left(\frac{D}{r}\right) \tag{2.12}$$

where $\varepsilon_0 = 8.854 \times 10^{-12}$ F/m is the permittivity of the free space. Then, the capacitance between the wires is given by

$$C_{ab} = \frac{q}{V_{ab}} = \frac{\pi\varepsilon_0}{\ln(D/r)} \text{ F/m} \tag{2.13}$$

Eq. (2.13) gives the capacitance between two conductors. For transmission line modeling, the capacitance is defined between the conductor and neutral. This is shown in Figure 2.7 (b). Therefore, the value of the capacitance C_{LN} is given as

$$C_{LN} = 2C_{ab} = \frac{2\pi\varepsilon_0}{\ln(D/r)} \text{ F/m} \tag{2.14}$$

For example, if two conductors of radius 0.01 m are placed at a distance of 4 m, then the capacitance is given by

$$C_{ab} = \frac{\pi \times 8.854 \times 10^{-12}}{\ln(4/0.01)} = 4.6425 \times 10^{-12} \text{ F/m}$$

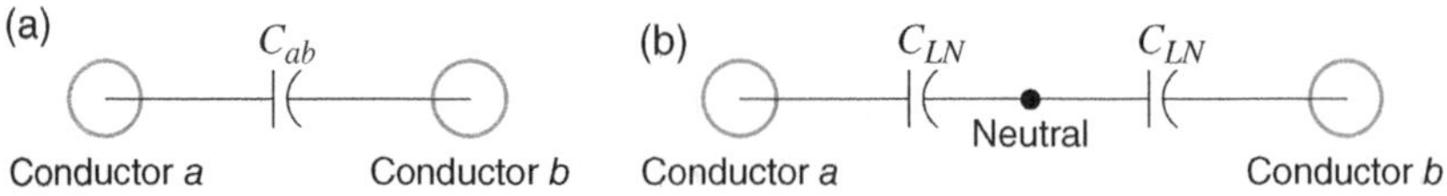

Figure 2.7 Capacitance between (a) two conductors and (b) conductor and neutral.

The line-to-neutral capacitance is then given by

$$C_{LN} = 2C_{ab} = 9.285 \times 10^{-12}\ \text{F/m}$$

For a three-phase line with symmetrical spacing, the capacitance C_{PN} between a phase and neutral is given by

$$C_{PN} = \frac{2\pi\varepsilon_0}{\ln(D/r)}\ \text{F/m} \tag{2.15}$$

Similarly, the capacitance between phase and neutral of an asymmetrically spaced three-phase line is given by

$$C_{PN} = \frac{2\pi\varepsilon_0}{\ln(GMD/r)}\ \text{F/m} \tag{2.16}$$

Using (2.16), we can also find the capacitance of a transposed line. For example, consider $D_{ab} = 4$ m, $D_{bc} = 5$ m, and $D_{ca} = 6$ m. Then, if the radius of each conductor is 0.01 m, the per phase capacitance is given by

$$C_{PN} = \frac{2\pi\varepsilon_0}{\ln\left(\sqrt[3]{4 \times 5 \times 6}/r\right)} = 8.971\ \text{F/m}$$

Earth affects the calculation of capacitance of three-phase lines as its presence alters the electric field lines. This is given by

$$C = \frac{2\pi\varepsilon_0}{\ln(2h/r)}\ \text{F/m} \tag{2.17}$$

where h is the height of the conductors placed on transmission towers. Obviously, as the height of the tower increases, the value of the capacitance decreases. Moreover, the height of the conductors is much larger than the spacing between the conductors. Therefore, the effect of Earth can be neglected for capacitance calculations, especially when balanced steady-state operation of the power system is considered. However, for underground cables, there is a strong dielectric effect between the cables and the ground and, therefore, the capacitance value will increase.

2.2 Synchronous Machine Model

A synchronous machine, as the name signifies, rotates at the rated (synchronous) speed, which is given in revolution per minute (rpm) by

$$\eta = \frac{120f}{p} \text{ rpm} \tag{2.18}$$

where f is the frequency in Hz, and p is the number of poles. Usually, large turboalternators have two poles. Then, if the system operates at 50 Hz, the alternator rotates at 3,000 rpm, while if the frequency is 60 Hz, it will rotate at 3,600 rpm.

A three-phase synchronous generator has a DC field winding that is placed on the rotor, while the phase windings are placed in the rotor, as shown in Figure 2.8 (a). The rotor of a synchronous machine can be of two types – salient pole or

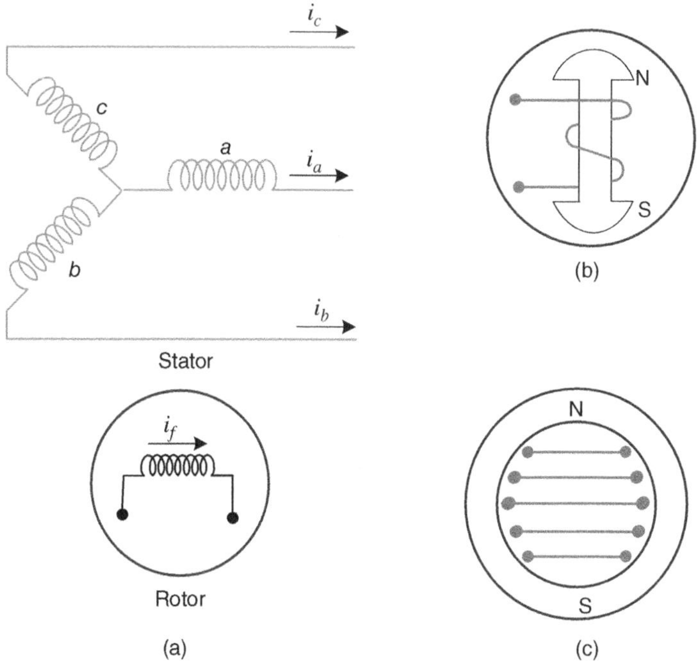

Figure 2.8 Synchronous machine configuration (a) stator and rotor, (b) salient pole rotor, and (c) nonsalient (cylindrical) rotor.

nonsalient pole. The term salient means projecting, i.e., the poles are projecting from the rotor surface, as shown in Figure 2.8 (b). Such rotors are used for low-speed applications, e.g., hydropower. However, the airgap is not uniform, and hence, the direct and quadrature axis reactances are not equal. For higher speed, the rotor is cylindrical (see Figure 2.8 c). Since the rotor is cylindrical, the airgap is uniform and the direct and quadrature axis reactances are equal.

For power system studies, synchronous machines are represented by their equivalent circuit model [4], which is discussed below. A more detailed derivation of model is given in [5]. In the equivalent circuit, it has been assumed that the machine operates in a balanced fashion, where the magnitude of the voltages are the same and they are phase displaced by 120° from each other. The field current I_f produces a flux ϕ_f in the airgap. The airgap flux produces an excitation voltage E_q, which is proportional to both machine speed and the flux. Again, the current I through phase produces a flux ϕ. This flux has two components. These are

- *Leakage flux*, ϕ_l, which links with stator winding and does not link with the field winding.
- *Armature reaction flux*, ϕ_m, that is established in the airgap and links with the field winding.

The armature reaction flux induces a voltage E_r in the stator winding. The resultant equivalent circuit is shown in Figure 2.9, where the terminal voltage is denoted by V_t. The reactances X_m and X_l are due to ϕ_m and ϕ_l, respectively, and their sum is called the *synchronous reactance* of the generator, i.e.,

$$X_s = X_m + X_l$$

The terminal voltage is then given by

$$V_t = E_q + (R + jX_S)I \tag{2.19}$$

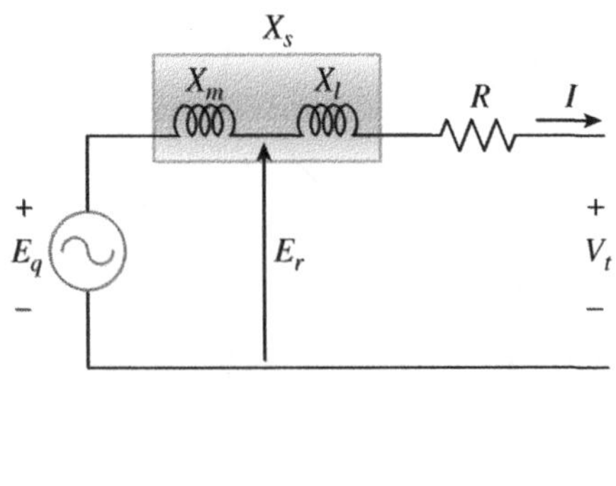

Figure 2.9 Equivalent circuit of a synchronous machine.

The resistance R is called the *effective resistance* and is often neglected in power system studies.

2.3 Transformer Model

The equivalent circuit of a single-phase transformer is shown in Figure 2.10. The following terms are defined with respect to this figure:

- V_1 and V_2, respectively, are the primary winding and the secondary winding voltages.
- I_1 is the current entering the primary terminal.
- R_c is the core loss component, which occurs due to hysteresis and eddy currents in the transformer core.
- X_m is the magnetizing reactance.
- X_{eq} is the leakage inductance of the transformer, which arises from the incomplete coupling of magnetic flux between the windings.
- R_{eq} is the transformer winding resistance.
- N_1:N_2 is the turns ratio of the transformer.

It is to be noted that all the quantities are referred to the primary side.

The impedance of the shunt branch is much larger compared to that of the series branch. Therefore, we neglect R_c and X_m. Again, of the series parameters, R_{eq} is much smaller than X_{eq}. Thus, the series impedance can be neglected as well, and, consequently, the transformer can be represented only by the leakage reactance X_{eq}. The single-phase transformer equivalent circuit, when referred to the primary side, is as shown in Figure 2.11 (a). The equivalent circuit, when referred to the secondary side, is shown in Figure 2.11 (b), where $a = N_1/N_2$.

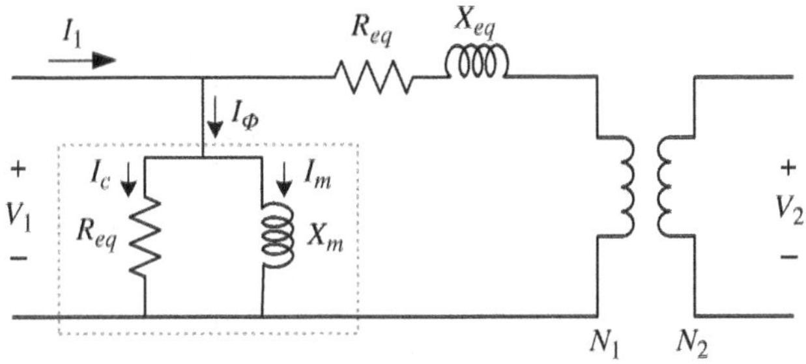

Figure 2.10 Equivalent circuit of a single-phase transformer.

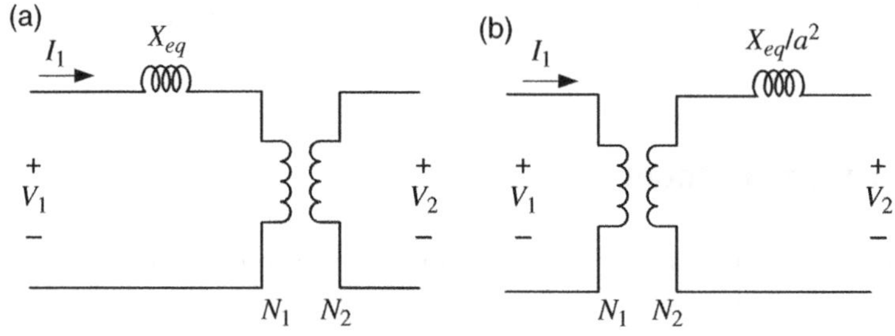

Figure 2.11 Simplified equivalent circuit of a single-phase transformer: (a) when referred to the primary side and (b) when referred to the secondary side.

2.4 Per Unit Representation

In the language of power systems, a three-phase circuit is said to be balanced if the following conditions are true.

1) If all the sources and loads are Y-connected and have equal impedances.
2) There is no mutual inductance between the phases.
3) All neutrals are at the same potential.
4) As a consequence of the points (2) and (3) above, all phases are decoupled.
5) All network variables are balanced sets in the same sequence as the sources.

Point (1) given above does not indicate that sources or loads that are Δ-connected cannot be balanced. Instead, all Δ-connected loads must be converted into an equivalent Y-connection. Consider, for example, the three-phase circuit shown in Figure 2.12 (a) that contains three balanced sources E_a, E_b, and E_c along with three balanced source impedances, denoted by Z_S. The sources supply two balanced loads—one Y-connected with impedance of Z_Y and the other Δ-connected with impedance of Z_Δ. Since this is a balanced network, the sum of the currents at the neutrals *N* and *n* is zero. Therefore, the neutrals are at the same potential. Transforming the Δ-connected load to an equivalent wye, we get the per phase equivalent circuit as shown in Figure 2.12 (b). In this fashion, an entire power system can be converted into its per phase equivalent. The line diagram showing a per phase equivalent circuit is called a *single-line diagram*.

The purpose of a per unit system is to convert all the system parameters into a uniform normalized platform. The use of the per unit system simplifies the comparison of components with different ratings, enabling seamless analysis across transformers, generators, and transmission lines. In this platform, each system quantity is normalized on its own base value. In a per unit system, all quantities

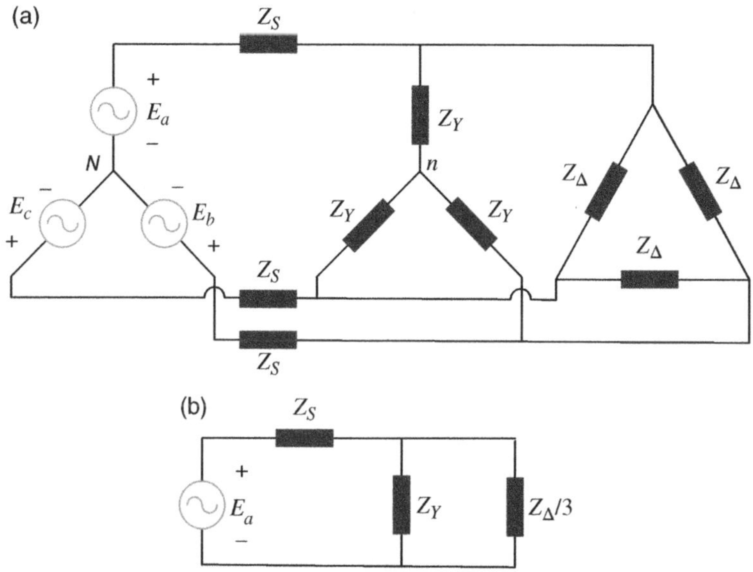

Figure 2.12 (a) Three balanced sources supplying two balanced loads through balanced source impedances and (b) its per phase equivalent circuit.

are denoted as per unit (or pu), and not in volt, amp, or ohm. The fundamental base quantities chosen are:

- VA base (P_{base}): This is the three-phase apparent power (Volt–Ampere) base that is *common to the entire circuit.*
- Voltage base (V_{base}): This is the line-to-line voltage of the different portions of a circuit. Note that this quantity is not uniform for the entire circuit but gets changed by the turns ratio of the transformer that joins the different portions of the circuit.

Based on the above two quantities, the impedance base is defined as

$$Z_{base} = \frac{(V_{base})^2}{P_{base}} \tag{2.20}$$

Even though P_{base} is defined for the three-phase power and V_{base} is defined for the line-to-line voltage, the current must be defined for the line current, i.e.,

$$I_{base} = \frac{P_{base}}{\sqrt{3} \times V_{base}} \tag{2.21}$$

It is interesting to note that all quantities are first defined at their own base, i.e., in terms of their P_{base} and V_{base}. They can then be converted into a base that is common for the entire circuit. For example, assume that an impedance Z is defined as Z_1 per unit in a base impedance of $Z_{base}^{(1)}$, i.e.,

$$Z_1(\text{pu}) = \frac{Z}{Z_{base}^{(1)}} \Rightarrow Z = Z_1 \times Z_{base}^{(1)} \tag{2.22}$$

The impedance now will be represented as Z_2 in a new base value denoted as $Z_{base}^{(2)}$. This is given by

$$Z_2 = \frac{Z}{Z_{base}^{(2)}} \Rightarrow Z_2 = \frac{Z_1 \times Z_{base}^{(1)}}{Z_{base}^{(2)}} \text{ pu} \tag{2.23}$$

From (2.20) and (2.23), Z_2 can be defined in terms of old and new values of VA base and voltage base as

$$Z_2 = Z_1 \times \left(\frac{V_{base}^{(1)}}{V_{base}^{(2)}}\right)^2 \times \frac{P_{base}^{(2)}}{P_{base}^{(1)}} \text{ pu} \tag{2.24}$$

Example 2.1 Let us consider the circuit shown in Figure 2.11 (a) which contains the equivalent circuit of a transformer. Let the transformer rating be 500 MVA, 400/33 kV with a leakage reactance of 10%.

The VA base of the transformer is 500 MVA and the voltage bases in the primary and secondary side are 400 kV and 33 kV, respectively. Therefore, the impedance bases of these two sides are

$$Z_{base}^{(1)} = \frac{(400 \times 10^3)^2}{500 \times 10^6} = 320\ \Omega \ \textit{and}\ Z_{base}^{(2)} = \frac{(33 \times 10^3)^2}{500 \times 10^6} = 2.178\ \Omega$$

where the subscripts 1 and 2 refer to the primary (high-tension) and secondary (low-tension) sides, respectively. Assume that the leakage reactance is referred to the primary side. Then for 10%, i.e., 0.1 per unit leakage reactance, we have

$$X_{eq}^{(1)} = 0.1 \times 320 = 32\ \Omega$$

Note that for the primary and secondary voltages given the turns ratio is $a = 400/33 = 12.1212$. The turns ratio is calculated here using the line-to-line voltages. However, the turns ratio will remain the same even when line-to-neutral voltages are considered since they are calculated by dividing line-to-neutral voltages by $\sqrt{3}$. Then the primary side reactance $X_{eq}^{(1)}$ is referred to the secondary side as

$$X_{eq}^{(2)} = \frac{32}{a^2} = \frac{32}{12.1212^2} = 0.2178\ \Omega$$

Dividing this by the base impedance of the secondary side, we have

$$X_{eq}^{(2)} = \frac{0.2178}{2.178} = 0.1 \text{ per unit.}$$

Therefore, we see that the per unit leakage reactance is the same for both sides of the transformer and, thus, the transformer can be represented by only its leakage reactance $X_{eq} = X_{eq}^{(1)} = X_{eq}^{(2)}$. The equivalent circuit of the transformer is then as shown in Figure 2.13. Since this diagram only shows the reactance (or impedance) of the circuit, this is called the *reactance* (or *impedance*) *diagram*.

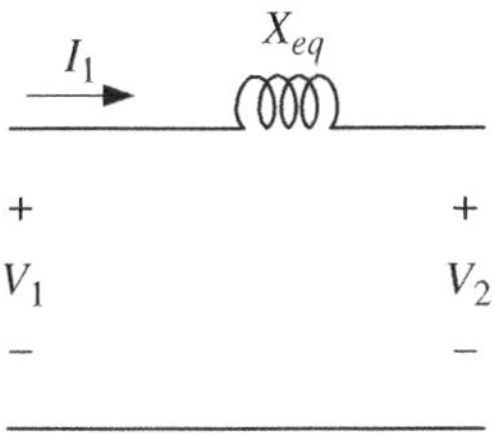

Figure 2.13 Per unit equivalent circuit of a transformer.

We have mentioned that for per unit calculations, we have assumed that all sources and loads are y-connected. This implies that all Δ-connected devices need to be converted to an equivalent y-connection. This does not, however, alter the per unit values. The following example illustrates this.

Example 2.2 Consider a Y/Δ connected transformer with the following parameters:

Rating 10 MVA, 33 kV(Y)/11 kV(Δ)
Leakage reactance on the LV side (Z_Δ) = 3 Ω
First, we convert the Δ-connected reactance to an equivalent star to get

$$Z_Y = \frac{Z_\Delta}{3} = 1\ \Omega$$

The base impedance on the secondary side is

$$Z_{base}^{(2)} = \frac{11^2}{10} = 12.1\ \Omega$$

Therefore, the equivalent reactance is

$$X_{eq}^{(2)} = \frac{1}{12.1} = 0.0826 \text{ per unit.}$$

Now, we refer the reactance to the primary side, which is given by

$$Z_Y^{(1)} = Z_Y \times \left(\frac{33}{11}\right)^2 = 9\ \Omega$$

Now the base impedance on the primary side is

$$Z_{base}^{(1)} = \frac{33^2}{10} = 108.9\ \Omega$$

Then, the per unit reactance on the primary side is

$$X_{eq}^{(1)} = \frac{9}{108.9} = 0.0826 \text{ per unit.}$$

Therefore, the per unit reactance remains the same irrespective of the connection (Y or Δ).

In all the subsequent discussions, the per unit impedance of an apparatus is given by assuming that the impedance is Y-connected or it is already converted into an equivalent Y. Furthermore, the per unit impedance of the apparatus is expressed in terms of its rated voltage and MVA. The following example illustrates this.

Example 2.3 Consider the 50 Hz power system the single-line diagram of which is shown in Figure 2.14. The system contains two buses, four generators, four transformers, and five transmission lines, the ratings of these are

Generator G_1: 400 MVA, 20 kV, $X_s = 15\%$
Generator G_2: 500 MVA, 18 kV, $X_s = 20\%$
Generator G_3: 500 MVA, 20 kV, $X_s = 20\%$
Generator G_4: 500 MVA, 20 kV, $X_s = 20\%$
Transformer T_1: 500 MVA, 220Y/22 kV, $X = 10\%$
Transformer T_2: Three single-phase transformers each rated 166.667 MVA, 130Y/25Δ kV, $X = 10\%$
Transformer T_3: 500 MVA, 220/22 kV, $X = 10\%$
Transformer T_4: 500 MVA, 200/22 kV, $X = 10\%$

The transmission line reactances are as indicated Figure 2.14. We shall draw the reactance diagram choosing the Generator G_3 circuit as the base.

Since the circuit of Generator G_3 is chosen as the base, the base MVA for the entire system is 500. Furthermore, the base voltage in the Generator G_3 side is 20 kV. The generator is connected to the transmission line through a 220 kV/22

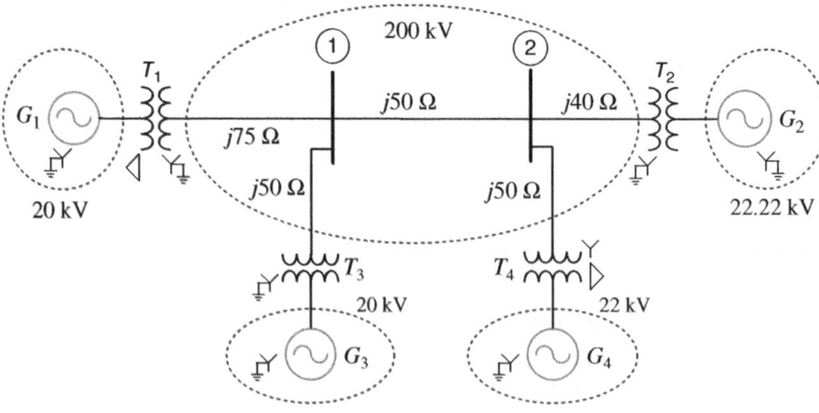

Figure 2.14 Single-line diagram of the power system of Example 2.2.

kV transformer T_3. Therefore, the base voltage of the high-tension side of the transformer is 200 kV. Thus, the bases of all four lines are 500 MVA and 200 kV. The base impedance of all the lines is then

$$Z_{base}^{TL} = \frac{200^2}{500} = 80\ \Omega$$

The per unit impedances of the transmission lines are then

$$X_{j75} = \frac{75}{80} = 0.938, X_{j50} = \frac{50}{80} = 0.625, X_{j40} = \frac{20}{80} = 0.25$$

The transformer T_3 has a base voltage of 220 kV on the high-tension side, whereas the base voltage of the lines is 200 kV. Therefore, the per unit impedance of the transformer will be changed as

$$X_{T3}^{(2)} = 0.1 \times \left(\frac{220}{200}\right)^2 = 0.121\ \text{pu}$$

Since transformers T_1 and T_3 are identical and both of them are connected to lines with a base of voltage of 200 kV, their per unit impedances will also be same, i.e.,

$$X_{T1}^{(2)} = X_{T3}^{(2)} = 0.121\ \text{pu}$$

Let us consider the circuit of Generator G_1. Since transformer T_1 also has a voltage ratio of 220 kV/22 kV, the base voltage for the generator side will be 20 kV, which is the same as base voltage of G_1. However, the base MVA for G_1 is 400. Therefore, its base MVA needs to be changed. The per unit reactance of the generator is then given by

$$X_{G1}^{(2)} = 0.15 \times \frac{500}{400} = 0.188\ \text{pu}$$

Transformer T_4 is rated at 500 kV and has a base voltage of 200 kV. Therefore, its per unit impedance remains the same, i.e.,

$$X_{T4}^{(2)} = 0.1\ \text{pu}$$

Similarly, generator G_4 has base MVA and voltage do not change. Therefore, its per unit impedance will also remain the same, i.e.,

$$X_{G4}^{(2)} = 0.2\ \text{pu}$$

Finally, we consider the circuit of Generator G_2. It is connected through three single-phase transformers rated 130Y/25Δ kV, 166.667 MVA. Then the base MVA of the transformer is $3 \times 166.667 = 500$ kV, which is the same as the base MVA of the circuit. The line-line voltage of the high-tension side of T_2 is

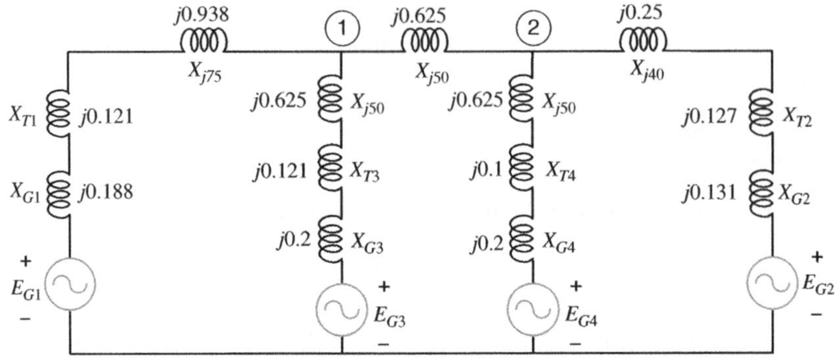

Figure 2.15 The impedance diagram of the power system of Figure 2.15.

$\sqrt{3} \times 130 = 225$ kV. Since the low-tension side is connected in Δ, its line-line voltage is 25 kV. This implies that the transformer T_2 has a turns ratio of 9 : 1. Therefore, the generator G_2 side voltage is 225/9 = 22.22 kV. Therefore, the per unit impedance of generator G_2 is

$$X_{G2}^{(2)} = 0.2 \times \left(\frac{18}{22.22}\right)^2 = 0.131 \text{ pu}$$

Furthermore, the base voltage of T_2 is also changed. Hence, its per unit impedance is

$$X_{T2}^{(2)} = 0.1 \times \left(\frac{25}{22.22}\right)^2 = 0.127 \text{ pu}$$

The base voltages at the different portions of the circuit are indicated by ellipses in Figure 2.14, and the impedance diagram of the circuit is shown in Figure 2.15.

2.5 Modeling Transmission Lines

As discussed in Section 2.2, transmission line parameters include series resistance and inductance and shunt capacitance. These are then used in transmission line models. These models are then used for various power system computations, which will be used in the remainder of the book. The line models are classified by their length. These classifications are

- Short line approximation for lines that are less than 80 km long.
- Medium line approximation for lines whose lengths are between 80 km and 250 km.
- Long line model for lines that are longer than 250 km.

The models, which will be discussed in this section, assume that the line is connected between a sending end and a receiving end. We start our discussion by introducing the ABCD parameters that are used for relating the sending end voltage and current to the receiving end voltage and currents.

2.5.1 ABCD Parameters

Consider the two-port representation of a transmission line shown in Figure 2.16. In this the sending and receiving end voltages are denoted by V_S and V_R, respectively. Also, the currents I_S and I_R are entering and leaving the network, respectively. The sending end voltage and current are then defined in terms of the ABCD parameters as

$$V_S = AV_R + BI_R \tag{2.25}$$

$$I_S = CV_R + DI_R \tag{2.26}$$

From (2.25), we see that

$$A = \left.\frac{V_S}{V_R}\right|_{I_R = 0} \tag{2.27}$$

This implies that A is the ratio of sending end voltage to the open circuit receiving end voltage. This quantity is dimensionless. Similarly,

$$B = \left.\frac{V_S}{I_R}\right|_{V_R = 0} \quad \Omega \tag{2.28}$$

i.e., B, given in Ohm, is the ratio of the sending end voltage and the short circuit receiving end current. Similarly, we can also define

$$C = \left.\frac{I_S}{V_R}\right|_{I_R = 0} \quad \mho \tag{2.29}$$

$$D = \left.\frac{I_S}{I_R}\right|_{V_R = 0} \tag{2.30}$$

The parameter D is dimensionless.

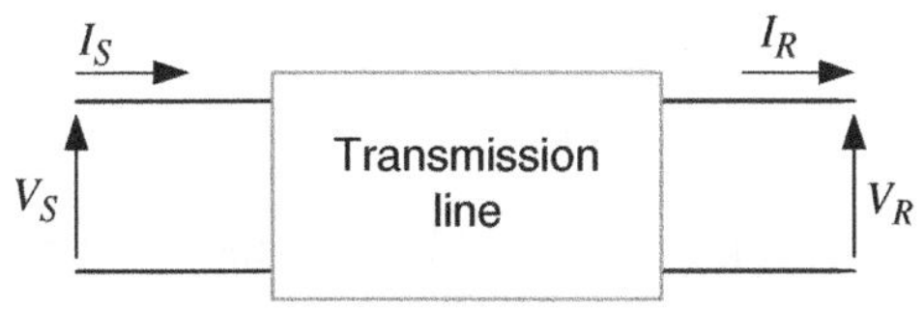

Figure 2.16 Two port representation of a power system.

2.5.2 Voltage Regulation

Voltage regulation (VR) is defined as the percentage change in the receiving end voltage when the load changes from full load to no load, i.e.,

$$\mathrm{VR} = \frac{|V_{RNL}| - |V_{RFL}|}{|V_{RFL}|} \times 100\% \tag{2.31}$$

where V_{RNL} is the no-load voltage, and V_{RFL} is the full-load voltage. At no load, $I_R = 0$, and therefore, from (2.27) we have

$$V_{RNL} = \frac{V_S}{A} \tag{2.32}$$

However, the full-load voltage will depend on the load current.

Example 2.4 A transmission line receives 50 MW, 30 MVAr at the receiving end voltage of 210 kV at full load. The system parameters are as follows:

$$V_S = 220\angle 0^\circ \text{ kv}, A = 0.98\angle 2^\circ\ \Omega \text{ and } B = 30\angle 75^\circ$$

We have to find the percentage voltage regulation.

The first step is to find the receiving end current. Let $V_R = 210\angle -\delta$ kV. Then from the receiving end complex power, we have

$$P_R - jQ_R = V_R^* I_R$$
$$\Rightarrow (50 - j30) \times 10^6 = 210 \times 10^3 I_R$$

Therefore, the receiving end current is

$$I_R = \frac{50 - j30}{210} \times 10^3 = 238.1 - j142.86 = 277.66\angle{-31^\circ} \text{ A}$$

From (2.25), the full-load voltage is calculated as

$$V_{RFL} = \frac{V_S - BI_R}{A} = \frac{220 \times 10^3 - 30\angle 75^\circ \times 277.66\angle{-31^\circ}}{0.98\angle 2^\circ}$$
$$= (218.04 - j13.53) \times 10^3 = 218.46\angle{-3.55^\circ} \text{ kV}$$

Now, from (2.32), the magnitude of the no-load receiving end voltage is given by

$$V_{RNL} = \frac{V_S}{|A|} = 224.49 \text{ kV}$$

Therefore, the percent voltage regulation is given from (2.31) as follows:

$$\mathrm{VR} = \frac{224.49 - 218.46}{218.46} \times 100\% = 2.76\%$$

2.5.3 Short Line Approximation

The shunt capacitance for a short line is almost negligible. The series impedance is assumed to be lumped as shown in Figure 2.17. Let us assume that line is l km long. If the impedance per km is $z_0 = r + jx$, then the total impedance of the line is

$$Z = R + jX = lz_0 = lr + jlx$$

The sending end voltage and current for this approximation are given by

$$V_S = V_R + ZI_R \tag{2.33}$$

$$I_S = I_R \tag{2.34}$$

Therefore, the ABCD parameters are given by

$$A = D = 1, B = Z\ \Omega \text{ and } C = 0 \tag{2.35}$$

2.5.4 Medium Line π Approximation

The line charging capacitance must be considered in a medium length line. There are two options available – one in which the total line charging capacitance is divided into two equal parts and is placed at the two ends of the line in π-like formation. In the other option, the capacitor is placed in the middle of the line in a T-like formation.

We first start our discussion with π-representation. In this representation, the lumped series impedance is placed in the middle while the shunt line charging capacitance is divided into two equal parts and placed at the two ends, as shown in Figure 2.18. Note that the capacitive admittance for a capacitance C is given by

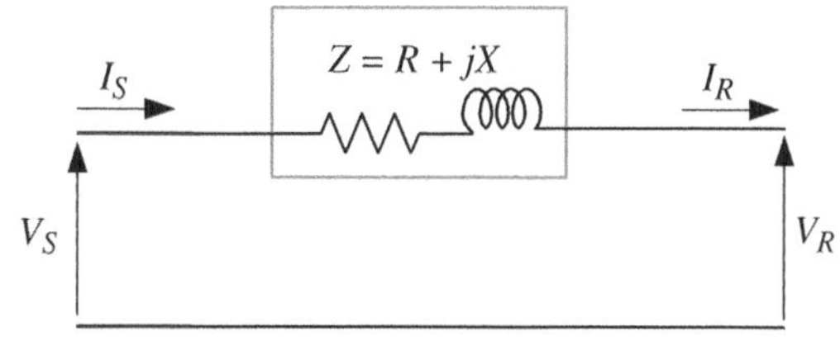

Figure 2.17 Short line approximation of a transmission line.

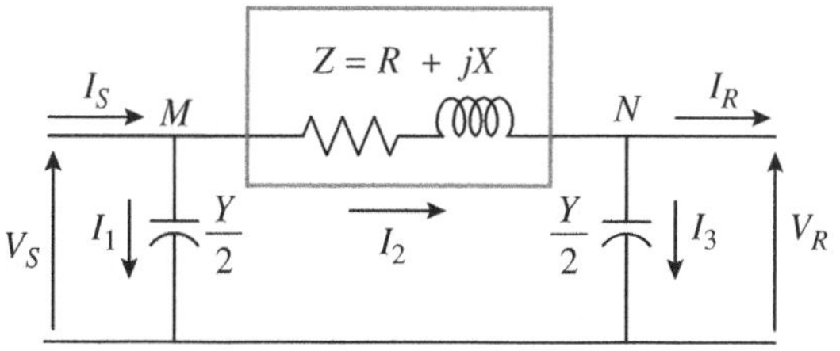

Figure 2.18 Nominal-π representation of a medium length transmission line.

$Y = j\omega C$. Therefore, dividing the line charging capacitance equally is akin to dividing the admittance into two equal parts in Figure 2.18. The nominal-π representation is typically used for load flow studies, as we shall see in Chapter 3.

Let us define three currents I_1, I_2, and I_3 as indicated in Figure 2.18. Applying Kirchhoff's current law (KCL) at node M, we get

$$I_S = I_1 + I_2 = \frac{Y}{2}V_S + I_2$$

Similarly, the KCL at node N gives

$$I_2 = I_3 + I_R = \frac{Y}{2}V_R + I_R$$

The last two equations are combined to form the following equation of the sending end current:

$$I_S = \frac{Y}{2}V_S + \frac{Y}{2}V_R + I_R \tag{2.36}$$

Now

$$V_S = ZI_2 + V_R = Z\left(V_R\frac{Y}{2} + I_R\right) + V_R = \left(\frac{YZ}{2} + 1\right)V_R + ZI_R \tag{2.37}$$

Substituting (2.36) in (2.37), we get

$$\begin{aligned} I_S &= \frac{Y}{2}\left[\left(\frac{YZ}{2} + 1\right)V_R + ZI_R\right] + \frac{Y}{2}V_R + I_R \\ &= Y\left(\frac{YZ}{4} + 1\right)V_R + \left(\frac{YZ}{2} + 1\right)I_R \end{aligned} \tag{2.38}$$

Therefore, from (2.37) and (2.38), we get the following ABCD parameters of the nominal-π representation

$$A = D = \frac{YZ}{2} + 1 \tag{2.39}$$

$$B = Z\ \Omega \tag{2.40}$$

$$C = Y\left(\frac{YZ}{4} + 1\right)\ \mho \tag{2.41}$$

2.5.5 Medium Line T Approximation

In this representation, the shunt admittance is placed in the middle and the series impedance is divided into two equal parts and these parts are placed on either side of the shunt admittance. The nominal-T representation is shown in Figure 2.19.

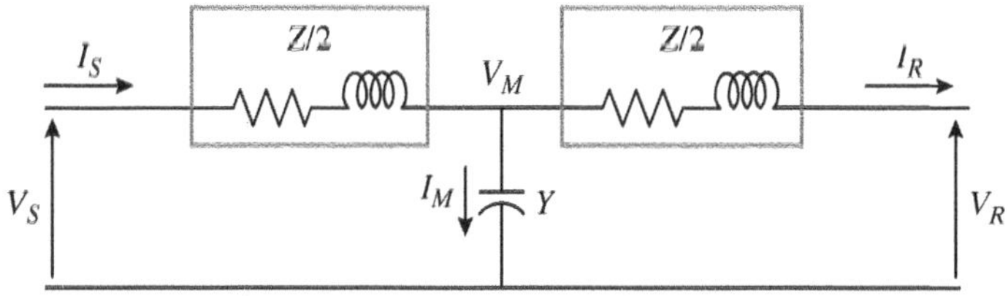

Figure 2.19 Nominal-T representation of a medium length transmission line.

Let us denote the midpoint voltage as V_M. Then the application of KCL at the midpoint results in

$$\frac{V_S - V_M}{Z/2} = YV_M + \frac{V_M - V_R}{Z/2}$$

Rearranging the above equation, the following expression is written as

$$V_M = \frac{2}{YZ + 4}(V_S + V_R) \tag{2.42}$$

Now the receiving end current is given by

$$I_R = \frac{V_M - V_R}{Z/2} \tag{2.43}$$

Substituting the value of V_M from (2.42) in (2.43) and rearranging we get

$$V_s = \left(\frac{YZ}{2} + 1\right)V_R + Z\left(\frac{YZ}{4} + 1\right)I_R \tag{2.44}$$

Furthermore, the sending end current is

$$I_S = YV_M + I_R \tag{2.45}$$

Then substituting the value of V_M from (2.42) in (2.45) and solving, we get

$$I_S = YV_R + \left(\frac{YZ}{2} + 1\right)I_R \tag{2.46}$$

Thus, the ABCD parameters of the T-network are

$$A = D = \left(\frac{YZ}{2} + 1\right) \tag{2.47}$$

$$B = Z\left(\frac{YZ}{4} + 1\right)\ \Omega \tag{2.48}$$

$$C = Y\ \mho \tag{2.49}$$

Example 2.5 Consider a medium length transmission line with the following parameters, given in per unit:

$$V_S = 1.05\angle 30°, V_R = 1\angle 0°, Z = j0.25 \text{ and } Y = j0.04$$

First, we shall consider the nominal-π representation. Then, from Figure 2.18, we have

$$I_2 = \frac{V_S - V_R}{Z} = 2.1 + j0.3627 = 2.1311\angle 9.8°$$

$$I_1 = \frac{Y}{2}V_S = -0.011 + j0.018 = 0.021\angle 120°$$

$$I_3 = \frac{Y}{2}V_R = 0.0 + j0.018 = 0.018\angle 90°$$

Noting that $I_S = I_1 + I_2$ and $I_R = I_2 - I_3$, the complex power dispatched from the sending end and received at the receiving end respectively are

$$S_S = P_S + jQ_S = V_S I_S^* = 2.1 + j0.751$$

$$S_R = P_R + jQ_R = V_R I_R^* = 2.1 - j0.343$$

The complex power flow through the different branches of the circuit are

$$S_1 = P_1 + jQ_1 = V_S I_1^* = 0 - j0.022$$

$$S_3 = P_3 + jQ_3 = V_R I_3^* = 0 - j0.02$$

It can be seen that S_1 and S_3 are purely reactive, which is not surprising. The real and reactive power distribution through the network is shown in Figure 2.20 (a). The reactive drop across the impedance is $0.751 + 0.022 + 0.343 + 0.02 = 1.135$. This can also be calculated from $|I_2|^2 \times |Z| = 1.135$.

We shall now consider the nominal-T representation of Figure 2.19. From this figure, we get

$$I_S = \frac{V_S - V_M}{Z/2}$$

$$I_R = \frac{V_M - V_R}{Z/2}$$

Also, using KCL at the midpoint, we have

$$I_M = V_M Y = I_S - I_R$$

$$\Rightarrow V_M Y = \frac{V_S - V_M}{Z/2} - \frac{V_M - V_R}{Z/2}$$

Solving the above equation, we have

$$V_M = \frac{V_S + V_R}{2 + YZ/2} = 0.957 + j0.263 = 0.993\angle 15.37°$$

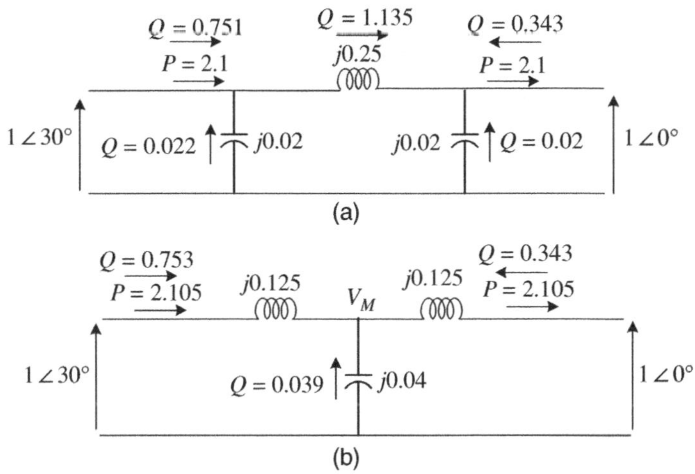

Figure 2.20 Real and reactive power flow of Example 2.5 for (a) π-representation and (b) T-representation.

Then, the complex power through the network is

$$S_S = P_S + jQ_S = V_S I_S^* = 2.105 + j0.753$$

$$S_R = P_R + jQ_R = V_R I_R^* = 2.105 - j0.344$$

$$S_M = P_M + jQ_M = V_S I_S^* = 0 - j0.039$$

The real and reactive power flow through the network is shown in Figure 2.20 (b).

Comparing Figure 2.20 (a) and (b), it can be observed that sending and receiving end active and reactive power are almost the same. However, the π-network has the advantage that an added midpoint node does not have to be considered, unlike the T-network. Moreover, the node voltage appears directly across the line charging capacitors. For these advantages, nominal π-representation is used in power systems, specifically for load flow studies, as discussed in Chapter 3.

2.5.6 Long Line Model

For accurate modeling of the transmission, the parameters are not assumed to be lumped but are distributed throughout the line. The single-line diagram of a long transmission line is shown in Figure 2.21. The length of the line is l. Let us consider a small strip Δx that is at a distance x from the receiving end. The voltage and current at the end of the strip are V and I, respectively, and the beginning of the strip are $V + \Delta V$ and $I + \Delta I$, respectively. The voltage drop across the strip is then ΔV.

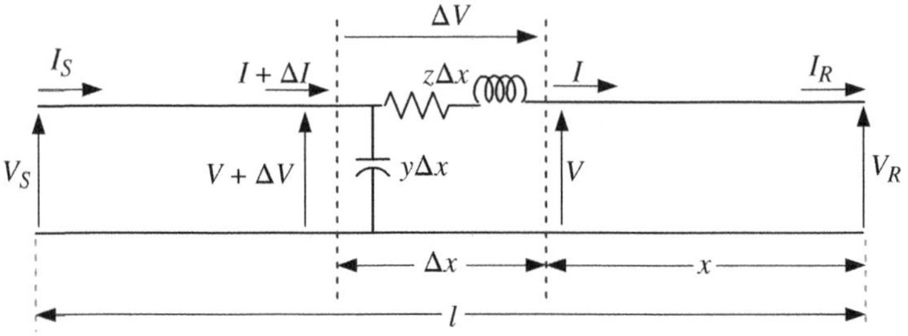

Figure 2.21 Long transmission line representation.

Since the length of the strip is Δx, the series impedance and shunt admittance are $z\Delta x$ and $y\Delta x$, respectively. It is to be noted here that the total impedance and admittance of the line are

$$Z = z \times l \text{ and } Y = y \times l \tag{2.50}$$

From the circuit of Figure 2.21, we see that

$$\Delta V = Iz\,\Delta x \Rightarrow Iz = \frac{\Delta V}{\Delta x} \tag{2.51}$$

Again as $\Delta x \rightarrow 0$, from (2.51), we get,

$$Iz = \frac{dV}{dx} \tag{2.52}$$

Now for the current through the strip, applying KCL, we have,

$$\Delta I = (V + \Delta V)y\,\Delta x = Vy\,\Delta x + \Delta Vy\,\Delta x \tag{2.53}$$

The second term on the right-hand side of the above equation is the product of two small quantities and therefore can be neglected. For $\Delta x \rightarrow 0$ we then have

$$\frac{dI}{dx} = Vy \tag{2.54}$$

Taking derivative with respect to x on both sides of (2.54), we get

$$\frac{d}{dx}\left(\frac{dV}{dx}\right) = z\frac{dI}{dx}$$

Substitution of (2.54) in the above equation results in

$$\frac{d^2V}{dx^2} - yzV = 0 \tag{2.55}$$

The roots of the above equation are located at $\pm\sqrt{yz}$.

Example 2.6 Consider a long transmission line for which the line impedance and admittance are given, respectively, by $z = 0.1 + j0.5145\ \Omega$ and $y = j3.1734 \times 10^{-6}\ \mho$ per kilometer. Then

$$z = 0.524\angle 79^\circ \text{ and } y = 3.1734 \times 10^{-6}\angle 90^\circ$$

Let us define $r^2 = yz$. Then

$$r = \sqrt{yz} = \sqrt{0.524 \times 3.1734 \times 10^{-6}\angle(79^\circ + 90^\circ)} = (-1.633 + j0.317) \times 10^{-6}\angle 169^\circ$$
$$= 0.0001 + j0.0013 = \alpha + j\beta$$

From Example 2.6, we find that the roots of (2.55) are $\alpha + j\beta$ and $\alpha - j\beta$. The solution of (2.55) is

$$V = A_1 e^{x(\alpha + j\beta)} + A_2 e^{-x(\alpha - j\beta)}$$

Then, the general solution of (2.55) is

$$V = A_1 e^{x\sqrt{yz}} + A_2 e^{-x\sqrt{yz}} \tag{2.56}$$

Taking derivative of (2.56) with respect to x, the following equation is obtained:

$$\frac{dV}{dx} = A_1\sqrt{yz}\, e^{x\sqrt{yz}} - A_2\sqrt{yz}\, e^{-x\sqrt{yz}} \tag{2.57}$$

Combining (2.54) with (2.52), the following equation is obtained.

$$I = \frac{1}{z}\left(\frac{dV}{dx}\right) = \frac{A_1}{\sqrt{z/y}}\, e^{x\sqrt{yz}} - \frac{A_2}{\sqrt{z/y}}\, e^{-x\sqrt{yz}} \tag{2.58}$$

Let us define the following two quantities:

$$Z_C = \sqrt{\frac{z}{y}}\ \Omega \text{ which is called the } \textbf{\textit{characteristic impedance}} \tag{2.59}$$

$$\gamma = \sqrt{yz} \text{ which is called the } \textbf{propagation constant} \tag{2.60}$$

Note that γ is unitless. Then, (2.56) and (2.58) can be written in terms of the characteristic impedance and propagation constant as

$$V = A_1 e^{\gamma x} + A_2 e^{-\gamma x} \tag{2.61}$$

$$I = \frac{A_1}{Z_C} e^{\gamma x} - \frac{A_2}{Z_C} e^{-\gamma x} \tag{2.62}$$

We now have to determine the constants A_1 and A_2 to find the solutions of (2.61) and (2.62). Note from Figure 2.21 that when $x = 0$, $V = V_R$, and $I = I_R$. Then, (2.61) and (2.62) can be rewritten as

$$V_R = A_1 + A_2 \tag{2.63}$$

$$I_R = \frac{A_1}{Z_C} - \frac{A_2}{Z_C} \Rightarrow Z_C I_R = A_1 - A_2 \tag{2.64}$$

Adding (2.63) with (2.64), we have

$$A_1 = \frac{V_R + Z_C I_R}{2} \tag{2.65}$$

Again, subtracting (2.64) from (2.63) we get

$$A_2 = \frac{V_R - Z_C I_R}{2} \tag{2.66}$$

Now, consider when $l = x$ in Figure 2.21. At this point, $V = V_S$ and $I = I_S$. Therefore, using the values of A_1 and A_2 obtained in (2.65) and (2.66), we can rewrite (2.61) and (2.62) as

$$V_S = \frac{V_R + Z_C I_R}{2} e^{\gamma l} + \frac{V_R - Z_C I_R}{2} e^{-\gamma l} \tag{2.67}$$

$$I_S = \frac{V_R/Z_C + I_R}{2} e^{\gamma l} - \frac{V_R/Z_C - I_R}{2} e^{-\gamma l} \tag{2.68}$$

Now, note that

$$\frac{e^{\gamma l} - e^{-\gamma l}}{2} = \sinh \gamma l \text{ and } \frac{e^{\gamma l} + e^{-\gamma l}}{2} = \cosh \gamma l$$

Then (2.67) and (2.68) can be rewritten as

$$V_S = V_R \cosh \gamma l + Z_C I_R \sinh \gamma l \tag{2.69}$$

$$I_S = V_R \frac{\sinh \gamma l}{Z_C} + I_R \cosh \gamma l \tag{2.70}$$

The ABCD parameters of the long transmission line can then be written as

$$A = D = \cosh \gamma l \tag{2.71}$$

$$B = Z_C \sinh \gamma l \ \Omega \tag{2.72}$$

$$C = \frac{\sinh \gamma l}{Z_C} \ \mho \tag{2.73}$$

Example 2.7 Consider the same line parameters that are given in Example 2.6. Assume that the line is 500 km. Then,

$$Z_C = \sqrt{\frac{z}{y}} = \sqrt{\frac{0.1 + j0.5145}{j3.1734 \times 10^{-6}}} = \sqrt{\frac{0.5241\angle 79^\circ}{3.1734 \times 10^{-6}\angle 90^\circ}}$$
$$= \sqrt{\frac{0.5241}{3.1734 \times 10^{-6}}}\angle\left(\frac{79^\circ - 90^\circ}{2}\right)$$
$$= 406.4024\angle -5.5^\circ \ \Omega$$

and

$$\gamma l = \sqrt{yz} \times l = \sqrt{0.5241 \times 3.1734 \times 10^{-6}} \times 500\angle\left(\frac{79^\circ + 90^\circ}{2}\right)$$
$$= 0.6448\angle 84.5^\circ = 0.0618 + j0.6419$$

We shall now use the following two formulas for evaluating the hyperbolic forms:

$$\cosh(\alpha + j\beta) = \cosh\alpha\cos\beta + j\sinh\alpha\sin\beta$$
$$\sinh(\alpha + j\beta) = \sinh\alpha\cos\beta + j\cosh\alpha\sin\beta$$

Using these two equations, the hyperbolic identities are evaluated as

$$\cosh\gamma l = 0.8025 + j0.037 \text{ and } \sinh\gamma l = 0.0495 + j0.5998$$

Therefore from (2.71) to (2.73), the ABCD parameters of the system can be written as

$$A = D = 0.8025 + j0.037, \quad B = 43.4 + j240.72\ \Omega \text{ and}$$
$$C = -2.01 \times 10^{-5} + j0.0015\ \mho$$

2.5.7 Equivalent-π Representation of a Long Line

The π-equivalent of a long transmission line is shown in Figure 2.22. In this, the series impedance is denoted by Z', while the shunt admittance is denoted by Y'. From (2.39) to (2.41) the ABCD parameters are defined as:

$$A = D = \left(\frac{Y'Z'}{2} + 1\right). \tag{2.74}$$

$$B = Z'\ \Omega, \tag{2.75}$$

$$C = Y'\left(\frac{Y'Z'}{4} + 1\right)\ \mho \tag{2.76}$$

Comparing (2.72) with (2.75), we can write

$$Z' = Z_C \sinh\gamma l = \sqrt{\frac{z}{y}}\sinh\gamma l = zl\frac{\sinh\gamma l}{l\sqrt{yz}} = Z\frac{\sinh\gamma l}{\gamma l}\ \Omega \tag{2.77}$$

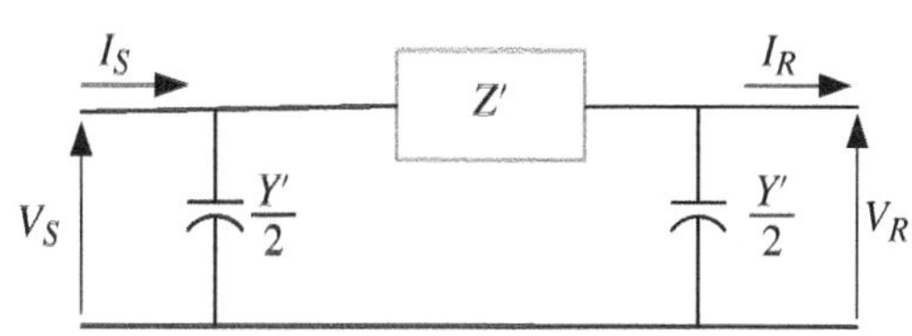

Figure 2.22 Equivalent-π representation of a long transmission line.

where $Z = zl$ is the total impedance of the line. Again comparing (2.71) with (2.74), we get

$$\cosh \gamma l = \frac{Y' Z'}{2} + 1 = \frac{Y'}{2} Z_C \sinh \gamma l + 1$$

The above equation is rearranged as

$$\frac{Y'}{2} = \frac{1}{Z_C}\frac{\cosh \gamma l - 1}{\sinh \gamma l} = \frac{1}{Z_C}\tanh(\gamma l/2) = \sqrt{\frac{y}{z}}\tanh(\gamma l/2) = \frac{yl}{2}\frac{\tanh(\gamma l/2)}{(l/2)\sqrt{yz}}$$
$$= \frac{Y}{2}\frac{\tanh(\gamma l/2)}{(\gamma l/2)} \tag{2.78}$$

where $Y = yl$ is the total admittance of the line. Note that for small values of l, $\sinh\gamma l = \gamma l$ and $\tanh(\gamma l/2) = \gamma l/2$. Therefore from (2.77), we get $Z = Z'$ and from (2.78), we get $Y = Y'$. This implies that when the length of the line is small, the nominal-π representation with lumped parameters is fairly accurate. However, the lumped parameter representation becomes erroneous as the length of the line increases. The following example illustrates this.

Example 2.8 Consider the transmission line given in Example 2.6. The equivalent system parameters for both lumped and distributed parameter representations are given in Table 2.1 for five different line lengths. It can be seen that the error between the parameters increases as the line length increases.

Table 2.1 The variations in the equivalent parameters with the changes in the line length.

Line length (km)	Lumped parameters		Distributed parameters	
	Z (Ω)	Y (℧)	Z' (Ω)	Y' (℧)
100	$52.41\angle 79°$	$3.17 \times 10^{-4}\angle 90°$	$52.27\angle 79°$	$3.17 \times 10^{-4}\angle 89.98°$
200	$104.83\angle 79°$	$6.35 \times 10^{-4}\angle 90°$	$103.69\angle 79.12°$	$6.38 \times 10^{-4}\angle 89.93°$
300	$157.24\angle 79°$	$9.52 \times 10^{-4}\angle 90$	$153.42\angle 79.28°$	$9.64 \times 10^{-4}\angle 89.88°$
400	$209.65\angle 79°$	$1.27 \times 10^{-3}\angle 90$	$200.64\angle 79.49°$	$1.29 \times 10^{-3}\angle 89.75°$
500	$262.06\angle 79°$	$1.59 \times 10^{-3}\angle 90$	$244.61\angle 79.78°$	$1.64 \times 10^{-3}\angle 89.6°$

2.5.8 Some Issues with Transmission Lines

Power transmission systems can face several challenges. One of them is the ohmic (I^2R) loss that increases with the line length and increased current. An increase in current can also cause thermal overloading that can decrease the lifespan of the line conductors. Overhead transmission lines span large areas and are subjected to interference by weather. One such phenomenon is called *corona discharge*, which occurs when the strong electric fields around a conductor ionize the surrounding air. This leads to partial breakdown as the ionization causes a continuous small amount of current flow through the ionized air. This results in energy loss. The corona discharge is prevalent in moist air. It is characterized by a hissing sound and ozone production and can cause radio interference. Higher voltages, which are desired for bulk power transmission, are likely to cause corona discharge. The effect of corona discharge can be mitigated by using thicker conductors; however, that might not be economically or technically viable.

A line overload can also cause a voltage drop. Furthermore, the lack of reactive power can cause voltage stability problems as well (see Chapter 8). It is interesting to note that a voltage rise can also occur in lightly loaded systems due to the presence of the line charging capacitors. Consider, for example, the nominal-π connected network of Fig. 2.18. Assume that the line is lossless, i.e., $R = 0$. Defining $Z_Y = -j2/Y$, the receiving end voltage for unloaded condition ($I_R = 0$) is given by

$$V_R = \frac{Z_Y}{Z + Z_Y} V_S$$

Consider now a transmission line with the per kilometer line reactance of 0.5145 Ω and admittance of 3.1734×10^{-6} ℧. Then, the receiving voltage is as follows:

- For a line length of 400 km: $V_R = 1.15 V_S$
- For a line length of 500 km: $V_R = 1.256 V_S$
- For a line length of 600 km: $V_R = 1.416 V_S$

Therefore, it can be seen that for an unloaded transmission line, the voltage rise at the receiving end increases as the line length increases. The phenomenon is called *Ferranti Effect*.

There are other potential problematic issues such as ageing infrastructures due to the high cost of line upgrade. Moreover, new line constructions may face public opposition due to land acquisition rights, environmental concerns, electromagnetic interference, or radiation. Therefore, the existing infrastructure needs to be operated optimally.

2.6 Lossless Transmission Lines

The resistance in a transmission line causes resistive I^2R power loss. Modern transmission lines are designed such that they have very minimal power losses, especially if the line is not very long. When the line resistance is assumed to be negligible (zero), the resistive drop across the line is zero, i.e., the power transmitted from the sending end is received at the receiving end. This is why such lines are lossless lines. A lossless line has the following characteristics:

- The characteristic impedance ($Z_C = x/y$) becomes a pure real number. This is often referred to as the *surge impedance.*
- The propagation constant ($\gamma = \sqrt{xy}$) becomes a pure imaginary number. Let us define the propagation constant as $\gamma l = j\beta$ and replace l by x. Noting that

$$\cosh(j\theta) = \cos(\theta) \text{ and } \sinh(j\theta) = j\sin(\theta)$$

Equations (2.69) and (2.70) can, respectively, be written as

$$V = V_R \cos\beta x + jZ_C I_R \sin\beta x \tag{2.79}$$

$$I = jV_R \frac{\sin\beta x}{Z_C} + I_R \cos\beta x \tag{2.80}$$

The term *surge impedance loading* or *SIL* is often used to indicate the nominal capacity of the line. The surge impedance is the ratio of voltage and current at any point along an infinitely long line. The term *SIL* or *natural power* (P_n) is a measure of power delivered by a transmission line when terminated by surge impedance and is given by

$$SIL = P_n = \frac{V_0^2}{Z_C} \tag{2.81}$$

where V_0 is the rated voltage of the line.

At *SIL*, $Z_C = V_R/I_R$ and hence from equations (2.79) and (2.80), we get

$$V = V_R \cos\beta x + jV_R \sin\beta x = V_R e^{-j\beta x} \tag{2.82}$$

$$I = jI_R \sin\beta x + I_R \cos\beta x = I_R e^{-j\beta x} \tag{2.83}$$

This implies that as the distance x changes, the magnitudes of the voltage and current in the above equations do not change. The voltage then has a flat profile all along the line. Also, as Z_C is real, V and I are in phase with each other all throughout the line.

Example 2.9 Consider a lossless, 500 km-long transmission line with the following parameters:

$$z = j0.5145\ \Omega \text{ per km and } y = j3.1734 \times 10^{-6}\ \mho \text{ per km.}$$

Then,

$$Z_C = \sqrt{\frac{z}{y}} = 402.652\ \Omega$$

$$\gamma = \sqrt{yz} = j0.0013$$

Assume that the receiving end voltage is 400 kV (L-L). Then, the natural power is

$$P_n = \frac{(400 \times 10^3)}{402.652} = 397.37 \text{ MW}$$

Therefore,

$$I_R = \frac{P_n}{3} \times \frac{\sqrt{3}}{V_R} = 573.547 \text{ A}$$

From (2.67) the sending end line-to-line voltage is computed as

$$V_S = 321.1 + j238.52 = 400\angle 36.6^\circ \text{ kV}$$

Since the magnitudes of the sending and receiving voltages are the same, we can surmise that the voltage profile is flat throughout the line. Again, from (2.68), the sending end current is given by

$$I_S = 460.42 + j342.01 = 573.547\angle 36.6^\circ \text{ A}$$

This not only indicates that the current profile is flat, but also that the sending end voltage and current are in phase. The phase angle difference between the sending end voltage and the receiving end voltage is equal to $\theta = \beta l$. This is shown in Figure 2.23.

It is interesting to note that if the line is loaded at *SIL*, it neither generates nor absorbs reactive power. However, if the line is loaded above *SIL*, there will be a voltage drop, then it will absorb reactive power. On the other hand, the line will

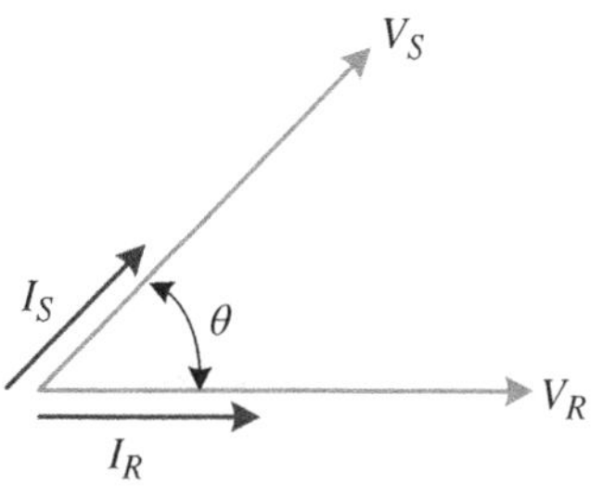

Figure 2.23 Voltage–current relationship in naturally loaded line.

supply reactive power causing a voltage rise if the line is loaded below *SIL*, as we have seen due to Ferranti effect.

2.6.1 Traveling Waves

A traveling wave in power lines refers to a disturbance, such as a sudden change in voltage or current, that can move along the transmission line at a high speed, typically close to the speed of light. These waves are usually caused by events like lightning strikes, switching operations, or faults in the system. When a traveling wave propagates, it carries energy and information along the line, and its behavior is influenced by the line's electrical properties, such as resistance, inductance, and capacitance. If the wave encounters changes in the line, like transformers or terminators, part of the wave may be reflected or transmitted, leading to complex wave patterns. Understanding traveling waves is crucial for designing protective systems, such as surge arresters and fault locators, to maintain the reliability and safety of the power network.

Let us consider a segment of a lossless line that is shown in Figure 2.24. Here also, we consider a small strip of length Δx as in Figure 2.21. The voltage drop across the strip is then given by

$$\Delta v = \frac{\partial v}{\partial x}\Delta x = -(L\Delta x)\frac{\partial i}{\partial t} \tag{2.84}$$

Also, the KCL at the right side of the strip is given by

$$\Delta i = \frac{\partial i}{\partial x}\Delta x = -(C\Delta x)\frac{\partial v}{\partial t} \tag{2.85}$$

Eqs. (2.84) and (2.85) can be simplified as

$$\frac{\partial v}{\partial x} = -L\frac{\partial i}{\partial t} \tag{2.86}$$

$$\frac{\partial i}{\partial x} = -C\frac{\partial v}{\partial t} \tag{2.87}$$

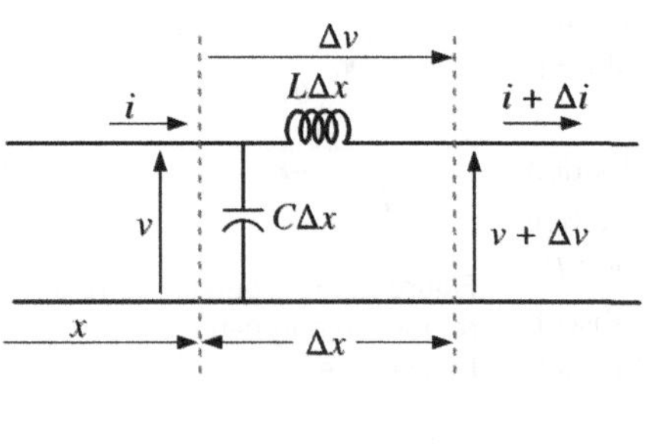

Figure 2.24 A small strip of a lossless distributed parameter line.

Taking a partial derivative (2.86) with respect to x, we get

$$\frac{\partial^2 v}{\partial x^2} = -L\frac{\partial}{\partial t}\left(\frac{\partial i}{\partial x}\right) \tag{2.88}$$

The traveling wave voltage equation of a lossless line is obtained by substituting (2.87) into (2.88) as

$$\frac{\partial^2 v}{\partial x^2} = LC\frac{\partial^2 v}{\partial t^2} \tag{2.89}$$

The solutions of (2.89) is of the following form:

$$v(x,t) = v^+\left(t - \frac{x}{\lambda}\right) + v^-\left(t + \frac{x}{\lambda}\right) \tag{2.90}$$

where $\lambda = 1/\sqrt{LC}$, the first term on the right-hand side is the forward traveling wave, while the second term represents the backward traveling wave. This is called d'Alembert's wave equation in one dimension, as shown in Figure 2.25.

The traveling wave current equation is calculated as follows:

$$\frac{\partial^2 i}{\partial x^2} = LC\frac{\partial^2 i}{\partial t^2} \tag{2.91}$$

with the solutions of

$$i(x,t) = i^+\left(t - \frac{x}{\lambda}\right) + i^-\left(t + \frac{x}{\lambda}\right) \tag{2.92}$$

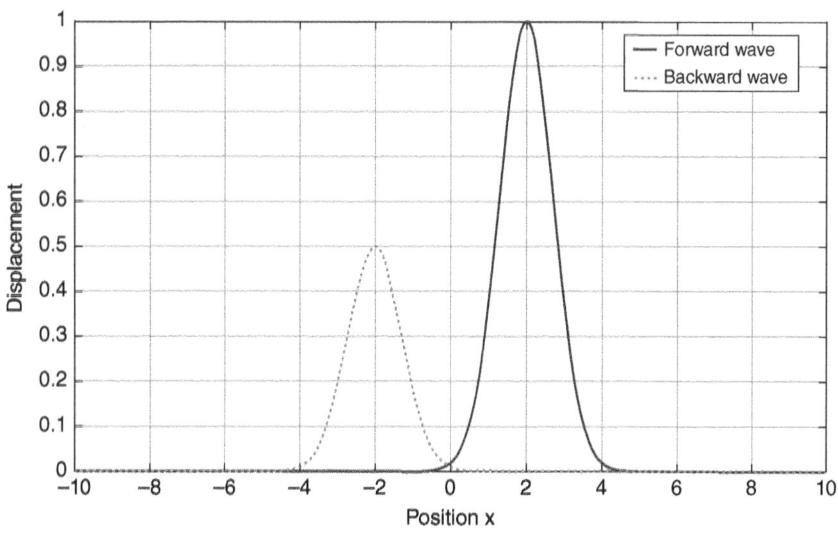

Figure 2.25 Typical forward and backward traveling waves.

The current equation can be represented using the characteristic impedance as

$$i(x,t) = \frac{1}{Z_C}\left[v^+\left(t-\frac{x}{\lambda}\right) + v^-\left(t+\frac{x}{\lambda}\right)\right] \tag{2.93}$$

2.6.2 Traveling Wave in Single-Phase, Two-Wire Line

Consider the single-phase, two-wire, lossless line shown in Figure 2.26. Sending end Thevenin voltage and impedance are, respectively, denoted in the Laplace domain by $E_G(s)$ and $Z_G(s)$, while $I(x, s)$ and $V(x, s)$, respectively, are the Laplace transform of the current and voltage at a position x. The line is terminated by an impedance $Z_R(s)$ and the characteristic impedance is Z_C. The total length of the line is l.

Taking Laplace transform, we can rewrite (2.90) and (2.93) as

$$V(x,s) = V^+(s)e^{-xs/\lambda} + V^-(s)e^{xs/\lambda} \tag{2.94}$$

$$I(x,s) = \frac{1}{Z_C}\left[V^+(s)e^{-xs/\lambda} + V^-(s)e^{xs/\lambda}\right] \tag{2.95}$$

These equations are solved using the boundary condition as [3]

$$V(x,s) = E_G(s)\left[\frac{Z_C}{Z_G(s)+Z_C}\right] \times \left[\frac{e^{-sx/\lambda} + \Lambda_R(s)e^{s(sx/\lambda-2\tau)}}{1-\Lambda_R(s)\Lambda_S(s)e^{-2s\tau}}\right] \tag{2.96}$$

$$I(x,s) = \left[\frac{E_G(s)}{Z_G(s)+Z_C}\right] \times \left[\frac{e^{-sx/\lambda} - \Lambda_R(s)e^{s(sx/\lambda-2\tau)}}{1-\Lambda_R(s)\Lambda_S(s)e^{-2s\tau}}\right] \tag{2.97}$$

where

$\Lambda_R(s) = \dfrac{Z_R(s)-Z_C}{Z_R(s)+Z_C}$ is the *receiving end voltage reflection coefficient* in per unit,

$\Lambda_S(s) = \dfrac{Z_G(s)-Z_C}{Z_G(s)+Z_C}$ is the sending *end voltage reflection coefficient* in per unit and

$\tau = \dfrac{1}{\lambda}$ is the *transit time of the line* in s.

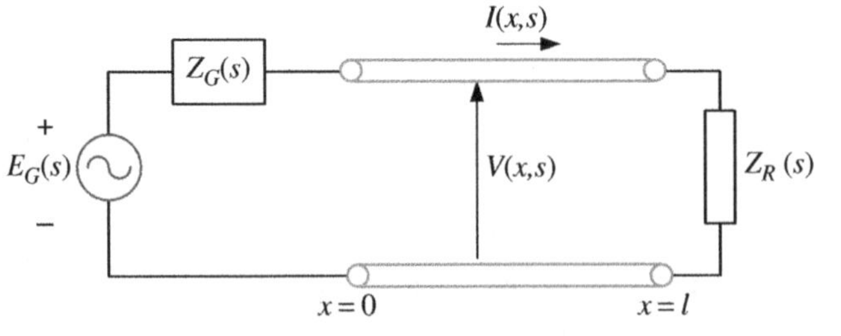

Figure 2.26 A single-phase, two-wire transmission line.

Example 2.10 Consider a single-phase, two-wire line with $Z_R = Z_C/4$ and $Z_G = 2Z_C$. We shall plot the voltage at the center of the line when a step function is applied at the sending end voltage, i.e., $e_G(t) = Vu_s(t)$, where $u_s(t)$ is a unit step function. This implies that

$$E_G(s) = \frac{V}{s}$$

The reflection coefficients are then

$$\Lambda_R(s) = \frac{Z_C/4 - Z_C}{Z_C/4 + Z_C} = -\frac{3}{5} \text{ and } \Lambda_S(s) = \frac{Z_C(2-1)}{Z_C(2+1)} = \frac{1}{3}$$

From (2.96), we have

$$V(x,s) = \frac{V}{3s} \times \left[\frac{e^{-sx/\lambda} - \frac{3}{5}e^{s(sx/\lambda - 2\tau)}}{1 + \frac{1}{5}e^{-2s\tau}}\right] = \frac{V}{3s} \times \left[\frac{5e^{-sx/\lambda} - 3e^{s(sx/\lambda - 2\tau)}}{5 + e^{-2s\tau}}\right]$$

Let us define $\alpha = x/\lambda$. Then, the above equation can be rewritten as follows:

$$V(x,s) = \frac{V}{3s} \times \left[\frac{5e^{-s\alpha} - 3e^{s(s\alpha - 2\tau)}}{5 + e^{-2s\tau}}\right]$$

The above equation is expanded as follows:

$$V(x,s) = \frac{V}{s} \times \left[\frac{e^{-s\alpha}}{3} - \frac{e^{s(s\alpha - 2\tau)}}{5} - \frac{e^{s(s\alpha + 2\tau)}}{15} + \cdots\right]$$

In this way, the voltage at any point is an infinite series of forward and backward traveling waves. Note that the voltage of the forward traveling wave arrives at the center of the line at $t = \tau/2$. Then after reflection, the backward traveling wave will arrive at the same point at $t = 3\tau/2$. The traveling wave will move backward and forward this way.

The expansion of $V(x, s)$ is a tedious process, as is evident from above. In 1933, L. V. Bewley developed the *lattice diagram* through which the forward and backward traveling waves can be calculated. Note that the forward and reverse reflection coefficients are given respectively by $\Lambda_R(s) = -3/5$ and $\Lambda_S(s) = 1/3$. Applying the initial value theorem on (2.96), we have

$$V_1 = \lim_{s \to \infty} sV(x,s) = \frac{V}{s} \times \left[\frac{Z_C}{Z_G(s) + Z_C}\right] = \frac{V}{3s}$$

Therefore, at the beginning, the wave will travel in forward direction with V_1. This is reflected backward with $V_2 = \Lambda_R V_1 = -V/5s$. The forward and backward voltages are listed in Table 2.2. The Bewley lattice diagram is shown in Figure 2.27.

Table 2.2 Forward and backward voltages.

Voltage	Direction	Value
V_1	Forward	$V/3s$
V_2	Backward	$-V/5s$
V_3	Forward	$\Lambda_S V_2 = -V/15s$
V_4	Backward	$\Lambda_R V_3 = V/25s$
V_5	Forward	$\Lambda_S V_4 = V/75s$
V_6	Backward	$\Lambda_R V_5 = -V/125s$

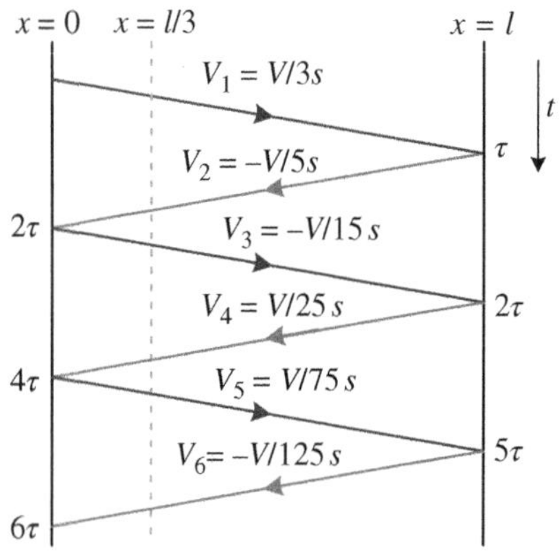

Figure 2.27 Bewley lattice diagram of the system of Example 2.9.

From the lattice diagram, we can calculate the voltage at any point on the line, say, at $x = l/3$ (as shown by the dotted line in Figure 2.27). Then, the first wave in the forward direction arrives at a time $t = \tau/3$. The reflected wave arrives at $t = \tau + 2\tau/3 = 5\tau/3$, and so on.

The voltages at the one-third point are calculated as follows:

- The first voltage waveform arrives at one-third point at $t = \tau/3$ is $V/3 = 0.3333V$.
- The second voltage waveform arrives at one-third point at $t = 5\tau/3$ is $V(1/3 - 1/5) = 0.1333V$.
- The third voltage waveform arrives at one-third point at $t = 2\tau + \tau/3 = 7\tau/3$ is $V(1/3 - 1/5 - 1/15) = 0.0667V$.
- The fourth voltage waveform arrives at one-third point at $t = 3\tau + 2\tau/3 = 11\tau/3$ is $V(1/3 - 1/5 - 1/15 + 1/25) = 0.1067V$.
- The fifth voltage waveform arrives at one-third point at $t = 4\tau + \tau/3 = 13\tau/3$ is $V(1/3 - 1/5 - 1/15 + 1/25 + 1/75) = 0.12V$.

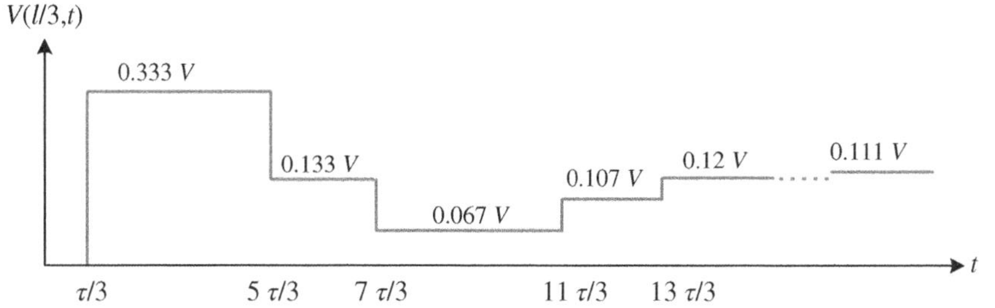

Figure 2.28 Voltage waveform at the one-third point of line obtained from the lattice diagram.

These are shown in Figure 2.28. The steady-state voltage is given by

$$V_{ss}(x) = V \times \left[\frac{Z_R}{Z_G + Z_R}\right] = V \times \left[\frac{Z_C/4}{2Z_C + Z_C/4}\right] = \frac{V}{9} = 0.1111V$$

If two lines are joined together and they have different characteristic impedances, the incident wave at the junction will behave in such a way that the first line is terminated by the characteristic impedance of the second line. The part of the incident wave that is not reflected will travel to the second line segment as a refracted wave. Furthermore, the reflected wave from the second line will be refracted back to the first line. Such a situation can arise when an overhead line is connected to an underground cable (see [3]).

In this section, we have demonstrated a simple two-wire system. However, the traveling wave phenomena are much more complicated for polyphase systems. One of the techniques uses modal decomposition where the network models are decomposed into modal equivalent circuits. A computer solution using finite element techniques is used for more complex cases. Some of the transient stability programs such as Electromagnetic Transients Program (EMTP) or EMTDC/PSCAD can also be utilized to investigate traveling wave phenomena.

It is interesting to note that if a line is terminated by its surge impedance Z_C, i.e., $Z_R(s) = Z_C$ the reflection coefficient will be

$$\Lambda_R(s) = \frac{Z_R(s) - Z_C}{Z_R(s) + Z_C} = 0 \tag{2.98}$$

This implies that there will be no reflected wave, and the transient will die out after reaching the receiving end. In general, however, the reflected wave will reduce in magnitude if the terminal impedance is close to but less than Z_C.

To protect the terminal equipment, surge arresters (often termed as lightning arresters) are used. One such device is a zinc oxide (ZnO) arrestor. It relies on highly nonlinear voltage–current characteristics of zinc oxide to effectively clamp high transient voltage. It will only allow safe voltage levels across it, while diverting excess energy to the ground.

Traveling waves in power systems are typically caused by lightning strikes, switching operations, or faults. Unchecked traveling waves can lead to equipment damage such as:

- High-voltage traveling waves can exceed the insulation strength of cables, transformers, generators, and other equipment, causing flashovers or punctures.
- Sensitive devices (like relays, CTs, PTs) can be overheated, damaged, or destroyed by sudden surges.
- Fast transients can cause internal winding failures due to uneven voltage distribution and mechanical stresses.
- Overhead line insulators can experience arc flashovers that can lead to the tripping of circuit breakers.

2.7 Concluding Remarks

In this chapter, the models of different components for power system studies are discussed. The basic concepts of transmission system parameters have been covered here. However, there are several other factors that need to be considered for accurate line parameter computations. Those interested in the topic can refer to a handbook on power systems (e.g., refer to [1]). More detailed descriptions are also given in [2]. In general, the simplified model of power lines will be discussed in the rest of the book. Similarly, only simplified models of generators are represented by a voltage source behind a reactance and transformers, which is represented by its leakage reactance will be considered. These models are also presented to highlight how the simplified representation is achieved.

References

1 K. O. Papailiou, ed., *Springer Handbook of Power Systems*, Springer Nature, Singapore, 2021.
2 W. D. Stevenson, *Elements of Power Systems Analysis*, 4th Ed., McGraw-Hill, New York, 1982.
3 J. D. Glover and M. S. Sarma, *Power System Analysis and Design*, 3rd Ed., Thomson Learning, Pacific Grove, CA, 2002.
4 P. C. Sen, *Principles of Electric Machine and Power Electronics*, 2nd Ed., John Wiley & Sons, New York, 1997.
5 P. Kundur, *Power System Stability and Control*, McGraw-Hill, New York, 1994.

Problems

P2.1 Consider a transmission line that has conductors with a radius of 1.59 cm and a spacing of 7 m.

a) Find the inductance and capacitance of the line.
b) Now assume that the conductors are used in a two-conductor bundle with a bundle spacing of 0.4 m. Then find the inductance and capacitance of the bundled conductor line.
c) From parts (a) and (b) comment on the effect of bundling.

P2.2 A balanced Δ-connected 5 MW resistive load is connected to the low-tension side of a Y-Δ connected transformer with negligible leakage reactance. The transformer is rated 132 Y/33 Δ kV and 10 MVA. Find the per phase load resistance, both in ohms and in per unit, when referred to the high-tension side of the transformer expressed in the base of the transformer circuit.

P2.3 A 50 Hz, three-phase transmission line is 280 km long. It has a total series impedance of $35 + j140\ \Omega$ and a total shunt admittance of $j930 \times 10^{-6}$ ℧. The line is connected to a 40 MW load at 220 kV. The load draws a current at a power factor of 0.9 (lagging). Find the sending end voltage using (a) short line approximation, (b) medium line π-approximation, and (c) long line model.

P2.4 A three-phase, 250 km line has the following parameters:

Resistance: 0.2 Ω/km, reactance: 0.8 Ω/km, and admittance: 5.3×10^{-6} ℧/km

The line is modeled by long line equation.

a) Find the ABCD constants of the line.
b) If the load in the receiving end of the line is 80 MVA and 220 kV at a unity power factor, find the line-to-line sending end voltage and the sending end current.

P2.5 A three-phase transmission line is 500 km long. It serves a 400 MVA and 440 kV load at 0.9 lagging power factor. The ABCD parameters are:

$$A = D = 0.8180\angle 1.3^\circ$$
$$B = 172.2\angle 84.2^\circ\ \Omega$$
$$C = 0.001933\angle 90.4^\circ\ ℧$$

a) Determine the sending-end line-to-neutral voltage and sending-end current at full load.

b) Determine the receiving-end line-to-neutral voltage and sending-end current at no load.

P2.6 Consider a 150 km long line. Its line parameters are as follows:

$$z = 0.2 + j0.4\ \Omega/\text{km}$$

$$y = j3 \times 10^{-6}\ ℧/\text{km}$$

The line supplies 100 MW at a lagging power factor of 0.9, while the receiving end voltage is 210 kV and the sending end voltage is 220 kV. Assuming a medium length π-approximation, determine the percent voltage regulation.

P2.7 Consider the transmission line of P2.4 with the same ABCD parameters. A series capacitor bank with the reactance of $146.6 \angle -90^\circ\ \Omega$ is installed at the middle of the line.

a) Determine the ABCD parameters of the series compensated line.
b) If the line supplies a 400 MVA and 440 kV load at 0.9 lagging power factor, determine the sending-end line-to-neutral voltage and sending-end current.

P2.8 A transmission line is represented by a nominal-T network, where the line impedance and admittance are

$$Z = 172.2\angle 84.2^\circ\ \Omega$$

$$Y = 0.0021\angle 90^\circ\ ℧$$

Find the ABCD parameters of the network.

P2.9 Consider two networks that are connected in series, as shown in Figure P2.9. Find the ABCD parameters of the combined network.

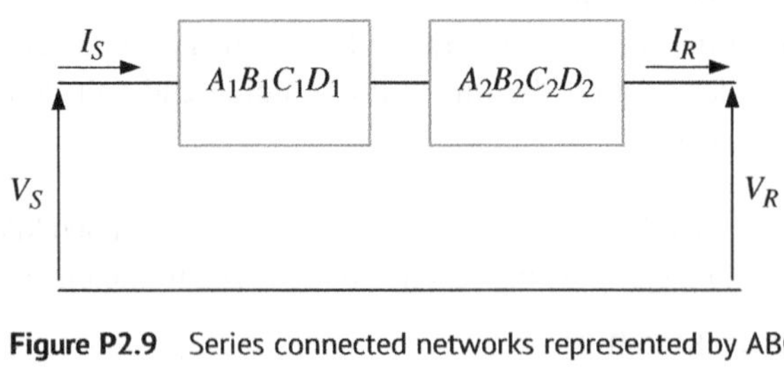

Figure P2.9 Series connected networks represented by ABCD parameters.

2.10 Find the ABCD parameters of the circuit shown in Figure P2.10. The system frequency is 50 Hz.

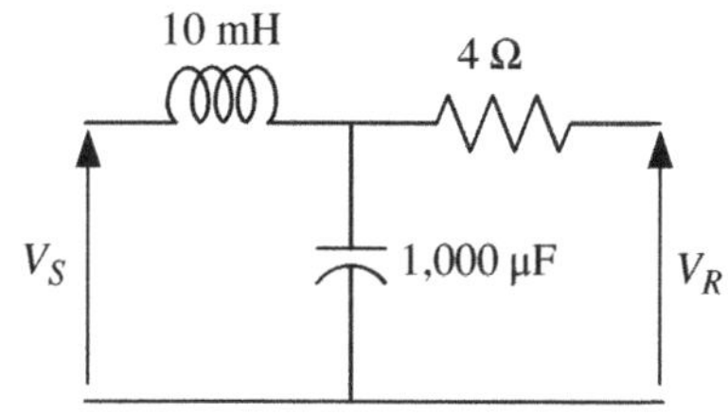

Figure P2.10 RLC network of Problem 2.10.

P2.11 For the power system shown in Figure P2.11, the generator and transformer ratings are

Generator-1: 200 MVA, 30 kV, $X = 20\%$
Generator-2: 100 MVA, 15 kV, $X = 20\%$
Transformer-1: 200 MVA, 200-Y/30-Δ kV, $X = 10\%$
Transformer-2: 100 MVA, 200-Y/30-Δ kV, $X = 10\%$

The transmission line impedances are given in ohms and are indicated in the figure. Choosing a base of 200 MVA, 30 kV in the circuit of Generator-1, draw an equivalent impedance diagram of the circuit.

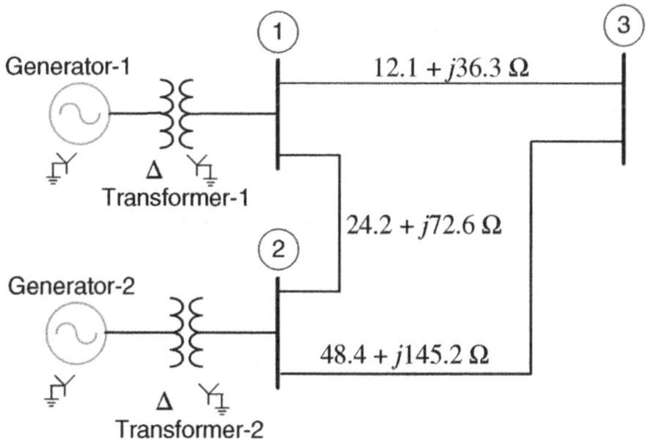

Figure P2.11 Three-bus power system of Problem 2.11.

P2.12 The ratings of the generators and transformers of the system shown in Figure P2.12 are given below

Generator G_1: 25 MVA, 20 kV, $X = 20\%$
Generator G_2: 25 MVA, 20 kV, $X = 20\%$

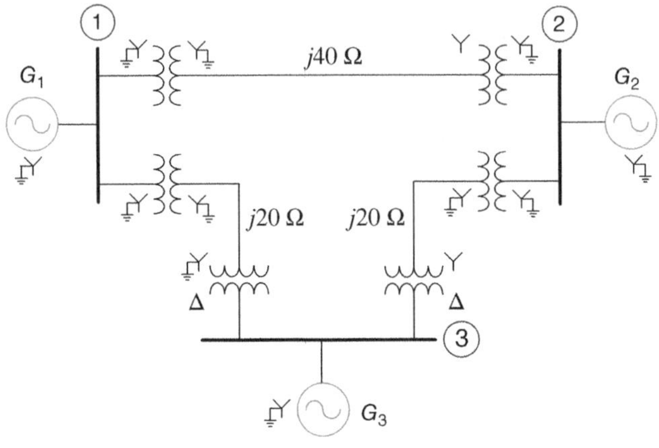

Figure P2.12 Three-bus power system of Problem 2.10.

Generator G_3: 25 MVA, 18 kV, $X = 20\%$
Y-Y connected transformer: 40 MVA, 120/20 kV, $X = 10\%$
Y-Δ connected transformer: 40 MVA, 120-Y/18-Δ kV, $X = 10\%$

The transmission line impedances are shown in the figure. The base MVA is chosen as 50. Choosing a base voltage of 20 kV in the circuit of the generator G1, draw the impedance diagram of the circuit.

P2.13 The voltages at the two ends of a lossless long transmission line are given by $V_S = 1.15\angle 22°$ and $V_R = 1\angle 0°$ per unit. Find the receiving end real power P_R and the reactive power Q_R when $Z_C = 1.0$ per unit and $\theta = 0.6985$ rad.

P2.14 For the system of Problem 2.13, find the sending end power P_S.

P2.15 Consider a medium length lossless line with the following parameters:

Reactance: 0.8 Ω/km and admittance: 5.3×10^{-6} ℧/km

It is represented by a nominal-π configuration of Figure 2.25. The system is unloaded. The sending voltage (V_S) is 440 kV. Find the receiving end voltage (V_R) if the length of the line is (a) 100 km and (b) 200 km.

P2.16 A single-phase, two-wire lossless line has a length l. It has a terminating impedance of $Z_R = 4Z_C$ and a source impedance of $Z_G = Z_C/3$. Draw the Bewley lattice diagram when a unit step voltage is of magnitude E. Furthermore, plot the voltage at the one-third point of the line. Calculate the steady-state voltage.

3

Power Flow Studies

Power flow or load flow studies are one of the most important aspects of power system planning and operation. The load flow gives us the sinusoidal steady state of the entire system – voltages, real and reactive power generated and absorbed, and line losses. Since the load is a static quantity and it is the power that flows through transmission lines, the purists prefer to call this power flow rather than load flow. We shall, however, stick to the more popular nomenclature of load flow.

Through the load flow studies, we can obtain the voltage magnitudes and angles at each bus in the steady state. This is rather important as the magnitudes of the bus voltages are required to be held within a specified limit. Once the bus voltage magnitudes and their angles are computed using the load flow, the real and reactive power flow through each line can be computed. Also based on the difference between power flow in the sending and receiving ends, the losses in a particular line can also be computed. Furthermore, from the line flow, we can also determine the overload and underload conditions.

The steady-state power and reactive power supplied by a bus in a power network are expressed in terms of nonlinear algebraic equations. We, therefore, would require iterative methods for solving these equations. In this chapter, we shall discuss different methods of load flow computations. We shall also delineate how to interpret the load flow results.

Power flow is an important concept that is generally carried out offline, usually during the planning stages. Performing load flow online is fraught with problems due to errors in measurements and telemetering data to the processing centers [1]. State estimation, on the other hand, is a mix of load flow with statistical estimation theory that can be applied online at the power flow control centers using supervisory control and data acquisition (SCADA). An introduction to state estimation is also presented in this chapter. Finally, both power flow and state estimation are performed through SCADA systems; a brief description of SCADA is provided at the end of this chapter.

Modern Power System, First Edition. Arindam Ghosh.

Published 2026 by John Wiley & Sons, Inc.
Companion website: www.wiley.com/go/ghoshpowersystem

3.1 Formation of Bus Admittance Matrix

The first step in a load flow process is the formation of bus admittance matrix, which is also known as $\mathbf{Y_{bus}}$ matrix. Consider the voltage source V_S with a source (series) impedance of Z_S, as shown in Figure 3.1 (a). Using Norton's theorem, this circuit can be replaced by a current source I_S with a parallel admittance of Y_S, as shown in Figure 3.1 (b). The relations between the original system and the Norton equivalent are as follows:

$$I_S = \frac{V_S}{Z_S} \text{ and } Y_S = \frac{1}{Z_S} \tag{3.1}$$

We shall use Norton's theorem for the formulation of the $\mathbf{Y_{bus}}$ matrix.

3.1.1 Without Line Charging Capacitors

Consider the four-bus power system shown in Figure 3.2. This contains two generators G_1 and G_2 that are connected through transformers T_1 and T_2 to buses 1 and 2. Let us denote the synchronous reactances of G_1 and by X_{G1} and X_{G2}, respectively, and the leakage reactances of T_1 and T_2 by X_{T1} and X_{T2}, respectively.

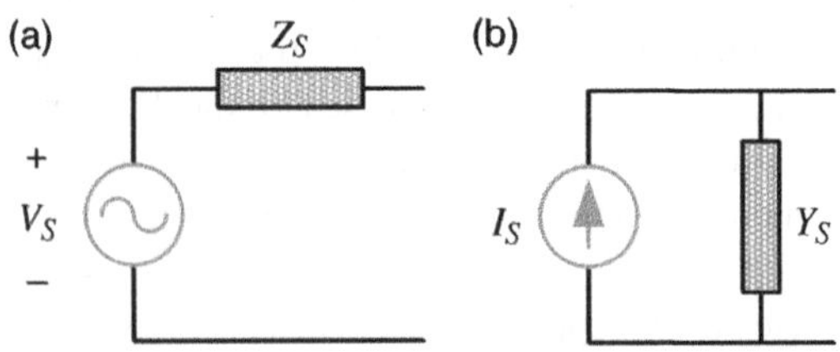

Figure 3.1 (a) Voltage source with a series impedance and (b) its Norton equivalent.

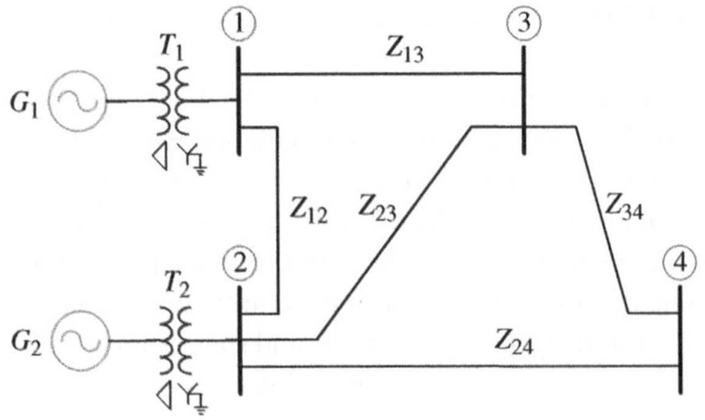

Figure 3.2 Single-line diagram of a four-bus power system.

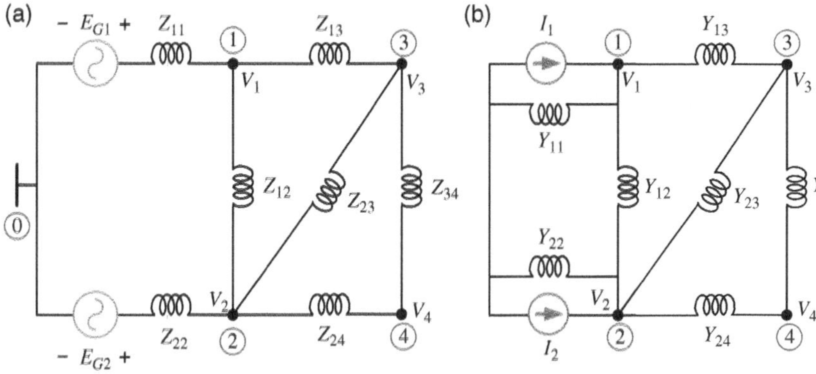

Figure 3.3 (a) Impedance diagram of the power network of Figure 3.2 and (b) and its equivalent admittance diagram.

The line impedances are denoted in the figure. The system impedance diagram is shown in Figure 3.3 (a), where $Z_{11} = j(X_{G1} + X_{T1})$ and $Z_{22} = j(X_{G2} + X_{T2})$. In this figure, the nodes with the node voltages V_1 to V_4 indicate the voltages of buses 1–4 respectively. Bus-0 indicates the reference node that is usually the neutral of the Y-connected system. The impedance diagram is converted into an equivalent admittance diagram shown in Figure 3.3 (b), where

$$Y_{ij} = \frac{1}{Z_{ij}}, \quad i = 1, \ldots, 4, \quad j = 1, \ldots, 4$$

The voltage sources E_{G1} and E_{G2} are converted into the equivalent current sources I_1 and I_2, respectively, using Norton's theorem discussed before.

We would like to determine the voltage–current relationships of the network shown in Figure 3.3 (b). Applying Kirchhoff's current law (KCL) at bus-1, we have

$$\begin{aligned} I_1 &= Y_{11}V_1 + Y_{12}(V_1 - V_2) + Y_{13}(V_1 - V_3) \\ &= (Y_{11} + Y_{12} + Y_{13})V_1 - Y_{12}V_2 - Y_{13}V_3 \end{aligned} \tag{3.2}$$

Similarly, the KCL at bus-2 is as follows:

$$I_2 = -Y_{12}V_1 + (Y_{22} + Y_{12} + Y_{23} + Y_{24})V_2 - Y_{23}V_3 - Y_{24}V_4 \tag{3.3}$$

Note that no current source is attached to either bus-3 or bus-4. Therefore, the KCL at these two buses are given by:

$$\begin{aligned} 0 &= -Y_{13}V_1 - Y_{23}V_2 + (Y_{13} + Y_{23} + Y_{34})V_3 - Y_{34}V_4 \\ 0 &= -Y_{24}V_2 - Y_{34}V_3 + (Y_{24} + Y_{34})V_4 \end{aligned} \tag{3.4}$$

Therefore, the current–voltage relations can be written in the following matrix–vector form:

$$\begin{bmatrix} I_1 \\ I_2 \\ 0 \\ 0 \end{bmatrix} = \begin{bmatrix} Y_{11}+Y_{12}+Y_{13} & -Y_{12} & -Y_{13} & 0 \\ -Y_{12} & Y_{22}+Y_{12}+Y_{23}+Y_{24} & -Y_{23} & -Y_{24} \\ -Y_{13} & -Y_{23} & Y_{13}+Y_{23}+Y_{34} & -Y_{34} \\ 0 & -Y_{24} & -Y_{34} & Y_{24}+Y_{34} \end{bmatrix} \begin{bmatrix} V_1 \\ V_2 \\ V_3 \\ V_4 \end{bmatrix} \tag{3.5}$$

The bus admittance matrix is then given by

$$\mathbf{Y_{bus}} = \begin{bmatrix} Y_{11}+Y_{12}+Y_{13} & -Y_{12} & -Y_{13} & 0 \\ -Y_{12} & Y_{22}+Y_{12}+Y_{23}+Y_{24} & -Y_{23} & -Y_{24} \\ -Y_{13} & -Y_{23} & Y_{13}+Y_{23}+Y_{34} & -Y_{34} \\ 0 & -Y_{24} & -Y_{34} & Y_{24}+Y_{34} \end{bmatrix} \tag{3.6}$$

From (3.6), it can be noted that the $\mathbf{Y_{bus}}$ matrix is symmetric, where the off-diagonal elements are the negative of the admittances connecting two buses, that is,

$$\mathbf{Y}_{\mathbf{bus}(jk)} = \mathbf{Y}_{\mathbf{bus}(kj)} = -Y_{jk} \tag{3.7}$$

Moreover, the diagonal elements of this matrix are given by

$$\mathbf{Y}_{\mathbf{bus}(kk)} = Y_{kk} + \sum_{j=1,\ j\neq k}^{n} \mathbf{Y}_{kj} \tag{3.8}$$

Equation (3.5) can be written in a compact form as follows:

$$\begin{bmatrix} I_1 \\ I_2 \\ 0 \\ 0 \end{bmatrix} = \mathbf{Y_{bus}} \begin{bmatrix} V_1 \\ V_2 \\ V_3 \\ V_4 \end{bmatrix}$$

Note that a bus impedance matrix ($\mathbf{Z_{bus}}$) can be obtained by inverting the bus admittance matrix. The voltage–current relations are then given by

$$\begin{bmatrix} V_1 \\ V_2 \\ V_3 \\ V_4 \end{bmatrix} = \mathbf{Y_{bus}^{-1}} \begin{bmatrix} I_1 \\ I_2 \\ 0 \\ 0 \end{bmatrix} = \mathbf{Z_{bus}} \begin{bmatrix} I_1 \\ I_2 \\ 0 \\ 0 \end{bmatrix} \tag{3.9}$$

Example 3.1 Consider the impedance diagram of Figure 3.3 (a). Let the impedances be given in per unit as:

$$Z_{11} = Z_{22} = 0.0125 + j0.25, \quad Z_{12} = 0.01 + j0.2, \quad Z_{13} = 0.0125 + j0.25$$
$$Z_{23} = Z_{34} = 0.02 + j0.4 \text{ and } Z_{24} = 0.025 + j0.5$$

Therefore, the admittances are computed as:

$$Y_{11} = Y_{22} = Y_{13} = \frac{1}{Z_{11}} = 0.2 - j3.99, \quad Y_{12} = \frac{1}{Z_{12}} = 0.25 - j4.99$$
$$Y_{23} = Y_{34} = \frac{1}{Z_{14}} = 0.12 - j2.49 \text{ and } Y_{24} = \frac{1}{Z_{24}} = 0.1 - j1.995$$

From (3.7) and (3.8), the $\mathbf{Y_{bus}}$ matrix is then formed as:

$$\mathbf{Y_{bus}} = \begin{bmatrix} 0.65 - j12.97 & -0.25 + j4.99 & -0.2 + j3.99 & 0 \\ -0.25 + j4.99 & 0.67 - j13.47 & -0.12 + j2.49 & -0.1 + j1.995 \\ -0.2 + j3.99 & -0.12 - j2.49 & 0.45 - j8.98 & -0.12 + j2.49 \\ 0 & -0.1 + j1.995 & -0.12 + j2.49 & 0.22 - j4.49 \end{bmatrix}$$

The bus impedance matrix is then

$$\mathbf{Z_{bus}} = \mathbf{Y_{bus}^{-1}} = \begin{bmatrix} 0.008 + j0.15 & 0.005 + j0.097 & 0.006 + j0.126 & 0.006 + j0.11 \\ 0.005 + j0.097 & 0.077 + j0.15 & 0.006 + j0.12 & 0.007 + j0.14 \\ 0.006 + j0.126 & 0.006 + j0.12 & 0.013 + j0.26 & 0.01 + j0.2 \\ 0.006 + j0.11 & 0.007 + j0.14 & 0.01 + j0.2 & 0.02 + j0.39 \end{bmatrix}$$

Notice that the $\mathbf{Z_{bus}}$ matrix does not contain the line impedances.

3.1.2 With Line Charging Capacitors

The $\mathbf{Y_{bus}}$ derivation in Section 3.1.1 assumes that the transmission lines are modeled with lumped series impedances without the shunt capacitances. This is valid for a short-line representation. In general, however, the $\mathbf{Y_{bus}}$ matrix may contain the shunt admittances, where the transmission lines are represented by its π-equivalent. Note that both medium-length and long-length lines can be represented by their π-equivalents.

Consider now the power system of Figure 3.2. Let us assume that all the lines are represented in an equivalent-π with the shunt admittance of the line between nodes i and j being denoted by Y_{chij}. Then the equivalent admittance at the two ends of this line will be $Y_{chij}/2$. For example, the shunt capacitance at the two ends

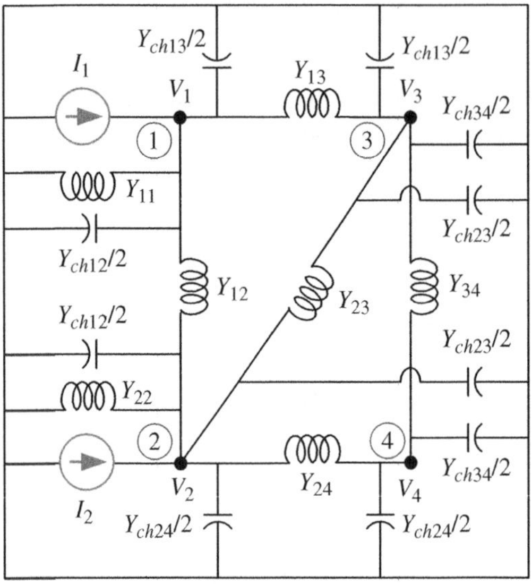

Figure 3.4 The admittance diagram of the power network in Figure 3.2, including the line charging capacitors.

of the line joining buses 1 and 3 will be $Y_{ch13}/2$. The admittance diagram in Figure 3.3 (b) is then modified, as shown in Figure 3.4. The $\mathbf{Y_{bus}}$ matrix of (3.4) is then modified as follows:

$$\mathbf{Y_{bus}} = \begin{bmatrix} Y_{11}+Y_{12}+Y_{13}+Y_{ch1} & -Y_{12} & -Y_{13} & 0 \\ -Y_{12} & Y_{22}+Y_{12}+Y_{23}+Y_{24}+Y_{ch2} & -Y_{23} & -Y_{24} \\ -Y_{13} & -Y_{23} & Y_{13}+Y_{23}+Y_{34}+Y_{ch3} & -Y_{34} \\ 0 & -Y_{24} & -Y_{34} & Y_{24}+Y_{34}+Y_{ch4} \end{bmatrix} \tag{3.10}$$

where

$$\begin{aligned} Y_{ch1} &= (Y_{ch12} + Y_{ch13})/2, \quad Y_{ch2} = (Y_{ch12} + Y_{ch23} + Y_{ch24})/2 \\ Y_{ch3} &= (Y_{ch13} + Y_{ch23} + Y_{ch34})/2, \quad Y_{ch4} = (Y_{ch24} + Y_{ch34})/2 \end{aligned} \tag{3.11}$$

3.2 Load Flow Preliminaries

Let the voltage of ith bus be denoted by

$$V_i = |V_i|\angle\delta_i = |V_i|(\cos\delta_i + j\sin\delta_i) \tag{3.12}$$

Let us denote the element-ij of the $\mathbf{Y_{bus}}$ matrix as Y_{bij}. Then, for an n-bus power system, these elements are written as:

$$\mathbf{Y}_{\mathbf{bus}(ij)} = Y_{bij} = |Y_{bij}|\angle\theta_{ij} = |Y_{bij}|(\cos\theta_{ij} + j\sin\theta_{ij}) = G_{ij} + jB_{ij},\ i = 1, \ldots, n, j = 1, \ldots, n \tag{3.13}$$

The current injected at a bus-i is then given by

$$I_i = Y_{bi1}V_1 + Y_{bi2}V_2 + \cdots + Y_{bin}V_n = \sum_{k=1}^{n} Y_{bik}V_k \tag{3.14}$$

We shall assume the following:

- The current entering a bus is positive and that leaving the bus is negative.
- Therefore, the real power and reactive power entering a bus will also be assumed to be positive.

The complex power at bus-i is then given by

$$\begin{aligned} P_i - jQ_i &= V_i^* I_i = V_i^* \sum_{k=1}^{n} Y_{bik}V_k \\ &= |V_i|(\cos\delta_i - j\sin\delta_i)\sum_{k=1}^{n}|Y_{bik}V_k|(\cos\theta_{ik} + j\sin\theta_{ik})(\cos\delta_k + j\sin\delta_k) \\ &= \sum_{k=1}^{n}|Y_{bik}V_iV_k|(\cos\delta_i - j\sin\delta_i)(\cos\theta_{ik} + j\sin\theta_{ik})(\cos\delta_k + j\sin\delta_k) \end{aligned} \tag{3.15}$$

Note that

$$\begin{aligned} &\cos(\theta_{ik} + \delta_k) + j\sin(\theta_{ik} + \delta_k) \\ &\quad = \cos\theta_{ik}\cos\delta_k - \sin\theta_{ik}\sin\delta_k + j[\sin\theta_{ik}\cos\delta_k + \cos\theta_{ik}\sin\delta_k] \\ &\quad = (\cos\theta_{ik} + j\sin\theta_{ik})(\cos\delta_k + j\sin\delta_k) \end{aligned}$$

Similarly,

$$\begin{aligned} &(\cos\delta_i - j\sin\delta_i)[\cos(\theta_{ik} + \delta_k) + j\sin(\theta_{ik} + \delta_k)] \\ &\quad = \cos(\theta_{ik} + \delta_k - \delta_i) + j\sin(\theta_{ik} + \delta_k - \delta_i) \end{aligned}$$

Substituting the above two identities in (3.15), we have

$$P_i - jQ_i = \sum_{k=1}^{n}|Y_{bik}V_iV_k|[\cos(\theta_{ik} + \delta_k - \delta_i) + j\sin(\theta_{ik} + \delta_k - \delta_i)] \tag{3.16}$$

Separating the real and imaginary components of (3.16), the real and reactive power injected to bus-i are given by

$$P_i = \sum_{k=1}^{n} |Y_{bik} V_i V_k| \cos(\theta_{ik} + \delta_k - \delta_i) \tag{3.17}$$

$$Q_i = -\sum_{k=1}^{n} |Y_{bik} V_i V_k| \sin(\theta_{ik} + \delta_k - \delta_i) \tag{3.18}$$

3.2.1 Classification of Buses

For load flow studies, it is assumed that the loads are constant, and they are defined by their real and reactive power consumption. It is further assumed that the generator terminal voltages are tightly regulated and therefore are constant. The main objective of the load flow is to find the voltage magnitude of each bus and its angle when the powers that are generated, and loads are prespecified. For these studies, three types of buses are considered. They are classified as follows:

- *Load Buses*: In these buses, no generators are connected; hence, the generated real power P_{Gi} and reactive power Q_{Gi} are taken as zero. The load drawn by these buses is defined by real power $-P_{Li}$ and reactive power $-Q_{Li}$ in which the negative sign indicates that the power flowing is out of the bus toward the load, with the assumption that the loads are mostly inductive. Since these buses are defined by their real and reactive power consumption, they are sometimes referred to as P–Q buses. The objective of the load flow is to find the voltage magnitude $|V_i|$ and its angle δ_i of these buses.
- *Voltage-Controlled Buses*: These are the buses where generators are connected. Therefore, the power generation in such buses is controlled through a prime mover while the terminal voltage is controlled through the generator excitation. Keeping the input power constant through turbine-governor control and keeping the bus voltage constant using an automatic voltage regulator, we can specify constant P_{Gi} and $|V_i|$ for these buses. This is why such buses are also referred to as P–V buses. It is to be noted that the reactive power supplied by the generator Q_{Gi} depends on the system configuration and cannot be specified in advance and will have to be determined. Furthermore, we have to find the unknown angle δ_i of the bus voltage.
- *Slack or Swing Bus*: Usually this bus is numbered as one for the load flow studies. This bus sets the angular reference for all the other buses. Since it is the angle difference between two voltage sources that dictates the real and reactive power flow between them, the particular value of the angle of the slack bus is not

important. However, it sets the reference against which angles of all the other bus voltages are measured. For this reason, the angle of this bus is usually chosen as 0°. Furthermore, it is assumed that the magnitude of the voltage of this bus is known.

Now consider a typical load flow problem in which all the load demands are known. Even if the generation matches the sum total of these demands exactly, the mismatch between generation and load will persist because of the I^2R losses in the lines. Since the I^2R loss of a line depends on the line current which, in turn, depends on the magnitudes and angles of voltages of the two buses connected to the line, it is rather difficult to estimate the line loss without calculating the voltages and angles. For this reason, a generator bus is usually chosen as the slack bus without specifying its real power. It is assumed that the generator connected to this bus will supply the balance of the real power required and the line losses.

3.2.2 Data Preparation

Let the real and reactive power generated at bus-i be denoted by P_{Gi} and Q_{Gi}, respectively. Moreover, let us denote the real and reactive power consumed at the ith bus by P_{Li} and Q_{Li}, respectively. Then the net real power injected in bus-i is as follows:

$$P_{i,inj} = P_{Gi} - P_{Li} \tag{3.19}$$

Let the injected power calculated by the load flow program be $P_{i,calc}$. Then, the mismatch between the actual injected and calculated values is given by

$$\Delta P_i = P_{i,inj} - P_{i,calc} = P_{Gi} - P_{Li} - P_{i,calc} \tag{3.20}$$

In a similar way, the mismatch between the reactive power injected and calculated values is given by

$$\Delta Q_i = Q_{i,inj} - Q_{i,calc} = Q_{Gi} - Q_{Li} - Q_{i,calc} \tag{3.21}$$

The purpose of the load flow is to minimize the above two mismatches. It is to be noted that (3.17) and (3.18) are used for the calculation of real and reactive power in (3.20) and (3.21). However, since the magnitudes of all the voltages and their angles are not known a priori, an *iterative procedure* must be used to estimate the bus voltages and their angles to calculate the mismatches. It is expected that mismatches ΔP_i and ΔQ_i reduce with each iteration. The load flow is said to have converged when the mismatches of all the buses become less than a very small number.

Example 3.2 For the load flow studies, we shall consider the system of Figure 3.5, which has two generators and three load buses. We define bus-1 as the slack bus while taking bus-5 as the P–V bus. Buses 2, 3, and 4 are P–Q buses. The per unit line impedances and the line charging admittances are given in Table 3.1, where the base MVA is chosen as 100. Based on this data, the $\mathbf{Y_{bus}}$ matrix is given by

$$\mathbf{Y_{bus}} = \begin{bmatrix} 2.69-j13.41 & -1.92+j9.62 & 0 & 0 & -0.77+j3.85 \\ -1.92+j9.62 & 3.65-j18.19 & -0.96+j4.81 & 0 & -0.77+j3.85 \\ 0 & -0.96+j4.81 & 2.21-j11.00 & -0.77+j3.85 & -0.48+j2.40 \\ 0 & 0 & -0.77+j3.85 & 1.15-j5.67 & -0.38+j1.92 \\ -0.77+j3.85 & -0.77+j3.85 & -0.48+j2.40 & -0.38+j1.92 & 2.40-j11.89 \end{bmatrix}$$

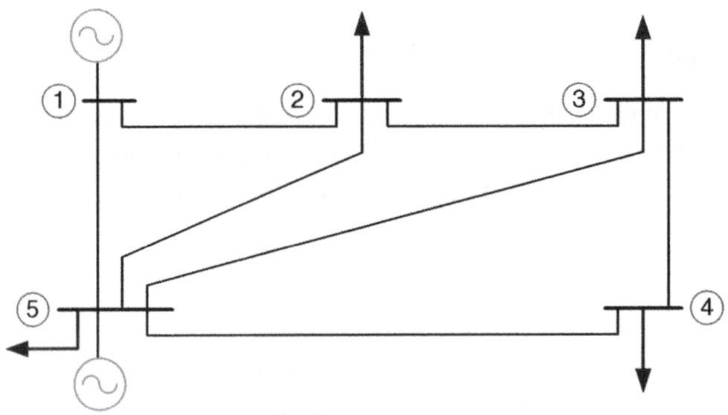

Figure 3.5 The single-line diagram of a five-bus power system used for load flow studies.

Table 3.1 Line impedance and line charging data of the system of Figure 3.5.

Line (bus to bus)	Impedance	Line charging (*Y*/2)
1–2	$0.02 + j0.10$	$j0.030$
1–5	$0.05 + j0.25$	$j0.020$
2–3	$0.04 + j0.20$	$j0.025$
2–5	$0.05 + j0.25$	$j0.020$
3–4	$0.05 + j0.25$	$j0.020$
3–5	$0.08 + j0.40$	$j0.010$
4–5	$0.10 + j0.50$	$j0.075$

Table 3.2 Bus voltages, power generated, and load – initial data.

Bus no. and type	Bus voltage		Power generated		Load	
	Magnitude (pu)	Angle (deg)	P (MW)	Q (MVAr)	P (MW)	Q (MVAr)
1 (Slack)	1.05	0	–	–	0	0
2 (P–Q)	①	⓪	0	0	96	62
3 (P–Q)	①	⓪	0	0	35	14
4 (P–Q)	①	⓪	0	0	16	8
5 (P–V)	1.02	⓪	48	–	24	11

Since the load flow procedure only produces the bus voltages, the sources and their internal impedances are not considered while forming the $\mathbf{Y}_{\mathbf{bus}}$ matrix.

The bus voltage magnitudes, their angles, and the power generated and consumed at each bus are given in Table 3.2. In this table, the initial assumed data of the voltages and their angles are circled. The power and reactive power generated at the slack bus and the reactive power generated at the P–V bus are unknown. Therefore, each of these quantities is indicated by a dash (–). As we shall see that we do not need these quantities for our load flow calculations. Hence, their initial estimates are not required. Moreover, note from Figure 3.5 that the slack bus does not contain any load while the P–V bus 5 has a local load, and this is indicated in the load column.

3.3 Load Flow Methods

The basic power flow Equations (3.17). (3.18) are nonlinear. Let us consider a power system that contains a total number of n buses. Of these,

- the total number of P–V (generator) buses is n_g,
- the total number of P–Q (load) buses is n_p, and
- one slack bus.

Therefore, $n = n_g + n_p + 1$. We make the following observations about the unknown quantities based on the bus types:

- *P–Q Buses*: Both voltage magnitudes and angles of these buses are unknown, resulting in $2n_p$ unknown quantities. However, their real and reactive powers are known.

- *P–V Buses*: Voltage angles and the reactive power of these buses are unknown, that is, further $2n_g$ unknown quantities.
- *Slack Bus*: The real and reactive powers are unknown.
- Therefore, the total number of unknown quantities is $2(n_p + n_g + 1) = 2n$.

The known quantities are as follows:

- *P–Q Buses*: Both real and reactive powers of these buses are known, that is, $2n_p$ known quantities.
- *P–V Buses*: Both real powers and voltage magnitudes of these buses are known, that is, further $2n_g$ known quantities.
- *Slack Bus*: The voltage magnitude and angle of the slack bus are known.
- Therefore, the total number of known quantities is $2(n_p + n_g + 1) = 2n$.

For example, consider Figure 3.5, where $n_g = 1$, $n_p = 3$, and $n = 5$. Now from Table 3.2, the unknown quantities are either circled or dashed. These are the voltage magnitudes of buses 2, 3, and 4, the angles of buses 2–5, the reactive powers of buses 1 and 4, and the real power of bus-1. Therefore, there are 10 unknown quantities. The known quantities in Table 3.2 are the real and reactive powers of the load buses 2–4, the voltage magnitude of buses 1 and 5, the network injected by bus-5, and the angle of bus-1. Therefore, there are 10 known quantities as well. Now, it can be seen from (3.20) and (3.21) that there are 10 equations, 5 each for real and reactive power. Therefore, there are as many known quantities as the number of equations. However, it is difficult to form a set of closed-form equations from (3.20) and (3.21). We, therefore, must resort to obtaining iterative solutions to the load flow problem. A noniterative technique, called the DC load flow (DCLF), determines angles of the bus voltages and the power flow between the lines under certain assumptions. This is discussed in Section 3.3.6.

3.3.1 Gauss–Seidel Load Flow Method

Consider the power system of Figure 3.5, for which the system data, including initial data, are given in Table 3.2. We define the initial bus voltage magnitudes as $|V_i|^{(0)}$, $i = 1, 2, ..., n - n_g - 1$ and their angles as $\delta_i^{(0)}, i = 1, 2, ..., n$. From these, and the active and reactive power injected, the bus voltage magnitudes and their angles for the next iteration $|V_i|^{(1)}$ and $\delta_i^{(1)}$ are computed. The process continues till both ΔP_i and ΔQ_i, given in (3.20) and (3.21) are below a small threshold value.

Equation (3.15) can be rewritten in terms of power injected as a bus as follows:

$$P_{i,inj} - jQ_{i,inj} = V_i^* \sum_{k=1}^{n} Y_{bik} V_k = V_i^* [Y_{bi1} V_1 + Y_{bi2} V_2 + \cdots + Y_{bii} V_i + \cdots + Y_{bin} V_n] \tag{3.22}$$

Therefore, the bus voltage can be written as follows:

$$V_i = \frac{1}{Y_{bii}}\left[\frac{P_{i,inj} - jQ_{i,inj}}{V_i^*} - Y_{bi1}V_1 - Y_{bi2}V_2 - \cdots - Y_{bin}V_n\right] \tag{3.23}$$

The iterative process of finding the bus voltage for the kth iteration is as follows:

$$V_i^{(k)} = \frac{1}{Y_{bii}}\left[\frac{P_{i,inj} - jQ_{i,inj}}{V_i^{*(k-1)}} - Y_{bi1}V_1 - Y_{bi2}V_2^{(k-1)} - \cdots - Y_{bin}V_{bin}^{(k-1)}\right] \tag{3.24}$$

Let us demonstrate the working principle of the Gauss–Seidel method with the help of the following example.

Example 3.3 Consider the power system of Example 3.2. The process starts at bus-2. Since this is a load bus, both real and reactive powers are known. We shall choose a base MVA of 100 to convert these quantities to per unit form. Therefore, from (3.24), we get

$$V_2^{(1)} = \frac{1}{Y_{b22}}\left[\frac{P_{2,inj} - jQ_{2,inj}}{V_2^{*(0)}} - Y_{b21}V_1 - Y_{b23}V_3^{(0)} - Y_{b24}V_4^{(0)} - Y_{b25}V_5^{(0)}\right]$$

From the $\mathbf{Y_{bus}}$ matrix of Example 3.2 and the data given in Table 3.2, we get

$$V_2^{(1)} = \frac{1}{Y_{b22}}\left[\frac{-0.96 + j0.62}{1} - 1.05Y_{b21} - Y_{b23} - Y_{b24} - 1.02Y_{b25}\right]$$
$$= 0.9927\angle -2.56^\circ$$

Note that since the real and reactive power are drawn from this bus, these quantities appear with negative signs in the equation above.

Since $V_2^{(1)}$ has already been computed, the voltage for bus-3 is given by

$$V_3^{(1)} = \frac{1}{Y_{b33}}\left[\frac{-0.35 + j0.14}{1} - 1.05Y_{b21} - Y_{b32}V_2^{(1)} - Y_{b34} - 1.02Y_{b35}\right]$$
$$= 0.9883\angle -2.83^\circ$$

Finally, the voltage of bus-4 is computed as

$$V_4^{(1)} = \frac{1}{Y_{b44}}\left[\frac{0.16 + j0.08}{V_4^{*(0)}} - 1.05Y_{b41} - Y_{b42}V_2^{(1)} - Y_{b43}V_3^{(1)} - 1.02Y_{45}\right]$$
$$= 0.9968\angle -3.48^\circ$$

Once the voltages of the P–Q buses are calculated, the voltages for the P–V buses will be calculated. The reactive power inject is expressed from (3.22) as:

$$\begin{aligned} Q_{i,inj} &= -\operatorname{Im}\left[V_i^* \sum\nolimits_{k=1}^{n} Y_{ik} V_k\right] \\ &= -\operatorname{Im}\left[V_i^* \{Y_{i1}V_1 + Y_{i2}V_2 + \cdots + Y_{ii}V_i + \cdots + Y_{in}V_n\}\right] \end{aligned} \tag{3.25}$$

For the voltage of bus-i, (3.25) is written for the kth iteration as:

$$Q_{i,inj}^{(k)} = -\operatorname{Im}\left[V_i^{*(k-1)}\left\{Y_{i1}V_1 + Y_{i2}V_2^{(k)} + \cdots + Y_{ii}V_i^{(k-1)} + \cdots + Y_{in}V_n^{(k-1)}\right\}\right] \tag{3.26}$$

The reactive power injected into bus-5 is then calculated as follows:

$$\begin{aligned} Q_{5,inj}^{(1)} &= -\operatorname{Im}\left[1.02\left\{Y_{b51}V_1 + Y_{b52}V_2^{(1)} + Y_{b53}V_3^{(1)} + Y_{b54}V_4^{(1)} + 1.02Y_{b55}\right\}\right] \\ &= -0.0899 \end{aligned}$$

The voltage of bus-5 is then evaluated from (3.24) is:

$$\begin{aligned} V_5'^{(1)} &= \frac{1}{Y_{55}}\left[\frac{P_{5,inj} - jQ_{5,inj}^{(1)}}{V_5^{*(0)}} - 1.05Y_{51} - Y_{52}V_2^{(1)} - Y_{53}V_3^{(1)} - 1.02Y_{54}\right] \\ &= 1.0169\angle -0.89^\circ \end{aligned}$$

Note that bus-5 injects 48 MW of power, while it has a local load of 24 MW. As a result, $P_{5,\,inj} = 24$ MW is chosen in the equation above. The magnitude of the bus-5 voltage is given as 1.02, which does not match the computed value. Therefore, the voltage magnitude is corrected using the following equation:

$$V_5^{(1)} = 1.02 \times \frac{V_5'^{(1)}}{\left|V_5'^{(1)}\right|} = 1.02\angle -0.89^\circ$$

This implies that the magnitude of bus-2 voltage is retained as 1.02, while the angle $\delta_5^{(1)}$ is computed from this process.

The values obtained from the first iteration are then used in the second iteration. At the end of each iteration, the mismatches ΔP_i and ΔQ_i, given in (3.20) and (3.21), respectively, are computed. The process continues till all the components of ΔP_i and ΔQ_i are below a prespecified small threshold value. For the system of this example, this threshold is chosen as 10^{-6}, and the process terminates in 28 iterations.

The Gauss–Seidel method usually takes a long time to converge and may sometimes have convergence issues. That is why it is not a preferred method used in industry. We shall now consider a more robust Newton–Raphson method.

3.3.2 Basics of Newton–Raphson Iterative Procedure

Consider a set of nonlinear equations, given by

$$\begin{aligned} f_1(x_1,\cdots,x_n) &= \eta_1 \\ f_2(x_1,\cdots,x_n) &= \eta_2 \\ &\vdots \\ f_n(x_1,\cdots,x_n) &= \eta_n \end{aligned} \tag{3.27}$$

where $x_1,\ x_2,\ ...,\ x_n$ are the unknown variables that will be estimated. We define another set of functions as follows:

$$\begin{aligned} g_1(x_1,\cdots,x_n) &= f_1(x_1,\cdots,x_n) - \eta_1 = 0 \\ g_2(x_1,\cdots,x_n) &= f_2(x_1,\cdots,x_n) - \eta_2 = 0 \\ &\vdots \\ g_n(x_1,\cdots,x_n) &= f_n(x_1,\cdots,x_n) - \eta_n = 0 \end{aligned} \tag{3.28}$$

Assuming the initial estimates of the n variables being $x_1^{(0)},\ x_2^{(0)},...,\ x_n^{(0)}$, and adding corrections $\Delta x_1^{(0)},\ \Delta x_2^{(0)},...,\ \Delta x_n^{(0)}$ to these variables such that we get the correct solution of these variables defined by

$$\begin{aligned} x_1^* &= x_1{}^{(0)} + \Delta x_1{}^{(0)} \\ x_2^* &= x_2{}^{(0)} + \Delta x_2{}^{(0)} \\ &\vdots \\ x_n^* &= x_n{}^{(0)} + \Delta x_n{}^{(0)} \end{aligned} \tag{3.29}$$

The functions in (3.28) then can be written in terms of the variables given in (3.29) as follows:

$$g_k\left(x_1^*,\cdots,x_n^*\right) = g_k\left(x_1{}^{(0)} + \Delta x_1{}^{(0)},\cdots,x_n{}^{(0)} + \Delta x_n{}^{(0)}\right), \quad k = 1,...,n \tag{3.30}$$

Equation (3.30) is expanded in Taylor's series around the nominal values of $x_1^{(0)},\ x_2^{(0)},...,\ x_n^{(0)}$. Neglecting the second- and higher-order terms of the series, the expansion of g_k, $k = 1,\ 2,\ ...,\ n$ is given as follows:

$$\begin{aligned} g_k\left(x_1^*,\cdots,x_n^*\right) = g_k\left(x_1{}^{(0)},\cdots,x_n{}^{(0)}\right) + \Delta x_1{}^{(0)}\left.\frac{\partial g_k}{\partial x_1}\right|^{(0)} + \Delta x_2{}^{(0)}\left.\frac{\partial g_k}{\partial x_2}\right|^{(0)} \\ + \cdots + \Delta x_n{}^{(0)}\left.\frac{\partial g_k}{\partial x_n}\right|^{(0)} \end{aligned} \tag{3.31}$$

where $\partial g_k/\partial x_i|^{(0)}$, $i = 1, 2, ..., n$ is the partial derivative of g_k evaluated at $x_1^{(0)}, x_2^{(0)}, ..., x_n^{(0)}$.

Equation (3.31) can be written in the vector–matrix form as:

$$\begin{bmatrix} \partial g_1/\partial x_1 & \partial g_1/\partial x_2 & \cdots & \partial g_1/\partial x_n \\ \partial g_2/\partial x_1 & \partial g_2/\partial x_2 & \cdots & \partial g_2/\partial x_n \\ \vdots & \vdots & \ddots & \vdots \\ \partial g_n/\partial x_1 & \partial g_n/\partial x_2 & \cdots & \partial g_n/\partial x_n \end{bmatrix}^{(0)} \begin{bmatrix} \Delta x_1{}^{(0)} \\ \Delta x_2{}^{(0)} \\ \vdots \\ \Delta x_n{}^{(0)} \end{bmatrix} = \begin{bmatrix} 0-g_1\left(x_1{}^{(0)},\cdots,x_n{}^{(0)}\right) \\ 0-g_2\left(x_1{}^{(0)},\cdots,x_n{}^{(0)}\right) \\ \vdots \\ 0-g_n\left(x_1{}^{(0)},\cdots,x_n{}^{(0)}\right) \end{bmatrix} \tag{3.32}$$

The square matrix of partial derivatives is called the Jacobian matrix $\mathbf{J}$ with $\mathbf{J}^{(0)}$ indicating that the matrix is evaluated for the initial values of $x_1^{(0)}, x_2^{(0)}, ..., x_n^{(0)}$. We can then write the solution of (3.32) as:

$$\begin{bmatrix} \Delta x_1{}^{(0)} \\ \Delta x_2{}^{(0)} \\ \vdots \\ \Delta x_n{}^{(0)} \end{bmatrix} = \left[\mathbf{J}^{(0)}\right]^{-1} \begin{bmatrix} \Delta g_1{}^{(0)} \\ \Delta g_2{}^{(0)} \\ \vdots \\ \Delta g_n{}^{(0)} \end{bmatrix} \tag{3.33}$$

Since Taylor's series is truncated by neglecting the second and higher-order terms, it is not expected to find the correct solution at the end of the first iteration, and hence, an iterative procedure will be required. In this process, the corrections are added to the initial estimates to form the next set of estimates as follows:

$$\begin{aligned} x_1{}^{(1)} &= x_1{}^{(0)} + \Delta x_1{}^{(0)} \\ x_2{}^{(1)} &= x_2{}^{(0)} + \Delta x_2{}^{(0)} \\ &\vdots \\ x_n{}^{(1)} &= x_n{}^{(0)} + \Delta x_n{}^{(0)} \end{aligned} \tag{3.34}$$

These are then used to find $\mathbf{J}^{(1)}$ and $\Delta g_k^{(1)}, i = 1, 2, ..., n$. From these, we can then calculate $\Delta x_1^{(1)}, \Delta x_2^{(1)}, ..., \Delta x_n^{(1)}$ from an equation like (3.33) and subsequently calculate $x_1^{(2)}, x_2^{(2)}, ..., x_n^{(2)}$. The process continues till all the mismatches Δg_k, $k = 1, 2, ..., n$ become less than a prespecified small quantity.

Example 3.4 Consider the following set of three nonlinear equations in three variables, that is, the number of equations is equal to the number of unknowns

$$\begin{aligned} g_1(x_1,x_2,x_3) &= x_1^2 - x_2^2 + x_3^2 - 11 = 0 \\ g_2(x_1,x_2,x_3) &= x_1x_2 + x_2^2 - 3x_3 - 3 = 0 \\ g_3(x_1,x_2,x_3) &= x_1 - x_1x_3 + x_2x_3 - 6 = 0 \end{aligned}$$

The Jacobian matrix of (3.32) is then

$$\mathbf{J} = \begin{bmatrix} 2x_1 & -2x_2 & 2x_3 \\ x_2 & x_1 + 2x_2 & -3 \\ 1 - x_3 & x_3 & -x_1 + x_2 \end{bmatrix}$$

The initial values to start the Newton–Raphson procedure must be carefully chosen. For example, if we choose $x_1^{(0)} = x_2^{(0)} = x_3^{(0)} = 0$, then all the elements of the first row will be zero making the matrix $\mathbf{J}^{(0)}$ singular, and therefore, non-invertible. To avoid this, let us choose $x_1^{(0)} = x_2^{(0)} = x_3^{(0)} = 1$. The Jacobian matrix is then given by

$$\mathbf{J}^{(0)} = \begin{bmatrix} 2 & -2 & 2 \\ 1 & 3 & -3 \\ 0 & 1 & 0 \end{bmatrix}$$

The mismatches are then computed as:

$$\begin{bmatrix} \Delta g_1^{(0)} \\ \Delta g_2^{(0)} \\ \Delta g_2^{(0)} \end{bmatrix} = \begin{bmatrix} 0 - g_1^{(0)} \\ 0 - g_2^{(0)} \\ 0 - g_3^{(0)} \end{bmatrix} = \begin{bmatrix} 10 \\ 4 \\ 5 \end{bmatrix}$$

Consequently, the corrections and updates calculated, respectively, from (3.33) and (3.34) and are given by:

$$\begin{bmatrix} \Delta x_1{}^{(0)} \\ \Delta x_2{}^{(0)} \\ \Delta x_3{}^{(0)} \end{bmatrix} = \begin{bmatrix} 4.75 \\ 5.00 \\ 5.25 \end{bmatrix} \text{ and } \begin{bmatrix} x_1{}^{(1)} \\ x_2{}^{(1)} \\ x_3{}^{(1)} \end{bmatrix} = \begin{bmatrix} 5.75 \\ 6.00 \\ 6.25 \end{bmatrix}$$

The process converges in seven iterations with the values of

$$x_1 = 2,\ x_2 = 3 \text{ and } x_3 = 4$$

3.3.3 Newton–Raphson Load Flow Method

Newton–Raphson load flow is similar in approach to that of solving a system of nonlinear equations using the Newton–Raphson method. However, the derivation of the Jacobian matrix is much more complicated than that discussed in

Section 3.3.2. For the load flow problem, the following equation needs to be solved:

$$\mathbf{J}\begin{bmatrix} \Delta\delta_2 \\ \vdots \\ \Delta\delta_n \\ \dfrac{\Delta|V_2|}{|V_2|} \\ \vdots \\ \dfrac{\Delta|V_{1+n_p}|}{|V_{1+n_p}|} \end{bmatrix} = \begin{bmatrix} \Delta P_2 \\ \vdots \\ \Delta P_n \\ \Delta Q_2 \\ \vdots \\ \Delta Q_{1+n_p} \end{bmatrix} \tag{3.35}$$

Note that the correction in the voltage magnitude is normalized with respect to the voltage magnitude, that is, $\Delta|V_i|/|V_i|$ has been used instead of $\Delta|V_i|$. This simplifies the calculation of the Jacobian matrix as discussed below.

The Jacobian matrix is divided into submatrices as [2]:

$$\mathbf{J} = \begin{bmatrix} \mathbf{J}_{11} & \mathbf{J}_{12} \\ \mathbf{J}_{21} & \mathbf{J}_{22} \end{bmatrix} \tag{3.36}$$

where

$$\mathbf{J}_{11} = \mathbf{L} = \begin{bmatrix} \dfrac{\partial P_2}{\partial \delta_2} & \cdots & \dfrac{\partial P_2}{\partial \delta_n} \\ \vdots & \ddots & \vdots \\ \dfrac{\partial P_n}{\partial \delta_2} & \cdots & \dfrac{\partial P_n}{\partial \delta_n} \end{bmatrix} = \begin{bmatrix} L_{22} & \cdots & L_{2n} \\ \vdots & \ddots & \vdots \\ L_{n2} & \cdots & L_{nn} \end{bmatrix} \in \Re^{(n-1)\times(n-1)} \tag{3.37}$$

$$\mathbf{J}_{12} = \mathbf{N} = \begin{bmatrix} |V_2|\dfrac{\partial P_2}{\partial |V_2|} & \cdots & |V_{1+n_p}|\dfrac{\partial P_2}{\partial |V_{1+n_p}|} \\ \vdots & \ddots & \vdots \\ |V_2|\dfrac{\partial P_n}{\partial |V_2|} & \cdots & |V_{1+n_p}|\dfrac{\partial P_n}{\partial |V_{1+n_p}|} \end{bmatrix} = \begin{bmatrix} N_{22} & \cdots & N_{2n_p} \\ \vdots & \ddots & \vdots \\ N_{n2} & \cdots & N_{nn_p} \end{bmatrix} \in \Re^{(n-1)\times n_p} \tag{3.38}$$

$$\mathbf{J}_{21} = \mathbf{M} = \begin{bmatrix} \dfrac{\partial Q_2}{\partial \delta_2} & \cdots & \dfrac{\partial Q_2}{\partial \delta_n} \\ \vdots & \ddots & \vdots \\ \dfrac{\partial Q_{1+n_p}}{\partial \delta_2} & \cdots & \dfrac{\partial Q_{1+n_p}}{\partial \delta_n} \end{bmatrix} = \begin{bmatrix} M_{22} & \cdots & M_{2n} \\ \vdots & \ddots & \vdots \\ M_{n_p2} & \cdots & M_{n_pn} \end{bmatrix} \in \Re^{n_p\times(n-1)} \tag{3.39}$$

$$\mathbf{J}_{22} = \mathbf{O} = \begin{bmatrix} |V_2|\frac{\partial Q_2}{\partial |V_2|} & \cdots & |V_{1+n_p}|\frac{\partial Q_2}{\partial |V_{1+n_p}|} \\ \vdots & \ddots & \vdots \\ |V_2|\frac{\partial Q_{1+n_p}}{\partial |V_2|} & \cdots & |V_{1+n_p}|\frac{\partial Q_{1+n_p}}{\partial |V_{1+n_p}|} \end{bmatrix} = \begin{bmatrix} O_{22} & \cdots & O_{2n_p} \\ \vdots & \ddots & \vdots \\ O_{n_p2} & \cdots & O_{n_pn_p} \end{bmatrix}$$
$$\in \Re^{n_p \times n_p} \tag{3.40}$$

The step-by-step Newton–Raphson algorithm is given below.

Step 1 (initialization): Choose the initial values of the voltage magnitudes of the P–Q buses and the voltage angles of all the buses except the slack bus.

Step 2: Using the initial values calculate the following:

a) the total $n - 1$ number of injected real power $P_{calc}^{(0)}$ and equal number of real power mismatch $\Delta P^{(0)}$ for all the buses except the slack bus,
b) the total n_p number of injected reactive power $Q_{calc}^{(0)}$ and equal number of reactive power mismatch $\Delta Q^{(0)}$ for P-Q buses, and
c) formulate the Jacobian matrix $\mathbf{J}^{(0)}$.

Step 3: Solve (3.35) for $\Delta\delta^{(0)}$ and $\Delta|V|^{(0)} \div |V|^{(0)}$.

Step 5: Obtain the updates from

$$\delta^{(1)} = \delta^{(0)} + \Delta\delta^{(0)}$$
$$|V|^{(1)} = |V|^{(0)}\left[1 + \frac{\Delta|V|^{(0)}}{|V|^{(0)}}\right] = |V|^{(0)} + \Delta|V|^{(0)} \tag{3.41}$$

Step 6: Check if all the mismatches (ΔP and ΔQ) are below a small number. Terminate the process if the answer is yes. Otherwise, go back to Step 2 to start the next iteration with the updates given by (3.41).

Example 3.5 The Newton–Raphson method is employed for the five-bus power system of Figure 3.5, for which the $\mathbf{Y_{bus}}$ matrix is computed in Example 3.2 and the initial data are listed in Table 3.2. As a first step, we shall formulate the Jacobian matrix. Let us rewrite (3.17) and (3.18) using (3.13) as follows:

$$P_i = |V_i|^2 G_{ii} + \sum_{k=1,\ k \neq i}^{n} |Y_{bik} V_i V_k| \cos(\theta_{ik} + \delta_k - \delta_i) \tag{3.42}$$

$$Q_i = -|V_i|^2 B_{ii} - \sum_{k=1,\ k \neq i}^{n} |Y_{bik} V_i V_k| \sin(\theta_{ik} + \delta_k - \delta_i) \tag{3.43}$$

Then the submatrices of the Jacobian matrix are computed in the following steps:

- Computing $\mathbf{J}_{11}$: Using (3.42), and from (3.36), the following equation is written:

$$\begin{aligned} L_{ik} &= \frac{\partial P_i}{\partial \delta_k} = -|Y_{bik} V_i V_k| \sin(\theta_{ik} + \delta_k - \delta_i), \quad i \neq k \\ &= -|Y_{bik} V_i V_k| \times [\sin(\theta_{ik}) \cos(\delta_k - \delta_i) + \cos(\theta_{ik}) \sin(\delta_k - \delta_i)] \\ &= -|V_i V_k| B_{ij} \cos(\delta_k - \delta_i) - |V_i V_k| G_{ij} \sin(\delta_k - \delta_i) \end{aligned} \tag{3.44}$$

Again, when $i = k$, we have

$$L_{ii} = \frac{\partial P_i}{\partial \delta_i} = \sum_{k=1,\ k \neq i}^{n} |Y_{bik} V_i V_k| \sin(\theta_{ik} + \delta_k - \delta_i)$$

Using (3.43), the above equation is rewritten as:

$$L_{ii} = \frac{\partial P_i}{\partial \delta_i} = -Q_i - |V_i|^2 B_{ii} \tag{3.45}$$

From the initial data given in Table 3.2, and from (3.44), we have

$$L_{23}{}^{(0)} = -\left|Y_{b23} V_2{}^{(0)} V_3{}^{(0)}\right| \sin(\theta_{23} + \delta_3 - \delta_2) = -|Y_{b23}| \sin\theta_{23} = -B_{23} = -4.8077$$

Again, from (4.43), we get

$$\begin{aligned} Q_2{}^{(0)} &= -\left|V_2{}^{(0)}\right|^2 B_{22} - \sum_{\substack{k=1 \\ k \neq 2}}^{n} \left|Y_{b2k} V_2{}^{(0)} V_k{}^{(0)}\right| \sin(\theta_{2k} + \delta_k - \delta_2) \\ &= -B_{22} - 1.05 B_{21} - B_{23} - B_{24} - 1.02 B_{25} = -0.6327 \end{aligned}$$

Therefore, from (3.45), the following is obtained:

$$L_{22}{}^{(0)} = -Q_2{}^{(0)} - \left|V_2{}^{(0)}\right|^2 B_{22} = -0.6327 - B_{22} = 18.8269$$

In a similar way, the rest of the components of the matrix $\mathbf{J}_{11}^{(0)}$ are obtained and the matrix is given by:

$$\mathbf{J}_{11}^{(0)} = \begin{bmatrix} 18.8269 & -4.8077 & 0 & -3.9231 \\ -4.8077 & 11.1058 & -3.8462 & -2.4519 \\ 0 & -3.8462 & 5.8077 & -1.9615 \\ -3.9231 & -2.4519 & -1.9615 & 12.4558 \end{bmatrix}$$

- Computing $\mathbf{J_{21}}$: From (3.43) and (3.38), the following equation can be written:

$$\begin{aligned} M_{ik} = \frac{\partial Q_i}{\partial \delta_k} &= -|Y_{bik} V_i V_k| \cos(\theta_{ik} + \delta_k - \delta_i), \quad i \neq k \\ &= -|Y_{bik} V_i V_k| \times [\cos(\theta_{ik}) \cos(\delta_k - \delta_i) - \sin(\theta_{ik}) \sin(\delta_k - \delta_i)] \\ &= -|V_i V_k| G_{ij} \cos(\delta_k - \delta_i) + |V_i V_k| B_{ij} \sin(\delta_k - \delta_i) \end{aligned} \tag{3.46}$$

On the other hand, by $i = k$, using (3.42) and (3.43), we get

$$M_{ii} = \frac{\partial Q_i}{\partial \delta_i} = \sum_{k=1, i \neq k}^{n} |Y_{bik} V_i V_k| \cos(\theta_{ik} + \delta_k - \delta_i) = P_i - |V_i|^2 G_{ii} \tag{3.47}$$

Then

$$M_{23}{}^{(0)} = -\left|Y_{b23} V_2{}^{(0)} V_3{}^{(0)}\right| \cos(\theta_{23} + \delta_2 - \delta_3) = -G_{23} = 0.9615$$

The real power injected into bus-2 is given by

$$\begin{aligned} P_2{}^{(0)} &= \left|V_2{}^{(0)}\right|^2 G_{22} + \sum_{\substack{k=1 \\ k \neq 2}}^{n} \left|Y_{b2k} V_2{}^{(0)} V_k{}^{(0)}\right| \cos(\theta_{2k} + \delta_k - \delta_2) \\ &= G_{22} + 1.05 G_{21} + G_{23} + G_{24} + 1.02 G_{25} = -0.1115 \end{aligned}$$

Therefore,

$$M_{22} = P_2{}^{(0)} - \left|V_2{}^{(0)}\right|^2 G_{22} = -3.7654$$

In a similar way, the rest of the elements of the $\mathbf{J}_{21}^{(0)}$ matrix are computed to form the matrix as follows:

$$\mathbf{J}_{21}^{(0)} = \begin{bmatrix} -3.7654 & 0.9615 & 0 & 0.7846 \\ 0.9615 & -2.2212 & 0.7692 & 0.4904 \\ 0 & 0.7692 & -1.1615 & 0.3923 \end{bmatrix}$$

- Computing $\mathbf{J_{21}}$: It is evident from (3.42) that the elements of this matrix are the derivatives of the real power P with respect to the magnitude of bus voltage $|V|$. Therefore, the off-diagonal elements are:

$$N_{ik} = |V_k| \frac{\partial P_i}{\partial |V_k|} = |Y_{bik} V_i V_k| \cos(\theta_{ik} + \delta_k - \delta_i) = -M_{ik} \quad i \neq k \tag{3.48}$$

On the other hand, the diagonal elements are given by

$$N_{ii} = |V_i|\frac{\partial P_i}{\partial |V_i|} = |V_i|\left[2|V_i|G_{ii} + \sum_{\substack{k=1\\k\neq i}}^{n}|Y_{bik}V_k|\cos(\theta_{ik}+\delta_k-\delta_i)\right]$$
$$= 2|V_i|^2G_{ii} + \sum_{\substack{k=1\\k\neq i}}^{n}|Y_{bik}V_iV_k|\cos(\theta_{ik}+\delta_k-\delta_i) = 2|V_i|^2G_{ii} + M_{ii} \tag{3.49}$$

For calculating the elements of the matrix $\mathbf{J_{12}}$, $j \neq k$, we note from (3.48) that they are negative of the elements of $\mathbf{J_{21}}$ when $j \neq k$. However, the size of $\mathbf{J_{21}}$ is (3×4) while the size of $\mathbf{J_{12}}$ is (4×3). Then, first, we need to make the following substitution:

$$\mathbf{J}_{12}^{(0)} = -\left[\mathbf{J}_{21}^{(0)}\right]^T$$

This, however, will lead to wrong diagonal elements. Therefore, the elements of $\mathbf{J_{21}}$ when $j = k$ are replaced by those computed using (3.49). The matrix is then given by

$$\mathbf{J}_{12}^{(0)} = \begin{bmatrix} 3.5423 & -0.9615 & 0 \\ -0.9615 & 2.2019 & -0.7692 \\ 0 & -0.7692 & 1.1462 \\ -0.7846 & -0.4904 & -0.3923 \end{bmatrix}$$

- Computing $\mathbf{J_{22}}$: For $j \neq k$, the following equation is written:

$$O_{ik} = |V_k|\frac{\partial Q_i}{\partial |V_k|} = -|Y_{bik}V_iV_k|\sin(\theta_{ik}+\delta_k-\delta_i) = L_{ik}, \quad i \neq k \tag{3.50}$$

Moreover, for $j = k$, the following expression is obtained:

$$O_{ii} = |V_i|\frac{\partial Q_i}{\partial |V_i|} = |V_i|\left[-2|V_i|B_{ii} - \sum_{\substack{k=1\\k\neq i}}^{n}|Y_{bik}V_k|\sin(\theta_{ik}+\delta_k-\delta_i)\right]$$
$$= -2|V_i|^2B_{ii} - \sum_{\substack{k=1\\k\neq i}}^{n}|Y_{bik}V_iV_k|\sin(\theta_{ik}+\delta_k-\delta_i) = -2|V_i|^2B_{ii} - L_{ii}$$
$$\tag{3.51}$$

For computing the matrix, we first set

$$\mathbf{J}_{22}^{(0)} = \mathbf{J}_{11}^{(0)}(1:3, 1:3)$$

Then the diagonal elements are computed using (3.51). This results in the following matrix:

$$\mathbf{J}_{22}^{(0)} = \begin{bmatrix} 17.5615 & -4.8077 & 0 \\ -4.8077 & 10.8996 & -3.8462 \\ 0 & -3.8462 & 5.5408 \end{bmatrix}$$

From the initial conditions, the active and reactive powers are computed as:

$$P_{calc}{}^{(0)} = [-0.1115 \quad -0.0096 \quad -0.0077 \quad -0.0098]^T$$

$$Q_{calc}{}^{(0)} = [-0.6327 \quad -0.1031 \quad -0.1335]^T$$

Consequently, the mismatches are found as:

$$\Delta P^{(0)} = [-0.8485 \quad -0.3404 \quad -0.1523 \quad 0.2302]^T$$

$$\Delta Q^{(0)} = [0.0127 \quad -0.0369 \quad 0.0535]^T$$

Thereafter, the updates are calculated. At the end of the first iteration, these are given by

$$\begin{bmatrix} \delta_2{}^{(0)} \\ \delta_3{}^{(0)} \\ \delta_3{}^{(0)} \\ \delta_4{}^{(0)} \end{bmatrix} = \begin{bmatrix} -4.91^\circ \\ -6.95^\circ \\ -7.19^\circ \\ -3.09^\circ \end{bmatrix} \text{ and } \begin{bmatrix} |V_2|^{(0)} \\ |V_3|^{(0)} \\ |V_4|^{(0)} \end{bmatrix} = \begin{bmatrix} 0.9864 \\ 0.9817 \\ 0.9913 \end{bmatrix}$$

Progressing in this fashion, the load flow converges in seven iterations. The results of the Newton–Raphson load flow are listed in Table 3.3. Here, the real and reactive powers are converted into per unit using a base MVA of 100. It is to be noted that the real and reactive power of the slack bus and the reactive power of the P–V bus are computed from (3.17) and (3.18) after the convergence of the load flow.

3.3.4 Fast Decoupled Load Flow

The Newton–Raphson load flow method is very robust – it has excellent convergence properties provided that the initial conditions are within reasonable bounds. The drawback of this method, however, is that all the elements of the Jacobian matrix must be recalculated at every iteration. Furthermore, at each iteration, the entire set of equations must also be recalculated to form the Jacobian matrix.

Table 3.3 Bus voltages, power generated, and load after load flow convergence using the Newton–Raphson method.

	Bus voltage		Power generated		Load	
Bus no.	**Magnitude (pu)**	**Angle (deg)**	**P (MW)**	**Q (MVAr)**	**P (MW)**	**P (MVAr)**
1	1.05	0	126.60	57.11	0	0
2	0.9826	−5.0124	0	0	96	62
3	0.9777	−7.1322	0	0	35	14
4	0.9876	−7.3705	0	0	16	8
5	1.02	−3.2014	48	15.59	24	11

Therefore, the Newton–Raphson method is very computationally intensive, which can be substantial for a large power system. The fast decoupled method is used to speed up computation. This is achieved by decoupling the elements of the Jacobian matrix under certain assumptions.

Consider the Jacobian matrix and its block elements given in (3.37) to (3.40). To simplify these matrices, we have to make a number of assumptions. These assumptions are listed below.

1) In a power system, the real power is mostly dependent on the angles of the sending and receiving ends. The bus voltage magnitudes at the two ends have very little influence on the power flow through the line, especially when both these voltages are operating near 1.0 per unit. We can, therefore, assume that

$$|V_k|\frac{\partial P_i}{\partial |V_k|} = 0, \ \forall i, k \tag{3.52}$$

2) Similarly, the reactive power flowing through a line is mostly dependent on the bus voltage magnitude and not on the angles of the bus voltages. This leads to the following assumption

$$\frac{\partial Q_i}{\partial \delta_k} = 0, \ \forall i, k \tag{3.53}$$

3) Since the difference between the bus voltage angles is small, we can assume that

$$\cos(\delta_i - \delta_k) = 1 \text{ and } \sin(\delta_i - \delta_k) = \delta_i - \delta_k, \ \forall i, k \tag{3.54}$$

4) In a power system, the line resistance is usually much smaller than the line reactance. This results in line susceptance B_{ik} being much larger than the line conductance G_{ik}. Therefore, we can make the following assumption

$$G_{ik}\sin(\delta_i-\delta_k) << B_{ik}\cos(\delta_i-\delta_k) \Rightarrow G_{ik}(\delta_i-\delta_k) << B_{ik},\ \forall i,k \tag{3.55}$$

5) Finally, the reactive power injected into a bus during normal operation is much less than the reactive power that would flow if all the lines from that bus were short circuited to the reference. That is

$$Q_i << B_{ii}|V_i|^2 \tag{3.56}$$

Based on the assumptions (1) and (2) above, we find that the submatrices $\mathbf{J}_{21}$ and $\mathbf{J}_{21}$ of the Jacobian matrix are null matrices. Again from (3.45), we find the off-diagonal elements of the matrix $\mathbf{J}_{11}$ are as follows:

$$\begin{aligned}\frac{\partial P_i}{\partial \delta_k} &= -|Y_{bik}V_iV_k|\sin(\theta_{ik}+\delta_k-\delta_i), \quad i\neq k\\ &= -|Y_{bik}V_iV_k|\{\sin\theta_{ik}\cos(\delta_k-\delta_i)+\cos\theta_{ik}\sin(\delta_k-\delta_i)\}\end{aligned}$$

Using (3.13), the above equation can be rewritten as:

$$\frac{\partial P_i}{\partial \delta_k} = -|V_iV_k|\{B_{ik}\cos(\delta_k-\delta_i)+G_{ik}\sin(\delta_k-\delta_i)\}$$

Based on the assumption (3.55), the above equation reduces to

$$\frac{\partial P_i}{\partial \delta_k} = -|V_iV_k|B_{ik},\ i\neq k \tag{3.57}$$

Moreover, from (3.49), we can write:

$$|V_k|\frac{\partial P_i}{\partial |V_k|} = -|V_iV_k|B_{ik},\ i\neq k \tag{3.58}$$

The diagonal elements of $\mathbf{J}_{11}$ and $\mathbf{J}_{22}$ have the same expressions that are given, respectively, by (3.46) and (3.52), that is,

$$\frac{\partial P_i}{\partial \delta_i} = -Q_i-|V_i|^2B_{ii}$$

$$|V_i|\frac{\partial Q_i}{\partial |V_i|} = -2|V_i|^2B_{ii}-\frac{\partial P_i}{\partial \delta_i}$$

We can then write the power flow adjustments as follows:

$$\Delta P_i = \left(\frac{\partial P_i}{\partial \delta_k}\right)\Delta\delta_k \tag{3.59}$$

$$\Delta Q_i = \left(\frac{\partial Q_i}{\partial |V_k|/|V_k|}\right)\frac{\Delta|V_k|}{|V_k|} \tag{3.60}$$

Substituting (3.57) in (3.59) and (3.58) in (3.60), the following two expressions are obtained:

$$\Delta P_i = -|V_i V_k| B_{ik} \Delta\delta_k \tag{3.61}$$

$$\Delta Q_i = -|V_i V_k| B_{ik} \frac{\Delta|V_k|}{|V_k|} \tag{3.62}$$

Since all the bus voltages are expected to be close to 1.0 per unit, we can assume $|V_k| = 1.0$ per unit. Then (3.61) and (3.62) can further be simplified as:

$$\frac{\Delta P_i}{|V_i|} = -B_{ik} \Delta\delta_k \tag{3.63}$$

$$\frac{\Delta Q_i}{|V_i|} = -B_{ik} \frac{\Delta|V_k|}{|V_k|} = -B_{ik} \Delta|V_k| \tag{3.64}$$

All these simplifications lead to the following two vector–matrix relations for the real and reactive power:

$$\begin{bmatrix} \Delta P_2/|V_2| \\ \Delta P_3/|V_3| \\ \vdots \\ \Delta P_n/|V_n| \end{bmatrix} = -\begin{bmatrix} B_{22} & B_{23} & \cdots & B_{2n} \\ B_{32} & B_{33} & \cdots & B_{3n} \\ \vdots & \vdots & \ddots & \vdots \\ B_{n2} & B_{n3} & \cdots & B_{nn} \end{bmatrix} \begin{bmatrix} \Delta\delta_2 \\ \Delta\delta_3 \\ \vdots \\ \Delta\delta_n \end{bmatrix} = \mathbf{B}' \begin{bmatrix} \Delta\delta_2 \\ \Delta\delta_3 \\ \vdots \\ \Delta\delta_n \end{bmatrix} \tag{3.65}$$

$$\begin{bmatrix} \Delta Q_2/|V_2| \\ \Delta Q_3/|V_3| \\ \vdots \\ \Delta Q_n/|V_n| \end{bmatrix} = -\begin{bmatrix} B_{22} & B_{23} & \cdots & B_{2n} \\ B_{32} & B_{33} & \cdots & B_{3n} \\ \vdots & \vdots & \ddots & \vdots \\ B_{n2} & B_{n3} & \cdots & B_{nn} \end{bmatrix} \begin{bmatrix} \Delta|V_2| \\ \Delta|V_3| \\ \vdots \\ \Delta|V_n| \end{bmatrix} = \mathbf{B}' \begin{bmatrix} \Delta|V_2| \\ \Delta|V_3| \\ \vdots \\ \Delta|V_n| \end{bmatrix} \tag{3.66}$$

Notice that both the matrices of the right-hand sides of (3.65) and (3.66) are the same, which is denoted by the matrix $\mathbf{B}'$. This matrix contains the susceptances of the $\mathbf{Y_{bus}}$ matrix. The elements of this matrix are constant and, hence, can be computed beforehand. The inverse of the matrix $\mathbf{B}'$ can be precomputed and stored. Moreover, note that at voltage-controlled (P–V) buses, Q is not specified and ΔV is zero. We can, therefore, eliminate these buses from the computation of (3.66). This is an advantage of this method. However, the algorithm may fail to converge if some of the assumptions are invalid. In particular, the assumption of $R < < X$ is very crucial. This assumption, however, is true for modern high-voltage transmission lines.

The step-by-step fast decoupled procedure is as follows:

Step 1: Choose the initial values of the voltage magnitudes $|V|^{(0)}$ of all n_p load buses and $(n-1)$ angles $\delta^{(0)}$ of the voltages of all the buses except the slack bus.
Step 2: Use the estimated $|V|^{(0)}$ and $\delta^{(0)}$ to calculate a total $(n-1)$ number of injected real power $P_{calc}^{(0)}$ and equal number of real power mismatch $\Delta P^{(0)}/|V|^{(0)}$.
Step 3: Solve (3.65) for $\Delta\delta$ and update δ.
Step 4: From the updated angles δ, calculate the mismatches $\Delta Q^{(0)}/|V|^{(0)}$.
Step 5: Solve (3.66) for $\Delta|V|$ and update $|V|$.
Step 6: Check if all the mismatches are below a small number. Terminate the process if yes. Otherwise, go back to Step 2 to start the next iteration.

Example 3.6 Let us consider the same system, as given in Example 3.3. The matrix $\mathbf{B}'$ given in (3.66) is computed as:

$$\mathbf{B}' = \begin{bmatrix} 18.1942 & -4.8077 & 0 & -3.8462 \\ -4.8077 & 11.0027 & -3.8462 & -2.4038 \\ 0 & -3.8462 & 5.6742 & -1.9231 \\ -3.8462 & -2.4038 & -1.9231 & 11.8942 \end{bmatrix}$$

With the initial data given and neglecting bus-1, the real power and real power mismatches are:

$$P_{calc} = [-0.1115 \quad -0.0096 \quad -0.0071 \quad 0.0098]^T$$
$$\Delta P = [-0.8455 \quad -0.3404 \quad -0.1523 \quad 0.2302]^T$$

The angle corrections are then obtained as:

$$\begin{bmatrix} \Delta\delta_2 \\ \Delta\delta_3 \\ \Delta\delta_4 \\ \Delta\delta_4 \end{bmatrix}^{(0)} = (\mathbf{B}')^{-1} \begin{bmatrix} -0.8455 \\ -0.3404 \\ -0.1523 \\ 0.2302 \end{bmatrix} = \begin{bmatrix} -0.0951 \\ -0.1344 \\ -0.1388 \\ -0.0614 \end{bmatrix} \text{rad}$$

The angles are then calculated as:

$$\begin{bmatrix} \delta_2 \\ \delta_3 \\ \delta_4 \\ \delta_4 \end{bmatrix}^{(1)} = \begin{bmatrix} \delta_2 \\ \delta_3 \\ \delta_4 \\ \delta_4 \end{bmatrix}^{(1)} + \begin{bmatrix} \Delta\delta_2 \\ \Delta\delta_3 \\ \Delta\delta_4 \\ \Delta\delta_4 \end{bmatrix}^{(0)} = \begin{bmatrix} -0.951 \\ -0.1344 \\ -0.1388 \\ -0.0614 \end{bmatrix} \text{rad}$$

With these values of the angles and neglecting bus-1, the reactive power and reactive power mismatches are calculated as:

$$Q_{calc} = [-0.4006 \quad -0.0226 \quad -0.0939 \quad -0.1006]^T$$
$$\Delta Q = [-0.2194 \quad -0.1174 \quad -0.0139 \quad -0.2494]^T$$

Therefore, using (3.66), we have

$$\begin{bmatrix} \Delta|V_2| \\ \Delta|V_3| \\ \Delta|V_4| \end{bmatrix}^{(0)} = \begin{bmatrix} -0.0182 \\ -0.0233 \\ -0.0133 \end{bmatrix} \text{ per unit}$$

Consequently, the voltage magnitudes are given by:

$$\begin{bmatrix} |V_2| \\ |V_3| \\ |V_4| \end{bmatrix}^{(1)} = \begin{bmatrix} |V_2| \\ |V_3| \\ |V_4| \end{bmatrix}^{(1)} + \begin{bmatrix} \Delta|V_2| \\ \Delta|V_3| \\ \Delta|V_4| \end{bmatrix}^{(0)} = \begin{bmatrix} 0.9818 \\ 0.9767 \\ 0.9867 \end{bmatrix} \text{ per unit}$$

The load flow algorithms converge in six iterations, and the results are identical to those given in Table 3.3.

3.3.5 Line Flows

The bus voltage magnitudes and angles of each bus along with the power generated and consumed at each bus are given in Table 3.3. It can be seen from this table that the total power generated is 174.6 MW, whereas the total load is 171 MW. This indicates that there is a total line loss of about 3.6 MW for all the lines put together.

The current flowing between the buses i and k can be written as:

$$I_{ik} = -Y_{bik}(V_i - V_k), \quad i \neq k \tag{3.67}$$

Therefore, the complex power leaving bus-i is given by

$$P_i + jQ_i = V_i I_i^* \tag{3.68}$$

Similarly, the complex power entering bus-k is

$$P_k + jQ_k = V_k I_k^* \tag{3.69}$$

Therefore, the I^2R losses in the different line segments are

$$P_{loss,i-k} = P_i - P_k \tag{3.70}$$

The real power flow over different lines is listed in Table 3.4. This table also gives the I^2Rloss along various segments. It is evident that all the losses add up to 3.6 MW, which is the net difference between power generation and load. Finally, we can compute the line I^2X drops in a similar fashion. This drop is given by

$$Q_{drop,i-k} = Q_i - Q_k \tag{3.71}$$

However, we must consider the effect of line charging separately.

Consider the line segment 1–2. The voltage of bus-1 is $V_1 = 1.05\angle 0°$ per unit, while that of bus-2 is $V_2 = 0.9826\angle -5.0124°$ per unit. From (3.67), we then have

Table 3.4 Real power flow over different lines.

Power dispatched		Power received		
From (bus)	Amount (MW)	In (bus)	Amount (MW)	Line loss (MW)
1	101.0395	2	98.6494	2.3901
1	25.5561	5	25.2297	0.3264
2	17.6170	3	17.4882	0.1288
3	0.7976	4	0.7888	0.0089
5	15.1520	2	14.9676	0.1844
5	18.6212	3	18.3095	0.3117
5	15.4566	4	15.2112	0.2454
				Total = 3.5956

$$I_{12} = 0.9623 - j0.5187 = 1.0932\angle -28.33^\circ \text{ per unit.}$$

The reactive power drop in the line segment 1–2 is

$$|I_{12}|^2 \times X_{12} \times 100 = 1.0932^2 \times 0.1 \times 100 = 10.9508 \text{ MVAr}$$

The same quantity can also be obtained by subtracting the reactive power absorbed by bus-2 from that supplied by bus-1. The above calculation however does not include the line charging. Note that, since the line is modeled by an equivalent-π, the voltage across the shunt capacitor is the bus voltage to which the shunt capacitor is connected. Therefore, the current I_{12} flowing through the line segment is not the current leaving bus-1 or entering bus-2; it is the current flowing in between the two charging capacitors. Since the shunt branches are purely reactive, the real power flow is not affected by the charging capacitors. Each charging capacitor is assumed to inject a reactive power that is the product of the half-line charging admittance and square of the magnitude of the voltage of that bus. The half-line charging admittance of this line is 0.03. Therefore, the line charging capacitor will inject the following reactive power at bus-1:

$$0.03 \times 100 \times |V_1|^2 = 3.3075 \text{ MVAr}$$

Similarly, the reactive injected at bus-2 will be

$$0.03 \times 100 \times |V_2|^2 = 2.8968 \text{ MVAr}$$

The power flow through line segments 1–2 and 1–5 is shown in Figure 3.6, where the dotted lines indicate the reactive power and the solid lines indicate the real power.

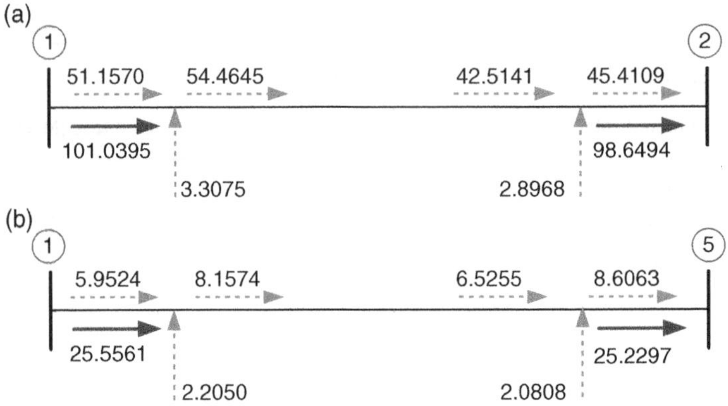

Figure 3.6 Real and reactive power flow through (a) line segment 1-2 and (b) line segment 1-5.

3.3.6 DC Load Flow

The Newton–Raphson and fast decoupled load flow techniques are usually sensitive to initial parameters – their convergence depends on the initial values chosen. One way to choose the initial starting point is to use what is called DCLF. This is a non-repetitive method, which obviously is less accurate than AC load flow techniques.

In this method, the nonlinear AC system model is simplified in a linear form using the following assumptions:

1) Line resistances are negligible. In modern power lines, the resistances are much less than the reactance ($R << X$), and therefore, the resistances can be ignored.
2) Voltage angle differences are small and, therefore, $\sin\delta = \delta$ and $\cos\delta = 1$.
3) Magnitudes of bus voltages are assumed to be 1.0 per unit. Since the automatic voltage regulators tightly hold the bus voltages constant, they can be assumed to be 1.0 per unit.
4) Reactive power flow in the network is neglected.

Since the voltage magnitudes are assumed to be 1 per unit, (3.65) can be rewritten as:

$$\begin{bmatrix} P_2 \\ P_3 \\ \vdots \\ P_n \end{bmatrix} = \mathbf{B}' \begin{bmatrix} \delta_2 \\ \delta_3 \\ \vdots \\ \delta_n \end{bmatrix} \tag{3.72}$$

Example 3.7 Consider the system of Figure 3.5. The Y-bus matrix is formed neglecting the line resistance and the line charging capacitors as given below:

$$\mathbf{Y_{bus}} = j\begin{bmatrix} -14 & 10 & 0 & 0 & 4 \\ 4 & -19 & 5 & 0 & 4 \\ 0 & 5 & -11.5 & 4 & 2.5 \\ 0 & 0 & 4 & -6 & 2 \\ 4 & 4 & 2.5 & 2 & -j12.5 \end{bmatrix}$$

Therefore, the matrix $\mathbf{B}^{'}$ is

$$\mathbf{B}' = \begin{bmatrix} 19 & -5 & 0 & -4 \\ -5 & 11.5 & -4 & -2.5 \\ 0 & -4 & 6 & -2 \\ -4 & -2.5 & -2 & 12.5 \end{bmatrix}$$

Then, the angles are given by

$$\begin{bmatrix} \delta_2 \\ \delta_3 \\ \delta_4 \\ \delta_5 \end{bmatrix} = [\boldsymbol{B}']^{-1}\begin{bmatrix} P_2 \\ P_3 \\ P_4 \\ P_5 \end{bmatrix} = \begin{bmatrix} -0.96 \\ -0.35 \\ -0.16 \\ 0.24 \end{bmatrix} \text{rad} = \begin{bmatrix} -5.65^\circ \\ -7.68^\circ \\ -7.81^\circ \\ -3.49^\circ \end{bmatrix}$$

Compare these values with those given in Table 3.3. Even if these values do not match, the errors are not significant.

Now the power transfer between two buses i and k that is separated by a reactance X_{ik} is given by

$$P_{ik} = \frac{V_i V_k}{X_{ik}} \sin(\delta_i - \delta_k)$$

Since the voltages are assumed to be 1 per unit, and the angle differences are assumed to be small, the above equation is rewritten as:

$$P_{ik} = \frac{1}{X_{ik}}(\delta_i - \delta_k) \tag{3.73}$$

The power flow over the lines is then calculated and is shown in Figure 3.7.

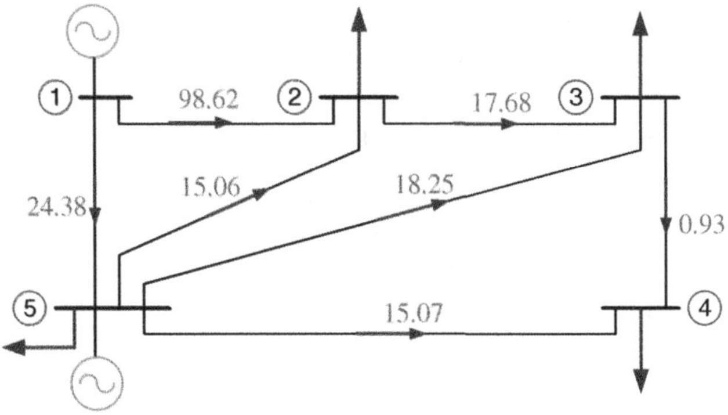

Figure 3.7 Power flow over the lines calculated through DC load flow.

3.4 State Estimation

State estimation is somewhat like load flow, except for the fact that state estimation can handle uncertainties that are associated with online load flow calculations [1]. These uncertainties can arise due to one or more of the following factors:

- Errors in measurement or meter errors.
- Errors in data communications.
- Errors in mathematical models.
- Unexpected system changes.

In general, load flow studies are performed offline for system planning, while state estimation is performed online for power system control [1]. However, online load flow can be a part of state estimation as well. State estimation can provide real-time information on power grids' operating conditions from redundant, noisy, and bad data, as will be discussed in this section. These and other measurements can be used to calculate the bus voltage magnitudes, angles, and power flow through different line segments using energy management systems (EMS).

3.4.1 Principles of Estimation

A linear dynamical system can be represented in the form

$$\dot{\mathbf{x}} = \mathbf{A}\mathbf{x} + \mathbf{B}\mathbf{u}, \quad y = \mathbf{C}\mathbf{x} \tag{3.74}$$

where $\mathbf{x}$ is the vector of state variable, $\mathbf{u}$ is the input vector or scalar, and $\mathbf{A}$ and $\mathbf{B}$ are matrices defining the system. The system output is denoted by y and the matrix

C relates the output with the states. The feedback control law is typically of the form

$$\mathbf{u} = -\mathbf{K}\mathbf{x} \tag{3.75}$$

where **K** is the feedback gain matrix. It is obvious that all the measurements of the states will be required for forming the control law. In the absence of full measurements, some of the states will have to be estimated from the measurements of the output and input. The estimator here is usually called an observer.

Now consider the following stochastic state space model:

$$\begin{aligned} \mathbf{x}(k+1) &= \mathbf{F}\mathbf{x}(k) + \mathbf{G}\mathbf{u}(k) + \mathbf{D}\mathbf{w}(k) \\ y(k) &= \mathbf{H}\mathbf{x}(k) + \mathbf{v}(k) \end{aligned} \tag{3.76}$$

where $\mathbf{w}(k)$ and $\mathbf{v}(k)$ are zero-mean, uncorrelated, Gaussian vectors. The main aim is to predict the state variable at a time $k+1$, given the measurements up to time k. The predicted quantity is denoted by $\hat{\mathbf{x}}(k+1|k)$. Due to the unpredictability of the Gaussian noise vectors, it is rather difficult to use an observer to obtain the estimates. A standard technique is to use Kalman Filter for this purpose, which first estimates $\hat{\mathbf{x}}(k|k)$ based on measurements, and then obtain the estimates, which are of the form

$$\hat{\mathbf{x}}(k+1|k) = \mathbf{F}\hat{\mathbf{x}}(k|k) + \mathbf{G}\mathbf{u}(k) \tag{3.77}$$

Given the impedance between two nodes, the real and reactive power flow through a line joining the nodes can be calculated if the magnitude and the phase angles of these node voltages are known. From this perspective, the voltage magnitudes and the phase angles are chosen as the state variables in power system state estimation, where the angles are measured with respect to a common reference point. The main idea is to produce the "best estimated" values of the magnitudes and angles from the measurements. This is obviously required when the measurements are not perfect and are prone to sensor errors. These errors can be small or large. Therefore, the estimator will have a built-in bad data detection scheme. Furthermore, the measured values are sent to a dispatch center through communication channels, which are prone to failures. Therefore, the operators at the dispatch centers require a robust procedure to obtain the required information about the bus voltages.

3.4.2 Maximum-Likelihood Estimation

Consider a set of measurements z, given by the batch model

$$\mathbf{z} = \begin{bmatrix} z_1 \\ z_2 \\ \vdots \\ z_m \end{bmatrix} = \begin{bmatrix} h_1(x_1, x_2, \ldots, x_n) \\ h_2(x_1, x_2, \ldots, x_n) \\ \vdots \\ h_m(x_1, x_2, \ldots, x_n) \end{bmatrix} + \begin{bmatrix} \varepsilon_1 \\ \varepsilon_2 \\ \vdots \\ \varepsilon_m \end{bmatrix} = \mathbf{h}(\mathbf{x}) + \boldsymbol{\varepsilon} \tag{3.78}$$

where $\mathbf{x} = \begin{bmatrix} x_1 & x_2 & \cdots & x_n \end{bmatrix}^T$ is a state vector and $\boldsymbol{\varepsilon} = \begin{bmatrix} \varepsilon_1 & \varepsilon_2 & \cdots & \varepsilon_m \end{bmatrix}^T$ is a vector of measurement errors. These errors are assumed to be zero-mean random variables that have Gaussian distribution. The properties of the variables are:

- The variables are zero mean, that is, their expected values are zero, $E(\varepsilon_i) = 0,\ i = 1, \cdots, m$.
- The variances of these variables are defined by $E\left(\varepsilon_i^2\right) = \sigma_i^2,\ i = 1, \cdots, m$.
- The variables are independent of each other, that is, $E(\varepsilon_i \varepsilon_j) = 0$.
- Since the variables are independent of each other, the off-diagonal elements of the covariance matrix will be zero. The matrix is then given by:

$$\mathbf{R} = E\left(\boldsymbol{\varepsilon}\boldsymbol{\varepsilon}^T\right) = \begin{bmatrix} \sigma_1^2 & 0 & \cdots & 0 \\ 0 & \sigma_2^2 & \cdots & 0 \\ \vdots & \vdots & \ddots & \vdots \\ 0 & 0 & \cdots & \sigma_m^2 \end{bmatrix} \tag{3.79}$$

A Gaussian or normal random variable z with a mean of μ_z and variance of σ_z^2 has a probability density function of

$$f(z) = \frac{1}{\sqrt{2\pi\sigma_z^2}} \exp\left[-\frac{\left(z - \mu_z\right)^2}{2\sigma_z^2}\right] \tag{3.80}$$

The logarithmic value of the above function is

$$\log f(z) = -\frac{\left(z - \mu_z\right)^2}{2\sigma_z^2} - \log\sqrt{2\pi\sigma_z^2} = -\frac{1}{2}\left(\frac{z - \mu_z}{\sigma_z}\right)^2 - \frac{1}{2}\log 2\pi - \log\sigma_z \tag{3.81}$$

If there are m Gaussian and independent (uncorrelated) random variables, then their joint probability density function can be given as the product of the probability density function of each variable, that is,

$$f_n(z) = f(z_1) f(z_2) \cdots f(z_m) \tag{3.82}$$

where $f_m(z)$ is called the likelihood function of z. Then we have

$$L = \log f_n(z) = \log f(z_1) + \log f(z_2) + \cdots + \log f(z_m) \tag{3.83}$$

Substituting (3.81) in (3.83), we get

$$L = -\frac{1}{2}\sum_{i=1}^{m}\left(\frac{z_i - \mu_{zi}}{\sigma_i}\right)^2 - \frac{m}{2}\log 2\pi - \sum_{i=1}^{m}\log\sigma_i \tag{3.84}$$

The maximum-likelihood estimate is obtained by maximizing L with respect to z. From (3.84), this is equivalent to minimizing the following function with respect to z:

$$\text{Minimize} \sum_{i=1}^{m} \left(\frac{z_i - \mu_{zi}}{\sigma_i} \right)^2 \tag{3.85}$$

Let the residual error of the measurement i is defined by

$$r_i = z_i - Ez_i, \; i = 1, ..., m \tag{3.86}$$

Since ε_i is zero-mean, (3.86) can be written from (3.78) as:

$$r_i = z_i - h_i(x), \; i = 1, ..., m \tag{3.87}$$

Equation (3.88) can be written in the floowing matrix–vector form:

$$\mathbf{r} = \begin{bmatrix} r_1 \\ r_2 \\ \vdots \\ r_m \end{bmatrix} = \begin{bmatrix} z_1 \\ z_2 \\ \vdots \\ z_m \end{bmatrix} - \begin{bmatrix} h_1(x_1, x_2, ..., x_n) \\ h_2(x_1, x_2, ..., x_n) \\ \vdots \\ h_m(x_1, x_2, ..., x_n) \end{bmatrix} = \mathbf{z} - \mathbf{h}(\mathbf{x}) \tag{3.88}$$

The estimation error can be minimized by minimizing **r** with respect to **x**. This is achieved using the weighted least squares (WLS) method, given by [3]

$$J(\mathbf{x}) = [\mathbf{z} - \mathbf{h}(\mathbf{x})]^T \mathbf{W} [\mathbf{z} - \mathbf{h}(\mathbf{x})] \tag{3.89}$$

where the matrix **W** contains the reciprocal of covariances given in (3.79), that is,

$$\mathbf{W} = \begin{bmatrix} 1/\sigma_1^2 & 0 & \cdots & 0 \\ 0 & 1/\sigma_2^2 & \cdots & 0 \\ \vdots & \vdots & \ddots & 0 \\ 0 & 0 & \cdots & 1/\sigma_m^2 \end{bmatrix} \tag{3.90}$$

Equation (3.89) is expanded as:

$$J(\mathbf{x}) = \mathbf{z}^T \mathbf{W} \mathbf{z} - \mathbf{z}^T \mathbf{W} \mathbf{h}(\mathbf{x}) - \mathbf{h}^T(\mathbf{x}) \mathbf{W} \mathbf{z} + \mathbf{h}^T(\mathbf{x}) \mathbf{W} \mathbf{h}(\mathbf{x}) \tag{3.91}$$

The derivative of J(**x**) is taken with respect to **x** and is equated to zero for obtaining the estimates of the state vector.

3.4.3 DC State Estimation

Assume that the angles $\delta_1, \delta_2, \ldots, \delta_n$ are assumed to be state variables, that is,

$$\mathbf{x} = [\delta_2 \quad \delta_3 \quad \cdots \quad \delta_n]^T$$

In the DC state estimation, the batch model of (3.78) is linear and is written as:

$$\mathbf{z} = \mathbf{Hx} + \boldsymbol{\varepsilon} \tag{3.92}$$

Therefore, the cost function of (3.91) is given by the following equation:

$$J(\mathbf{x}) = \mathbf{z}^T\mathbf{Wz} - \mathbf{z}^T\mathbf{WH} - \mathbf{H}^T\mathbf{Wz} + \mathbf{H}^T\mathbf{WH} \tag{3.93}$$

Taking the derivative of $J(\mathbf{x})$ with respect to $\mathbf{x}$ and equating it equal to zero, we have

$$-2\mathbf{H}^T\mathbf{Wz} + \mathbf{H}^T\Omega\mathbf{H}\hat{\mathbf{x}} = 0 \tag{3.94}$$

where $\hat{\mathbf{x}}$ is the estimate of the state vector $\mathbf{x}$. The estimates are then given by

$$\hat{\mathbf{x}} = \left(\mathbf{H}^T\mathbf{WH}\right)^{-1}\mathbf{H}^T\mathbf{Wz} \tag{3.95}$$

Example 3.8 Consider the four-bus power system shown in Figure 3.8. This has two generator buses – 1 and 4, of which bus-1 is taken as the slack bus. The load buses 2 and 3 draw 100 MW and 50 MW of power, respectively. The per unit line reactances are chosen as:

$$X_{12} = 0.15,\ X_{14} = 0.3,\ X_{23} = 0.3,\ X_{24} = 0.5 \text{ and } X_{34} = 0.15$$

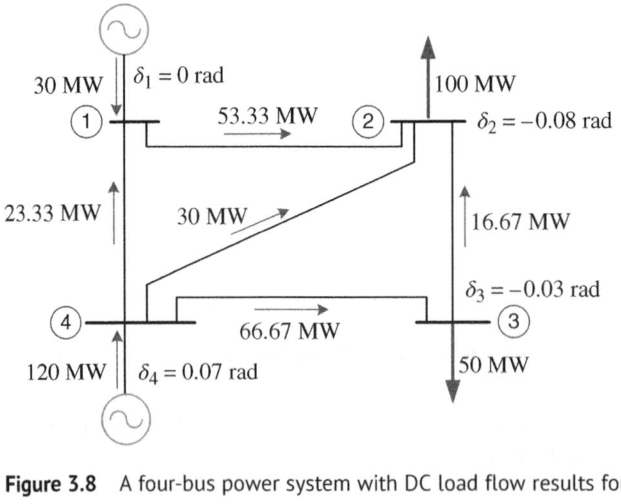

Figure 3.8 A four-bus power system with DC load flow results for Example 3.8.

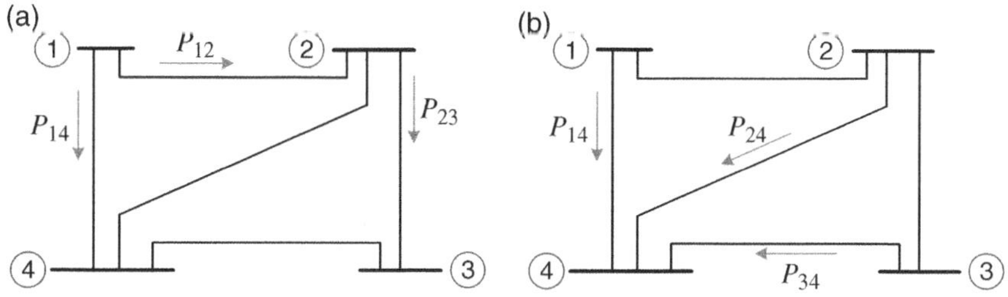

Figure 3.9 The four-bus power system with two different sets of measurements for Example 3.8. (a) measurement set-a and (b) measurement set-b.

The base MVA is chosen as 100. The results of DCLF, that is, the power flow through the lines and voltage angles are shown in Figure 3.8.

Given $\delta_1 = 0°$, the angles δ_2, δ_3, and δ_4 need to be estimated based on the line flows. It can be seen from Figure 3.8 that five data of line flow are available – 1-2, 1-4, 2-3, 2-4, and 3-4. However, only three of the line flow measurements will be sufficient for the estimation of the angles. Two sets of measurements are shown in Figure 3.9. For Figure 3.9 (a), the following equations are written for a base MVA of 100. These are measurement set-a.

$$z_1 = P_{12} = 0.5333 = \frac{1}{Z_{12}}(\delta_1 - \delta_2) = -6.67\delta_2$$

$$z_2 = P_{32} = -0.1667 = \frac{1}{Z_{23}}(\delta_2 - \delta_3) = 3.33\delta_2 - 3.33\delta_3$$

$$z_3 = P_{14} = -0.2333 = \frac{1}{Z_{14}}(\delta_1 - \delta_4) = -3.33\delta_4$$

We then have the following batch model of (3.92):

$$\mathbf{z} = \begin{bmatrix} P_{12} \\ P_{32} \\ P_{14} \end{bmatrix} = \begin{bmatrix} 0.53 \\ -0.16 \\ -0.23 \end{bmatrix}, \mathbf{x} = \begin{bmatrix} \delta_2 \\ \delta_3 \\ \delta_4 \end{bmatrix} \text{ and } \mathbf{H} = \begin{bmatrix} -6.67 & 0 & 0 \\ 3.33 & -3.33 & 0 \\ 0 & 0 & -3.33 \end{bmatrix}$$

It is assumed that all the meters have a standard deviation (σ) of 0.01 and hence a covariance (σ^2) of 0.0001. The weighting matrix is then given by

$$\mathbf{W} = \begin{bmatrix} 1 & 0 & 0 \\ 1 & 1 & 0 \\ 0 & 0 & 1 \end{bmatrix} \times 10^4$$

The solution of (3.95) then produces the following estimated angles:

$$\hat{\delta}_2 = -0.0795 \text{ rad} = -4.555^\circ$$
$$\hat{\delta}_3 = -0.0315 \text{ rad} = -1.805^\circ$$
$$\hat{\delta}_4 = 0.069 \text{ rad} = 3.953^\circ$$

Again, from Figure 3.9 (b), we have the measurement set-b

$$z_1 = P_{24} = 0.53 = \frac{1}{Z_{12}}(\delta_1 - \delta_2) = -6.67\delta_2$$

$$z_2 = P_{32} = -0.16 = \frac{1}{Z_{23}}(\delta_2 - \delta_2) = 3.33\delta_2 - 3.33\delta_3$$

$$z_3 = P_{14} = -0.23 = \frac{1}{Z_{14}}(\delta_1 - \delta_4) = -3.33\delta_4$$

Therefore,

$$\mathbf{z} = \begin{bmatrix} P_{24} \\ P_{34} \\ P_{14} \end{bmatrix} = \begin{bmatrix} -0.29 \\ -0.66 \\ -0.23 \end{bmatrix} \text{ and } \mathbf{H} = \begin{bmatrix} 2.0 & 0 & -2.0 \\ 0 & 6.67 & -6.67 \\ 0 & 0 & -3.33 \end{bmatrix}$$

The resulting estimates are the same as given above.

3.4.4 AC State Estimation

The optimization of (3.89) is written as follows:

$$\frac{\partial J(\mathbf{x})}{\partial x} = \frac{\partial [\mathbf{z} - \mathbf{h}(\mathbf{x})]^T \mathbf{W} [\mathbf{z} - \mathbf{h}(\mathbf{x})]}{\partial x} = 0 \tag{3.96}$$

The solution of (3.96) is given by [4]

$$\mathbf{H}^T \mathbf{W} [\mathbf{z} - \mathbf{h}(\mathbf{x})] = \mathbf{0} \tag{3.97}$$

where

$$\mathbf{H} = \frac{\partial \mathbf{h}(\mathbf{x})}{\partial \mathbf{x}} \tag{3.98}$$

Unlike the DC state estimation, the vector $\mathbf{h}(\mathbf{x})$ will contain nonlinear terms for AC state estimation. Each element of this vector can be linearized using Taylor's series expansion, where the second- and higher-order terms are neglected to obtain

$$h_i(x + \Delta x) = h_i(x) + \left.\frac{\partial h_i(x)}{\partial x_i}\right| \Delta x_i$$

This is then written in the vector–matrix form as:

$$\mathbf{h}(\mathbf{x} + \Delta\mathbf{x}) = \mathbf{h}(\mathbf{x}) + \mathbf{H}\Delta\mathbf{x} \tag{3.99}$$

The state estimation problem is solved in an iterative manner, which is given by [4]

$$\mathbf{G}(\mathbf{x}^k)\Delta\mathbf{x}^k = \mathbf{H}^T(\mathbf{x}^k)\mathbf{W}[\mathbf{z} - \mathbf{h}(\mathbf{x}^k)] \tag{3.100}$$

where

$$\mathbf{G}(\mathbf{x}^k) = \mathbf{H}^T(\mathbf{x}^k)\mathbf{W}\mathbf{H}(\mathbf{x}^k) \tag{3.101}$$

The state vector is updated in the same manner as the Newton–Raphson load flow program from:

$$\mathbf{x}^{k+1} = \mathbf{x}^k + \Delta\mathbf{x}^k \tag{3.102}$$

Example 3.9 Consider the three-bus power system shown in Figure 3.10, which contains two voltmeters to measure the voltage magnitudes of buses 1 and 2 and three power meters to measure the active and reactive power at locations indicated in the figure. It is obvious from Figure 3.10 that we can have a total of eight measurements – six from the power meters and two from the voltmeters. The measurement vector is then given by

$$\mathbf{z} = [\,|V_1| \quad |V_2| \quad P_{12} \quad P_{13} \quad P_2 \quad Q_{12} \quad Q_{13} \quad Q_2\,]^T$$

The states that need to be estimated for the problem are the magnitude of all three bus voltages and angles of buses 2 and 3. The state vector is then given by

$$\mathbf{x} = [\,\delta_2 \quad \delta_3 \quad |V_1| \quad |V_2| \quad |V_3|\,]^T$$

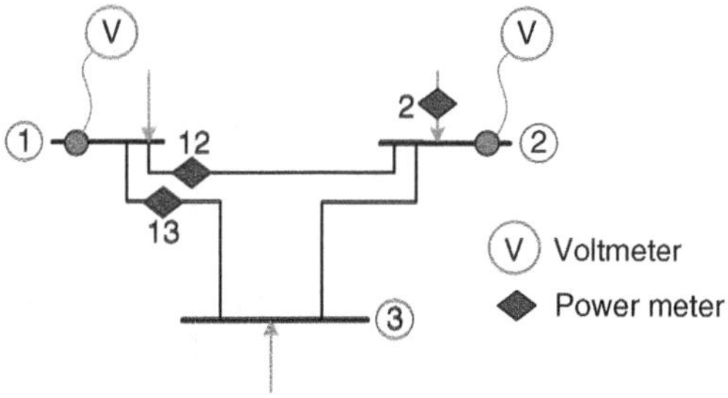

Figure 3.10 The three-bus power system of Example 3.9.

The Jacobian matrix of (3.98) is then given by

$$\mathbf{H} = \begin{bmatrix} \frac{\partial|V_1|}{\partial\delta_2} & \frac{\partial|V_1|}{\partial\delta_3} & \frac{\partial|V_1|}{\partial|V_1|} & \frac{\partial|V_1|}{\partial|V_2|} & \frac{\partial|V_1|}{\partial|V_3|} \\ \frac{\partial|V_2|}{\partial\delta_2} & \frac{\partial|V_2|}{\partial\delta_3} & \frac{\partial|V_2|}{\partial|V_1|} & \frac{\partial|V_2|}{\partial|V_2|} & \frac{\partial|V_2|}{\partial|V_3|} \\ \frac{\partial P_{12}}{\partial\delta_2} & \frac{\partial P_{12}}{\partial\delta_3} & \frac{\partial P_{12}}{\partial|V_1|} & \frac{\partial P_{12}}{\partial|V_2|} & \frac{\partial P_{12}}{\partial|V_3|} \\ \frac{\partial P_{13}}{\partial\delta_2} & \frac{\partial P_{13}}{\partial\delta_3} & \frac{\partial P_{13}}{\partial|V_1|} & \frac{\partial P_{13}}{\partial|V_2|} & \frac{\partial P_{13}}{\partial|V_3|} \\ \frac{\partial P_2}{\partial\delta_2} & \frac{\partial P_2}{\partial\delta_3} & \frac{\partial P_2}{\partial|V_1|} & \frac{\partial P_2}{\partial|V_2|} & \frac{\partial P_2}{\partial|V_3|} \\ \frac{\partial Q_{12}}{\partial\delta_2} & \frac{\partial Q_{12}}{\partial\delta_3} & \frac{\partial Q_{12}}{\partial|V_1|} & \frac{\partial Q_{12}}{\partial|V_2|} & \frac{\partial Q_{12}}{\partial|V_3|} \\ \frac{\partial Q_{13}}{\partial\delta_2} & \frac{\partial Q_{13}}{\partial\delta_3} & \frac{\partial Q_{13}}{\partial|V_1|} & \frac{\partial Q_{13}}{\partial|V_2|} & \frac{\partial Q_{13}}{\partial|V_3|} \\ \frac{\partial Q_2}{\partial\delta_2} & \frac{\partial Q_2}{\partial\delta_3} & \frac{\partial Q_2}{\partial|V_1|} & \frac{\partial Q_2}{\partial|V_2|} & \frac{\partial Q_2}{\partial|V_3|} \end{bmatrix}$$

The process of the formation of the Jacobian matrix is somewhat similar to that discussed for Newton–Raphson load flow in Section 3.3.3. There are, however, some differences. For example, in the load flow, the real and reactive power flows between the buses are not considered. However, as evident from Figure 3.10, these are required for state estimation. Moreover, bus voltages can also be measurement variables. We must, therefore, include the derivatives of these quantities for the formation of the Jacobian matrix.

The submatrices $\mathbf{J_{11}}$, $\mathbf{J_{12}}$, $\mathbf{J_{21}}$, and $\mathbf{J_{22}}$ given in (3.37) to (3.40) can be used to compute $\partial P/\partial\delta$, $\partial P/\partial|V|$, $\partial Q/\partial\delta$, and $\partial Q/\partial|V|$, respectively. (3.67) describes the current flow between two buses. Then, the complex power flow between these buses is given by

$$P_{ik} - jQ_{ik} = V_i^* I_{ik} = V_i^* \times [-Y_{bik}(V_i - V_k)] = -Y_{bik}|V_i|^2 + Y_{bik} V_i^* V_k \tag{3.103}$$

Using (3.13), (3.103) is rewritten as:

$$P_{ik} - jQ_{ik} = -(G_{ik} + jB_{ik})\left(|V_i|^2 + V_i^* V_k\right) \tag{3.104}$$

Since $V_i = |V_i| \angle \delta_i$ and $V_k = |V_k| \angle \delta_k$, we have

$$\begin{aligned} V_i^* V_k &= |V_i||V_k|\angle(\delta_k - \delta_i) = |V_i||V_k|\{\cos(\delta_k - \delta_i) + j\sin(\delta_k - \delta_i)\} \\ &= |V_i||V_k|\{\cos(\delta_i - \delta_k) - j\sin(\delta_i - \delta_k)\} \end{aligned}$$

Separating the real and imaginary components of (3.103), the following real and reactive powers are obtained:

$$P_{ik} = -G_{ik}|V_i|^2 - G_{ik}|V_i||V_k|\cos(\delta_i - \delta_k) - B_{ik}|V_i||V_k|\sin(\delta_i - \delta_k) \tag{3.105}$$

$$Q_{ik} = B_{ik}|V_i|^2 - G_{ik}|V_i||V_k|\sin(\delta_i - \delta_k) + B_{ik}|V_i||V_k|\cos(\delta_i - \delta_k) \tag{3.106}$$

Therefore,

$$\frac{\partial P_{ik}}{\partial \delta_i} = |V_i||V_k|\{G_{ik}\sin(\delta_i - \delta_k) - B_{ik}\cos(\delta_i - \delta_k)\} \tag{3.107}$$

$$\frac{\partial P_{ik}}{\partial \delta_k} = -|V_i||V_k|\{G_{ik}\sin(\delta_i - \delta_k) - B_{ik}\cos(\delta_i - \delta_k)\} \tag{3.108}$$

$$\frac{\partial P_{ik}}{\partial |V_i|} = -2G_{ik}|V_i| - |V_k|\{G_{ik}\cos(\delta_i - \delta_k) + B_{ik}\sin(\delta_i - \delta_k)\} \tag{3.109}$$

$$\frac{\partial P_{ik}}{|V_k|} = -|V_i|\{G_{ik}\cos(\delta_i - \delta_k) + B_{ik}\sin(\delta_i - \delta_k)\} \tag{3.110}$$

$$\frac{\partial Q_{ik}}{\partial \delta_i} = -|V_i||V_k|\{G_{ik}\sin(\delta_i - \delta_k) + B_{ik}\cos(\delta_i - \delta_k)\} \tag{3.111}$$

$$\frac{\partial Q_{ik}}{\partial \delta_k} = |V_i||V_k|\{G_{ik}\cos(\delta_i - \delta_k) + B_{ik}\sin(\delta_i - \delta_k)\} \tag{3.112}$$

$$\frac{\partial Q_{ik}}{\partial |V_i|} = 2B_{ik}|V_i| - |V_k|\{G_{ik}\sin(\delta_i - \delta_k) + B_{ik}\cos(\delta_i - \delta_k)\} \tag{3.113}$$

$$\frac{\partial Q_{ik}}{\partial |V_k|} = |V_i|\{-G_{ik}\sin(\delta_i - \delta_k) + B_{ik}\cos(\delta_i - \delta_k)\} \tag{3.114}$$

Moreover, note that

$$\frac{\partial |V_i|}{\partial |V_i|} = 1 \text{ and } \frac{\partial |V_i|}{\partial |V_k|} = 0 \tag{3.115}$$

The iterative process is then carried out through the following steps:

- *Step 1*: Choose an initial state vector $\mathbf{x}^{(0)}$.
- *Step 2*: Compute the Jacobian matrix $\mathbf{H}^{(0)}$ using the values chosen in Step 1.
- *Step 3*: Compute the gain matrix from $\mathbf{G}^{(0)}(\mathbf{x}) = [\mathbf{H}^{(0)}]^T\mathbf{W}\mathbf{H}^{(0)}$.
- *Step 4*: Compute $\Delta\mathbf{x}^{(0)}$ from (3.100).
- *Step 5*: Compute $\mathbf{x}^{(1)}$ from (3.102).
- *Step 6*: If all the elements of $\Delta\mathbf{x}^{(0)}$ are below a certain threshold, terminate the process. Otherwise, go back to Step 2.

For worked out examples refer to [4, 5]. There are several other methods and variations of state estimation, as discussed in [3–5].

3.4.5 Bad Data Detection

A WLS state estimator is very sensitive to bad data due to measurement errors [6]. One or more bad data can be present after the convergence of the state estimator algorithm. After the convergence, a two-step procedure is employed, first to determine the presence of bad data and then to identify the source of the bad data [7]. A chi-squared test is a procedure that can be used for the detection of the presence of bad data.

Let x_i, $i = 1, \ldots, n$ be a set of independent normal random variables. Then, the distribution of

$$X = \sum_{i=1}^{n} x_i^2 \tag{3.116}$$

is called the chi-squared distribution with a degree of freedom (df) n. For $x > 0$ and $n > 0$, the chi-squared probability density function is given by [8]

$$f(x|n) = \frac{e^{-x/2} x^{(n-2)/2}}{2^{n/2}\Gamma(n/2)} \tag{3.117}$$

where Γ is the gamma function, which for an integer $x > 0$ is defined as $\Gamma(x) = (x-1)!$ The chi-squared random variable with the df on n is defined as χ_n^2.

The MATLAB function "chi2pdf(x,n)" in "Statics and Machine Learning Toolbox" can compute the chi-squared probability density function, where x is the series of observations and n is the df. For example, consider the following code:

```
x = 0:0.001:30;
y1 = chi2pdf(x,5);
y2 = chi2pdf(x,10);
```

The probability density function plots of these two outputs are shown in Figure 3.11.

The procedure for the detection of bad data using chi-squared test has been discussed in [3, 7]. It depends on the measurement residuals, given by

$$\hat{\mathbf{z}} = \mathbf{H}\hat{\mathbf{x}} \tag{3.118}$$

The chi-squared distribution is then computed as follows:

$$f(m,n) = \sum_{i=1}^{n} \frac{(z_i - \hat{z}_i)^2}{\sigma_i^2} \tag{3.119}$$

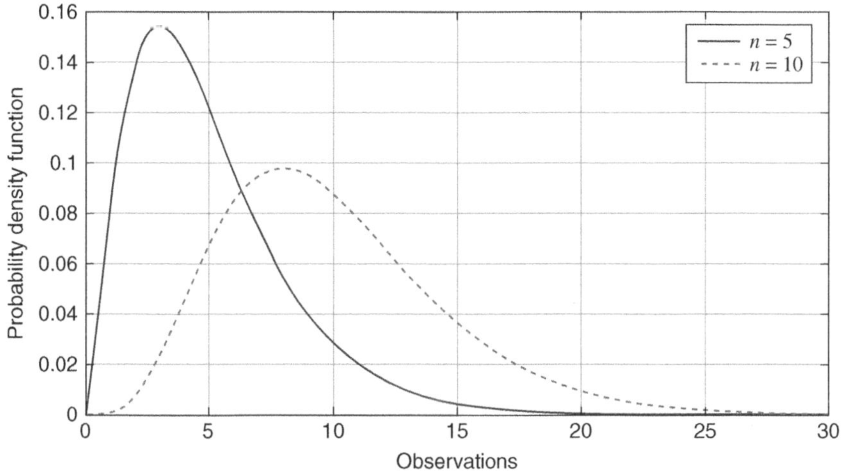

Figure 3.11 Chi-squared probability density functions.

where m is the number of state variables, n is the number of measurements, and σ_i^2 is variance of the sensors. The quantity $(n-m)$ is called the degree of freedom. The presence of bad data is detected for the following condition:

$$f(m,n) > p \tag{3.120}$$

where p is a threshold value that is normally chosen as 95%. Example 3.10 illustrates the procedure.

Example 3.10 Consider the three-bus power system shown in Figure 3.12. Assume that the load in bus-2 is 1.5 per unit and that in bus-3 is 1.9 per unit. Then the DCLF results obtained are:

$$\delta_1 = 0°, \delta_2 = -0.3341 \text{ rad}, \delta_2 = -0.3095 \text{ rad}$$

$$P_1 = 3.4 \text{ pu}, P_{12} = 1.3364 \text{ pu}, P_{13} = 2.0636 \text{ pu and } P_{23} = -0.1636 \text{ pu}$$

Figure 3.12 Three-bus power system of Example 3.10.

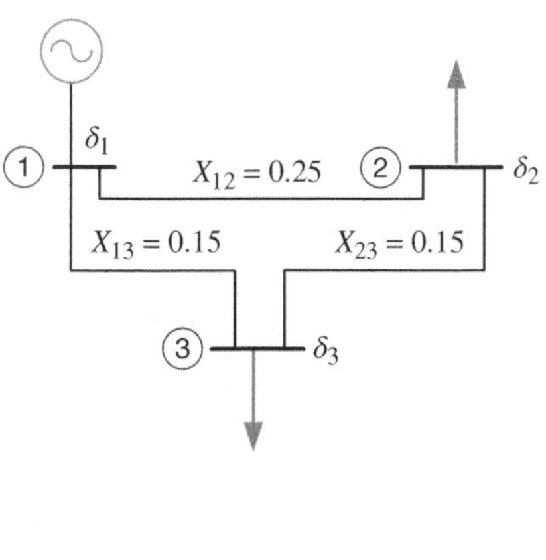

Assume that there are four measurements. The measurement sensor may not be perfect and there may be measurement errors. These erroneous measurements are given in per unit below:

$$P_1 = 3.45,\ P_2 = -1.55,\ P_3 = -1.8, \text{ and } P_{23} = -0.15$$

The measurement vector is:

$$z_m = \begin{bmatrix} z_{1m} & z_{2m} & z_{3m} & z_{4m} \end{bmatrix}^T = \begin{bmatrix} P_1 & P_2 & P_3 & P_{23} \end{bmatrix}^T$$

Assuming $\delta_1 = 0$, the power flow equations in the measurements in terms of the angles are written as:

$$P_1 = P_{12} + P_{13} = -4\delta_2 - 6.667\delta_3$$

$$P_2 = P_{21} + P_{23} = 4\delta_2 + 6.667(\delta_2 - \delta_3) = 10.667\delta_2 - 6.667\delta_3$$

$$P_3 = P_{31} + P_{32} = 6.667\delta_3 + 6.667(\delta_3 - \delta_2) = 13.333\delta_3 - 6.667\delta_2$$

$$P_{23} = 6.667\delta_2 - 6.667\delta_3$$

Then, we have the following components of the batch model:

$$\mathbf{z_m} = \begin{bmatrix} 3.45 \\ -1.55 \\ -1.8 \\ -0.15 \end{bmatrix}, \mathbf{H} = \begin{bmatrix} -4 & -6.667 \\ 10.667 & -6.667 \\ -6.667 & 13.333 \\ 6.667 & -6.667 \end{bmatrix}, \mathbf{x} = \begin{bmatrix} \delta_2 \\ \delta_3 \end{bmatrix}$$

Let the covariances of the sensor error be:

$$\sigma_1^2 = 0.001, \sigma_2^2 = 0.002, \sigma_3^2 = 0.002, \sigma_4^2 = 0.001$$

The weighting matrix is then

$$W = \begin{bmatrix} 1/\sigma_1^2 & & & \\ & 1/\sigma_2^2 & & \\ & & 1/\sigma_3^2 & \\ & & & 1/\sigma_4^2 \end{bmatrix} = \begin{bmatrix} 10 & & & \\ & 5 & & \\ & & 5 & \\ & & & 2 \end{bmatrix} \times 10^2$$

Then, the estimates are computed as:

$$\hat{x} = \begin{bmatrix} -0.3415 \\ -0.3094 \end{bmatrix} \text{rad}$$

The estimated measurements are then calculated as given below.

$$\hat{\mathbf{z}} = \mathbf{H}\hat{\mathbf{x}} = \begin{bmatrix} 3.4291 \\ -1.5802 \\ -1.8459 \\ -0.214 \end{bmatrix}$$

The measurement residuals are then given by

$$\mathbf{z}_\mathrm{m} - \hat{\mathbf{z}} = \begin{bmatrix} 0.0209 \\ 0.0302 \\ 0.0489 \\ 0.0640 \end{bmatrix}$$

Then, from (3.119), we have

$$f(m, n) = \frac{0.0209^2}{0.001} + \frac{0.0302^2}{0.002} + \frac{0.0489^2}{0.002} + \frac{0.064^2}{0.005} = 2.9083$$

Since $m = 3$ and $n = 4$, the df is 1. For this degree of freedom, the threshold value is found from the chi-squared distribution table as 3.841. Since $f(m, n) < p$, no bad data is suspected. Note that the table of chi-squared distribution with respect to degrees of freedom is available online, for example, refer to [9].

Example 3.11 Consider again Example 3.10, where the measurements now are as given below:

$$P_1 = 3.45,\ P_2 = -1.55,\ P_3 = -2.1, \text{ and } P_{23} = -0.1$$

The estimates are then

$$\hat{\mathbf{x}} = \begin{bmatrix} -0.3350 \\ -0.32074 \end{bmatrix} \text{ rad}$$

Then,

$$\hat{\mathbf{z}} = \begin{bmatrix} 3.4895 \\ -1.4676 \\ -2.0222 \\ -0.1156 \end{bmatrix}$$

The chi-squared distribution is then:

$$f(m, n) = 8.0539$$

Obviously, this is greater than 3.841, and hence, bad data are present for these measurements.

Once the presence of bad data is detected, the next step is to identify the measurements for which sensors are erroneous. One of the methods of identifying bad data is the use of normalized residuals. The residuals of (3.118) are normalized in such a way that they have a Gaussian distribution with a mean of 0 and a variance of 1 [4]. The largest normalized residual will indicate the bad data, provided that only one component of the dataset is erroneous. This procedure can be applied to multiple bad data so long as these do not interact with each other. The other method of bad data identification is called hypothesis testing [10]. This method is much more complicated than the normalized residual method in which the measurement errors are computed directly. Multiple bad data can be estimated using this method even when they interact.

3.5 SCADA and EMS

Supervisory control and data acquisition (SCADA) is not a controller but provides supervisory commands to the operators. As such, SCADA systems are used not only in power systems but also in other industries such as oil and gas pipelines, factory automation, water treatment, and chemical processes. It is essentially a software process that sits on top of a hardware. The hardware is interfaced with the outside world both for data collection and user interface. The schematic diagram of SCADA is shown in Figure 3.13. The basic components of SCADA are as follows:

- Remote terminal unit (RTU): It collects data from the field devices. It can also process the data before sending them to the master station (MS) through a communication channel. At the same time, an RTU can also receive data from MS and communicate them to the field devices.

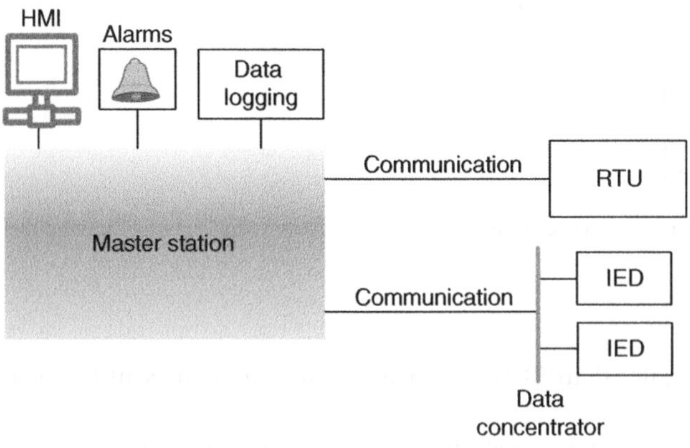

Figure 3.13 A SCADA system.

- Intelligent electronic device (IED): It is basically a programmable logic controller (PLC) that is connected to sensors to collect data. Even though a PLC has a more sophisticated control capability than an RTU, it does not have telemetry capabilities. Thus, all the IED data are concentrated and sent to the MS through communication channels.
- The MS: It performs various supervisory functions. It can display data through human–machine interface (HMI), can trigger alarms, and can log the data for future processing.

In the modern power system, data from a phasor measurement unit (PMU) can also be included in SCADA systems. The SCADA can monitor bus voltages, power, reactive and current flow through different lines, and breaker status. The software residing in the MS can then perform different tasks such as power flow, state estimation, fault detection, and isolation. Therefore, the SCADA system is an integral part of what has been discussed in this chapter. There are various functions of SCADA and its application to smart grid that can be found in [11].

An EMS has more advanced features than SCADA. It can collect data from the SCADA system to execute advanced algorithms such as optimization, planning, load forecasting, economic dispatch, and generator scheduling. The main purpose of an EMS is to facilitate the efficient use of energy by reducing the operating cost. Using the real-time operational status from SCADA, EMS makes optimized adjustments to make a power system more efficient. Thus, even though both SCADA and EMS perform similar functions, EMS is more versatile as it can be used for near-instantaneous matching load with demand. A review of EMS can be found in [12].

3.6 Concluding Remarks

This chapter presents the methods to calculate a power system bus voltages and their angles in a network using iterative methods. Both load flow techniques and state estimation are discussed. Usually, load flow studies are performed for system planning for future expansion and to maintain operational efficiency. These studies are also performed to determine the size of the equipment required to maintain the power flow during normal operations and withstand contingencies. Even though the state estimation procedure uses the load flow data, their purpose is different than load flow. The purpose of state estimation is real-time monitoring to help the system operators make informed decisions. State estimation also improves the system security by detecting and responding to abnormal conditions, and predicting the potential failure due to system outages. With the advent of the smart grid, both these computer-based techniques are gaining more importance due to the increased penetration of renewable energy generators in power systems, both for planning and operation. In addition, the threat of cyberattacks through bad data injection can cause havoc to a power systems normal operation. One such attack scenario is discussed in Section 3.4.5. A brief description of cyberattacks and their countermeasures will be discussed in Chapter 12.

References

1 F. C. Schweppe and J. Wildes, "Power system static-state estimation, Part I: exact model,", *IEEE Transactions on Power Apparatus and Systems*, Vol. PAS-89, No. 1, pp. 120–125, 1970.

2 J. J. Grainger and W. D. Stevenson, *Power System Analysis*, McGraw-Hill, New York, 1994.

3 A. J. Wood and B. F. Wollenberg, *Power Generation, Operation and Control*, 2nd Ed., John Wiley, New York, 2003.

4 A. Gomez-Exposito, A. J. Conejo and C. Canizares, Ed., *Electric Energy System: Analysis and Operation*, CRC Press, Boca Raton, 2009.

5 M. Ahmad, *Power System State Estimation*, Archer House, Norwood, MA, 2012.

6 H. M. Merrill and F. C. Schweppe, "Bad data suppression in power system static state estimation," *IEEE Transactions on Power Apparatus and Systems*, Vol. PAS-90, No. 6, pp. 2718–2725, 1971.

7 M. Ahmad, *Power System State Estimation*, Artech House, Norwood, MA, 2013.

8 K. Krishnamoorthy, *Handbook of Statistical Distribution with Applications*, Chapman & Hall/CRC, Boca Raton, FL, 2006.

9 The University of Arizona, *Chi-Squared Distribution Table*. Available. https://math.arizona.edu/~jwatkins/chi-square-table.pdf

10 Milli, L. and Van Custem, T. (1988). Implementation of hypothesis testing identification in power system state estimation. *IEEE Trans. on Power Systems* **3** (4): 887–893.

11 M. S. Thomas and J. D. McDonald, *Power System SCADA and Smart Grids*, CRC Press, Boca Raton, 2015.

12 S. K. Rathor and D. Saxena, "Energy management system for smart grid: an overview and key issues," *Wiley International Journal of Energy Research*, Vol. 44, No. 6, pp. 4063–5034, 2020.

Problems

P3.1 For the $\mathbf{Y_{bus}}$ matrix given in per unit below, draw the impedance diagram.

$$\mathbf{Y_{bus}} = j\begin{bmatrix} -6 & 2 & 2.5 & 0 \\ 2 & -10 & 2.5 & 4 \\ 2.5 & 2.5 & -9 & 4 \\ 0 & 4 & 4 & -8 \end{bmatrix}$$

P3.2 Compute the bus admittance ($\mathbf{Y_{bus}}$) matrices for the systems given in Problems (a) P2.11 and (b) P2.12.

P3.3 Consider the bus admittance matrix that is portioned as follows:

$$\mathbf{Y_{bus}} = \begin{bmatrix} \mathbf{K} & \mathbf{L} \\ \mathbf{L}^T & \mathbf{M} \end{bmatrix}$$

The voltage–current relationship for the matrix is given by

$$\begin{bmatrix} \mathbf{I_A} \\ \mathbf{0} \end{bmatrix} = \begin{bmatrix} \mathbf{K} & \mathbf{L} \\ \mathbf{L}^T & \mathbf{M} \end{bmatrix}\begin{bmatrix} \mathbf{V_A} \\ \mathbf{V_B} \end{bmatrix}$$

where

$$\mathbf{I_A} \in \Re^{n\times 1}, \quad \mathbf{V_A} \in \Re^{n\times 1}, \quad \mathbf{V_B} \in \Re^{m\times 1},$$
$$\mathbf{K} \in \Re^{n\times n}, \quad \mathbf{L} \in \Re^{n\times m}, \quad \mathbf{M} \in \Re^{m\times m}$$

This implies that there are n number of buses where currents get injected and another m number of buses where no current is injected. We can then eliminate buses contained in the vector $\mathbf{V}_B$ and write

$$I_A = Y_{bus}^{new} V_A \text{ where } Y_{bus}^{new} = K - LML^T$$

Consider the impedance diagram shown in Figure P3.3. Find the bus voltages V_1 and V_2.

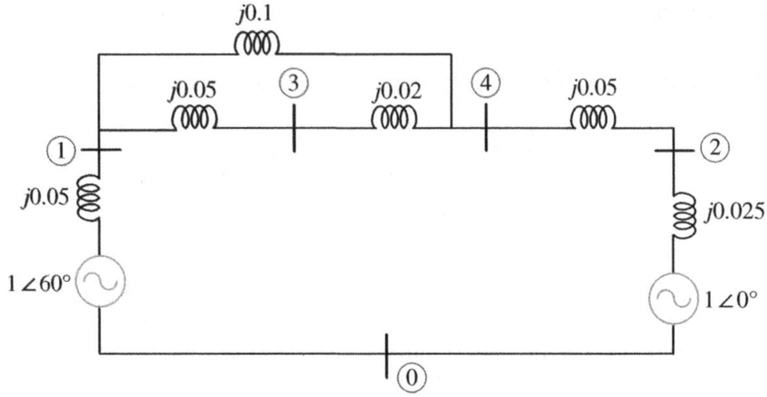

Figure P3.3 Impedance diagram of the system for Problem 3.3.

P3.4 From the results obtained in Problem P3.3, determine the bus voltages V_3 and V_4.

P3.5 Consider the three-bus power system shown in Figure P3.5. The network is lossless and the line charging capacitors are neglected. Bus-1 is the slack bus, while the other two are load buses. The system data are as given below:

$$V_1 = 1.05\angle 0°, P_2 + jQ_2 = 0.1952 + j0.1366, P_3 + jQ_3 = 0.7074 + j0.1908$$

$$\mathbf{Y_{bus}} = j\begin{bmatrix} -15 & 4 & 3 \\ 4 & -6 & 2 \\ 3 & 2 & -5 \end{bmatrix}$$

After the convergence of a load flow program, the voltage of bus-2 is found to be $V_2 = 1\angle -5°$. Find the magnitude and angle of the voltage of bus-3.

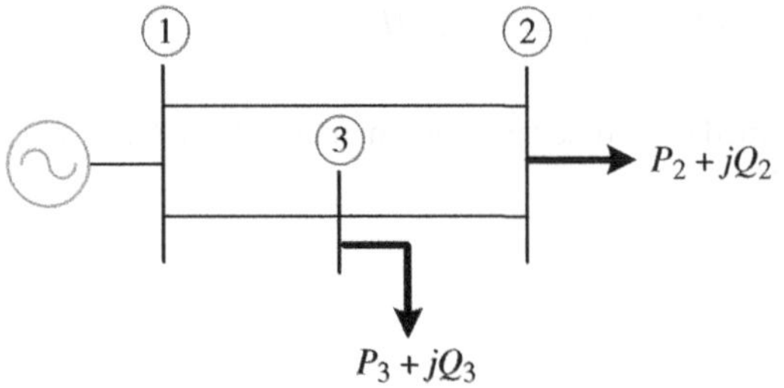

Figure P3.5 Three-bus power system of Problem 3.5.

P3.6 Thc following data are given in per unit for a four-bus power system

$$Y_{b21} = j6.0, Y_{b22} = -j10.5, Y_{b23} = j1.5 \text{ and } Y_{b24} = j3.0$$

Bus-2 is a load bus which draws $P_2 = 1.0$ and $Q_2 = 0.75$. The initial values of the bus voltages are

Bus-1 (slack): $V_1 = 1.05 \angle 0^\circ$ pu
Bus-2 (P–Q): $V_2 = 1.0 \angle 0^\circ$ pu
Bus-3 (P–V): $V_3 = 1.02 \angle 0^\circ$ pu
Bus-4 (P–Q): $V_4 = 1.0 \angle 0^\circ$ pu

Find V_2 produced after the first iteration of Gauss–Seidel method.

P3.7 A system of nonlinear equations is given by

$$g_1(x_1, x_2, x_3) = 4x_2 \sin x_1 + 6x_3 + 4x_2 = 0$$
$$g_2(x_1, x_2, x_3) = x_3^2 + \cos x_1 + 1 = 0$$
$$g_3(x_1, x_2, x_3) = 3x_3 \cos x_1 + x_2 x_3 + 4 = 0$$

The following two possible initial conditions are chosen:

$$x_1^{(0)} = 0 \text{ rad}, x_2^{(0)} = 2 \text{ and } x_3^{(0)} = 0$$

$$x_1^{(0)} = 0 \text{ rad}, x_2^{(0)} = x_3^{(0)} = 1$$

Explain which of these two choices are feasible.

P3.8 A system of nonlinear equations is given by

$$g_1(x_1, x_2) = x_1 + 2x_2^2 - 5 = 0$$
$$g_2(x_1, x_2) = 2x_1 x_2 + x_2 - 7 = 0$$

Choosing initial conditions of $x_1^{(0)} = 2$ and $x_2^{(0)} = 0$, perform one iteration of Newton–Raphson method to determine $x_1^{(1)}$ and $x_2^{(1)}$.

P3.9 In a five-bus power system, the elements of the fourth row of the $\mathbf{Y_{bus}}$ matrix, computed on a base of 220 kV and 100 MVA, are

$$Y_{4b1} = Y_{b42} = 0, \quad Y_{b43} = 2.887\angle 104.04^\circ,$$
$$Y_{b44} = 3.850\angle 104.04^\circ, \quad Y_{b45} = 6.737\angle -75.96^\circ$$

The initial load flow data are given in Table P3.9.

Table P3.9 Initial conditions of the five-bus system of Problem 3.9.

	Generation		Load		
Bus no. and type	**P (MW)**	**Q (MVAr)**	**P (MW)**	**Q (MAVr)**	**Voltage (pu) and angle**
1. Slack	–	–	65	30	$1.04\angle 0°$
2. Load	0	0	115	60	$1.00\angle 0°$
3. PV	180	–	70	40	$1.02\angle 0°$
4. Load	0	0	70	30	$1.00\angle 0°$
5. Load	0	0	85	40	$1.00\angle 0°$

(a) Evaluate $\Delta P_4^{(0)}$ with the data given In Table P3.9. Choose a base of 100 MVA.
(b) Determine the value of $(\partial P_4/\partial \delta_3)^{(0)}$.

P3.10 (Note this problem will need the use of a computer with suitable software, e.g., MATLAB). Consider the four-bus power system shown in Figure P3.10, which contains a P–V and two P–Q buses in addition to the slack bus. The initial bus voltages are also indicated in the figure. Moreover, the power and reactive consumed power at the load buses and the power injected at the P–V bus are also indicated in the figure, where all quantities are in per unit. The $\mathbf{Y_{bus}}$ matrix for the system is computed as:

$$\mathbf{Y_{bus}} = \begin{bmatrix} 3-j9 & -2+j6 & 0 & -1+j3 \\ -2+j6 & 3.5-j10.5 & -1+j3 & -0.5+j1.5 \\ 0 & -1+j3 & 3-j9 & -2+j6 \\ -1+j3 & -0.5+j1.5 & -2+j6 & 3.5-j10.5 \end{bmatrix}$$

After the first iteration of Newton–Raphson method, compute
(a) the Jacobian matrix,
(b) power and reactive power mismatches, and
(c) voltage magnitudes and their angles.

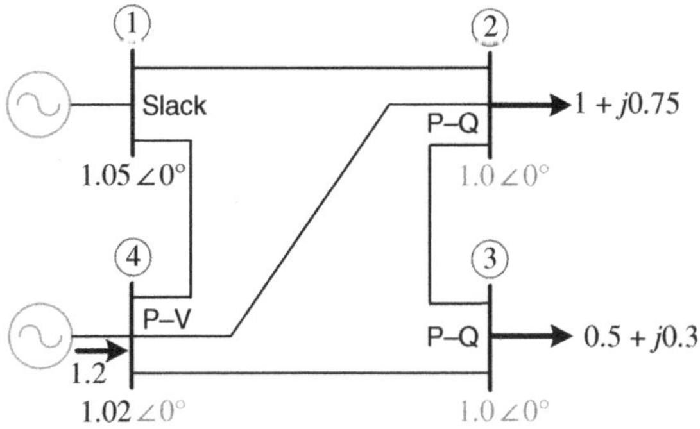

Figure P3.10 Four-bus power system with initial data of Problem 3.10.

P3.11 Write a computer program to solve the Newton–Raphson load flow of Problem P3.10, and then calculate the line losses. Tabulate the results in the same fashion as given in Tables 3.3 and 3.4.

P3.12 Consider the system of Figure P3.10, where the $\mathbf{Y_{bus}}$ matrix is given by

$$\mathbf{Y_{bus}} = \begin{bmatrix} 1-j9 & -0.75+j6 & 0 & -0.25+j3 \\ -0.75+j6 & 2-j14 & -1+j5 & -0.25+j3 \\ 0 & -1+j5 & 2-j11 & -1+j6 \\ -0.25+j3 & -0.25+j3 & -1+j6 & 1.5-j12 \end{bmatrix}$$

The initial data and the power injected and consumed remain the same as that given in Figure 3.10. Find the voltage magnitudes and the angles after the first iteration of the fast decoupled load flow program.

P3.13 Write a suitable program to compute the fast decoupled load flow and tabulate the data in the voltage magnitudes, their angle, and power generated in the same fashion, as shown in Table 3.3.

P3.14 For the system of Problems 3.10 and 3.11, use DCLF to determine the angles and the line flows.

P3.15 The partial load flow solution data are given in Table P3.15. The per unit base is chosen as 100 MVA. The line charging admittance in the line connecting buses 4 and 5 is 0.041 per unit. Using the structure shown in Figure P3.15, determine the real and reactive power flow at the various parts of the network. Moreover, calculate the power loss in lines 4 and 5 and the reactive power drop in the line.

Table P3.15 Partial load flow result of Problem P3.15.

				Line flow		
Bus	**Type**	**Voltage (pu)**	**Angle (deg)**	**To bus**	**MW**	**MVAr**
4	PQ	0.920	−10.9°	5	−31.25	−11.09
5	PQ	0.968	−6.2°	4	32.03	8.77

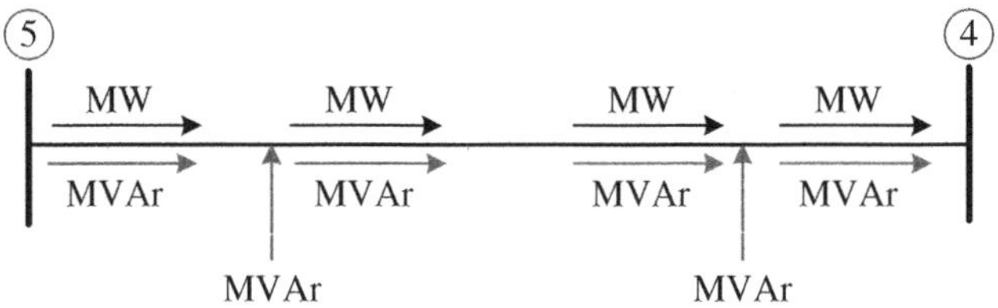

Figure P3.15 Partial line flow of Problem 3.15.

P3.16 Consider a lossless SMIB system with the sending end voltage being $V_S = V_1 \angle 0°$ per unit and the receiving end voltage being $V_R = V_2 \angle \delta_2$ per unit. The line reactance is given by 0.2 per unit. The power flow through the line is measured at two points – one from the sending end and the other from the receiving end. These are given by $P_{12} = 0.8$ per unit and $P_{21} = -0.82$ per unit. The variances of both measurement sensors are 0.01 per unit. Determine the estimate of angle δ_2 from DC state estimation.

P3.17 Consider the system shown in Figure P3.17, where there are three meters placed at the locations shown. The meter measurements and the sensor variances in per unit are as follows:

$$P_{12} = 0.55, \quad \sigma_{12}^2 = 0.02$$

$$P_{13} = 0.29, \quad \sigma_{13}^2 = 0.05$$

$$P_{32} = 0.54, \quad \sigma_{32}^2 = 0.002$$

The line reactances are given by:

$$X_{12} = 0.2, \quad X_{13} = 0.25, \quad X_{23} = 0.1$$

Using DC state estimation, determine if there is any bad data in the measurement for a threshold of 95%.

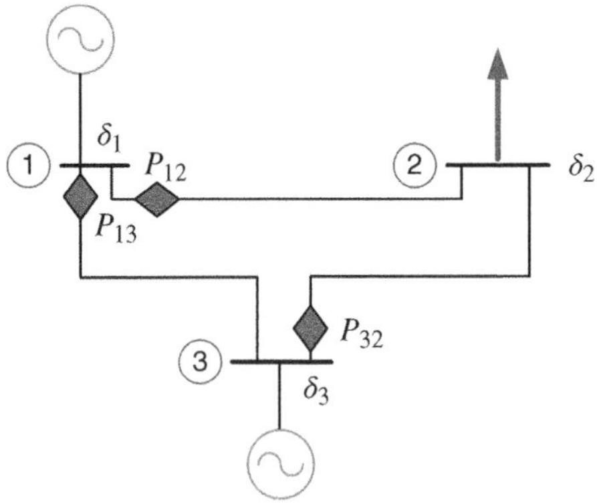

Figure P3.17 DC state estimation of Problem 3.15.

4

Economic Operation of Power System

A good business practice is the one in which the production cost is minimized without sacrificing the quality. This is not any different in the power sector as well. The main aim here is to reduce the production cost while maintaining the voltage magnitudes at each bus. In this chapter, we shall discuss the economic operation strategy, which will take into account the generation limits. The optimization problem will be solved using Lagrange multipliers. First, it will be demonstrated how several units of a plant can serve a load. Next, this will be extended to include line losses, where several plants that are connected through transmission lines combine to serve loads.

The economic operation assumes that several generators are already connected to the system. However, the daily load consumption is not fixed and varies with the time of the day and time of the year. The power generation must vary according to the load pattern, which may in turn vary with season. Therefore, it is not reasonable to think that all the units of a plant will run simultaneously and change their power output as and when required. In the unit commitment method, a subset of the available generators is scheduled to serve loads as they change over a day. This is also introduced in this chapter.

A power system must operate in a manner such that the frequency at every part of the system must be the same. Ideally, the frequency should be the nominal frequency of 50 or 60 Hz. However, the frequency can change depending on the load. Consider, for example, that there are N number of units supplying P_L amount of power. Now suppose the power demand is increased by ΔP_L. Then, all the units must share the power in such a way that the change in frequency caused by the load increase is the same for all the units. This is accomplished through the so-called droop equations through which the turbine–governors are controlled. It is to be noted that both the frequency control and the economic dispatch issue commands to change the power setting of each turbine–governor unit. It would

Modern Power System, First Edition. Arindam Ghosh.

Published 2026 by John Wiley & Sons, Inc.
Companion website: www.wiley.com/go/ghoshpowersystem

seem that their demands are conflicting. However, in an automatic generation control (AGC) strategy, both the objectives are coordinated. These aspects are discussed at the end of the chapter.

4.1 Economic Operation of a Power Plant

In an early attempt at economic operation of a power plant, it was decided to supply power from the most efficient plant at light load conditions. As the load increased, the power was supplied by this most efficient plant till the point of maximum rating of this plant was reached. With further increase in load, the next most efficient plant would supply power till its maximum capacity is reached. In this way, the power would be supplied by the most efficient to the least efficient plant to reach the peak demand. Unfortunately, however, this method failed to minimize the total cost of electricity generation. Therefore, there is a need for an alternative method which considers the total generation cost of all the units of a plant that is supplying a load.

4.1.1 Economic Distribution of Loads Between Two Units of a Plant

To determine the economic distribution of a load amongst the different units of a plant, the variable operating costs of each unit must be expressed in terms of its power output. Note that the fuel price is the main cost in a thermal or nuclear unit. Then the fuel cost must be expressed in terms of the power output. Other costs, such as the operation and maintenance costs, can also be expressed in terms of the power output. Fixed costs, such as the capital cost and depreciation, are not included in the fuel cost.

Several countries around the world use dollars ($) as their currencies. Since the fuel cost is volatile and changes over time, we shall define the cost in terms of a fictitious dollar, which can be replaced by the currency of a country for actual calculations.

The fuel requirement of each generator is given in terms of this fictitious \$/hour. Let us define the input cost of a unit-i, f_i in \$/h and the power output of the unit as P_i. Then the input cost can be expressed in terms of the power output as

$$f_i = \frac{a_i}{2} P_i^2 + b_i P_i + c_i \ \$/\text{h} \tag{4.1}$$

As evident from (4.1) that total cost has three components. The fixed cost can include maintenance, salaries of personnel, depreciations, etc. given by c_i in (4.1). The term b_i defines the cost component that increases linearly with the power generation, and a_i is the quadratic cost component, which defines the nonlinear behavior of fuel consumption and efficiency. For example, when the generator approaches its maximum capacity, its efficiency decreases. The question is how these coefficients are calculated. The following example illustrates this.

Example 4.1 Notice that (4.1) has three unknowns. Therefore, to estimate the coefficients, we need three equations. We also need to know the total cost from historical data or from measurements. Let the generation cost of a unit be \$22, \$47, and \$82 for power generation levels of 10 MW, 20 MW, and 30 MW, respectively. Then we have the following three equations:

$$22 = \alpha 10^2 + b10 + c$$
$$47 = \alpha 20^2 + b20 + c$$
$$82 = \alpha 30^2 + b30 + c$$

where $\alpha = a/2$. We can then write the following vector–matrix equation:

$$\begin{bmatrix} 22 \\ 47 \\ 82 \end{bmatrix} = \begin{bmatrix} 100 & 10 & 1 \\ 400 & 20 & 1 \\ 900 & 30 & 1 \end{bmatrix} \begin{bmatrix} \alpha \\ b \\ c \end{bmatrix} \Rightarrow \mathbf{A} = \mathbf{B} \begin{bmatrix} \alpha \\ b \\ c \end{bmatrix}$$

The coefficients then can be found by using the least square technique, for example, using the following MATLAB function:

$$\begin{bmatrix} \alpha = a/2 \\ b \\ c \end{bmatrix} = \mathbf{B} \backslash \mathbf{A} = \begin{bmatrix} 0.05 \\ 1 \\ 7 \end{bmatrix}$$

Therefore, $a = 0.05 \times 2 = 0.1$. Then, the cost function for the unit is given by

$$f = 0.05P^2 + P + 7 \,\$/\text{h}$$

Figure 4.1 plots the quadratic power output versus the cost curve.

The economic dispatch problem considers the incremental fuel cost, which is the additional cost that may be incurred to produce an additional 1 MW of power. This is given by

$$\lambda_i = \frac{df_i}{dP_i} = a_i P_i + b_i \,\$/\text{MWh} \tag{4.2}$$

From the values calculated in Example 4.1, the cost of generating 25 MW of power that is obtained from (4.1) is \$63.26. On the other hand, the incremental cost of generating 1 MW, obtained from (4.2) is \$1.1. Note that the incremental cost does not include the fixed cost. Therefore, it cannot be calculated by taking the difference between generating 26 MW of power (which is \$66.8) and 25 MW of power.

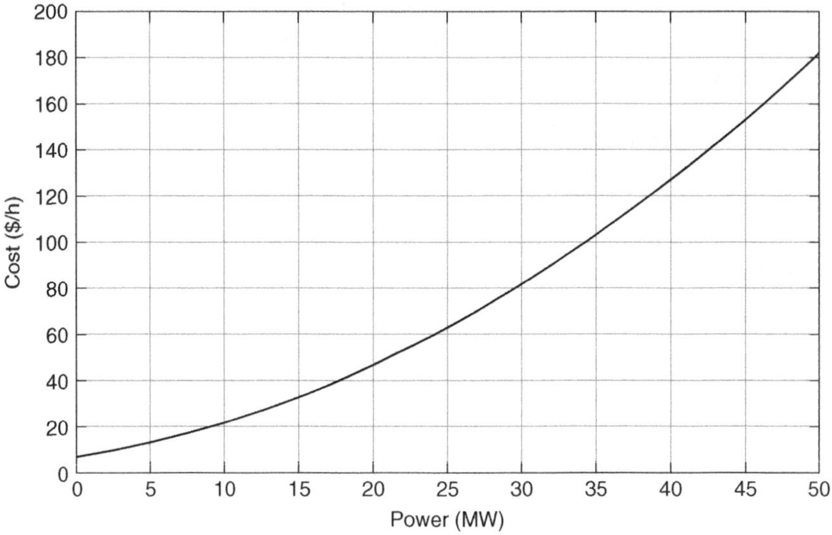

Figure 4.1 The power output versus cost curve of Example 4.1.

Let us now assume that only two units having different incremental costs supply a load. There will be a reduction in cost if some amount of load is transferred from the unit with higher incremental cost to the unit with lower incremental cost. In this fashion, the load is transferred from the less efficient unit to the more efficient unit thereby reducing the total operation cost. The load transfer will continue till the incremental costs of both units are the same. This will be the optimum point of operation for both units.

Example 4.2 Consider two units of a power plant that have the following fuel costs:

$$f_1 = \frac{0.08}{2}P_1^2 + 10P_1 + 25\,\$/\text{h} \text{ and } f_2 = \frac{0.06}{2}P_2^2 + 8P_2 + 20\,\$/\text{h}$$

The incremental costs of these two units are then

$$\lambda_1 = 0.08P_1 + 10\,\$/\text{MWh} \text{ and } \lambda_2 = 0.06P_2 + 8\,\$/\text{MWh}$$

Let us assume that the units together are supplying 500 MW of power. Suppose Unit-1 supplies 200 MW of power, while Unit-2 supplies the rest, that is, 300 MW. Then we have the following incremental cost of generating 1 MW of power by the units as

$$\lambda_1 = 0.08 \times 200 + 10 = 26\,\$/\text{MWh} \text{ and } \lambda_2 = 0.06 \times 300 + 8 = 26\,\$/\text{MWh}$$

Therefore, we find the incremental cost of generation of both units are the same. We shall call this the optimum setting, as explained in Section 4.1.2.

Now what happens when both units supply the same amount of power of 250 MW? The changes in the costs of the two units are:

$$\text{Unit-1:} \int_{200}^{250} (0.08P_1 + 10)dP_1 = \left(0.04P_1^2 + 10P_1\right)\Big|_{200}^{250} = 1{,}400\ \$/\text{h}$$

$$\text{Unit-2:} \int_{300}^{250} (0.04P_2 + 8)dP_2 = \left(0.03P_2^2 + 8P_2\right)\Big|_{300}^{250} = -1{,}225\ \$/\text{h}$$

Thus, we find that an excess of $f_{ex} = 1400 - 1225 = 175$ \$/h is required for this setting.

We shall calculate the excess cost (f_{ex}) from the optimal setting of $P_1 = 200$ MW and $P_2 = 300$ MW as the power generation of the two units changes from 0 MW to 500 MW for the total load of 500 M. For example, when Unit-1 supplies the entire 500 MW of power and Unit-2 supplies no power, we have

$$\text{Uniti-1:} \int_{200}^{500} (0.08P_1 + 10)dP_1 = \left(0.04P_1^2 + 10P_1\right)\Big|_{200}^{500} = 11{,}400\ \$/\text{h}$$

$$\text{Unit-2:} \int_{300}^{0} (0.04P_2 + 8)dP_2 = \left(0.03P_2^2 + 8P_2\right)\Big|_{300}^{0} = -5{,}700\ \$/\text{h}$$

Therefore, the excess cost is

$$f_x = 11{,}400 - 5{,}100 = 6{,}300\ \$/\text{h}$$

On the other hand, when Unit-2 supplies the entire 500 MW of power and Unit-1 supplies no power, we have

$$\text{Unit-1:} \int_{200}^{0} (0.08P_1 + 10)dP_1 = \left(0.04P_1^2 + 10P_1\right)\Big|_{200}^{0} = -3{,}600\ \$/\text{h}$$

$$\text{Unit-2:} \int_{300}^{500} (0.04P_2 + 8)dP_2 = \left(0.03P_2^2 + 8P_2\right)\Big|_{300}^{0} = 6{,}400\ \$/\text{h}$$

Therefore, the excess cost is

$$f_x = 6400 - 3600 = 2800\ \$/\text{h}$$

Figure 4.2 shows the variation in the excess cost when the P_1 changes from 0 to 500 MW. Consequently, P_2 changes from $P_2 - P_1$ MW to 0 MW. The excess cost is zero when $P_1 = 200$ MW, as expected.

With the current trend of including renewable energy generator, the economic operation problem becomes more complicated since they do not have any fuel cost. This implies that their operating does not increase with the power output.

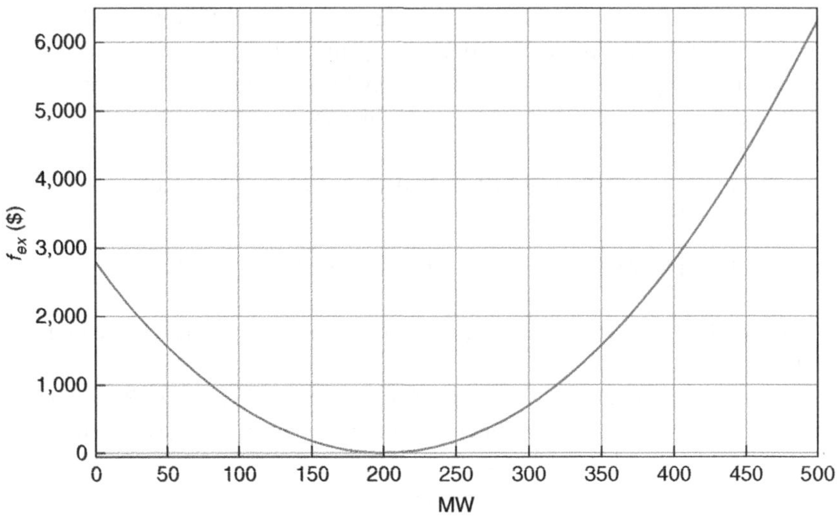

Figure 4.2 Variation in excess cost from optimal setting as P_1 changes from 0 to 500 MW in Example 4.2.

However, their output depends on resource availability (wind speed, sunlight), not on dispatch instructions. The cost equation for wind (or solar) power is given by

$$f = \text{LCOE} \times P + c\,\$/\text{h} \tag{4.3}$$

where the levelized cost of energy (LCOE) is given by

$$\text{LCOE} = \frac{\text{Total Lifetime Cost (\$)}}{\text{Total Lifetime Energy Output (MW)}} \tag{4.4}$$

Therefore, the incremental fuel cost is constant and is equal to LOCE. However, a penalty term can be used for uncertainty and variability of wind speed or solar irradiance. Sometimes a battery energy storage system (BESS) can be used to smoothen out the variability of wind or solar. Then the BESS cost can be included in the economic dispatch problem. Note that LCOE can vary from county to county. The typical value of LCOE in the United States is around US$45/MWh for onshore windfarms and around US$120 $/MWh for offshore windfarms.

4.1.2 Economic Distribution of Loads Between Multiple Units of a Plant

Let us consider a plant that has a total of N number of units. The total fuel cost will then be the summation of the individual fuel cost f_i, $i = 1, \ldots, N$ of each unit, that is,

$$f_T = f_1 + f_2 + \cdots + f_N = \sum_{k=1}^{N} f_k \tag{4.5}$$

Let us denote that the total power that the plant is required to supply by P_T, such that

$$P_T = P_1 + P_2 + \cdots + P_N = \sum_{k=1}^{N} P_k \tag{4.6}$$

where P_i, $i = 1, \ldots, N$ are the power supplied by the N different units.

The objective is to minimize f_T for a given P_T. This can be achieved when the total difference df_T becomes zero, that is,

$$df_T = \frac{\partial f_T}{\partial P_1} dP_1 + \frac{\partial f_T}{\partial P_2} dP_2 + \cdots + \frac{\partial f_T}{\partial P_N} dP_N = 0 \tag{4.7}$$

Now, since the power supplied is assumed to be constant, we have

$$dP_T = dP_1 + dP_2 + \cdots + dP_N = 0 \tag{4.8}$$

Multiplying (4.8) by λ and subtracting from (4.7), we get

$$\left(\frac{\partial f_T}{\partial P_1} - \lambda\right) dP_1 + \left(\frac{\partial f_T}{\partial P_2} - \lambda\right) dP_2 + \cdots + \left(\frac{\partial f_T}{\partial P_N} - \lambda\right) dP_N = 0 \tag{4.9}$$

In (4.9), the term λ is called the *Lagrange multiplier.*

The equality in (4.9) is satisfied when each individual term given in brackets is zero. This gives us

$$\frac{\partial f_T}{\partial P_i} - \lambda = 0, \quad i = 1, \ \ldots, \ N \tag{4.10}$$

Note that f_T is the summation of the costs of each individual unit, where, as per (4.1), the cost f_i is only dependent on the power output P_i, that is,

$$\frac{\partial f_T}{\partial P_i} = \frac{df_i}{dP_i}, \quad i = 1, \ldots, N$$

Therefore, the partial derivative becomes a full derivative for all N units, that is,

$$\lambda = \frac{df_1}{dP_1} = \frac{df_2}{dP_2} = \cdots = \frac{df_N}{dP_N} \tag{4.11}$$

Example 4.3 Consider a power plant that has three units. The incremental fuel costs are

$$\text{Unit-1: } \frac{df_1}{dP_1} = 0.12P_1 + 12.0\,\$/\text{MWh}$$

$$\text{Unit-2: } \frac{df_2}{dP_2} = 0.08P_2 + 8.0\,\$/\text{MWh}$$

$$\text{Unit-3: } \frac{df_3}{dP_3} = 0.096P_3 + 9.0\,\$/\text{MWh}$$

The units together are supplying 900 MW of power. We have to determine $P_1, P_2,$ and P_3.

To determine three unknowns, we need three equations. The first one is total power supplied, that is,

$$P_1 + P_2 + P_3 = 900\text{ MW}$$

Now, the incremental costs must be the same. Therefore, we have the following two equations:

$$\lambda = \frac{df_1}{dP_1} = \frac{df_2}{dP_2}$$
$$\Rightarrow 0.12P_1 + 12.0 = 0.08P_2 + 8.0 \Rightarrow 0.12P_1 - 0.08P_2 = -4$$

$$\lambda = \frac{df_2}{dP_2} = \frac{df_3}{dP_3}$$
$$\Rightarrow 0.08P_2 + 8.0 = 0.096P_3 + 9.0 \Rightarrow 0.08P_2 - 0.096P_3 = -1$$

Writing these two equations in a matrix–vector form, we have

$$\begin{bmatrix} 0.12 & -0.08 & 0 \\ 0 & 0.08 & -0.096 \\ 1 & 1 & 1 \end{bmatrix} \begin{bmatrix} P_1 \\ P_2 \\ P_3 \end{bmatrix} = \begin{bmatrix} -4 \\ -1 \\ 900 \end{bmatrix}$$

Solving the above equation, we have

$$P_1 = 212.78\text{ MW } P_2 = 369.17\text{ MW and } P_3 = 318.05\text{ MW}$$

The incremental fuel cost then is

$$\lambda = \frac{df_1}{dP_1} = 0.12P_1 + 12.0 = 37.53\,\$/\text{MWh}$$

Let us now assume that the plant also contains a wind farm nearby, which has a maximum capacity of 200 MW. However, the wind farm can generate between 0 and 200 MW of power. Then, if the total power requirement still remains 900 MW, we have

$$P_1 + P_2 + P_3 = 900 - P_W\text{ MW}$$

where P_W is the power generated by the wind farm. The variations in the optimal settings of the generating units and the incremental cost are shown in Figure 4.3 when the wind power changes from 0 to 200 MW.

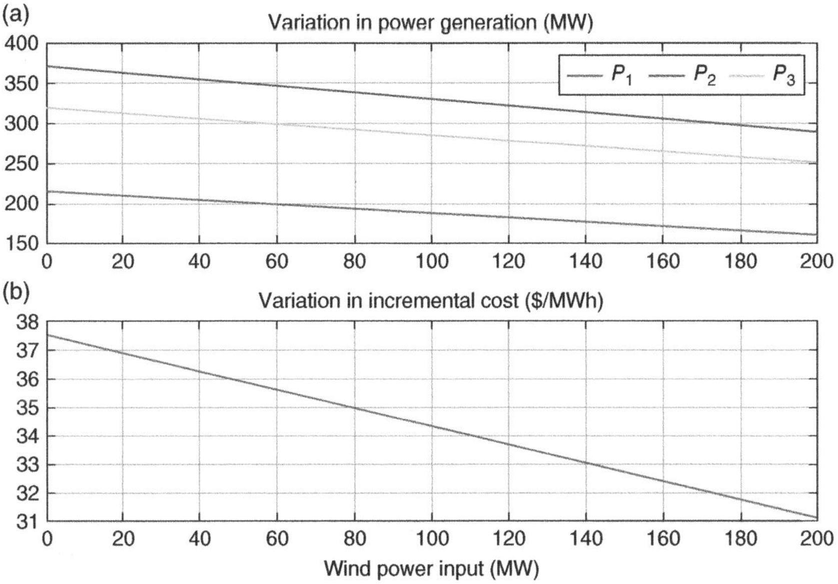

Figure 4.3 Variations in (a) generating power and (b) incremental cost with the changes in wind power input.

4.1.3 Consideration of Generator Limits

It is not always necessary that all the units of a plant are available to share a load. Some of the units may be taken off due to scheduled maintenance. Also, it is not necessary that the less efficient units are switched off during off-peak hours. There is a certain amount of shutdown and startup costs associated with shutting down a unit during the off-peak hours and servicing it back online during the peak hours. To complicate the problem further, it may take about eight hours or more to restore the boiler of a unit and synchronize the unit with the bus. Therefore, to meet the sudden change in the power demand, it may be necessary to keep more units connected to the system than is necessary to meet the load demand during that time. This safety margin in generation is called *spinning reserve*. The optimal load dispatch problem must then incorporate this startup and shutdown cost for not endangering the system security.

The power generation limit of each unit is then given by the inequality constraints

$$P_{min,i} \leq P_i \leq P_{max,i}, \quad i = 1, \ldots, N \tag{4.12}$$

The maximum limit P_{max} is the upper limit of the power generation capacity of each unit. On the other hand, the lower limit P_{min} pertains to the thermal consideration of operating a boiler in a thermal or nuclear generating station. An operational unit must produce a minimum amount of power such that the boiler thermal components are stabilized at the minimum designed operating temperature.

Example 4.4 Consider again the power plant of Example 4.3 without the wind farm. Let the limits of the three generators be given by

$$\text{Unit-1: } 30\,\text{MW} \leq P_1 \leq 300\,\text{MW}$$

$$\text{Unit-2: } 30\,\text{MW} \leq P_2 \leq 300\,\text{MW}$$

$$\text{Unit-3: } 30\,\text{MW} \leq P_3 \leq 500\,\text{MW}$$

From Example 4.3, it can be seen that when the generators are required to supply 900 MW of power, P_2 needs to supply 369.17 MW, which exceeds its generation limit. Furthermore, each generator cannot supply less than 30 MW of power. Therefore, the economic dispatch problem must consider the generation limits in a constrained optimization problem. We shall discuss the economic dispatch problem starting from the minimum load of 90 MW to the full load of 1,100 MW.

At the minimal load of 90 MW, all the three units must generate 30 MW each. Then, the incremental fuel cost of each unit is

$$\text{Unit-1: } \frac{df_1}{dP_1} = 0.12 \times 30 + 12.0 = 15.6\,\$/\text{MWh}$$

$$\text{Unit-2: } \frac{df_2}{dP_2} = 0.08 \times 30 + 8.0 = 10.4\,\$/\text{MWh}$$

$$\text{Unit-3: } \frac{df_3}{dP_3} = 0.096 \times 30 + 9 = 11.88\,\$/\text{MWh}$$

Since Units 1 and 3 have higher incremental costs, they must operate at 30 MW each when the power demand increases above 90 MW. The incremental cost will be due to Unit-2, which is equal to 10.4 \$/MWh. Then, the optimal dispatch problem is separated into the following segments.

- *Segment-1*: With the generation of Units 1 and 3 remaining constant at 30 MW, the generation of Unit-2 is increased due to a load increase till its incremental cost is equal to that of Unit-3, that is, 11.88 \$/MWh. This is achieved when P_2 is equal to 48.5 MW, at a total power of 108.5 MW. Beyond this power level, the incremental costs of both Units 2 and 3 will be the same.
- *Segment*-2: Units 2 and 3 together will supply the increased load demand with the contribution of Unit-1 remaining at 30 MW till the incremental costs of all the three units are equal to that of Unit-1, that is, 15.6 \$/MWh. To calculate the power supplied by Units 2 and 3, we have

$$15.6 = 0.08P_2 + 8.0\,\$/\text{MWh} \Rightarrow P_2 = 95\,\text{MW}$$
$$15.6 = 0.096P_3 + 9.0\,\$/\text{MWh} \Rightarrow P_3 = 68.75\,\text{MW}$$

Therefore, the total power that will be supplied during this time is 30 + 95 + 68.75 = 193.75 MW.

- *Segment-3*: For a power supply level above 193.75 MW, all the three units will supply power together with the same incremental fuel cost. However, Unit-2, being the most efficient unit, reaches its generation limits of $P_2 = 300$ MW. To calculate the operating condition for which Unit-2 reaches its maximum limit, we have

$$\lambda = \frac{df_2}{dP_2} = 0.08 \times 300 + 8.0 = 32\,\$/\text{MWh}$$

Using this value of λ, we have the power generated by the other two units as

$$\lambda = 32 = 0.12P_1 + 12.0 \Rightarrow P_1 = 166.67\,\text{MW}$$
$$\lambda = 32 = 0.096P_3 + 9.0 \Rightarrow P_3 = 239.58\,\text{MW}$$

The total power then is 166.7 + 300 + 239.58 = 706.25 MW.

- *Segment-4*: Let us assume that the plant is required to supply 900 MW of power. Since Unit-2 has reached its maximum power level of 300 MW, Units 1 and 3 must have the same incremental fuel cost. Then, we have

$$P_1 + P_3 = 900 - 300 = 600\,\text{MW}$$

To supply 600 MW of power, the incremental fuel cost will be

$$\lambda = \frac{df_1}{dP_1} = \frac{df_3}{dP_3}$$
$$\Rightarrow 0.12P_1 + 12.0 = 0.096P_3 + 9.0 \Rightarrow 0.12P_1 - 0.096P_3 = -3$$

The solutions of the last two equations are

$$P_1 = 252.78\,\text{MW and } P_3 = 347.22\,\text{MW}$$

The fuel cost for Unit-2 is fixed. Therefore, the incremental fuel of 1 MW excess generation will be shared by Units 1 and 3. The incremental cost is then equal to

$$\lambda = \frac{df_1}{dP_1} = 0.12P_1 + 12.0 = 42.33\,\$/\text{MWh}$$

- *Segment-5*: Finally, Units 1 and 3 will supply power till Unit-1 reaches its maximum limit of 300 MW. The incremental fuel cost at this point is

Table 4.1 Load distribution and incremental cost for the units of Example 4.4.

P_T (MW)	P_1 (MW)	P_2 (MW)	P_3 (MW)	λ ($/MWh)
90	30	30	30	10.4
100	30	40	30	11.2
108.5	30	48.5	30	11.88
150	30	71.14	48.86	13.69
193.75	30	95	68.75	15.6
300	58.33	137.5	104.17	19
400	85	177.5	137.5	22.2
500	111.67	217.5	170.83	25.4
600	138.33	257.5	204.17	28.6
700	165	297.5	237.5	31.8
706.25	166.67	300	239.58	32
800	208.33	300	291.67	37
900	252.78	300	347.22	42.33
1006.25	300	300	406.25	48
1,100	300	300	500	57

$$\frac{df_1}{dP_1} = 0.12 \times 300 + 12 = 48 \text{ \$/MWh}$$

The power generated by Unit-3 for this incremental cost is

$$\lambda = 48 = 0.096P_3 + 9.0 \Rightarrow P_3 = 406.25 \text{ MW}$$

The total power generation at this time is 300 + 300 + 406.25 = 1,000.25 MW. From this point onward till the maximum power level is reached, only Unit-3 will supply power, and the incremental fuel cost will be that of Unit-3 alone. Table 4.1 lists the load distribution and the incremental fuel cost for the generators. The load distribution is pictorially depicted in Figure 4.4.

4.2 Economic Operation of a Power System

In Section 4.1, the economic operation of a single plant has been discussed where it has been shown how a particular amount of load is shared between the different units of a plant. In this problem, the transmission line losses are not considered

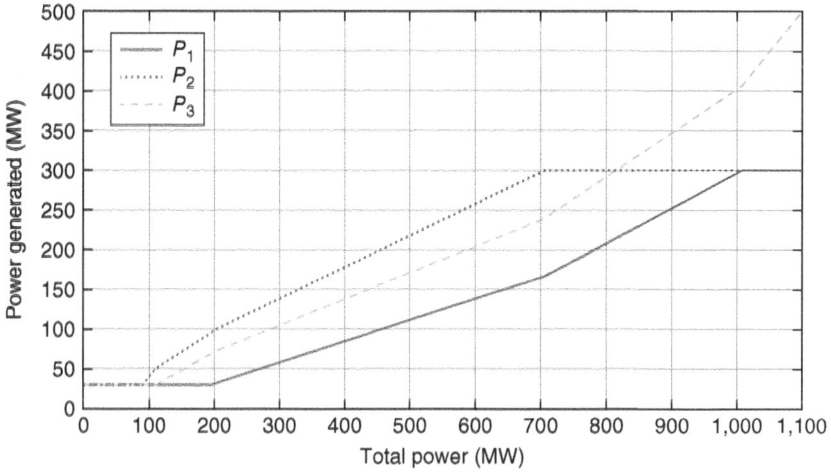

Figure 4.4 Power distribution between the units in Example 4.4.

assuming that the losses were a part of the load supplied. However, for the distribution of loads between the different plants that are joined by transmission lines, the line losses must be explicitly included in the economic dispatch problem. In this section, this problem is discussed.

When the transmission losses are included in the economic dispatch problem, we can modify (4.6) as

$$P_T = P_1 + P_2 + \cdots + P_N - P_{LOSS} \tag{4.13}$$

where P_{LOSS} is the total line loss. Since P_T is assumed to be constant, we have

$$0 = dP_1 + dP_2 + \cdots + dP_N - dP_{LOSS} \tag{4.14}$$

In the above equation, dP_{LOSS} includes power loss due to every generator, that is,

$$dP_{LOSS} = \frac{\partial P_{LOSS}}{\partial P_1} dP_1 + \frac{\partial P_{LOSS}}{\partial P_2} dP_2 + \cdots + \frac{\partial P_{LOSS}}{\partial P_N} dP_N \tag{4.15}$$

Also, the minimum generation cost implies $df_T = 0$ as given in (4.7). Multiplying both (4.14) and (4.15) by λ and combining them, we get

$$0 = \left(\lambda \frac{\partial P_{LOSS}}{\partial P_1} - \lambda\right) dP_1 + \left(\lambda \frac{\partial P_{LOSS}}{\partial P_2} - \lambda\right) dP_2 + \cdots + \left(\lambda \frac{\partial P_{LOSS}}{\partial P_N} - \lambda\right) dP_N \tag{4.16}$$

Adding (4.16) with (4.7), we obtain

$$0 = \sum_{i=1}^{N} \left(\frac{\partial f_T}{\partial P_i} + \lambda \frac{\partial P_{LOSS}}{\partial P_i} - \lambda \right) dP_i \tag{4.17}$$

The above equation is satisfied when

$$\frac{\partial f_T}{\partial P_i} + \lambda \frac{\partial P_{LOSS}}{\partial P_i} - \lambda = 0, \quad i = 1, \ldots, N \tag{4.18}$$

Again since

$$\frac{\partial f_T}{\partial P_i} = \frac{df_i}{\partial P_i}, \quad i = 1, \ldots, N$$

from (4.18) we get

$$\lambda = \frac{df_1}{dP_1} L_1 = \frac{df_2}{dP_2} L_2 = \cdots = \frac{df_N}{dP_N} L_N \tag{4.19}$$

where L_i is called the *penalty factor* and is given by

$$L_i = \frac{1}{1 - \partial P_{LOSS}/\partial P_i}, \quad i = 1, \ldots, N \tag{4.20}$$

Example 4.5 Let us consider the two generating units of Example 4.2. It has been assumed that the transmission loss is defined in terms of these two units as

$$P_{LOSS} = 0.75 \times 10^{-4} P_1^2 + 1.4 \times 10^{-5} P_1 P_2 + 0.8 \times 10^{-4} \times P_2^2$$

Therefore, from (4.20), we have

$$\begin{aligned} L_1 &= \frac{1}{1 - \partial P_{LOSS}/\partial P_1} = \frac{1}{1 - 1.5 \times 10^{-4} P_1 - 1.4 \times 10^{-5} P_2} \\ L_2 &= \frac{1}{1 - \partial P_{LOSS}/\partial P_2} = \frac{1}{1 - 1.4 \times 10^{-5} P_1 - 1.6 \times 10^{-4} P_2} \end{aligned}$$

Let the incremental fuel cost be $\lambda = \$26/\text{MWh}$. Then, from (4.19), we get

$$26.9 = \frac{0.08 P_1 + 10}{1 - 1.5 \times 10^{-4} P_1 - 1.4 \times 10^{-5} P_2}$$

$$\Rightarrow 0.0847 P_1 + 3.77 \times 10^{-4} P_2 = 16.9$$

$$26.9 = \frac{0.06 P_2 + 8}{1 - 1.4 \times 10^{-5} P_1 - 1.6 \times 10^{-4} P_2}$$

$$\Rightarrow 3.77 \times 10^{-4} P_1 + 0.0643 P_2 = 18.9$$

Rearranging the above two equations, we get

$$\begin{bmatrix} 0.0847 & 3.77\times10^{-4} \\ 3.77\times10^{-4} & 0.0643 \end{bmatrix}\begin{bmatrix} P_1 \\ P_2 \end{bmatrix} = \begin{bmatrix} 16.9 \\ 18.9 \end{bmatrix}$$

The solution of the equation gives $P_1 = 199.79$ MW and $P_2 = 292.75$ MW. Then, the power loss is calculated as 10.77 MW.

The power loss equation in the example above can be rewritten as

$$P_{LOSS} = [P_1 \;\; P_2]\begin{bmatrix} 0.75\times10^{-4} & 0.7\times10^{-5} \\ 0.7\times10^{-5} & 0.8\times10^{-4} \end{bmatrix}\begin{bmatrix} P_1 \\ P_2 \end{bmatrix}$$

In general, consider a power system with N number of plants. The power generated is defined by the vector

$$\mathbf{P} = [P_1 \;\; P_2 \;\; \cdots \;\; P_N]^T$$

The transmission losses can then be written as

$$P_{LOSS} = \mathbf{P}^T\mathbf{B}\mathbf{P} \tag{4.21}$$

where the matrix $\mathbf{B}$ is a symmetric matrix given by

$$\mathbf{B} = \begin{bmatrix} B_{11} & B_{12} & \cdots & B_{1N} \\ B_{12} & B_{22} & \cdots & B_{2N} \\ \vdots & \vdots & \ddots & \vdots \\ B_{1N} & B_{2N} & \cdots & B_{NN} \end{bmatrix}$$

The elements B_{ij} of the matrix $\mathbf{B}$ are called the *loss coefficients*. These coefficients are not constant but vary with plant loading. However, for the simplified calculation of the penalty factor L_i, these coefficients are often assumed to be constant.

When the incremental cost equations are linear, we can use analytical equations to find out the economic settings. However, in practice, the incremental costs are given by nonlinear equations that may even contain nonlinearities. In that case, iterative solutions are required to find the optimal generator settings. The following example shows one approach of the iterative procedure [1].

Example 4.6 We shall consider the system of Examples 4.2 and 4.4. The incremental costs are given by

$$\frac{df_1}{dP_1} = a_1P_1 + b_1 \text{ and } \frac{df_2}{dP_2} = a_2P_2 + b_2$$

where

$$a_1 = 0.08, b_1 = 10, a_2 = 0.06, \text{and } b_2 = 8$$

The power loss is defined from (4.21) as

$$P_{LOSS} = \begin{bmatrix} P_1 & P_2 \end{bmatrix} \begin{bmatrix} B_{11} & B_{12} \\ B_{12} & B_{22} \end{bmatrix} \begin{bmatrix} P_1 \\ P_2 \end{bmatrix}$$
$$= B_{11}P_1^2 + 2B_{12}P_1P_2 + B_{22}P_2^2$$

where

$$B_{11} = 0.75 \times 10^{-4}, B_{12} = 0.7 \times 10^{-5}, \text{and } B_{22} = 0.8 \times 10^{-4}$$

The steps of the iterative process are given below.

Step 1: First, choose the total load level, P_T, which is chosen as $P_T = 500$ MW.
Step 2: Choose the initial values of the power supplied by the two units. From Example 4.2, these are chosen as

$$P_1 = 200\text{ MW and } P_2 = 300\text{ MW}$$

The power loss is computed from (4.21) as $P_{LOSS} = 11.04$ MW
Step 3: Compute the derivatives of the power loss with respect to the chosen values

$$\frac{\partial P_{LOSS}}{\partial P_1} = 1.5 \times 10^{-4}P_1 + 1.4 \times 10^{-5}P_2 = 0.0342$$
$$\frac{\partial P_{LOSS}}{\partial P_2} = 1.4 \times 10^{-5}P_1 + 1.6 \times 10^{-4}P_2 = 0.0508$$

Step 4: From (4.18), we obtain the following two equations:

$$\frac{df_1}{dP_1} + \lambda\frac{\partial P_{LOSS}}{\partial P_1} - \lambda = 0 \Rightarrow (a_1P_1 + b_1) + \lambda\frac{\partial P_{LOSS}}{\partial P_1} - \lambda = 0$$

$$\frac{df_2}{dP_2} + \lambda\frac{\partial P_{LOSS}}{\partial P_2} - \lambda = 0 \Rightarrow (a_2P_2 + b_2) + \lambda\frac{\partial P_{LOSS}}{\partial P_2} - \lambda = 0$$

In addition, the power balance equation is given by

$$P_1 + P_2 = P_T + P_{LOSS}$$

These three equations are written in the matrix–vector form as

$$\begin{bmatrix} a_1 & 0 & \dfrac{\partial P_{LOSS}}{\partial P_1} - 1 \\ 0 & a_2 & \dfrac{\partial P_{LOSS}}{\partial P_2} - 1 \\ 1 & 1 & 0 \end{bmatrix} \begin{bmatrix} P_1 \\ P_2 \\ \lambda \end{bmatrix} = \begin{bmatrix} -b_1 \\ -b_2 \\ P_T + P_{LOSS} \end{bmatrix}$$

The solution of the above equation is

$$\begin{bmatrix} P_1 \\ P_2 \\ \lambda \end{bmatrix} = \begin{bmatrix} 208.002\text{ MW} \\ 303.038\text{ MW} \\ 27.5835\text{ \$/MWh} \end{bmatrix}$$

The error in power computation is then

$$P_{err} = P_T + P_{LOSS} - (P_1 + P_2) = 11.04\text{ MW}$$

Since the error is above a certain threshold, we go back to Step 2 and calculate the quantities again. The iterative process converges in five iterations, where the power error is found to be $P_{err} = 4.2 \times 10^{-5}$ MW. The converged parameters are:

$$P_1 = 208.0839\text{ MW}, P_2 = 303.4121\text{ MW and } \lambda = 27.6264\text{ \$/MWh}$$

4.3 Unit Commitment

In the previous two sections, we have discussed economic operation of power systems, which assumes that a total number of N units are already connected to the systems. The goal of the economic dispatch is to optimally schedule these N units such that the total power can be served. The unit commitment problem, however, is more complex. Assume that N units are available to serve a power system. However, all these units need not be operational all the time. In the unit commitment problem, the optimal scheduling of a subset of these N units is performed based on the forecasted load demand.

The daily load consumption is not fixed and varies with the time of the day and the time of the year. For example, in an industrial region, the daytime load will be high due to power consumption by the industries, irrespective of the time of the year. The domestic load can vary depending on the time of the year. In Australia, the summer months are hot, and air conditioners can consume a large amount of energy during the evenings to cool houses. In winter months, the air conditioner may run in reverse cycle, but usually late in the evenings. On the other hand, the lighting load is higher in winter months than in the summer months. We know that the power system requires an almost instantaneous balance between power generation and power consumption. Therefore, is it reasonable to think that all the units of a plant will run simultaneously and change their power output as required? If a power system is operated based solely on thermal generators, a great deal of money can be saved by turning off some of the units during lighter load conditions. Let us consider the following example.

Example 4.7 Consider the load pattern for a particular day that is shown in Figure 4.5 that is to be supplied by the following three generating units:

Unit-1: $$f_1 = \frac{0.12}{2}P_1^2 + 12P_1 + 25\,\$/\text{h}$$
$$200\text{ MW} \leq P_1 \leq 500\text{ MW}$$

Unit-2: $$f_2 = \frac{0.08}{2}P_2^2 + 8P_2 + 15\,\$/\text{h}$$
$$100\text{ MW} \leq P_1 \leq 300\text{ MW}$$

Unit-3: $$f_3 = \frac{0.096}{2}P_3^2 + 9P_3 + 20\,\$/\text{h}$$
$$100\text{ MW} \leq P_3 \leq 300\text{ MW}$$

The incremental fuel costs of these units are given as follows:

$$\frac{df_1}{dP_1} = 0.12P_1 + 12.0\,\$/\text{MWh}$$

$$\frac{df_2}{dP_2} = 0.08P_2 + 8.0\,\$/\text{MWh}$$

$$\frac{df_3}{dP_3} = 0.096P_3 + 9.0\,\$/\text{MWh}$$

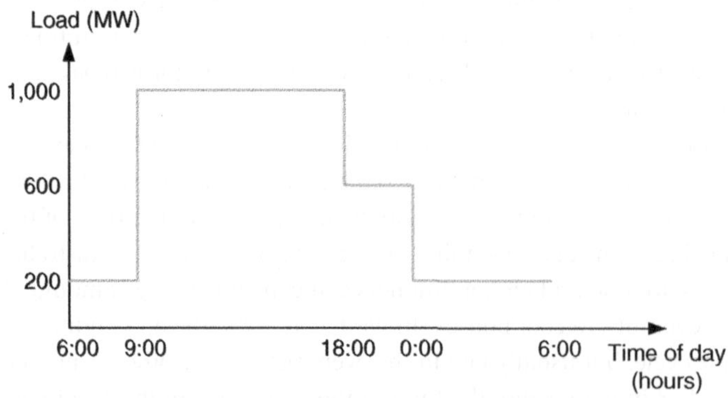

Figure 4.5 Hourly distribution loads for a particular day.

We shall consider three load levels: 200 MW, 600 MW, and 1,000 MW. These are considered individually as discussed below.

200 MW Load: The minimum load level of 200 MW can be (a) supplied by Unit-1 or (b) shared by Units 2 and 3. Note that this minimum level needs to be supplied between 0:00 and 9:00 hours, that is, a total of nine hours.

Option-a: The cost of running Unit-1 supplying 200 MW for nine hours is

$$\left(\frac{0.12}{2} \times 200^2 + 12 \times 200 + 25\right) \times 9 = \$43,425$$

Option-b: The cost of running Units 2 and 3 supplying 100 MW each is

$$\left(\frac{0.08}{2} \times 100^2 + 8 \times 100 + 15 + \frac{0.096}{2} \times 100^2 + 9 \times 100 + 20\right) \times 9 = \$23,535$$

Obviously shutting Unit-1 down during the off-peak hours is financially beneficial. This result is not surprising since the generating cost of Unit-1 is higher than the other two units.

600 MW Load: This load level is supplied for six hours (between 18:00 and 24:00 hours). The unconstrained scheduling as discussed in Example 4.3 will give us the following values of the power share:

$$P_1 = 138.33 \text{ MW } P_2 = 257.5 \text{ MW and } P_3 = 204.17 \text{ MW}$$

This obviously is not feasible since it violates the minimum power condition of Unit-1. Therefore, the choice is (a) to supply 600 MW by Units 2 and 3 equally or (b) let Unit-1 supply 200 MW, while the rest being shared by Units 2 and 3 economically.

Option-a: The cost of this option is

$$\left(\frac{0.08}{2} \times 300^2 + 8 \times 300 + 15 + \frac{0.096}{2} \times 300^2 + 9 \times 300 + 20\right) \times 6 = \$78,330$$

Option-b: If Unit-1 supplies 200 MW, then the optimal setting for Units 2 and 3 are

$$P_2 = 223.86 \text{ MW and } P_3 = 176.14 \text{ MW}$$

The total cost is then

$$\left(\frac{0.12}{2} \times 200^2 + 12 \times 200 + 25\right) \times 6$$
$$+ \left(\frac{0.08}{2} \times 223.86^2 + 8 \times 223.86 + 15 + \frac{0.096}{2} \times 176.14^2 + 9 \times 176.14 + 20\right)$$
$$\times 6 = \$70,379$$

Obviously, Option-b is the cheaper option.

1,000 MW Load: The load at this level must be supplied for nine hours (between 9:00 and 18:00 hours). There are two options. These are (a) Units 2 and 3 supplying their maximum rated power of 300 MW each and Unit-1 supplying the rest 400 MW and (b) Unit-1 supplying (say) 500 MW and the rest is shared by the other two units economically.

Option-a: The cost of this option is

$$\left(\frac{0.12}{2}\times 400^2+12\times 400+25\right)\times 9$$
$$+\left(\frac{0.08}{2}\times 300^2+8\times 300+15+\frac{0.096}{2}\times 300^2+9\times 300+20\right)$$
$$\times 9=\$247,320$$

Option-b: The economic operation of Units 2 and 3 supplying 500 MW of power lead to

$$P_2=278.41\text{ MW and }P_3=221.59\text{ MW}$$

Therefore, the total cost for this option is

$$\left(\frac{0.12}{2}\times 500^2+12\times 500+25\right)\times 9$$
$$+\left(\frac{0.08}{2}\times 278.41^2+8\times 278.41+15+\frac{0.096}{2}\times 221.59^2+9\times 221.59+20\right)$$
$$\times 9=\$276,651$$

Obviously, Option-a is cheaper in this case. The unit commitment problem is summarized in Table 4.2.

The above example shows that unit commitment is a complex problem even for this simple three-unit system. Usually, this problem is solved through a suitable optimization technique. There are several other factors that need to be considered in the optimization process. Some of these are discussed below.

Table 4.2 Unit commitment for Example 4.6.

Hours	Total load	Unit-1	Unit-2	Unit-3
0:00–9:00	200 MW	Off	100 MW	100 MW
9:00–18:00	1,000 MW	400 MW	300 MW	300 MW
18:00–24:00	600 MW	200 MW	223.86 MW	176.14 MW

4.3.1 Spinning Reserve

At any given time, there are a certain number of generators that are synchronized with an ac power system. At the same time, the generators together supply a certain amount of load. The spinning reserve is defined as the difference between the generation capacity of the generators minus the loads. The term spinning is used to signify that the generators are spinning with the system, i.e., synchronized with the ac system.

As will be explained in Chapter 7, an increase in load will cause a drop in frequency. Conversely, the failure of one of the generating units can also cause a drop in frequency. A large mismatch between the generation and load can have a catastrophic effect on the power system. Therefore, power systems must be equipped with fast responding generators so that the frequency can be restored using automatic generation control.

Modern-day power systems are equipped with large wind and solar farms. These are called intermittent in nature. A passing cloud can cause the solar power output to drop. Similarly, wind speed can drop suddenly. Some wind turbine technologies are known to be susceptible to tripping even if the voltage transiently falls to levels as high as 70%. Such plants would jeopardize the grid stability. A large number of wind turbines use induction generators. During voltage dips, induction generators tend to significantly increase their reactive power demand to the extent that the system voltage may be further depressed. This causes slower recovery of the voltage once the fault has been removed.

If several wind generators are suddenly disconnected, there will be a mismatch between generated and consumed power, and as a result, the frequency will drop. If there is not enough spinning reserve in the system, a black-out may occur. If wind farms are not able to ride-through voltage dips, the transmission systems need a higher spinning reserve.

4.3.2 Thermal Limit Constraints

Thermal units cannot be brought online quickly. Their temperature needs to be changed gradually. The boilers need to cool down completely before they can be fired again. These may require hours of cooling down and hours of heating up, plus the time required for synchronization with the ac grid. However, a unit that has been shut down recently can be restarted relatively quickly since the boiler temperature may not have dropped significantly. The thermal limits constraints are [1]:

- *Minimum uptime*: A running unit cannot be turned off instantaneously.
- *Minimum downtime*: As has been mentioned earlier, a certain amount of time is required to bring the boiler to operating temperature and synchronization.

- *Crew constraint*: The starting up of a unit or multiple units may require a number of crew members. It may not always be possible for the crews to attend to more than one unit at any given time.

There are two starting modes of a boiler. The first one is called the *cold start*, where the boiler has cooled down completely. The other is called *banking*, where the boiler operating temperature is maintained by providing sufficient heat. The costs for the two operations are given as [1]

$$f_c = f_{cs}\left(1 - \varepsilon^{-t/\alpha}\right) \times F + c \tag{4.22}$$

where f_c is the startup cost when cooling; f_{cs} is the cold start cost in kJ; F is the fuel cost in \$; α is the thermal time constant of the unit; t is the number of hours the unit was cooled.

$$f_b = f_t \times t \times F + c \tag{4.23}$$

where f_b is the startup cost when banking; f_t is the cost of maintaining the temperature of the unit in kJ/h.

It is to be noted that the cost of banking is not necessarily less than that of the cold start cost. Up to a certain time, the banking cost is lower than the cold start cost. However, it is logical that if a unit has to maintain the operating temperature for a longer period of time, then its banking cost will be higher.

4.3.3 Solution Methods for Unit Commitment Problem

As has been mentioned before, the load varies throughout the day. At the same time, there may be several units available to cater to the load. A typical load variation curve is shown in Figure 4.6 (a). Figure 4.6 (b) shows the different generation levels that are required to supply this load variation. Suppose there are N number of units in a power system. Then, there are $2^N - 1$ combinations that are available to supply the load throughout the day. It is evident from Figure 4.6 (b) that it is not economical to run all the units throughout the day. Therefore, in the unit commitment problem, an optimization process must be applied to find the best possible economic combinations that can supply the daily load.

Unit commitment problem is usually very difficult to tackle. There are (1) binary variables, for example, generator on or off, (2) constraints such supply must meet demand, and (3) large systems with several units that need to be coordinated. There are three methods that can be applied to the unit commitment method. These are [1]:

- Priority list scheme
- Lagrange relaxation
- Dynamic programming

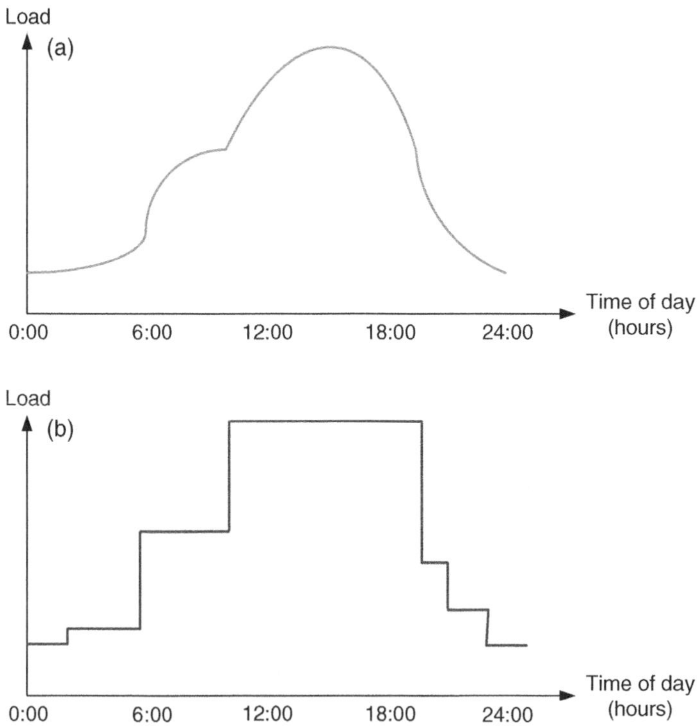

Figure 4.6 (a) A typical load variation curve and (b) power requirement to serve the load.

Furthermore, machine learning methods can also be used. We have demonstrated the basic principles of the priority list scheme in Example 4.7. In the Lagrange relaxation method, the power balance constraint, that is, $\sum_i P_i = P_T$ is relaxed and is replaced by a penalty term using Lagrange multiplier. Let us assume that there are three units with the following power constraints in MW:

$$20 \le P_1 \le 100$$
$$30 \le P_2 \le 150$$
$$10 \le P_3 \le 50$$

The costs of generation for the three units are given by

$$f_1 = 0.02P_1^2 + 20P_1 + 100\,\$$$
$$f_2 = 0.01P_2^2 + 25P_2 + 120\,\$$$
$$f_3 = 0.015P_3^2 + 30P_3 + 80\,\$$$

Then, from (4.1), we get

$$a_1 = 0.04, b_1 = 20, c_1 = 100$$
$$a_2 = 0.02, b_2 = 25, c_2 = 120$$
$$a_3 = 0.03, b_3 = 30, c_3 = 80$$

The total demand from the three units is 180 MW.

Note from (4.2), the Lagrange multiplier is given by

$$\lambda = \frac{df_i}{dP_i} = a_i P_i + b_i \; \$/\text{MWh}$$

This can be rewritten as

$$P_i = \frac{\lambda - b_1}{a_i}$$

The Lagrange relaxation is an iterative method, where we first choose $\lambda = 30$. Then, in the first iteration, we have

Unit-1: $P_1 = \dfrac{30-20}{0.04} = 250$ MW. This is more than the maximum rating. Then the generation is clipped at $P_1 = 100$ MW.

Unit-2: $P_2 = \dfrac{30-25}{0.02} = 250$ MW. This is more than the maximum rating. Then the generation is clipped at $P_2 = 150$ MW.

Unit-3: $P_2 = \dfrac{30-30}{0.03} = 0$ MW. This is less than the maximum rating. Then the generation is clipped at $P_3 = 10$ MW.

Therefore, the total generation is $100 + 150 + 10 = 260$ MW. With the load demand of 180 MW, we have an over generation of 80 MW. Therefore, λ must be reduced in the next iteration. The problem with this method is that the mismatch can oscillate between two values, without converging. For a comprehensive analysis of unit commitment problem, see [1].

4.4 Automatic Generation Control

Electric power is generated by converting mechanical energy into electrical energy. The rotor mass, which contains turbine and generator units, stores kinetic energy due to its rotation. This stored kinetic energy accounts for sudden increase in the load. Let us denote the mechanical torque input by T_m and the output electrical torque by T_e. Neglecting the rotational losses, a generator unit is said to be operating in the steady state at a constant speed when the difference between these two elements of torque is zero. In this case, we say that the accelerating torque is zero.

$$T_a = T_m - T_e \tag{4.24}$$

When the electric power demand increases suddenly, the electric torque increases. However, without any feedback mechanism to alter the mechanical torque, T_m remains constant. Therefore, the accelerating torque given by (4.24) becomes negative causing a deceleration of the rotor mass. During this time, the system frequency, which is proportional to the rotor speed, also decreases. We can thus infer that any deviation in the frequency for its nominal value of 50 or 60 Hz is indicative of the imbalance between T_m and T_e. The frequency drops when $T_m < T_e$ (T_a is negative) and rises when $T_m > T_e$ (T_a is positive).

The steady-state power–frequency relation is shown in Figure 4.7. In this figure, the slope of the ΔP_{ref} line is negative and is given by

$$-R = \frac{\Delta f}{\Delta P_m} \tag{4.25}$$

where R is called the regulating constant. Note that from Figure 4.7 that this quantity is given in per unit. From this figure, we can write the steady-state power–frequency relation as

$$\Delta P_m = \Delta P_{ref} - \frac{1}{R}\Delta f \tag{4.26}$$

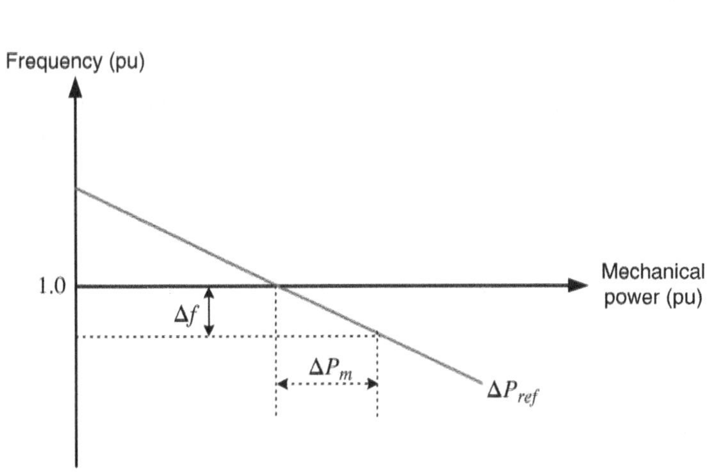

Figure 4.7 A typical steady-state power–frequency curve.

Suppose an interconnected power system contains N turbine–generator units. Then the steady-state power–frequency relation is given by the summation of (4.26) for each of these units as

$$\begin{aligned}\Delta P_m &= \Delta P_{m1} + \Delta P_{m2} + \cdots + \Delta P_{mN}\\ &= \left(\Delta P_{ref1} + \Delta P_{ref2} + \cdots + \Delta P_{refN}\right) - \left(\frac{1}{R_1} + \frac{1}{R_2} + \cdots + \frac{1}{R_N}\right)\Delta f\\ &= \Delta P_{ref} - \left(\frac{1}{R_1} + \frac{1}{R_2} + \cdots + \frac{1}{R_N}\right)\Delta f\end{aligned} \tag{4.27}$$

In the above equation, ΔP_m is the total change in turbine–generator mechanical power and ΔP_{ref} is the total change in the reference power settings in the power system. Also note that since all the generators are supposed to work in synchronism, the change in frequency of each of the units is the same and is denoted by Δf. Then the frequency response characteristics is defined as

$$\beta = \frac{1}{R_1} + \frac{1}{R_2} + \cdots + \frac{1}{R_N} \tag{4.28}$$

We can therefore modify (4.28) as

$$\Delta P_m = \Delta P_{ref} - \beta\Delta f \tag{4.29}$$

Example 4.8 Consider an interconnected power system that contains three turbine–generator units rated at 750 MW, 500 MW, and 250 MW. The regulating constant of each unit is 5% based on its own rating. Each unit is operating at 50% of its own rating at 50 Hz when the load suddenly increases by 250 MW. We shall choose a common base of 100 MW to calculate the drop in frequency and the rise in the mechanical power output of each unit.

We first convert the regulating constant to a common base from the per unit base of each generator. Since the frequency remains constant for both the cases, this is given by

$$R_{new} = R_{old} \times \frac{S_{base}^{new}}{S_{base}^{old}} \tag{4.30}$$

Therefore, we have

$$\begin{aligned}R_1 &= 0.05 \times \frac{100}{750} = 0.0067\\ R_2 &= 0.05 \times \frac{100}{500} = 0.01\\ R_3 &= 0.05 \times \frac{100}{250} = 0.02\end{aligned}$$

From (4.28), the constant β is obtained as

$$\beta = \frac{1}{R_1} + \frac{1}{R_2} + \frac{1}{R_3} = 300$$

The total change in the frequency is calculated from (4.29) $\Delta P_{ref} = 0$, that is, for no change in the reference setting. Since the per unit change in load $-250/100 = -2.5$, the change in frequency is given by

$$\Delta f = \frac{\Delta P_m}{\beta} = -\frac{2.5}{300} = -0.0083 \text{ per unit}$$

This means that frequency drops by $0.0083 \times 50 = 0.4167$ Hz.

Then, from (4.26), we have

$$\begin{aligned}
\Delta P_{m1} &= -\frac{1}{R_1}\Delta f \times 100 = \frac{0.833}{0.0067} = 125.50 \text{ MW} \\
\Delta P_{m2} &= -\frac{1}{R_2}\Delta f \times 100 = \frac{0.833}{0.01} = 83.33 \text{ MW} \\
\Delta P_{m3} &= -\frac{1}{R_3}\Delta f \times 100 = \frac{0.833}{0.01} = 41.67 \text{ MW}
\end{aligned}$$

It can be seen that

$$\Delta P_{m1}\colon \Delta P_{m2}\colon \Delta P_{m3} = 3\colon 2\colon 1$$

This implies that the increase in power of each generator is in proportion to its own ratings.

Since the generators were supplying half their rated power before the load change, the new power setting of the three generators is then given by

$$\begin{aligned}
P_{m1} &= 375 + 150 = 500.0 \text{ MW} \\
P_{m2} &= 250 + 83.33 = 333.33 \text{ MW} \\
P_{m3} &= 125 + 41.67 = 166.67 \text{ MW}
\end{aligned}$$

It can also be seen that

$$P_{m1}\colon P_{m2}\colon P_{m3} = 3\colon 2\colon 1$$

This implies that the generators supply power according to their ratings as well.

There is an alternate formulation that is called droop control, which has generated considerable interest in recent times due to microgrid applications (see [2]). Consider a power system that contains N number of generators. Then the frequency of generator-i is given by

$$f_i = f_{\min} + n_i(P_{\max,i} - P_i), \quad i = 1, 2, \ldots, N \tag{4.31}$$

where $P_{\max,i}$ and P_i, respectively, are the maximum power and generated power of generator-i and n_i is its droop gain in Hz/MW. The frequency $f_{\min}$ is the minimum allowable frequency in Hz. This implies that the frequency is $f_{\min}$ when the generators are supplying their maximum rated power. Eq. (4.31) is rewritten as

$$\Delta f_i = f_i - f_{\min} = n_i \Delta P_i \Rightarrow n_i = \frac{\Delta f_i}{\Delta P_i} \tag{4.32}$$

Note from (4.32) that, for each generator, the product of droop gain and the drop in power generated is the drop in frequency, that is,

$$\Delta f = n_1 \Delta P_1 = n_2 \Delta P_2 = \cdots = n_N \Delta P_N \tag{4.33}$$

The following example illustrates this concept.

Example 4.9 Consider the same system as discussed in Example 4.8. The system operates in such a way that the minimum frequency ($f_{\min}$) is 49.5 Hz. The droop gains are chosen such that the generators supply half their rated power when the frequency drops by 0.25 Hz. Then, from (4.32), we have

$$n_1 = \frac{49.75 - 49.5}{750 - 375} = \frac{0.25}{375} = 6.6667 \times 10^{-4}\ \text{Hz/MW}$$

$$n_2 = \frac{0.25}{250} = 1 \times 10^{-3}\ \text{Hz/MW}$$

$$n_3 = \frac{0.25}{125} = 2 \times 10^{-3}\ \text{Hz/MW}$$

Then, at half the rated power, the frequency is computed from (4.31) is for all the three generators is the same and is equal to 49.75 Hz.

When the load increases by 250 MW, we expect that the output powers of the generators will increase according to their own ratings, that is,

$$\Delta P_{m1} = \frac{250 \times 750}{1500} = 125\ \text{MW}$$

$$\Delta P_{m2} = \frac{250 \times 500}{1500} = 83.33\ \text{MW}$$

$$\Delta P_{m3} = \frac{250 \times 250}{1500} = 41.67\ \text{MW}$$

These are the same as those obtained in Example 4.8. The operating frequencies of these units are then computed from (4.31) as

$$f_1 = f_{\min} + n_1(750 - 375 - 125) = 49.6667\ \text{Hz}$$
$$f_2 = f_{\min} + n_2(500 - 250 - 83.33) = 49.6667\ \text{Hz}$$
$$f_3 = f_{\min} + n_3(250 - 125 - 41.67) = 49.6667\ \text{Hz}$$

It can thus be seen that all the generators operate at the same frequency and the frequency drop is 50 − 49.667 = 0.333 Hz.

4.4.1 Load Frequency Control (LFC)

Modern-day power systems are divided into various areas. For example, the National Electricity Market (NEM) in Australia is one of the longest interconnected power systems – from Port Douglas in Queensland to Port Lincoln in South Australia, covering a distance of about 5,000 km. The NEM is composed of five physically connected regions on the eastern seaboard of Australia. These are:

- Queensland
- New South Wales (including Australian Capital Territory)
- Victoria
- Tasmania
- South Australia

Except for Tasmania, which is interconnected with Victoria with undersea HVDC line, all others are connected to their neighboring areas. The transmission lines that connect an area to its neighboring area are called *tie-lines*. Power sharing between two areas occurs through these tie-lines. LFC, as the name signifies, regulates the power flow between different areas while holding the frequency constant.

As evident from (4.29), the system frequency rises when the load decreases if ΔP_{ref} is kept at zero. Similarly, the frequency may drop if the load increases. However, it is desirable to maintain the frequency constant such that $\Delta f = 0$. The power flows through different tie-lines are scheduled – for example, area-i may export a pre-specified amount of power to area-j, while importing another pre-specified amount of power from area-k. However, it is expected that to fulfill this obligation, area-i absorbs its own load change, that is, increase generation to supply extra load in the area or decrease generation when the load demand in the area has reduced. While doing this, area-i must however maintain its obligation to areas j and k as far as importing and exporting power is concerned. A conceptual diagram of the interconnected areas is shown in Figure 4.8.

The LFC has the following two objectives:

- Hold the frequency constant ($\Delta f = 0$) against any load change. Each area must contribute to absorb any load change such that frequency does not deviate.
- Each area must maintain the tie-line power flow to its pre-specified value.

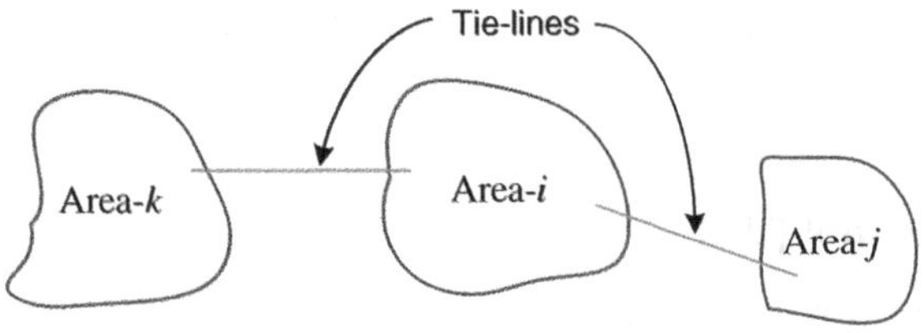

Figure 4.8 Interconnected areas in a power system.

The first step in the LFC is to form the *area control error* (ACE) that is defined as

$$\mathrm{ACE} = (P_{tie} - P_{sch}) + B_f\Delta f = \Delta P_{tie} + B_f\Delta f \tag{4.34}$$

where P_{tie} and P_{sch} are the tie-line power and the scheduled power through tie-line, respectively, and the constant B_f is called the frequency bias constant.

The change in the reference of the power setting $\Delta P_{ref,\ i}$ of the area-i is then obtained by the feedback of the ACE through an integral controller of the form

$$\Delta P_{ref,i} = -K_i\int \mathrm{ACE}\ dt \tag{4.35}$$

where K_i is the integral gain. The ACE is negative if the net power flow out of an area is low or if the frequency has dropped or both. In this case, the generation must be increased. This can be achieved by increasing $\Delta P_{ref,\ i}$. This negative sign accounts for the inverse relation between $\Delta P_{ref,\ i}$ and ACE. The tie-line power flow and frequency of each area are monitored in its control center. Once the ACE is computed and $\Delta P_{ref,\ i}$ is obtained from (4.35), commands are given to various turbine–generator controls to adjust their reference power settings.

Example 4.10 Consider a two-area power system in which Area-1 generates a total of 5,000 MW, while Area-2 generates 4,000 MW. The bias constant of Area-1 (β_1) is 850 MW/Hz and that of Area-2 (β_2) is 680 MW/Hz, where the fundamental frequency is 50 Hz. Both the areas are operating at 60% of the rated power, when the load of Area-2 suddenly increases by 200 MW.

First, consider the case when there is no LFC. Since the two areas are interconnected, they must operate at the same frequency and hence Δf must be the same for both the areas. Then, from (4.34), we have

$$\Delta P_{m1} + \Delta P_{m2} = \left(\Delta P_{ref1} + \Delta P_{ref2}\right) - (\beta_1 + \beta_2)\Delta f$$

Now without LFC, $\Delta P_{ref1} = \Delta P_{ref2} = 0$. Therefore, for 200 MW load increase, the following relation is obtained from the equation above:

$$200 = -(850 + 680)\Delta f \Rightarrow \Delta f = -\frac{200}{1530} = -0.1307\ \mathrm{Hz}$$

Then, the power share between the areas is given by

$$\Delta P_{m1} = -\beta_1 \Delta f = -850 \times (-0.1307) = 111.11 \text{ MW}$$
$$\Delta P_{m2} = -\beta_2 \Delta f = -680 \times (-0.1307) = 88.89 \text{ MW}$$

Therefore, of the 200 MW load increase in Area-2, this area picks up 88.89 MW of power and Area-1 supplies 111.11 MW of power through the tie-line, such that

$$\Delta P_{tie1} = -\Delta P_{tie2} = 111.11 \text{ MW}$$

The total loads in the two areas are then

$$P_{m1} = 5000 \times 0.6 + 111.11 = 3111.11 \text{ MW}$$
$$P_{m2} = 4000 \times 0.6 + 88.89 = 2489.89 \text{ MW}$$

With the LFC, it is desirable that Area-2 absorbs its own load change while not allowing the frequency to drift. The area control errors of the two areas are given by

$$\text{ACE}_1 = \Delta P_{tie1} + B_1 \Delta f_1 \text{and } \text{ACE}_2 = \Delta P_{tie2} + B_2 \Delta f_2$$

Since the net change in the power flow through tie-lines connecting these two areas must be zero, we have

$$\Delta P_{tie1} + \Delta P_{tie2} = 0 \Rightarrow \Delta P_{tie1} = -\Delta P_{tie2}$$

Also, as the transients die out, the drift in the frequency of both these areas is assumed to be constant, that is,

$$\Delta f_1 = \Delta f_2 = \Delta f$$

If the load frequency controller (4.35) can set the power reference of Area-2 properly, the ACE of the two areas will be zero, that is, $\text{ACE}_1 = \text{ACE}_2 = 0$. Then we have

$$\text{ACE}_1 + \text{ACE}_2 = (B_1 + B_2)\Delta f = 0$$

This will imply that Δf will be equal to zero while maintaining $\Delta P_{tie1} = \Delta P_{ti2} = 0$. This means that Area-2 picks up the additional load in the steady state.

4.4.2 Coordination Between LFC and Economic Operation

Both the LFC and the economic dispatch issue command to the power setting of each turbine–governor unit. At first glance, it may seem that these two commands can be conflicting. This, however, is not true. A typical automatic generation control (AGC) strategy is shown in Figure 4.9 in which both the objectives are coordinated. First, the area control error is computed. A share of this ACE,

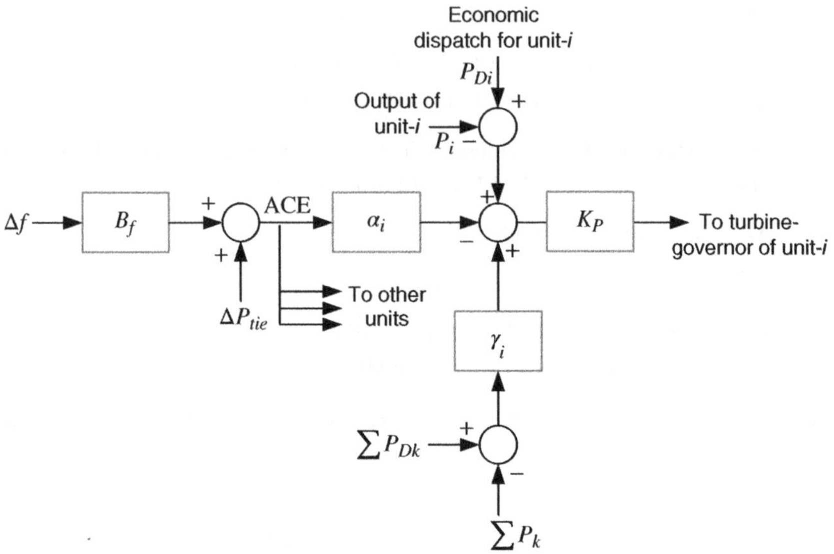

Figure 4.9 Automatic generation control of unit-*i*.

proportional to α_i, is allocated to each of the turbine–generator unit of an area. Also, the share of unit-*i*, $\gamma_i \times \sum (P_{Dk} - P_k)$, for the deviation of total generation from actual generation, is computed. Furthermore, the error between the economic power setting and actual power setting of unit-*i* is computed. All these signals are then combined and passed through a proportional gain K_p to obtain the turbine–governor control signal.

4.5 Concluding Remarks

In this chapter, we have discussed how a power system can schedule the power supply economically. Furthermore, the turbine–governor control that may be required to maintain frequency also has been discussed. Economic dispatch is a complex problem when the scheduling of all the units is considered together through unit commitment optimization problem. The basic background of the procedure has also been presented.

There are several books that cover some of these topics. Reference [1] provides the most comprehensive analysis of all the topics covered in this chapter. There are also excellent references, such as the books [3, 4] that discuss some of the topics in detail. With the present-day increased penetration of renewable energy generators, the unit commitment problem is becoming very crucial as these generators are intermittent in nature. The storage elements are considered to smoothen the non-periodical power fluctuations of wind and solar power. References [5, 6] discuss the unit commitment problem in the presence of renewable energy.

References

1 G. J. Wood and B. F. Wollenberg, *Power Generation, Operation, and Control*, 2nd Ed., John Wiley & Sons, New York, 1996.
2 A. Ghosh and F. Zare, *Control of Power Electronic Converters with Microgrid Applications*, IEEE Press-Wiley, New Jersey, 2022.
3 J. J. Grainger and W. D. Stevenson, *Power System Analysis*, McGraw-Hill, New York, 1994.
4 J. D. Glover, M. S. Sarma and T. J. Overbye, *Power System Analysis and Design*, 5th Ed., Cengage Learning, Stamford, CT, 2012.
5 S. Abujarad, M. W. Mustafa and J. J. Jamian, "Recent approaches of unit commitment in the presence of intermittent renewable energy resources: A review," *Renewable and Sustainable Energy Reviews*, Vol. 70, pp. 215–223, 2017.
6 Z. Zhang, E. Du, F. Teng, et al., "Modeling frequency dynamics in unit commitment with a high share of renewable energy," *IEEE Transactions on Power Systems*, Vol. 35, No. 6, pp. 4383–4395, 2020.

Problems

P4.1 A generating unit has the following fuel input expressed in terms of million Joules per hour

$$0.025P^2 + 5P + 100$$

where P is the power output in MW. Write the equation of incremental fuel cost in \$/MWh based on the fuel cost of \$5 per million Joules.

P4.2 The incremental fuel costs of two units of a plant are given by

$$\frac{df_1}{dP_1} = 0.01P_1 + 11.0\,\$/\text{MWh and } \frac{df_2}{dP_2} = 0.012P_2 + 8.0\,\$/\text{MWh}$$

Find the saving in \$/h for economical allocation of a load between the units compared to their sharing a total load demand of 750 MW.

P4.3 The incremental fuel costs of three units of a plants are

$$\frac{df_1}{dP_1} = 0.012P_1 + 9.0\,\$/\text{MWh}$$
$$\frac{df_2}{dP_2} = 0.008P_2 + 8.0\,\$/\text{MWh}$$
$$\frac{df_3}{dP_3} = 0.0096P_3 + 6.0\,\$/\text{MWh}$$

(a) Determine the incremental fuel cost (λ) when these three units together are supplying a load of 784 MW if they do not have any generating limits.
(b) Determine λ when these they supply 784 MW with P3 ≤ 400 MW.

P4.4 A system contains two generating units with the incremental fuel costs of

$$\frac{df_1}{dP_1} = 0.008P_1 + 8.0\,\$/\text{MWh and } \frac{df_2}{dP_2} = 0.012P_2 + 9.0\,\$/\text{MWh}$$

The system is operating in economic dispatch with P1 = P2 = 500 MW with $\partial P_{Loss}/\partial P_2 = 0.2$. Find $\partial P_{Loss}/\partial P_1$.

P4.5 Consider the power system of Figure P4.5 with the following data in per unit:

$$I_1 = 1\angle 0°, I_2 = 0.8\angle 0°, V_3 = 1\angle 0°$$
$$Z_{13} = 0.04 + j0.16$$
$$Z_{23} = 0.03 + j0.12$$
$$Z_{34} = 0.02 + j0.08$$

Determine if the following loss coefficients describe the power loss correctly.

$$B_{11} = 0.055, B_{22} = 0.0476 \text{ and } B_{12} = 0.0188$$

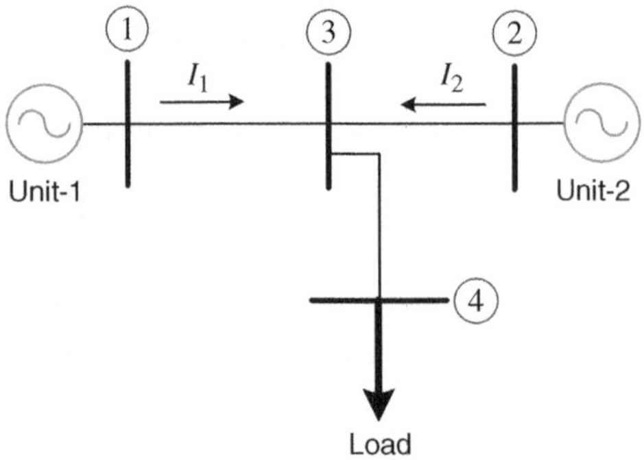

Figure P4.5 Four-bus power system of Problem 4.5.

P4.6 Assume that the loss coefficients derived in Problem P4.5 are in per unit with a base of 100 MVA. The incremental fuel costs of the two units of Problem P4.5 are given by

$$\frac{df_1}{dP_1} = 0.010P_1 + 8.5\,\$/\text{MWh and } \frac{df_2}{dP_2} = 0.015P_2 + 9.5\,\$/\text{MWh}$$

Find P_1 and P_2 when the incremental fuel cost (λ) is 15 \$/MWh.

P4.7 A power system has two generating units. They dispatch power economically with $P_1 = 140$ MW and $P_2 = 250$ MW. The loss coefficients are

$$B_{11} = 0.1 \times 10^{-2}\,\text{MW}^{-1}$$
$$B_{12} = -0.02 \times 10^{-2}\,\text{MW}^{-1}$$
$$B_{13} = 0.13 \times 10^{-2}\,\text{MW}^{-1}$$

To raise the total load of the system will require an additional \$14/h. Find the penalty factor of Unit-1 and the additional cost per hour to increase the output of this unit by 1 MW.

P4.8 Two power systems are delivering power to a load such that $P_1 + P_2$ - $P_{LOSS} = 400$ MW. The incremental costs and the line losses are given by

$$\frac{df_1}{dP_1} = 0.035P_1 + 20.5\,\$/\text{MWh}$$
$$\frac{df_2}{dP_2} = 0.035P_1 + 20.5\,\$/\text{MWh}$$
$$P_{LOSS} = 0.001(P_2 - 50)^2\,\text{MW}$$

Determine the power output P_1 and P_2 when $\lambda = 30.1$ \$/MWh.

Suppose now the total power of 400 MW is supplied without considering the line losses. Then find the P_{LOSS} and the cost per hour due to this non-optimal setting.

P4.9 The fuel costs of two generating units are given by

$$f_1 = 0.1P_1^2 + 40P_1 + 120\,\$/\text{h}$$
$$f_2 = 0.2P_2^2 + 30P_2 + 100\,\$/\text{h}$$

The minimum and maximum ratings of the units are

$$30\,\text{MW} \le P_1 \le 175\,\text{MW}$$
$$20\,\text{MW} \le P_2 \le 125\,\text{MW}$$

The load pattern of a particular day is shown in Figure P4.9. An additional cost of \$400 is incurred by taking one unit off and bringing it back to service after 12 hours. Determine if it is economical to run both

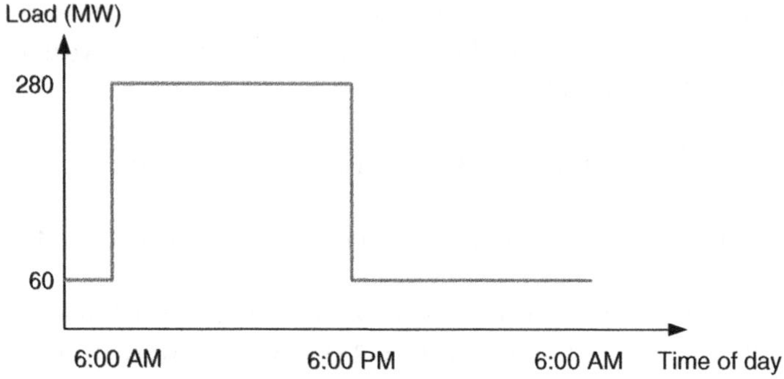

Figure P4.9 Load pattern of a particular day.

units all 24 hours of the day or remove one unit during the light load condition (6 PM to 6 AM).

P4.10 An interconnected 50 Hz power system of one area consists of three turbine–generator units rated 1,500 MVA, 1,000 MVA, and 500 MVA. Their respective regulation constants (R) for all the generators are chosen as 0.05. Each generator is operating at 75% of their own rating when the load increases by 500 MW. Determine the steady-state drop in the frequency and the increase in the turbine power output of each generator neglecting system losses. Choose a base MVA of 1,000.

P4.11 Repeat Problem 4.10 using the method discussed in Example 4.9, assuming the maximum allowable frequency drop is 1 Hz from the base frequency of 50 Hz.

5

Power System Fault Analysis

Short circuits occur in power systems due to various reasons like equipment failure, lightning strikes, falling off branches or trees on transmission lines, switching surges, insulation failures, and other electrical or mechanical causes. All these are collectively called *faults* in power systems. A fault usually results in a high current flowing through the lines and if adequate protection is not taken, it may result in damage to the power apparatus. In this chapter, we shall discuss the effects of both symmetrical and unsymmetrical faults on the power system. Here the term symmetrical fault refers to those conditions in which all three phases of a power system are grounded at the same point. For this reason, the symmetrical faults sometimes are also called three-line-to-ground (3LG) faults. The unsymmetrical faults are of three types. These are:

- Single-line-to-ground (1LG) fault
- Line-to-line (LL) fault
- Double-line-to-ground (2LG) fault

We shall first discuss the behavior of electrical circuits and synchronous generators under faulted conditions. We shall then proceed to calculate fault currents in a network under balanced fault conditions. Unsymmetrical faults are difficult to analyze because the current flow in the three phases may be different under unbalanced fault conditions. To alleviate this problem, a three-phase circuit is usually resolved into three balanced circuits, called zero, positive, and negative sequences using symmetrical component transformation. However, the behavior of power system loads and transformers under symmetrical component transformation is different for star (grounded or ungrounded neutral) and delta connections. Therefore, a power circuit must be resolved into three (zero, positive, and negative) networks. It will be shown how these networks can be joined together for the three types of unbalanced faults mentioned above for the calculation of unbalanced fault currents.

Modern Power System, First Edition. Arindam Ghosh.

Published 2026 by John Wiley & Sons, Inc.
Companion website: www.wiley.com/go/ghoshpowersystem

As we have discussed in Chapter 2, a power transmission line has series resistors and inductors and line charging capacitors. In general, the capacitors are neglected for fault calculations since the capacitive effect is small compared to the inductive effects. Moreover, the fault analysis mainly focuses on subtransient and transient time frames of synchronous generators (see Section 5.2), where the inductive effect dominates. However, for unbalanced 1LG faults, capacitors can affect zero-sequence networks which can result in overvoltages. Nevertheless, we shall neglect the charging capacitors for fault current calculations.

5.1 Transients in an RL Circuit

In this section, we will consider transients in a circuit that contains a resistor and inductor (RL circuit). Consider the circuit shown in Figure 5.1 which contains an ideal source (v_s), a resistor (R), an inductor (L), and a switch (S). It is assumed that the switch is open and is closed at an instant of time $t = 0$. This implies that the current i is zero before the closing of the switch.

5.1.1 DC Source

Let us assume that the source voltage is DC and is given by $v_s = V_{dc}$. Then the line current is given by the differential equation

$$\frac{di}{dt} = -\frac{R}{L}i + \frac{1}{L}v_s \tag{5.1}$$

The solution of the above equation is written in the following form:

$$i(t) = e^{-Rt/L}i(0) + \frac{1}{L}\int_0^t e^{-(R/L)(t-\tau)}v_s(\tau)d\tau \tag{5.2}$$

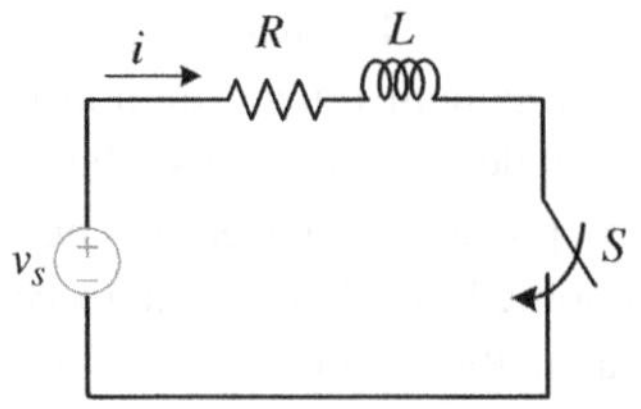

Figure 5.1 A simple RL circuit.

Since the initial current $i(0) = 0$ and since $v_s(\tau) = V_{dc}$ for $0 \le \tau < \infty$, we can rewrite (5.2) as

$$i(t) = \frac{V_{dc}}{R}\left(1 - e^{-Rt/L}\right) = \frac{V_{dc}}{R}\left(1 - e^{-t/T}\right) \tag{5.3}$$

where $T = L/R$ is the time constant of the circuit.

Example 5.1 Consider the system of Figure 5.1, which is supplied by DC source of voltage $V_{dc} = 100$ V, $R = 1\ \Omega$, and $L = 10$ mH. The time response of the current is as shown in Figure 5.2, where it is assumed that the switch is closed at $t = 0$ s. Since in a DC circuit, an inductor is short circuited in the steady state, the current reaches a steady-state value of 100 A. The time constant of the circuit is 0.01 s. This is defined by the time in which the current $i(t)$ reaches 63.2% of its final value and is obtained by substituting $t = \tau$. Note that the slope of the curve is given by

$$\frac{di}{dt} = \frac{V_{dc}}{R\tau}e^{-t/T}\bigg|_{t=0} = \frac{V_{dc}}{RT} = 10^4 \text{A/s} \tag{5.4}$$

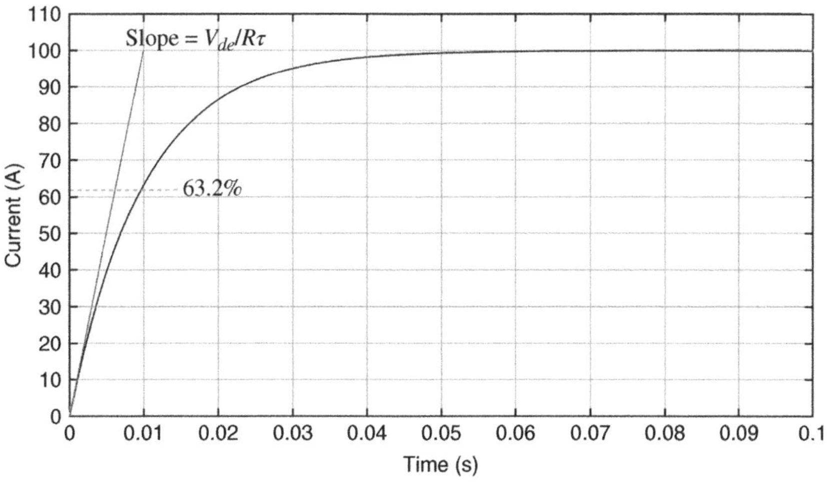

Figure 5.2 Transient current in an RL circuit when excited by a DC source.

5.1.2 AC Source

Let us assume that the source voltage in Figure 5.1 is an AC and is given by

$$v_s = \sqrt{2}V_m \sin(\omega t + \alpha) \tag{5.5}$$

where α is the phase angle of the applied voltage. Substituting (5.5) into (5.2), we have the solution of the form

$$i(t) = e^{-Rt/L}i(0) + \sqrt{2}V_m[A\sin(\omega t + \alpha) + B\cos(\omega t + \alpha)] \tag{5.6}$$

where

$$A = \frac{R}{R^2 + (\omega L)^2} \text{ and } B = -\frac{\omega L}{R^2 + (\omega L)^2}$$

Again,

$$A\sin(\omega t + \alpha) + B\cos(\omega t + \alpha) = F\sin(\omega t + \alpha - \theta)$$

where

$$F = \sqrt{A^2 + B^2} = \frac{1}{\sqrt{R^2 + (\omega L)^2}} \text{ and } \theta = \tan^{-1}\left(\frac{B}{A}\right) = \tan^{-1}\left(\frac{\omega L}{R}\right)$$

Then, denoting $Z = \sqrt{R^2 + (\omega L)^2}$, (5.6) can be rewritten as

$$i(t) = e^{-Rt/L}i(0) + \frac{\sqrt{2}V_m}{Z}\sin(\omega t + \alpha - \theta) \tag{5.7}$$

The initial condition $i(0)$ is given by

$$i(0) = -\frac{\sqrt{2}V_m}{Z}\sin(\alpha - \theta)$$

Then, (5.7) is written as

$$i(t) = i_{ac}(t) + i_{dc}(t) \tag{5.8}$$

where

$$i_{ac} = \frac{\sqrt{2}V_m}{Z}\sin(\omega t + \alpha - \theta)$$

$$i_{dc} = -\frac{\sqrt{2}V_m}{Z}\sin(\alpha - \theta)e^{-t/T}$$

Equation (5.8) implies that the current has an AC component with a frequency that is the same as the frequency of applied voltage and a DC component that is dependent on the initial condition. The DC component exponentially dies out, leaving only the AC component. In fact, the phasor AC component can be obtained from the phasor AC voltage using the sinusoidal steady-state analysis.

We shall investigate the response of the current with the changes with a change in α. The current response for $V_m = 100$ V and $\alpha = 45°$ is shown in Figure 5.3 when the switch is closed at $t = 0$ s. The system frequency is $\omega = 100\pi$ rad/s, and the impedances are chosen as $R = 1\ \Omega$ and $L = 10$ mH. It is obvious that the current becomes equal to i_{ac} once i_{dc} dies out once the initial transient phase is over. Figure 5.4 shows the response of the current for different values of α. Since the current is almost inductive (due to small resistance), the transient is minimal when $\alpha = 90°$, that is, the circuit is switched on almost at the zero-crossing of the current. On the other hand, the transient is maximum when $\alpha = 0°$, that is, almost at the peak of the current.

5.1.3 Fault in an AC Circuit

Consider the AC circuit of Figure 5.5. Let us choose the following parameters for the circuit:

$$V_s = 240\,(\text{rms}),\ \text{frequency} = 50\ \text{Hz},\ R = 0.864\ \Omega,\ \text{and}\ L = 11\ \text{mH}$$

The line impedance is then given $Z = R + j100\pi L = 0.864 + j3.46\ \Omega$. The RL load has a resistance of 8.64 Ω and an inductance of 49.5 mH, such that its impedance is $8.64 + j15.55\ \Omega$. With the system operating in the steady state, switch S is closed suddenly, creating a short circuit.

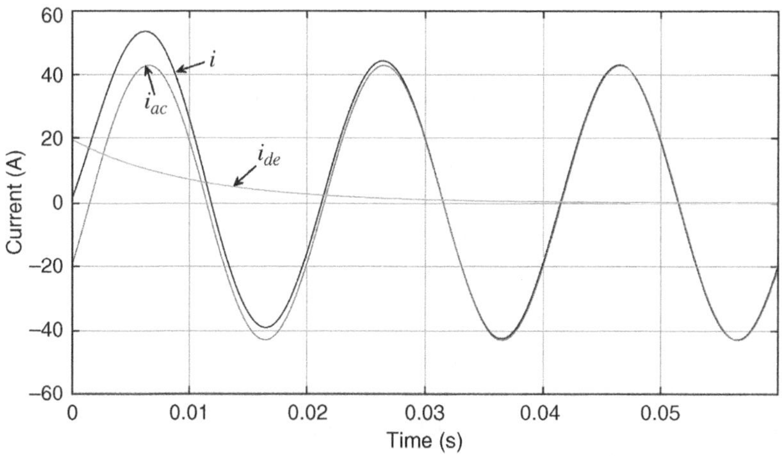

Figure 5.3 Transient current in an AC circuit with both its AC and DC components.

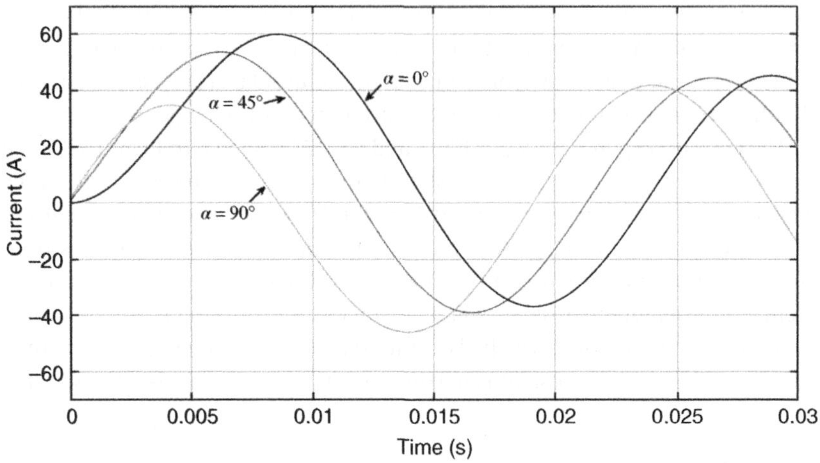

Figure 5.4 Transient current in an AC circuit for different values of the voltage phase angle α.

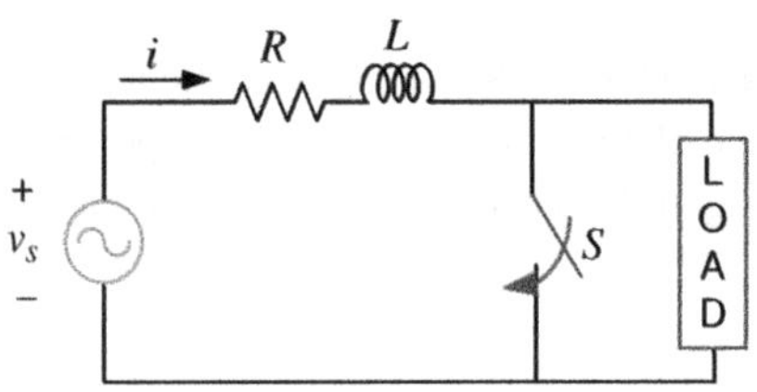

Figure 5.5 Single-phase AC circuit for fault study.

The phasor current that is flowing through the circuit is given by

$$I = \frac{240}{9.504 + j19.01} = 5.05 - j10.10 = 11.29\angle - 63.43^\circ$$

This means that the pre-fault current has a peak value of 15.97 ($= 11.29 \times \sqrt{2}$) A. When the system reaches a steady state following the fault, the current phasor is given by

$$I = \frac{240}{0.864 + j3.46} = 16.34 - j65.36 = 67.37\angle - 75.96^\circ \text{ A}$$

This current has a peak value of 95.28 A. The current waveform is shown in Figure 5.6 for two different values of the incipient fault times. The current rises

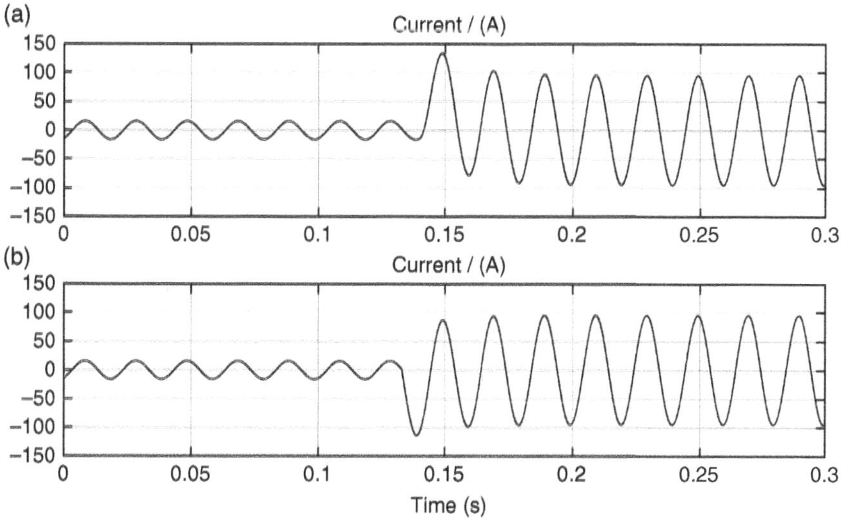

Figure 5.6 The current waveform of the circuit of Figure 5.5 before and after the closing of the switch S. The switch is closed at (a) 0.14 s and (b) 0.133 s.

(or falls) sharply at the occurrence of the fault before it reaches its steady-state waveform. The peak of the current rises in the positive direction, as shown in Figure 5.6 (a) is about 135 A when the switch is closed at 0.14 s, while the maximum current is negative as shown in Figure 5.6 (b) is about 115 A when the switch is closed at 0.133 s. This implies that the current peak is about nine times the peak of the steady-state pre-fault current. In general, depending on the ratio of source and load impedances, the faulted current may shoot up anywhere between 10 and 20 times the pre-fault current.

5.2 Short Circuit in an Unloaded Synchronous Generator

To improve dynamic stability, synchronous generators have short-circuited bars embedded in the rotor core. These are called *damper windings* or amortisseur windings. Their purpose is to provide an additional pathway for the circulating current to flow during a short circuit or fault. The effect of damper winding is to reduce the machine's reactance. This reactance is called transient reactance X'_d and the time constant for the decay of the transient current is denoted by T'_d. Additionally, the current immediately following a fault is called subtransient current which is dependent on the field circuit.

Figure 5.7 shows a typical response of the armature current when a three-phase symmetrical short circuit occurs at the terminals of an unloaded synchronous

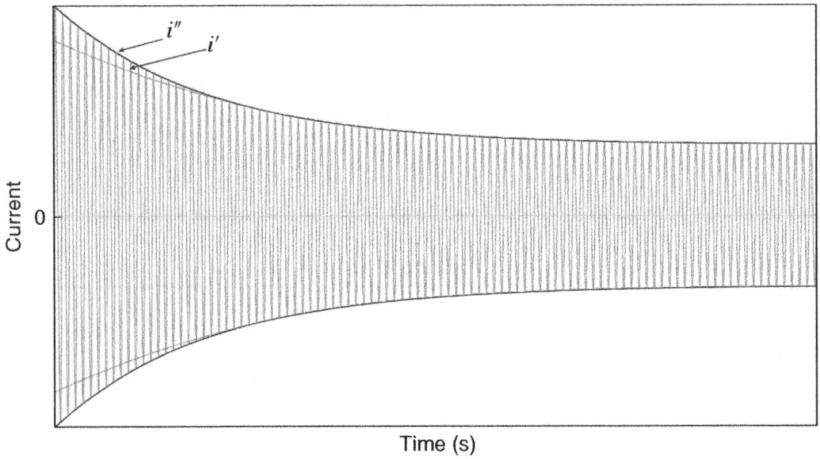

Figure 5.7 The armature current of a synchronous generator as a short circuit occurs at its terminals.

generator. It is assumed that there is no DC offset in the armature current. The magnitude of the current decreases exponentially from a high initial value. Neglecting the effects of armature resistance, the instantaneous expression for the fault current is given by [1]

$$i_f(t) = \sqrt{2}V_t\left[\left(\frac{1}{X''_d}-\frac{1}{X'_d}\right)e^{-t/T''_d}+\left(\frac{1}{X'_d}-\frac{1}{X_d}\right)e^{-t/T'_d}+\frac{1}{X_d}\right]\sin(\omega t+\alpha-\pi/2) \tag{5.9}$$

where V_t is the magnitude of the terminal voltage, α is its phase angle, X''_d is the direct axis subtransient reactance, X'_d is the direct axis transient reactance, X_d is the direct axis synchronous reactance, with $X''_d < X'_d < X_d$. The time constants are the following:

T''_d is the direct axis subtransient time constant.
T'_d is the direct axis transient time constant.

Assuming that the fault occurs at time $t = 0$, the RMS value of the current at that instant is

$$I_f(0) = I''_f = \frac{V_t}{X''_d} \tag{5.10}$$

This is called the subtransient fault current. The duration of the subtransient current is dictated by the time constant T''_d. Between $T''_d < t < T'_d$, the first exponential term of (6.9) will start decaying and will eventually vanish. However, since t is still nearly equal to zero, we have the following RMS value of the current:

$$I'_f = \frac{V_t}{X'_d} \tag{5.11}$$

This is called the transient fault current. Now as time progresses further, the second exponential term will also diminish, and the RMS value of the current in the sinusoidal steady state will be

$$I_f = \frac{V_t}{X_d} \tag{5.12}$$

In addition to the AC, the fault currents will also contain the DC offset. Note that a symmetrical fault occurs when three different phases are in three different locations in the AC cycle. Therefore, the DC offsets in the three phases are different. The maximum value of the DC offset is given by

$$i_{\text{dc}}^{\text{max}} = \sqrt{2} I''_f \, e^{-t/T_A} \tag{5.13}$$

where T_A is the armature time constant.

Example 5.2 Consider a 500 MVA, 22 kV, 50 Hz alternator with the following parameters given in per unit:

$$X_d = 2.5, X'_d = 1 \text{ and } X''_d = 0.5$$

The time constants given in seconds are

$$T'_d = 1.55, T''_d = 0.0225 \text{ and } T_A = 0.25$$

The generator operates in a steady state at the rated voltage at no load, when a bolted three-phase fault occurs at the generator terminals.

The base current that is given by

$$I_{base} = \frac{500 \times 10^6}{\sqrt{3} \times 22 \times 10^3} = 13.12 \text{ kA}$$

Then from (5.10) to (5.12), we have

$$I''_f = \frac{1}{0.5} \times 13.12 = 26.24 \text{ kA}$$

$$I'_f = \frac{1}{1} \times 13.12 = 13.12 \text{ kA}$$

$$I_f = \frac{1}{2.5} \times 13.12 = 5.25 \text{ kA}$$

Assuming α in (5.9) is equal to zero, the fault current is now computed exactly at the completion of three cycles. The time at that instant is $20 \times 3 = 60$ ms. Therefore, from (5.9), we have

$$|i_f(t)| = \left|\sqrt{2}\left[\left(\frac{1}{0.5}-\frac{1}{1}\right)e^{-0.06/0.0225} + \left(\frac{1}{1}-\frac{1}{2.5}\right)e^{-0.06/1.55} + \frac{1}{2.5}\right]\sin(100\pi \times 0.06 - \pi/2)\right|$$
$$= 1.0723 \times 13.12 = 14.07\text{kA}$$

This implies that if a circuit breaker must interrupt this fault current, it must operate to open the circuit at least in this fault current level. However, the fault current can be higher due to the presence of the DC offset. The circuit breaker operation method will be discussed in Chapter 6.

5.3 Symmetrical Fault in a Power System

As has been mentioned earlier, a symmetrical fault means that the fault occurs simultaneously in all three phases, when all the phases fall to the ground at the same location. There are two ways of calculating the fault currents. Both these methods are presented in this section.

5.3.1 Calculation of Fault Current Using Impedance Diagram

Let us consider the following example.

Example 5.3 Consider the power system of Figure 5.8 in which a synchronous generator supplies a synchronous motor. The motor is operating at the rated voltage and rated mega volt-ampere while drawing a load current at a power factor of 0.9 (lagging) when a three-phase symmetrical short circuit occurs at its terminals, as shown in Figure 5.8. We shall calculate the fault current that flows from both the generator and the motor.

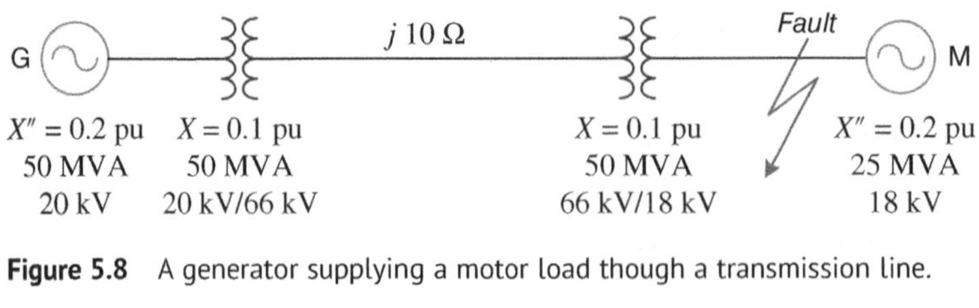

Figure 5.8 A generator supplying a motor load though a transmission line.

The base is chosen as 50 MVA and 20 kV in the circuit of the generator. Then the motor subtransient reactance is given by

$$X''_m = 0.2 \times \frac{50}{25} = 0.4 \text{ per unit.}$$

Also, the base impedance in the circuit of the transmission line is

$$Z_{base} = \frac{66^2}{50} = 87.12\,\Omega$$

Therefore, the impedance of the transmission line is

$$X_{line} = j\frac{10}{87.12} = j0.1148 \text{ per unit.}$$

The impedance diagram for the circuit is shown in Figure 5.9 in which switch S indicates the fault.

The motor draws a load at rated voltage and rated MVA with a lagging power factor of 0.9, and hence,

$$I_L = 1\angle - \cos^{-1}(0.9) = 0.9 - j0.4359 = 1\angle -25.84^\circ \text{ per unit}$$

Then the subtransient voltages of the motor and the generator are

$$E''_m = 1.0 - j0.4 \times i_L = 0.8256 - j0.36 \text{ per unit}$$

$$E''_g = 1.0 + j0.5148 \times i_L = 1.2244 + j0.4633 \text{ per unit}$$

Then, the subtransient fault currents fed by the motor and the generator are

$$I''_m = \frac{E''_m}{j0.4} = -0.9 - j2.0641 \text{ per unit}$$

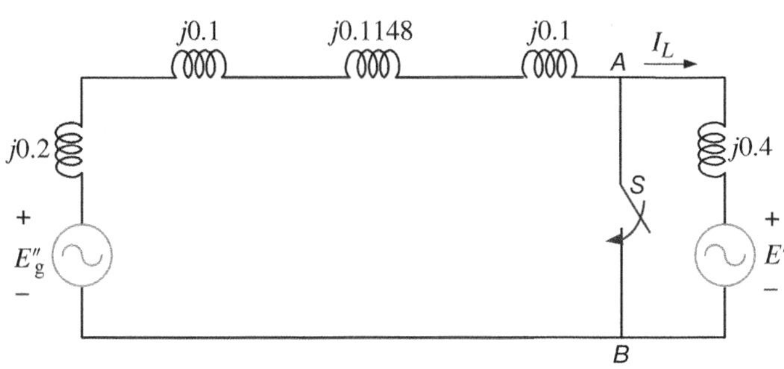

Figure 5.9 Impedance diagram of the circuit of Figure 6.8.

$$I''_g = \frac{E''_g}{j0.5148} = 0.9 - j2.3784 \text{ per unit}$$

and the total current flowing to the fault is

$$I''_f = I''_g + I''_m = -j4.4425 \text{ per unit}$$

Note that the base current in the circuit of the motor is

$$I_{base} = \frac{50 \times 10^3}{\sqrt{3} \times 18} = 1603.8 \text{ A}$$

Therefore, while the load current before the fault was 1603.8 A, the fault current is 7124.7 A.

Example 5.4 We shall now solve Problem 5.3 differently. The Thevenin impedance at the circuit between terminals A and B of the circuit of Figure 5.9 is the parallel combination of the impedances $j0.4$ and $j0.5148$. This is then given as

$$Z_{th} = j\frac{0.4 \times 0.5148}{0.4 + 0.5148} = j0.2251 \text{ per unit}$$

Since the voltage at the motor terminals before the fault is 1.0 per unit, the fault current is

$$I''_f = \frac{1.0}{Z_{th}} = -j4.4425 \text{ per unit}$$

If we neglect the pre-fault current flowing through the circuit, then the fault current fed by the motor and the generator can be determined using the current divider principle, that is,

$$I''_{m0} = \frac{I''_f}{j0.9148} \times j0.5148 = -j2.5 \text{ per unit}$$

$$I''_{g0} = \frac{I''_f}{j0.9148} \times j0.4 = -j1.9425 \text{ per unit}$$

If, on the other hand, the pre-fault current is not neglected, then the fault current supplied by the motor and the generator are

$$I''_m = I''_{m0} - I_L = -0.9 - j2.0641 = 2.96\angle 180^\circ \text{ per unit}$$

$$I''_g = I''_{g0} + I_L = 0.9 - j2.3784 = 2.54\angle -69.27^\circ \text{ per unit}$$

5.3.2 Calculation of Fault Current Using Bus Impedance Matrix

Consider the impedance diagram of Figure 3.3 (a). This is redrawn in Figure 5.10. Assume that a symmetrical fault has occurred at Bus-4 such that it is now connected to the reference bus. Let us assume that the pre-fault voltage of this bus is V_f. To denote a short circuit at Bus-4, we add two voltage sources V_f and $-V_f$ together in series between Bus-4 and the reference bus. Also, note that the subtransient fault current I_f'' flows from Bus-4 to the reference bus. This implies that a current that is equal to $-I_f''$ is injected into Bus-4. This current, which is due to the source $-V_f$, will flow through the various branches of the network and will cause a change in the bus voltages. If the two voltage sources and V_f are short circuited, then $-V_f$ is the only source left in the network that injects a current, $-I_f''$, into Bus-4. The changes in the different node voltages that are caused by $-V_f$ and the current, $-I_f''$, are then given by

$$\begin{bmatrix} \Delta V_1 \\ \Delta V_2 \\ \Delta V_3 \\ -V_f \end{bmatrix} = \mathbf{Z_{bus}} \begin{bmatrix} 0 \\ 0 \\ 0 \\ -I_f'' \end{bmatrix} \tag{5.14}$$

where the prefix Δ indicates the changes in the bus voltages due to the current, $-I_f''$.

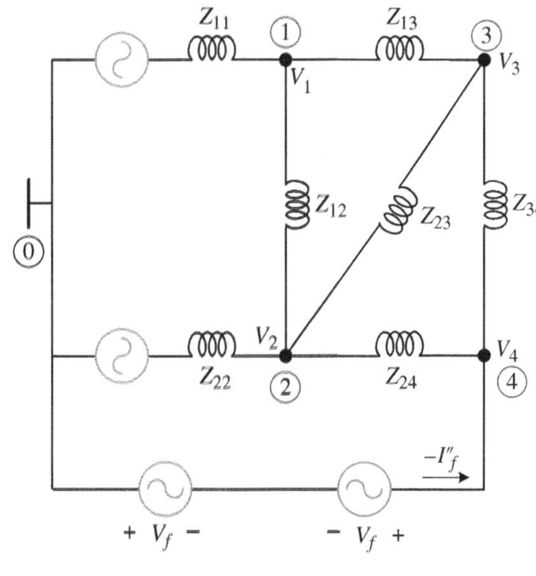

Figure 5.10 Network depicting a symmetrical fault at Bus-4.

Let the elements of a $\mathbf{Z_{bus}} \in \Re^{n \times n}$ matrix be define by Z_{bij}, $i = 1, \ldots, n, j = 1, \ldots n$. From the fourth row of (5.14), the following equation can be written:

$$V_f = Z_{b44} I_f'' \tag{5.15}$$

Combining (5.14) and (5.15), we get

$$\Delta V_i = -Z_{i4} I_f'' = -\frac{Z_{i4}}{Z_{44}} V_f, i = 1, 2, 3 \tag{5.16}$$

We further assume that the system is unloaded before the fault occurs and that the magnitude and phase angles of all the generator internal EMFs are the same. Then, there will be no current circulating anywhere in the network and the bus voltages of all the nodes before the fault will be same and equal to V_f. Then the new altered bus voltages due to the fault will be given from (5.16) by

$$V_i = V_f + \Delta V_i = \left(1 - \frac{Z_{i4}}{Z_{44}}\right) V_f, i = 1, \cdots, 4 \tag{5.17}$$

Example 5.5 Let us consider the same system as discussed in Example 3.1 except that we assume that the internal voltages of both generators are equal to $1 \angle 0°$. Then the current injected in both Bus-1 and Bus-2 will be given by $1.0/j0.25 = -j4.0$ per unit. We, therefore, get the pre-fault bus voltages using the $\mathbf{Z_{bus}}$ matrix given in Example 3.1 as

$$\begin{bmatrix} V_1 \\ V_2 \\ V_3 \\ V_4 \end{bmatrix} = j \begin{bmatrix} 0.1531 & 0.0969 & 0.1264 & 0.1133 \\ 0.0969 & 0.1531 & 0.1236 & 0.1367 \\ 0.1264 & 0.1236 & 0.2565 & 0.1974 \\ 0.1133 & 0.1367 & 0.1974 & 0.3926 \end{bmatrix} \begin{bmatrix} 4\angle -90° \\ 4\angle -90° \\ 0 \\ 0 \end{bmatrix} = \begin{bmatrix} 1.0 \\ 1.0 \\ 1.0 \\ 1.0 \end{bmatrix}$$

Now the altered bus voltages for a symmetrical fault in Bus-4 are given from (6.17) as

$$V_1 = 1 - \frac{0.1133}{0.3926} = 0.7114 \text{ per unit}$$

$$V_2 = 1 - \frac{0.1367}{0.3926} = 0.6518 \text{ per unit}$$

$$V_3 = 1 - \frac{0.1974}{0.3926} = 0.4972 \text{ per unit}$$

$$V_4 = 1 - \frac{0.3926}{0.3926} = 0 \text{ per unit}$$

Therefore, from (5.15), we have

$$I_f'' = \frac{1}{j0.3926} = -j2.5471 \text{ per unit}$$

It is interesting to note that the Thevenin impedance between two buses k and m in a power network is given by [1]

$$Z_{th,km} = Z_{bmm} + Z_{bkk} - 2Z_{bkm} \tag{5.18}$$

5.4 Symmetrical Components

Let us assume that a balanced y-connected load of resistance 10 Ω is connected to an unbalanced y-connected voltage source through an inductive reactance of 3 Ω. The neutrals of both the voltage source and the load are solidly connected. Let the voltages of the three phases be given by

$$V_a = 100\angle 0^\circ \text{ V}, V_b = 100\angle -100^\circ \text{ V and } V_c = 100\angle 110^\circ \text{ V}$$

Note that a set of voltages (currents) is balanced only when the magnitudes of all three phases are the same and the angles are displaced by 120°. Then the line currents are computed as

$$I_a = \frac{V_a}{10 + j3} = 9.17 - j2.75 = 9.58\angle -16.7^\circ \text{ A}$$
$$I_b = \frac{V_b}{10 + j3} = -4.3 - j8.56 = 9.58\angle -116.7^\circ \text{ A}$$
$$I_c = \frac{V_c}{10 + j3} = -0.55 + j9.56 = 9.58\angle -93.3^\circ \text{ A}$$

We can then compute the complex power supplied by the source as

$$S = V_a I_a^* + V_b I_b^* + V_c I_c^* = 2752.3 + j825.69$$

This is a simple system where we can compute the currents, including real and reactive power of each phase by simple phasor analysis. However, when the circuit contains a Y–Δ transformer, or the load is Δ-connected, we will have to use Y–Δ transformation to solve the circuit. Moreover, a power system may contain several generators, transformers, transmission lines, and loads. In that case, it will be almost impossible to perform per-phase computation, even with powerful software tools like MATLAB. Fortunately, the solution to this problem was proposed by Canadian-American electrical engineer Charles LeGeyt Fortescue, who was born in Ontario, Canada, graduated from the University of Toronto, and spent his working life at Westinghouse Electric Company in Pittsburg.

Fortescue derived how an unbalanced three-phase system can be resolved into three balanced systems in the sinusoidal steady state. This is called *resolving symmetrical components of the original phasors* or simply *symmetrical components*. A system of three unbalanced phasors can be resolved in the following three symmetrical components:

- *Positive Sequence*: A balanced three-phase system with the same phase sequence as the original sequence.
- *Negative Sequence*: A balanced three-phase system with the opposite phase sequence as the original sequence.
- *Zero Sequence*: Three phasors that are equal in magnitude and phase.

Figure 5.11 depicts a set of three unbalanced phasors that are resolved into the three sequence components mentioned above. In this the original set of three phasors are denoted by V_a, V_b and V_c, while their positive-, negative-, and zero-sequence components are denoted by the subscripts 1, 2, and 0 respectively. This implies that the positive-, negative-, and zero-sequence components of phase-a are denoted by V_{a1}, V_{a2} and V_{a0}, respectively. Note that just like the voltage phasors given in Figure 5.11, we can also resolve three unbalanced current phasors into three symmetrical components.

5.4.1 Symmetrical Component Transformation

The symmetrical component is defined in terms of an operator a, which is given by

$$a = e^{j\,120^0} = -\frac{1}{2} + j\frac{\sqrt{3}}{2} \tag{5.19}$$

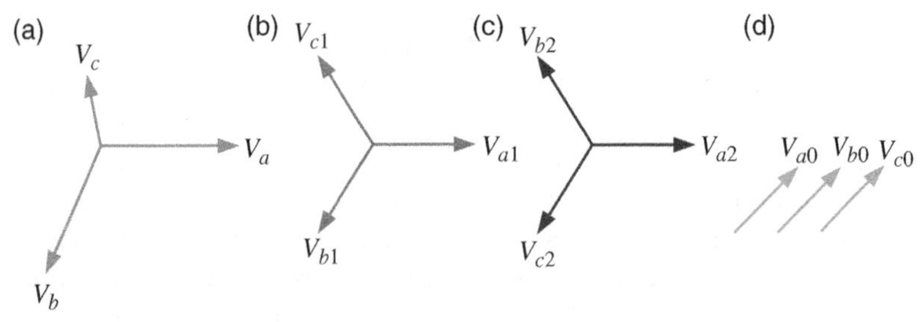

Figure 5.11 Representation of (a) an unbalanced network, its (b) positive sequence, (c) negative sequence, and (d) zero sequence.

This operator has the following properties:

$$a^2 = e^{j\,240^0} = e^{-j\,120^0} - \frac{1}{2} - j\frac{\sqrt{3}}{2} = a^* \tag{5.20a}$$

$$a^3 = e^{j\,360^0} = 1, a^4 = e^{j\,480^0} = a, a^5 = e^{j\,600^0} = a^2 \text{ and so on} \tag{5.20b}$$

$$1 + a + a^2 = 1 - \frac{1}{2} + j\frac{\sqrt{3}}{2} - \frac{1}{2} - j\frac{\sqrt{3}}{2} = 0 \tag{5.20c}$$

Using the operator a, the symmetrical components are written in Figure 5.11 as *Positive Sequence*

$$V_{b1} = a^2 V_{a1} \text{ and } V_{c1} = aV_{a1} \tag{5.21}$$

Negative Sequence

$$V_{b2} = aV_{a2} \text{ and } V_{c2} = a^2 V_{a2} \tag{5.22}$$

Zero Sequence

$$V_{a0} = V_{b0} = V_{c0} \tag{5.23}$$

The voltage of each phase is the sum of all its three symmetrical components, that is,

$$V_a = V_{a0} + V_{a1} + V_{a2} \tag{5.24}$$

$$V_b = V_{b0} + V_{b1} + V_{b2} = V_{a0} + a^2 V_{a1} + aV_{a2} \tag{5.25}$$

$$V_c = V_{c0} + V_{c1} + V_{c2} = V_{a0} + aV_{a1} + a^2 V_{a2} \tag{5.26}$$

Let us define the following vectors:

$$\mathbf{V_{a012}} = \begin{bmatrix} V_{a0} \\ V_{a1} \\ V_{a2} \end{bmatrix} \text{ and } \mathbf{V_{abc}} = \begin{bmatrix} V_a \\ V_b \\ V_c \end{bmatrix} \tag{5.27}$$

Then, combining (5.24) to (5.27), we have

$$\mathbf{V_{abc}} = \begin{bmatrix} V_a \\ V_b \\ V_c \end{bmatrix} = \mathbf{C}^{-1} \begin{bmatrix} V_{a0} \\ V_{a1} \\ V_{a2} \end{bmatrix}, \mathbf{C}^{-1} = \begin{bmatrix} 1 & 1 & 1 \\ 1 & a^2 & a \\ 1 & a & a^2 \end{bmatrix} \tag{5.28}$$

This is called the *inverse symmetrical component transformation*. The *symmetrical component transformation* is given from (5.28) as

$$\mathbf{V_{a012}} = \mathbf{C}\mathbf{V_{abc}}, \mathbf{C} = \frac{1}{3} \begin{bmatrix} 1 & 1 & 1 \\ 1 & a & a^2 \\ 1 & a^2 & a \end{bmatrix} \tag{5.29}$$

For an unbalanced set of current phasors $\mathbf{I_{abc}}$, their symmetrical components $\mathbf{I_{a012}}$ are written as

$$\begin{aligned}\mathbf{I_{a012}} &= \mathbf{C}\mathbf{I_{abc}}\\ \mathbf{I_{abc}} &= \mathbf{C}^{-1}\mathbf{I_{a012}}\end{aligned} \tag{5.30}$$

Example 5.6 Consider a balanced set of voltages, given in per unit by

$$V_a = |V|,\ V_b = |V|\angle -120^\circ \ \text{ and } \ V_c = |V|\angle 120^\circ$$

These imply

$$V_b = |V|a^2 \ \text{ and } \ V_c = |V|a$$

Then, from (5.29), we have

$$V_{a0} = \frac{|V|}{3}\left(1 + a^2 + a\right) = 0$$

$$V_{a1} = \frac{|V|}{3}\left(1 + a^3 + a^3\right) = |V|$$

$$V_{a2} = \frac{|V|}{3}\left(1 + a^4 + a^2\right) = 0$$

Therefore, it can be inferred that, for a set of balanced voltages, the negative- and zero-sequence components are zero. Furthermore, the positive-sequence voltages are the same as those of the original three-phase voltages, that is,

$$V_{a1} = V_a,\ V_{b1} = V_b \text{ and } V_{c1} = V_c \tag{5.31}$$

Example 5.6 Consider an unbalanced set of voltages, given in per unit by

$$V_a = 1.0,\ V_b = 1.2\angle -110^\circ \ \text{ and } \ V_c = 0.9\angle 120^\circ$$

Then, using (5.29), the symmetrical components of the voltages are given by

$$\begin{bmatrix} V_{a0}\\ V_{a1}\\ V_{a2}\end{bmatrix} = C\begin{bmatrix} 1.0\\ 1.2\angle -110^\circ\\ 0.9\angle 120^\circ\end{bmatrix} = \begin{bmatrix} 0.0465 + j0.1161\\ 1.0273 + j0.0695\\ -0.0738 + j0.0466\end{bmatrix} = \begin{bmatrix} 0.1250\angle -68.16^\circ\\ 1.0296\angle 3.87^\circ\\ 0.0873\angle 147.72^\circ\end{bmatrix}$$

Therefore, the symmetrical components of the voltages are given by

$$V_{a0} = V_{b0} = V_{c0} = 0.125\angle -68.16^\circ$$

$$V_{b1} = 1.0296\angle -116.13^\circ, \quad V_{c1} = 1.0296\angle 123.87^\circ$$

$$V_{b2} = 0.0873\angle 267.72^\circ, \quad V_{c2} = 0.0973\angle 27.72^\circ$$

The original phase voltages can be computed from their symmetrical components using (5.24) to (5.26) as

$$V_a = V_{a0} + V_{a1} + V_{a2} = 1.0$$

$$V_b = V_{b0} + V_{b1} + V_{b2} = 1.2\angle -110^\circ$$

$$V_c = V_{c0} + V_{c1} + V_{c2} = 0.9\angle 120^\circ$$

5.4.2 Real and Reactive Power

The three-phase complex power in the a-b-c frame is given by

$$P_{abc} + jQ_{abc} = V_a I_a^* + V_b I_b^* + V_c I_c^* = \mathbf{V}_{\mathbf{abc}}^T \mathbf{I}_{\mathbf{abc}}^* \tag{5.32}$$

Then from (5.28) and (5.30), we have

$$P_{abc} + jQ_{abc} = \left(\mathbf{C}^{-1}\mathbf{V}_{\mathbf{a012}}\right)^T \mathbf{C}^{-1*}\mathbf{I}_{\mathbf{a012}}^* = \mathbf{V}_{\mathbf{a012}}^T \mathbf{C}^{-T}\mathbf{C}^{-1*}\mathbf{I}_{\mathbf{a012}}^* \tag{5.33}$$

Now

$$C^{-T}C^{-1*} = 3\begin{bmatrix} 1 & 0 & 0 \\ 0 & 1 & 0 \\ 0 & 0 & 1 \end{bmatrix}$$

Using the above equation, (5.33) is rewritten as

$$P_{abc} + jQ_{abc} = \mathbf{3V}_{\mathbf{a012}}^T \mathbf{I}_{\mathbf{a012}}^* = 3\left(V_{a0}I_{a0}^* + V_{a1}I_{a1}^* + V_{a2}I_{a2}^*\right) \tag{5.34}$$

Thus, the complex power in a three-phase system in a-b-c frame is three times the summation of the complex power of the three-phase sequences.

Example 5.7 Let us consider the voltages given in Example 5.6. Let us further assume that these voltages supply a balanced Y-connected load whose per-phase impedance is $Z_Y = 0.2 + j0.8$ per unit. Then the per-unit currents in the three phases are

$$I_a = \frac{V_a}{Z_Y} = 1.2127\angle -75.96^\circ \text{ pu}$$

$$I_b = \frac{V_b}{Z_Y} = 1.4552\angle 174.04^\circ \text{ pu}$$

$$I_c = \frac{V_c}{Z_Y} = 1.0914\angle 44.04^\circ \text{ pu}$$

Then the real and reactive power consumed by the load is given by

$$P_{abc} + jQ_{abc} = V_a I_a^* + V_b I_b^* + V_c I_c^* = 0.9559 + j3.8235 \text{ pu}$$

Now using the transformation (5.30), we get

$$\begin{bmatrix} I_{a0} \\ I_{a1} \\ I_{a2} \end{bmatrix} = \begin{bmatrix} -0.1229 - j0.0889 \\ 0.3839 - j1.1881 \\ 0.0331 + j0.1005 \end{bmatrix} = \begin{bmatrix} 0.1516\angle -144.12° \\ 1.2486\angle -72.10° \\ 0.1058\angle 71.75° \end{bmatrix} \text{pu}$$

From the results given in Example 5.6, and from the values of the currents given above, the zero-sequence complex power is computed as

$$P_0 + jQ_0 = V_{a0} I_{a0}^* = 0.0046 + j0.0184 \text{ pu}$$

The positive-sequence complex power is

$$P_1 + jQ_1 = V_{a1} I_{a1}^* = 0.3118 + j1.2472 \text{ pu}$$

Finally, the negative-sequence complex power is

$$P_2 + jQ_2 = V_{a2} I_{a2}^* = 0.0022 + j0.009 \text{ pu}$$

Adding the three complex powers together, we get the total complex power consumed by the load as

$$P_{012} + jQ_{012} = (P_0 + P_1 + P_2) + j(Q_0 + Q_1 + Q_2) = 0.3186 + j1.2745 \text{ pu}$$

Verify the following:

$$P_{abc} + jQ_{abc} = 3(P_{012} + jQ_{012})$$

5.5 Sequence Circuits and Networks

As discussed in Section 5.4, the unbalanced voltages and currents can be resolved into their symmetrical components. Consider the power system shown in Figure 5.12, in which the network is supplying a three-phase passive load with grounded Y–Y-connected transformers. Suppose a fault has occurred in phase-c at the location shown. Then, after the occurrence of the fault, the voltages of the other two phases (V_{fa} and V_{fb}) will not be affected. Similarly, the currents of these two phases (I_{fa} and I_{fb}) will also remain unchanged. However, the voltage at the fault location of phase-c will be zero, that is, $V_{fc} = 0$. Consequently, a large amount of current I_{cf} will flow to the ground and the current flowing to the load will be zero, that is, $I_{cL} = 0$. Therefore, it can be surmised that both sides of the network will be unbalanced.

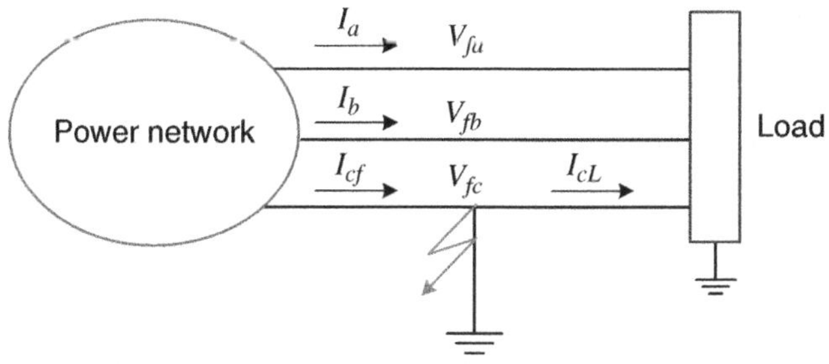

Figure 5.12 A power network during a fault.

A power system contains components such as loads, generators transformers, and transmission lines. Some of them may be connected in wye, and others can be connected in delta. They behave differently depending on the fault location and type. In this section, we shall discuss how different power system components can be resolved into their positive-, negative-, and zero-sequence circuits. When all components are resolved into their sequence circuits, they are joined to form three separate networks, which are called sequence networks.

5.5.1 Sequence Circuit for a Y-Connected Load

Consider the balanced Y-connected load that is shown in Figure 5.13. The neutral point (n) of the windings is grounded through an impedance Z_n. The load in each phase is denoted by Z_Y. Let us consider phase-a of the load. With respect to Figure 5.13, the following voltages are defined:

- V_a is the phase voltage, that is, between the line and ground,
- V_{an} is the line-to-neutral voltage, and
- V_n is the voltage between the neutral and ground.

Figure 5.13 Schematic diagram of a balanced Y-connected load.

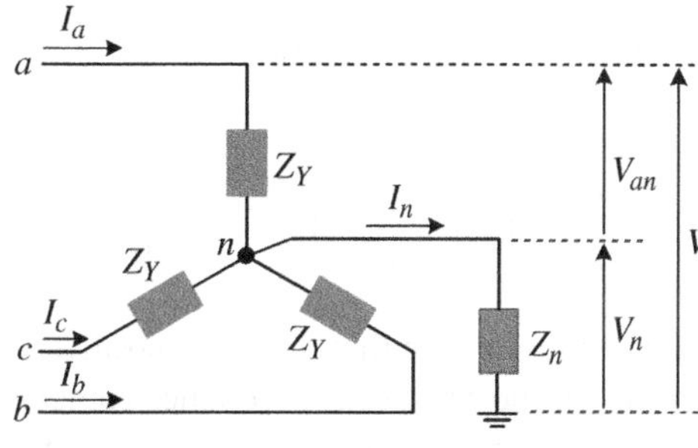

The neutral current is then given by

$$I_n = I_a + I_b + I_c = 3I_{a0} + (I_{a1} + I_{b1} + I_{c1}) + (I_{a2} + I_{b2} + I_{c2})$$

Using the property given in (5.20c), it can be stated that the summations of the positive- and negative-sequence components will be zero. The above equation can then be rewritten as

$$I_n = 3I_{a0} \tag{5.35}$$

Therefore, only zero-sequence current will flow out of the neutral and there will not be any positive or negative-sequence current flowing out of the neutral point.

The voltage drop between the neutral and ground is

$$V_n = 3Z_n I_{a0} \tag{5.36}$$

Now,

$$V_a = V_{an} + V_n = V_{an} + 3Z_n I_{a0} \tag{5.37}$$

Similar expressions can also be written for the other two phases, such that

$$\begin{bmatrix} V_a \\ V_b \\ V_c \end{bmatrix} = \begin{bmatrix} V_{an} \\ V_{bn} \\ V_{cn} \end{bmatrix} + \begin{bmatrix} V_n \\ V_n \\ V_n \end{bmatrix} = Z_Y \begin{bmatrix} I_a \\ I_b \\ I_c \end{bmatrix} + 3Z_n I_{a0} \begin{bmatrix} 1 \\ 1 \\ 1 \end{bmatrix} \tag{5.38}$$

Pre-multiplying both sides of (5.38) by the matrix **C**, we get

$$\mathbf{V_{a012}} = Z_Y \mathbf{I_{a012}} + 3Z_n I_{a0} C \begin{bmatrix} 1 \\ 1 \\ 1 \end{bmatrix} \tag{5.39}$$

Now since

$$\mathbf{C} \begin{bmatrix} 1 \\ 1 \\ 1 \end{bmatrix} = \begin{bmatrix} 1 \\ 0 \\ 0 \end{bmatrix}$$

Equation (5.39) is rewritten as

$$\begin{bmatrix} V_{a0} \\ V_{a1} \\ V_{a2} \end{bmatrix} = Z_Y \begin{bmatrix} I_{a0} \\ I_{a1} \\ I_{a2} \end{bmatrix} + 3Z_n \begin{bmatrix} I_{a0} \\ 0 \\ 0 \end{bmatrix} \tag{5.40}$$

It can thus be seen that the zero-, positive-, and negative-sequence voltages only depend on their respective sequence component currents. The sequence component equivalent circuits are shown in Figure 5.14. While the positive and the

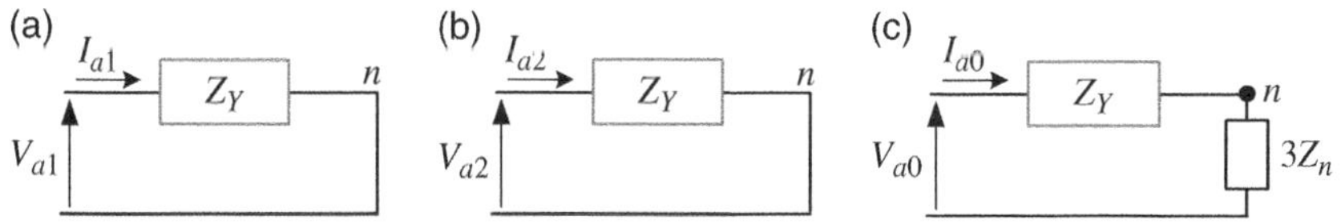

Figure 5.14 Sequence circuits of Y-connected load: (a) positive, (b) negative, and (c) zero sequence.

negative-sequence impedances are both equal to Z_Y, the zero-sequence impedance is equal to

$$Z_0 = Z_Y + 3Z_n \tag{5.41}$$

If the neutral is grounded directly (i.e., $Z_n = 0$), then $Z_0 = Z_Y$. On the other hand, if the neutral is kept floating (i.e., $Z_n = \infty$), then there will not be any zero-sequence current flowing in the circuit at all.

5.5.2 Sequence Circuit for a Delta-Connected Load

Consider the balanced Δ-connected load shown in Figure 5.15 in which the load in each phase is denoted by Z_Δ. The line-to-line voltages are given by

$$\begin{aligned} V_{ab} &= Z_\Delta I_{ab} \\ V_{bc} &= Z_\Delta I_{bc} \\ V_{ca} &= Z_\Delta I_{ca} \end{aligned} \tag{5.42}$$

Adding these three voltages, we get

$$V_{ab} + V_{bc} + V_{ca} = Z_\Delta(I_{ab} + I_{bc} + I_{ca}) \tag{5.43}$$

Figure 5.15 Schematic diagram of a balanced Δ-connected load.

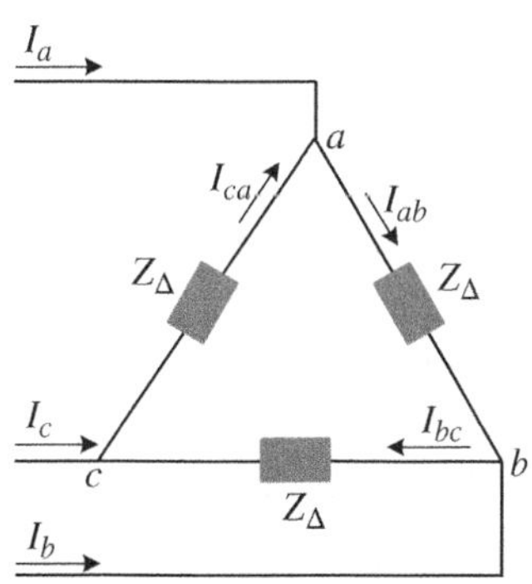

Denoting the zero-sequence component of the voltages as $V_{ab0} = V_{bc0} = V_{ca0}$ and that of the currents as $I_{ab0} = I_{bc0} = I_{ca0}$, (5.43) is rewritten as

$$V_{ab0} = Z_\Delta I_{ab0} \tag{5.44}$$

Again since

$$V_{ab0} = V_{ab} + V_{bc} + V_{ca} = V_a - V_b + V_b - V_c + V_c - V_a = 0,$$

we find from (5.44) that $I_{ab0} = 0$. Hence, in a Δ-connected load without any mutual coupling will not have any zero-sequence circulating current. Note that the positive- and negative-sequence impedance for this load will be equal to Z_Δ, and their equivalent circuits will be similar to those shown in Figure 5.14 (a) and (b), respectively, where Z_Y is replaced by Z_Δ.

Example 5.8 Consider the circuit shown in Figure 5.16 in which a Δ-connected load is connected in parallel with a Y-connected load. The neutral point of the Y-connected load is grounded through an impedance. Applying Kirchhoff's current law at the point-a in the circuit, we get

$$\begin{aligned} I_a &= \frac{V_a - V_b}{Z_\Delta} + \frac{V_a - V_c}{Z_\Delta} + \frac{V_a - V_n}{Z_Y} \\ &= \left(\frac{2}{Z_\Delta} + \frac{1}{Z_Y}\right) V_a - \frac{1}{Z_\Delta}(V_b + V_c) - \frac{V_n}{Z_Y} \end{aligned}$$

The above expression can be written in terms of the vector $\mathbf{V_{abc}}$ as

$$I_a = \left(\frac{3}{Z_\Delta} + \frac{1}{Z_Y}\right) V_a - \frac{1}{Z_\Delta}[1\ \ 1\ \ 1]\mathbf{V_{abc}} - \frac{V_n}{Z_Y}$$

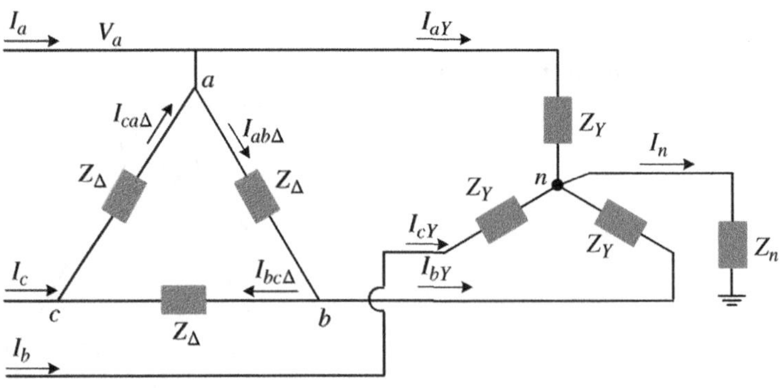

Figure 5.16 Parallel connected balanced Δ and Y-connected load.

Since the load is balanced, we can write

$$\mathbf{I_{abc}} = \left(\frac{3}{Z_\Delta} + \frac{1}{Z_Y}\right)\mathbf{V_{abc}} - \frac{1}{Z_\Delta}\begin{bmatrix}1 & 1 & 1\\ 1 & 1 & 1\\ 1 & 1 & 1\end{bmatrix}\mathbf{V_{abc}} - \frac{V_n}{Z_Y}\begin{bmatrix}1\\ 1\\ 1\end{bmatrix}$$

Pre-multiplying both sides of the above expression by the transformation matrix **C**, we get

$$\mathbf{I_{a012}} = \left(\frac{3}{Z_\Delta} + \frac{1}{Z_Y}\right)\mathbf{V_{a012}} - \frac{1}{Z_\Delta}\mathbf{C}\begin{bmatrix}1 & 1 & 1\\ 1 & 1 & 1\\ 1 & 1 & 1\end{bmatrix}\mathbf{C}^{-1}\mathbf{V_{a012}} - \frac{V_n}{Z_Y}\mathbf{C}\begin{bmatrix}1\\ 1\\ 1\end{bmatrix}$$

Now since

$$\mathbf{C}\begin{bmatrix}1 & 1 & 1\\ 1 & 1 & 1\\ 1 & 1 & 1\end{bmatrix}\mathbf{C}^{-1} = \begin{bmatrix}3 & 0 & 0\\ 0 & 0 & 0\\ 0 & 0 & 0\end{bmatrix}$$

$\mathbf{I_{a012}}$ can be rewritten as

$$\mathbf{I_{a012}} = \left(\frac{3}{Z_\Delta} + \frac{1}{Z_Y}\right)\mathbf{V_{a012}} - \frac{3}{Z_\Delta}\mathbf{V_{a0}} - \frac{V_n}{Z_Y}\begin{bmatrix}1\\ 0\\ 0\end{bmatrix}$$

Separating the three components, the following three equations are obtained:

$$I_{a1} = \left(\frac{3}{Z_\Delta} + \frac{1}{Z_Y}\right)V_{a1}$$

$$I_{a2} = \left(\frac{3}{Z_\Delta} + \frac{1}{Z_Y}\right)V_{a2}$$

$$I_{a0} = \frac{1}{Z_Y}V_{a0} - \frac{1}{Z_Y}V_n$$

If we convert the Δ-connected load into an equivalent Y, then the composite load will be a parallel combination of two Y-connected circuits – one with an impedance of Z_Y and the other with an impedance of $Z_\Delta/3$. Therefore, the positive- and the negative-sequence impedances are given by the parallel combination of these two impedances, that is,

$$Z_1 = Z_2 = \frac{Z_Y Z_\Delta}{3Z_Y + Z_\Delta}$$

Now refer to Figure 5.16. The voltage V_n is given by

$$V_n = Z_n(I_{aY} + I_{bY} + I_{cY}) = 3Z_n I_{aY0}$$

From Figure 5.16, we can also write $I_a = I_{a\Delta} + I_{aY}$, where $I_{a\Delta} = I_{ab\Delta} - I_{ca\Delta}$. Therefore,

$$\begin{aligned} I_a + I_b + I_c &= I_{a\Delta} + I_{b\Delta} + I_{c\Delta} + I_{aY} + I_{bY} + I_{cY} \\ &= I_{ab\Delta} - I_{ca\Delta} + I_{bc\Delta} - I_{ab\Delta} + I_{ca\Delta} - I_{bc\Delta} + I_{aY} + I_{bY} + I_{cY} \end{aligned}$$

This implies that the zero-sequence current is

$$\frac{I_{a0}}{3} = I_{aY} + I_{bY} + I_{cY} = \frac{I_{aY0}}{3}$$

Therefore, we have

$$I_{a0} = \frac{1}{Z_Y} V_{a0} - \frac{3Z_n}{Z_Y} I_{a0}$$

The zero-sequence impedance is then given by

$$Z_0 = \frac{V_{a0}}{I_{a0}} = Z_Y + 3Z_n$$

It can be seen that the Z_Δ term is absent from the zero-sequence impedance.

5.5.3 Sequence Circuit for a Synchronous Generator

The three-phase equivalent circuit of a synchronous generator is shown in Figure 5.17, where the neutral point is grounded through a reactor with the impedance of Z_n. The neutral current is then given by

$$I_n = I_a + I_b + I_c \tag{5.45}$$

For a balanced generator operation, the sum of the three currents is zero, that is, $I_a + I_b + I_c = 0$. However, for unbalanced operation, the summation is not zero but is equal to I_n as per (5.45). With respect to Figure 5.17, the phase-a line-to-neutral voltage expression is written as

$$\begin{aligned} V_{an} &= -(R + j\omega L_s)I_a + j\omega M_s(I_b + I_c) + E_{an} \\ &= -(R + j\omega L_s + j\omega M_s)I_a + j\omega M_s(I_a + I_b + I_c) + E_{an} \end{aligned} \tag{5.46}$$

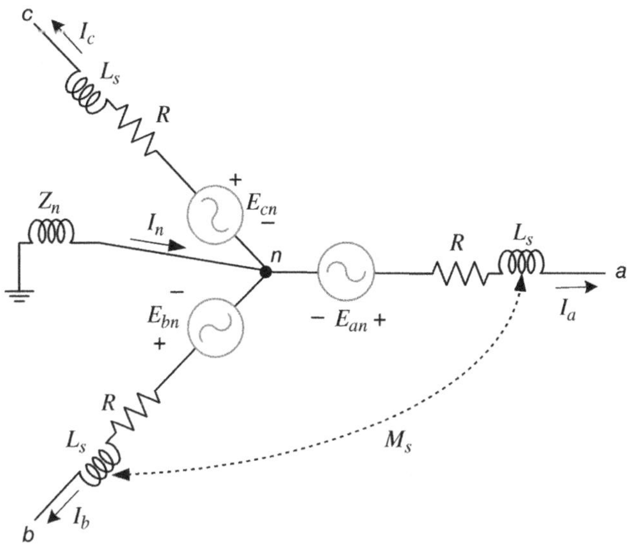

Figure 5.17 Equivalent circuit of a synchronous generator with grounded neutral.

Similar expressions can also be written for the other two phases. Therefore, the line-to-neutral voltages of the three phases are written in the matrix–vector form as

$$\begin{bmatrix} V_{an} \\ V_{bn} \\ V_{cn} \end{bmatrix} = -\left[R + j\omega\left(L_s + M_s\right)\right] \begin{bmatrix} I_a \\ I_b \\ I_c \end{bmatrix} + j\omega M_s \begin{bmatrix} 1 & 1 & 1 \\ 1 & 1 & 1 \\ 1 & 1 & 1 \end{bmatrix} \begin{bmatrix} I_a \\ I_b \\ I_c \end{bmatrix} + \begin{bmatrix} E_{an} \\ E_{bn} \\ E_{cn} \end{bmatrix} \tag{5.47}$$

Pre-multiplying both sides of (5.47) by the transformation matrix **C** we get

$$\begin{bmatrix} V_{an0} \\ V_{an1} \\ V_{an2} \end{bmatrix} = -\left[R + j\omega\left(L_s + M_s\right)\right] \begin{bmatrix} I_{a0} \\ I_{a1} \\ I_{a2} \end{bmatrix} + j\omega M_s \mathbf{C} \begin{bmatrix} 1 & 1 & 1 \\ 1 & 1 & 1 \\ 1 & 1 & 1 \end{bmatrix} \mathbf{C}^{-1} \begin{bmatrix} I_{a0} \\ I_{a1} \\ I_{a2} \end{bmatrix} + \begin{bmatrix} E_{an0} \\ E_{an1} \\ E_{an2} \end{bmatrix} \tag{5.48}$$

Since the synchronous generator is operated to supply only balanced voltages, it can be assumed that $E_{an0} = E_{an2} = 0$ and $E_{an1} = E_{an}$. Equation (5.48) is then modified as

$$\begin{bmatrix} V_{an0} \\ V_{an1} \\ V_{an2} \end{bmatrix} = -\left[R + j\omega\left(L_s + M_s\right)\right] \begin{bmatrix} I_{a0} \\ I_{a1} \\ I_{a2} \end{bmatrix} + j\omega M_s \begin{bmatrix} 3 & 0 & 0 \\ 0 & 0 & 0 \\ 0 & 0 & 0 \end{bmatrix} \begin{bmatrix} I_{a0} \\ I_{a1} \\ I_{a2} \end{bmatrix} + \begin{bmatrix} 0 \\ E_{an} \\ 0 \end{bmatrix} \tag{5.49}$$

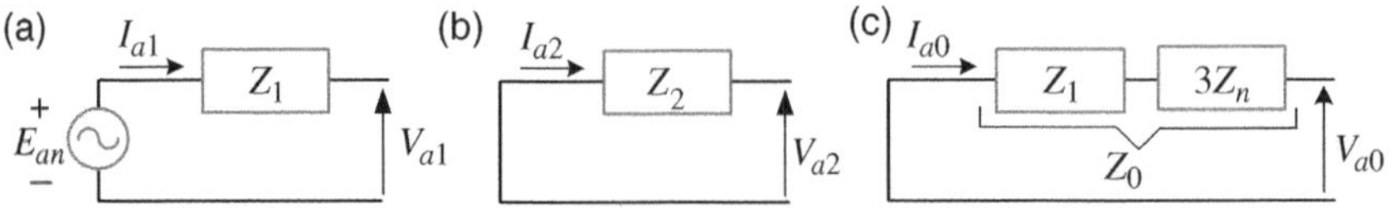

Figure 5.18 Equivalent circuit of a synchronous generator with grounded neutral: (a) positive, (b) negative and (c) zero sequence.

The zero-, positive-, and negative-sequence terms are then expressed from (5.49) as

$$V_{an0} = -[R + j\omega(L_s - 2M_s)]I_{a0} = -Z_{g0}I_{a0} \tag{5.50}$$

$$V_{an1} = -[R + j\omega(L_s + M_s)]I_{a1} + E_{an} = E_{an} - Z_1 I_{a1} \tag{5.51}$$

$$V_{an2} = -[R + j\omega(L_s + M_s)]I_{a2} = -Z_2 I_{a2} \tag{5.52}$$

Furthermore, we have seen for a Y-connected load that $V_{a1} = V_{an1}$ and $V_{a2} = V_{an2}$ since the neutral current does not affect these voltages. However, $V_{a0} = V_{an0} + V_n$ and $V_n = -3Z_n I_{a0}$. Therefore, (5.50) is rewritten as

$$V_{a0} = -(Z_{g0} + 3Z_n)I_{a0} = -Z_0 I_{a0} \tag{5.53}$$

The sequence diagrams for a synchronous generator are shown in Figure 5.18.

5.5.4 Sequence Circuit for a Symmetrical Transmission Line

The schematic diagram of a transmission line is shown in Figure 5.19. In this diagram, the self-impedances of the three phases are denoted by Z_{aa}, Z_{bb} and Z_{cc}, while that of the neutral wire is denoted by Z_{nn}. Let us assume that the self-impedances of the conductors are the same, that is,

$$Z_{aa} = Z_{bb} = Z_{cc}$$

Since the transmission line is assumed to be symmetric, we further assume that the mutual inductances between the conductors are the same and so are the mutual inductances between the conductors and the neutral, that is,

$$Z_{ab} = Z_{bc} = Z_{ca} \text{ and } Z_{an} = Z_{bn} = Z_{cn}$$

The directions of the currents flowing through the lines are indicated in Figure 5.19, where the voltages between the different conductors are indicated.

Kirchhoff's voltage law between points a and n gives the following equation:

$$V_{an} = V_{aa'} + V_{a'n'} + V_{n'n} = V_{aa'} + V_{a'n'} - V_{nn'} \tag{5.54}$$

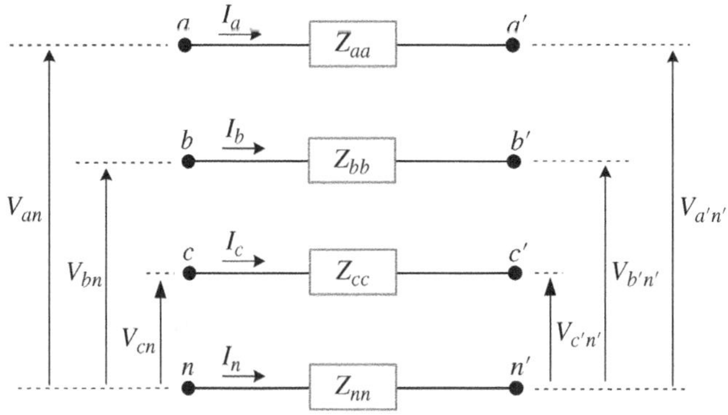

Figure 5.19 Lumped parameter representation of a symmetrical transmission line.

Again

$$V_{aa'} = Z_{aa}I_a + Z_{ab}(I_b + I_c) + Z_{an}I_n \tag{5.55}$$

$$V_{nn'} = Z_{nn}I_a + Z_{an}(I_a + I_b + I_c) \tag{5.56}$$

Substituting (5.55) and (5.56) in (5.54), we get

$$V_{an} - V_{a'n'} = (Z_{aa} - Z_{an})I_a + (Z_{ab} - Z_{an})(I_b + I_c) + (Z_{an} - Z_{nn})I_n \tag{5.57}$$

Since the neutral provides a return path for the currents I_a, I_b, and I_c, the neutral current is written as

$$I_n = -(I_a + I_b + I_c) \tag{5.58}$$

Substituting (5.58) in (5.57), the following equation for phase-a of the circuit is obtained:

$$V_{an} - V_{a'n'} = (Z_{aa} + Z_{nn} - 2Z_{an})I_a + (Z_{ab} + Z_{nn} - 2Z_{an})(I_b + I_c) \tag{5.59}$$

Denoting

$$Z_s = Z_{aa} + Z_{nn} - 2Z_{an} \text{ and } Z_m = Z_{ab} + Z_{nn} - 2Z_{an}$$

Equation (5.59) is rewritten as

$$V_{an} - V_{a'n'} = Z_sI_a + Z_m(I_b + I_c) \tag{5.60}$$

Since (5.60) does not explicitly include the neutral conductor, we can define the voltage drop across the phase-a conductor as

$$V_{aa'} = V_{an} - V_{a'n'} \tag{5.61}$$

Combining (5.60) and (5.61), we get

$$V_{aa'} = Z_s I_a + Z_m (I_b + I_c) \tag{5.62}$$

A similar expression can also be written for the other two phases, to obtain

$$\begin{bmatrix} V_{aa'} \\ V_{bb'} \\ V_{cc'} \end{bmatrix} = \begin{bmatrix} Z_s & Z_m & Z_m \\ Z_m & Z_s & Z_m \\ Z_m & Z_m & Z_s \end{bmatrix} \begin{bmatrix} I_a \\ I_b \\ I_c \end{bmatrix} \tag{5.63}$$

Pre-multiplying both sides of (5.63) by the transformation matrix **C**, we get

$$\mathbf{V_{aa'012}} = \mathbf{C} \begin{bmatrix} Z_s & Z_m & Z_m \\ Z_m & Z_s & Z_m \\ Z_m & Z_m & Z_s \end{bmatrix} \mathbf{C}^{-1} \mathbf{I_{a012}} \tag{5.64}$$

Now

$$\begin{bmatrix} Z_s & Z_m & Z_m \\ Z_m & Z_s & Z_m \\ Z_m & Z_m & Z_s \end{bmatrix} \mathbf{C}^{-1} = \begin{bmatrix} Z_s & Z_m & Z_m \\ Z_m & Z_s & Z_m \\ Z_m & Z_m & Z_s \end{bmatrix} \begin{bmatrix} 1 & 1 & 1 \\ 1 & a^2 & a \\ 1 & a & a^2 \end{bmatrix}$$

$$= \begin{bmatrix} Z_s + 2Z_m & Z_s - Z_m & Z_s - Z_m \\ Z_s + 2Z_m & a^2 Z_s + (1+a) Z_m & a Z_s + \left(1 + a^2\right) Z_m \\ Z_s + 2Z_m & a Z_s + \left(1 + a^2\right) Z_m & a^2 Z_s + (1+a) Z_m \end{bmatrix}$$

Hence,

$$\mathbf{C} \begin{bmatrix} Z_s & Z_m & Z_m \\ Z_m & Z_s & Z_m \\ Z_m & Z_m & Z_s \end{bmatrix} \mathbf{C}^{-1} = \frac{1}{3} \begin{bmatrix} 1 & 1 & 1 \\ 1 & a & a^2 \\ 1 & a^2 & a \end{bmatrix} \begin{bmatrix} Z_s + 2Z_m & Z_s - Z_m & Z_s - Z_m \\ Z_s + 2Z_m & a^2 Z_s + (1+a) Z_m & a Z_s + \left(1 + a^2\right) Z_m \\ Z_s + 2Z_m & a Z_s + \left(1 + a^2\right) Z_m & a^2 Z_s + (1+a) Z_m \end{bmatrix}$$

$$= \frac{1}{3} \begin{bmatrix} 3Z_s + 6Z_m & 0 & 0 \\ 0 & 3Z_s - 3Z_m & 0 \\ 0 & 0 & 3Z_s - 3Z_m \end{bmatrix}$$

Therefore, (5.64) is rewritten as

$$\begin{bmatrix} V_{aa'0} \\ V_{aa'1} \\ V_{aa'2} \end{bmatrix} = \begin{bmatrix} Z_s + 2Z_m & & \\ & Z_s - Z_m & \\ & & Z_s - Z_m \end{bmatrix} \begin{bmatrix} I_{a0} \\ I_{a1} \\ I_{a2} \end{bmatrix} \tag{5.65}$$

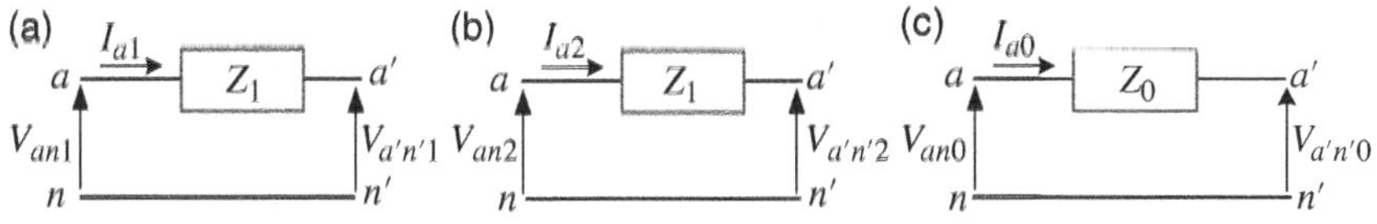

Figure 5.20 Sequence circuits of symmetrical transmission line: (a) positive, (b) negative, and (c) zero sequence.

The positive-, negative-, and zero-sequence equivalent circuits of the transmission line are shown in Figure 5.20, where the sequence impedances are

$$Z_1 = Z_2 = Z_s - Z_m = Z_{aa} - Z_{ab}$$
$$Z_0 = Z_s + 2Z_m = Z_{aa} + 2Z_{ab} + 3Z_{nn} - 6Z_{an}$$

5.5.5 Sequence Circuits for Transformers

In this section, we shall discuss the sequence circuits of transformers, which can have several possible combinations. Since the sequence circuits are different for Y- and Δ-connected loads, the sequence circuits are also different for Y- and Δ-connected transformers. Therefore, the different transformer connections are treated separately.

5.5.5.1 Y–Y-Connected Transformer

Figure 5.21 shows the schematic diagram of a Y–Y-connected transformer in which both neutrals are grounded. The primary and secondary side quantities are denoted by subscripts in uppercase letters and lowercase letters, respectively. The turns ratio of the transformer is given by $\alpha = N_1 : N_2$.

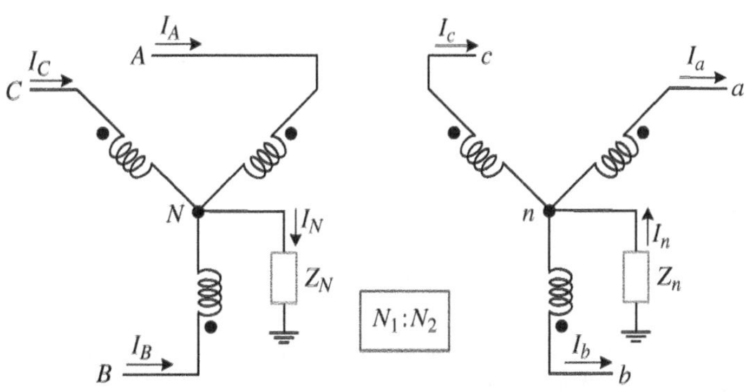

Figure 5.21 Schematic diagram of a grounded neutral Y–Y-connected transformer.

The primary side voltage of phase-a is

$$V_A = V_{AN} + V_N = V_{AN} + 3Z_N I_{A0}$$

Expanding V_A and V_{AN} in terms of their positive-, negative-, and zero-sequence components, the above equation can be rewritten as

$$V_{A0} + V_{A1} + V_{A2} = V_{AN0} + V_{AN1} + V_{AN2} + 3Z_N I_{A0} \tag{5.66}$$

Noting that the direction of the neutral current I_n is opposite to that of I_N, an expression similar to that of (5.66) for the secondary side is written as

$$V_{a0} + V_{a1} + V_{a2} = V_{an0} + V_{an1} + V_{an2} - 3Z_n I_{a0} \tag{5.67}$$

Now since the turns ratio of the transformer is $\alpha = N_1 : N_2$, the following expressions can be written for an ideal transformer:

$$\begin{aligned} \alpha &= \frac{N_1}{N_2} = \frac{V_{AN}}{V_{an}} \Rightarrow V_{an} = \frac{V_{AN}}{\alpha} \\ N_1 I_A &= N_2 I_a \Rightarrow I_a = \alpha I_A \end{aligned} \tag{5.68}$$

Substituting (5.58) in (5.67), we get

$$V_{a0} + V_{a1} + V_{a2} = \frac{1}{\alpha}(V_{AN0} + V_{AN1} + V_{AN2}) - 3Z_n \alpha I_{A0}$$

Multiplying both sides of the above equation by α results in

$$\alpha(V_{a0} + V_{a1} + V_{a2}) = V_{AN0} + V_{AN1} + V_{AN2} - 3Z_n \alpha^2 I_{A0} \tag{5.69}$$

Finally, combining (5.88) with (5.69), we get

$$\alpha(V_{a0} + V_{a1} + V_{a2}) = V_{A0} + V_{A1} + V_{A2} - 3(Z_N + Z_n \alpha^2) I_{A0} \tag{5.70}$$

The positive-, negative-, and zero-sequence components are separated from (5.70) to obtain

$$\alpha V_{a1} = \frac{N_1}{N_2} V_{a1} = V_{A1} \tag{5.71}$$

$$\alpha V_{a2} = \frac{N_1}{N_2} V_{a2} = V_{A2} \tag{5.72}$$

$$\alpha V_{a0} = \frac{N_1}{N_2} V_{a0} = V_{A0} - 3\left[Z_N + (N_1/N_2)^2 Z_n\right] I_{A0} \tag{5.73}$$

From (5.71) and (5.72), it is evident that the positive- and negative-sequence relations are the same as that has been used for representing transformer circuits given in Figure 2.20. Hence, the positive- and negative-sequence impedances are the

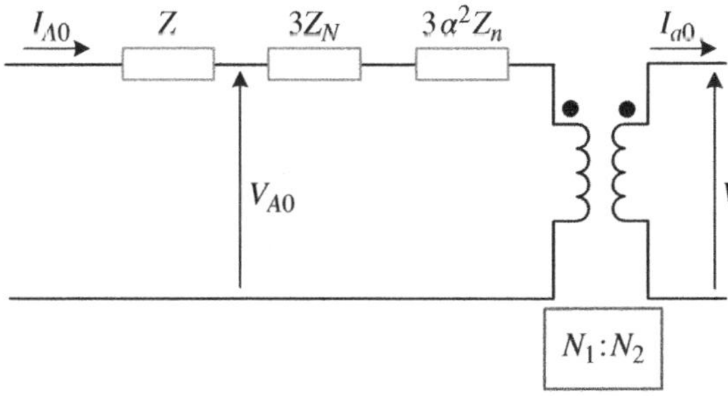

Figure 5.22 Zero-sequence equivalent circuit of a grounded neutral Y–Y-connected transformer.

same as the transformer leakage impedance Z. The zero-sequence equivalent circuit is shown in Figure 5.22. The total zero-sequence impedance is given by

$$Z_0 = Z + 3Z_N + 3(N_1/N_2)^2 Z_n \tag{5.74}$$

The following points must be noted for the zero-sequence circuit vis-à-vis their grounding.

- The zero-sequence diagram of the grounded neutral Y–Y-connected transformer is shown in Figure 5.23 (a) in which the impedance Z_0 is as given in (5.74).
- If both the neutrals are solidly grounded, that is, $Z_n = Z_N = 0$, then $Z_0 = Z$. The single-line diagram is still the same as that shown in Figure 5.23 (a).
- If one of the two neutrals or both neutrals are ungrounded, then we have either $Z_n = \infty$ or $Z_N = \infty$ or both. The zero-sequence diagram is then shown in Figure 5.23 (b), where the value of Z_0 will depend on which neutral is kept ungrounded.

5.5.5.2 Δ–Δ-Connected Transformer

The schematic diagram of a Δ–Δ-connected transformer is shown in Figure 5.24. The line-to-line voltage between phases A and B is given by

$$\begin{aligned} V_{AB} &= V_A - V_B \\ &= V_{A0} + V_{A1} + V_{A2} - V_{B0} - V_{B1} - V_{B2} = V_{AB1} + V_{AB2} \end{aligned} \tag{5.75}$$

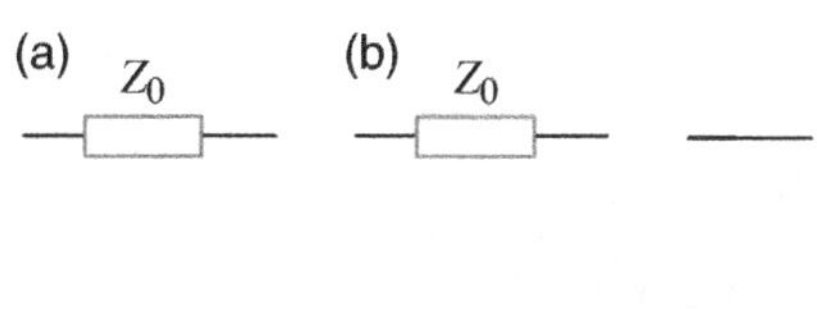

Figure 5.23 Zero-sequence diagram of (a) grounded neutral and (b) ungrounded neutral Y–Y-connected transformer.

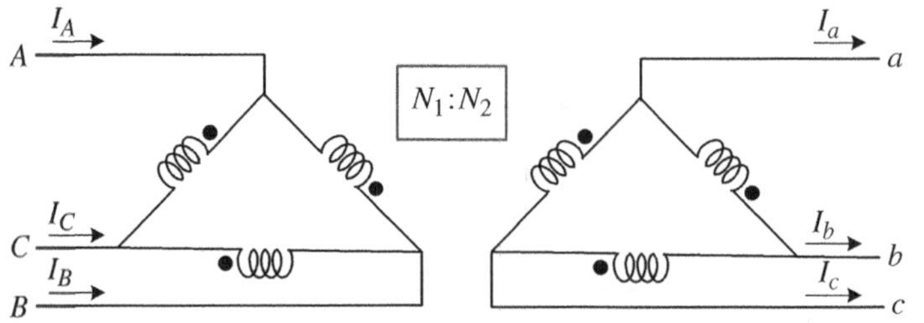

Figure 5.24 Schematic diagram of a Δ–Δ-connected transformer.

Again

$$V_{AB} = \frac{N_1}{N_2} V_{ab} = \alpha V_{ab}$$

Therefore, from (5.75), we get

$$V_{AB} = V_{AB1} + V_{AB2} = \alpha(V_{ab1} + V_{ab2}) \tag{5.76}$$

The sequence components of the line-to-line voltage V_{AB} can be written in terms of the sequence components of the line-to-neutral voltage as

$$V_{AB1} = \sqrt{3} V_{AN1} \angle 30° \tag{5.77}$$

$$V_{AB2} = \sqrt{3} V_{AN2} \angle -30° \tag{5.78}$$

Combining (5.76) to (5.78), we get

$$\sqrt{3} V_{AN1} \angle 30° + \sqrt{3} V_{AN2} \angle -30° = \alpha\left(\sqrt{3} V_{an1} \angle 30° + \sqrt{3} V_{an2} \angle -30°\right) \tag{5.79}$$

Hence, the positive- and negative-sequence voltages are given by

$$V_{AN1} = \alpha V_{an1} \text{ and } V_{AN2} = \alpha V_{an2} \tag{5.80}$$

Thus, the positive- and negative-sequence equivalent circuits are represented by a series impedance that is equal to the leakage impedance of the transformer. Since the Δ-connected winding does not provide any path for the zero-sequence current to flow, we have

$$I_{A0} = I_{a0} = 0$$

However, the zero-sequence current can sometimes circulate within the Δ windings. We can then draw the zero-sequence equivalent circuit as shown in Figure 5.25.

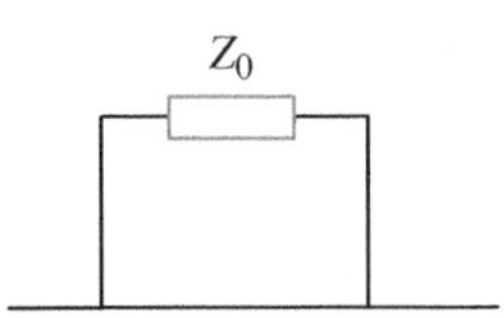

Figure 5.25 Zero-sequence diagram of a Δ–Δ-connected transformer.

5.5.5.3 Y–Δ-Connected Transformer

The schematic diagram of a Y–Δ-connected transformer is shown in Figure 5.26. It is assumed that the Y-connected side is grounded with the impedance Z_N. Even though the zero-sequence current in the primary Y-connected side has a path to the ground, the zero-sequence current flowing in the Δ-connected secondary winding has no path to flow in the line. Hence, the zero-sequence current will be zero, that is, $I_{a0} = 0$. However, the circulating zero-sequence current in the Δ winding magnetically balances the zero-sequence current of the primary winding.

Phase-a voltage in both sides of the transformer is related by

$$V_{AN} = \frac{N_1}{N_2} V_{ab} = \alpha V_{ab}$$

Furthermore, the following relation is also true

$$V_A = V_{AN} + V_N$$

The phase-a voltage on the primary side is then given by

$$\begin{aligned} V_{A0} + V_{A1} + V_{A2} &= V_{AN0} + V_{AN1} + V_{AN2} + 3Z_N I_{A0} \\ &= \alpha(V_{ab0} + V_{ab1} + V_{ab2}) + 3Z_N I_{A0} \end{aligned} \tag{5.81}$$

Separating zero-, positive-, and negative-sequence components, we can write

$$V_{A0} - 3Z_N I_{A0} = \alpha V_{ab0} = 0 \tag{5.82}$$

$$V_{A1} = \alpha V_{ab1} = \sqrt{3}\alpha V_{a1} \angle 30^\circ \tag{5.83}$$

$$V_{A2} = \alpha V_{ab2} = \sqrt{3}\alpha V_{a2} \angle -30^\circ \tag{5.84}$$

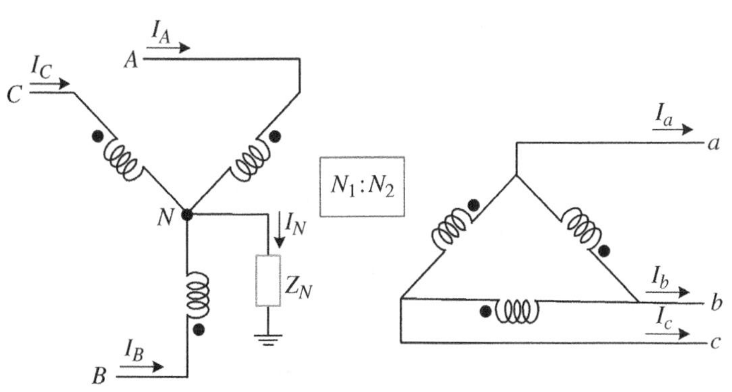

Figure 5.26 Schematic diagram of a Y–Δ-connected transformer.

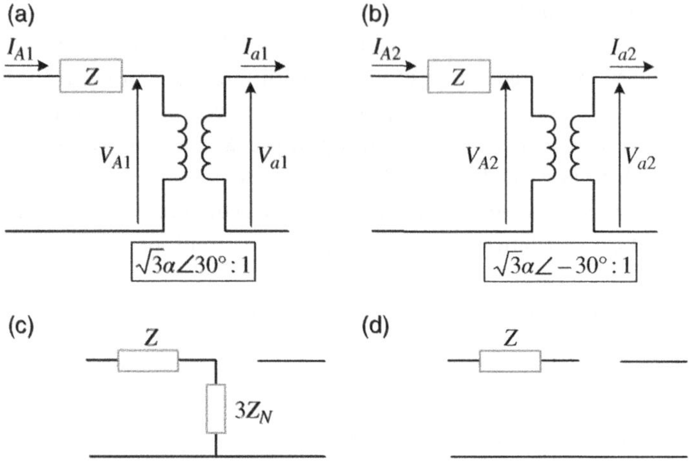

Figure 5.27 Sequence diagram of a Y–Δ-connected transformer: (a) positive sequence, (b) negative sequence, (c) zero sequence with grounded Y-connection, and (d) zero sequence with ungrounded Y-connection.

The positive-sequence equivalent circuit is shown in Figure 5.27 (a). The negative-sequence circuit is the same as that of the positive-sequence circuit except for the phase shift in the induced emf. This is shown in Figure 5.27 (b). The zero-sequence equivalent circuit is shown in Figure 5.27 (c), where $Z_0 = Z + 3Z_N$. Note that the primary and the secondary sides are not connected and hence there is an open circuit between them. However, since the zero-sequence current flows through primary windings, a return path is provided through the ground. If, however, the neutral in the primary side is not grounded, that is, $Z_N = \infty$, then the zero-sequence current cannot flow in the primary side as well. The sequence diagram is then shown in Figure 5.27 (d), where $Z_0 = Z$.

5.5.6 Sequence Networks

The sequence circuits developed in Sections 5.5.1 and 5.5.5 are combined to form the sequence networks. The sequence networks for the positive, negative, and zero sequences are formed separately by combining the sequence circuits of all the individual elements. Certain assumptions are made while forming the sequence networks. These are listed below:

- Apart from synchronous machines, the network is made of static elements.
- The voltage drop caused by the current in a particular sequence depends only on the impedance of that part of the network.

- The positive- and negative-sequence impedances are equal for all static circuit components.
- The subtransient positive- and negative-sequence impedances of a synchronous machine are equal.
- Voltage sources are connected to the positive-sequence circuits of the rotating machines.
- No positive- or negative-sequence current flows between neutral and ground.

Example 5.9 Let us consider the network shown in Figure 5.28. The values of the various reactances are not important here and hence are not given in this figure. However, various points of the circuit are denoted by the letters A to G. This has been done to identify the impedances of various circuit elements. For example, the leakage reactance of the transformer T_1 is placed between points A and B and that of transformer T_2 is placed between D and E.

The sequence networks are shown in Figure 5.29. These are summarized as follows:

- The positive-sequence network is shown in Figure 5.29 (a).
- The negative-sequence diagram is shown in Figure 5.29 (b). It is almost identical to the positive-sequence diagram except that the voltage sources are absent in this circuit.
- The zero-sequence network is shown in Figure 5.29 (c). This is divided into several segments. These are as follows:
  - The neutral point of generator G_1 is grounded. Hence, a path from Point A to the ground is provided through the zero-sequence reactance of the generator.
  - The primary side of the transformer T_1 is Δ-connected and hence there is discontinuity in the circuit after Point A.

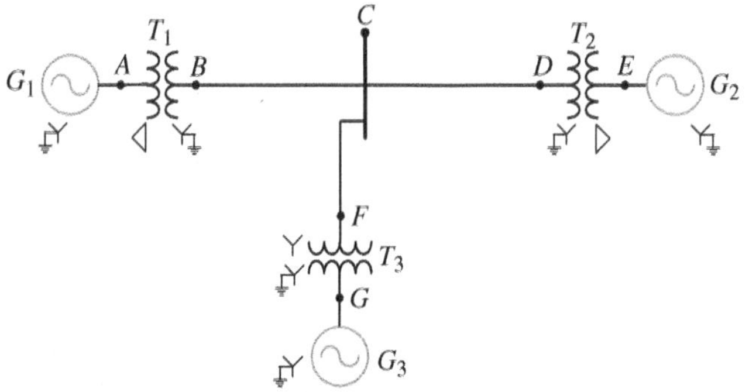

Figure 5.28 Single-line diagram of a three-machine power system.

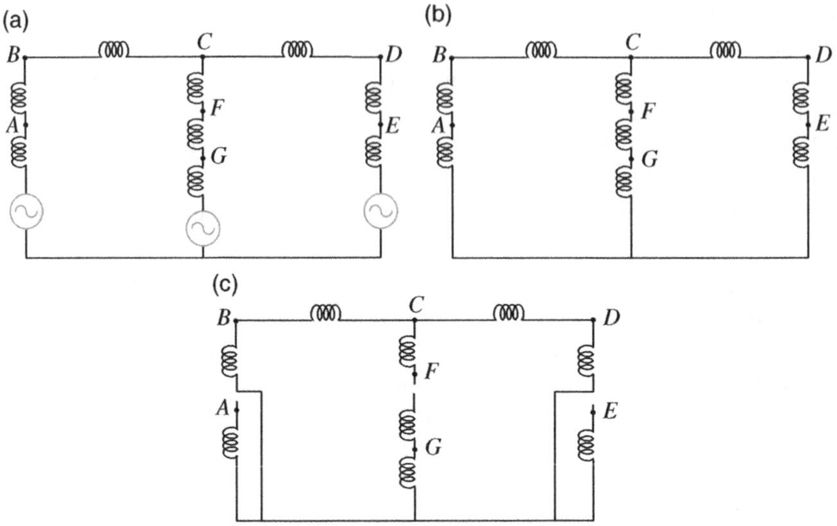

Figure 5.29 Sequence networks of the power system of Figure 5.8: (a) positive, (b) negative, and (c) zero sequence.

- The generator G_2 and the transformer T_2 also have similar configurations.
- The transmission line impedances are placed between the points *B-C*, *C-D*, and *C-F*.
- The secondary side of transformer T_3 is ungrounded and hence there is a break in the circuit after Point *F*.
- However, the primary side of T_3 is grounded and so is the neutral point of generator G_3. Hence, the zero-sequence components of these two apparatus are connected to the ground.

5.6 Unsymmetrical Faults

The sequence circuits and the sequence networks developed in Section 5.5 will now be used for determining fault currents during unsymmetrical faults. For the calculation of fault currents, the following assumptions are made:

- The power system is balanced before the fault occurs such that of the three sequence networks only the positive-sequence network is active. Also, as a fault occurs, the sequence networks are connected only through the fault location.
- The load current is negligible such that the pre-fault positive-sequence voltages are the same at all nodes and at the fault location.

- All network resistances and line charging capacitances are negligible.
- All loads are passive except the rotating loads which are represented by synchronous machines.

Based on the assumptions stated above, the faulted network will be as shown in Figure 5.30, where the voltage at the faulted point will be denoted by V_f and the current in the three faulted phases are denoted by I_{fa}, I_{fb} and I_{fc}. We shall now discuss how the three sequence networks are connected when the three different types of faults occur. These faults are single-line-to-ground faults, line-to-line faults, and double-line-to-ground faults.

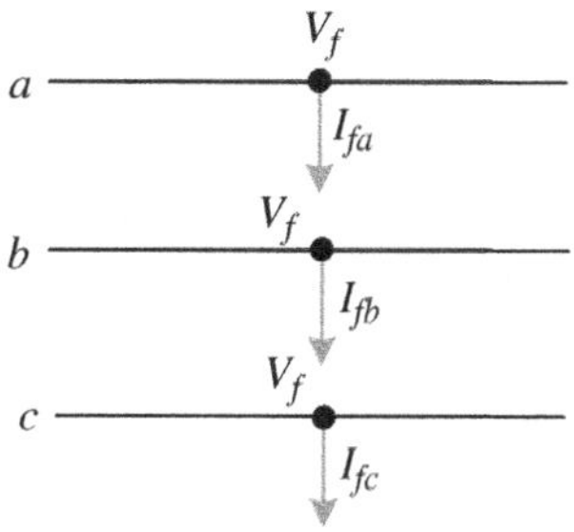

Figure 5.30 Representation of a faulted segment.

5.6.1 Single-Line-to-Ground (1LG) Fault

Let us suppose a 1LG fault has occurred at node k of a network. The faulted segment is then as shown in Figure 5.31 where it is assumed that phase-a has touched the ground through an impedance Z_f. Since the system is assumed to be unloaded before the occurrence of the fault, we have

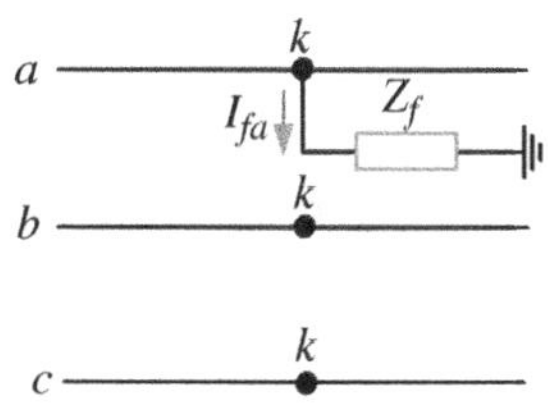

Figure 5.31 Representation of 1LG fault.

$$I_{fb} = I_{fc} = 0 \tag{5.85}$$

Also, the phase-a voltage at the fault point is given by

$$V_{ka} = Z_f I_{fa} \tag{5.86}$$

The symmetrical components of the fault current can then be written using (5.85) as

$$I_{fa012} = \frac{1}{3}\begin{bmatrix} 1 & 1 & 1 \\ 1 & a & a^2 \\ 1 & a^2 & a \end{bmatrix}\begin{bmatrix} I_{fa} \\ 0 \\ 0 \end{bmatrix} \tag{5.87}$$

Solution of (5.87) gives the following relation:

$$I_{fa0} = I_{fa1} = I_{fa2} = \frac{I_{fa}}{3} \tag{5.88}$$

This implies that the three sequence currents are in series for a 1LG fault. Let us denote the zero, positive, and negative sequence Thevenin impedances at the faulted point as Z_{kk0}, Z_{kk1}, and Z_{kk2}, respectively. Also, since the Thevenin voltage at the faulted phase is V_f, the following three sequence voltages at the faulted point are obtained:

$$
\begin{aligned}
V_{ka0} &= -Z_{kk0} I_{fa0} \\
V_{ka1} &= V_f - Z_{kk1} I_{fa1} \\
V_{ka2} &= -Z_{kk2} I_{fa2}
\end{aligned}
\tag{5.89}
$$

Then, the phase-a voltage at the faulted bus is written from (5.88) and (5.89) as

$$
\begin{aligned}
V_{ka} &= V_{ka0} + V_{ka1} + V_{ka2} \\
&= V_f - (Z_{kk0} + Z_{kk1} + Z_{kk2}) I_{fa0}
\end{aligned}
\tag{5.90}
$$

From Figure 5.31, the voltage V_{ka} is written as

$$V_{ka} = Z_f I_{fa} = Z_f \left(I_{fa0} + I_{fa1} + I_{fa2}\right) = 3Z_f I_{fa0} \tag{5.91}$$

Phase-a fault current is then obtained by combining (5.90) and (5.91) as

$$I_{fa0} = \frac{V_f}{Z_{kk0} + Z_{kk1} + Z_{kk2} + 3Z_f} \tag{5.92}$$

The Thevenin equivalent of the sequence network is shown in Figure 5.32.

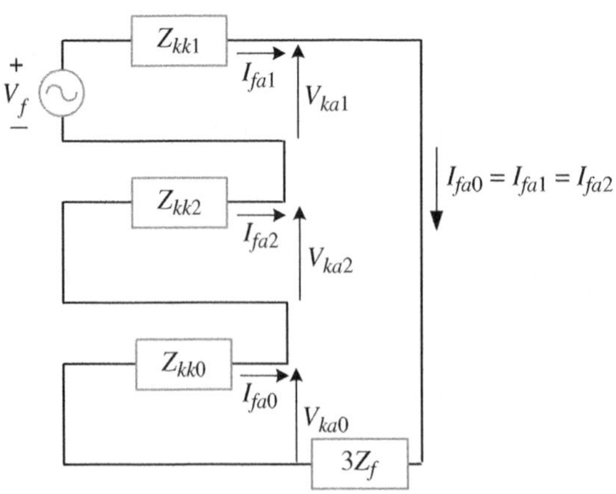

Figure 5.32 Thevenin equivalent circuit of a 1LG fault.

Example 5.10 A three-phase Y-connected unloaded synchronous generator is running at its rated voltage when a 1LG fault occurs at its terminals. The generator is rated 20 kV, 220 MVA, with a subsynchronous reactance of 0.2 per unit. Assume that the subtransient mutual reactance between the windings is 0.025 per unit. The neutral of the generator is grounded through a 0.05 per unit reactance. The equivalent circuit of the generator is shown in Figure 5.33. We have to find out the fault current.

Since the generator is unloaded the internal EMFs are

$$E_{an} = 1.0 \quad E_{bn} = 1.0\angle -120^\circ \quad E_{cn} = 1.0\angle 120^\circ$$

Since no current flows in phases b and c, the phase-a fault current is given in Figure 5.33 as

$$I_{fa} = \frac{1}{j(0.2+0.05)} = -j4.0$$

The neutral point voltage is given by

$$V_n = -X_n I_{fa} = -0.2$$

From Figure 5.33 and (5.47), we get

$$V_a = 0$$
$$V_b = E_{bn} + V_n + j0.025 I_{fa} = -0.6 - j0.866 = 1.0536\angle -124.72^\circ$$
$$V_c = E_{cn} + V_n + j0.025 I_{fa} = -0.6 + j0.866 = 1.0536\angle 124.72^\circ$$

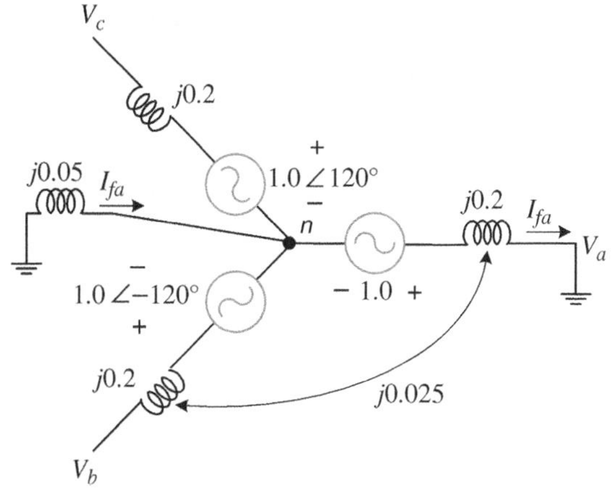

Figure 5.33 Unloaded generator of Example 5.10.

Therefore,

$$\mathbf{V_{a012}} = \mathbf{C}\begin{bmatrix} 0 \\ 1.0536\angle-124.72^\circ \\ 1.0536\angle124.72^\circ \end{bmatrix} = \begin{bmatrix} -0.4 \\ 0.7 \\ -0.3 \end{bmatrix}$$

From (5.51), the positive-sequence impedance is calculated as $Z_1 = j\omega(L_s + M_s) = j0.225$. Then from Figure 5.18 (a), we have

$$I_{fa1} = \frac{E_{an} - V_{a1}}{Z_1} = \frac{1-0.7}{j0.225} = -j1.3333$$

Also note from (5.88) that

$$I_{fa0} = I_{fa1} = I_{fa2}$$

Therefore, from Figure 5.18, the zero- and negative-sequence impedances are given by

$$Z_{g0} = -\frac{V_{a0}}{I_{a0}} - 3Z_n = j(0.3-0.15) = j0.15$$

$$Z_2 = -\frac{V_{a2}}{I_{a2}} = j0.225$$

Comparing the above two values with (5.50) and (5.52), we find that Z_0 indeed is equal to $j\omega(L_s - 2M_s)$ and $Z_2 = j\omega(L_s + M_s)$. The fault current is computed from (5.92) as

$$I_{fa0} = \frac{1}{j(0.225 + 0.225 + 0.15 + 3\times 0.05)} = -j1.3333$$

5.6.2 Line-to-Line (LL) Fault

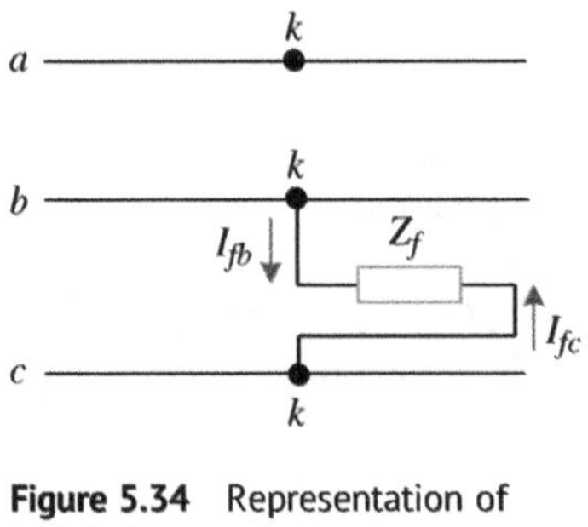

Figure 5.34 Representation of L-L fault.

The faulted segment for an LL fault is shown in Figure 5.34 where it is assumed that the fault has occurred at node k of the network. For this fault, it has been assumed that the phases b and c got shorted through the impedance Z_f. Since the system is unloaded before the occurrence of the fault, the phase-a current is zero, that is,

$$I_{fa} = 0 \tag{5.93}$$

Also, since phases b and c are shorted, the currents of these phases are negative of each other, that is,

$$I_{fb} = -I_{fc} \tag{5.94}$$

Therefore, from (5.93) and (5.94), we have

$$\mathbf{I_{fa012}} = \mathbf{C}\begin{bmatrix} 0 \\ I_{fb} \\ -I_{fb} \end{bmatrix} = \frac{I_{fb}}{3}\begin{bmatrix} 0 \\ (a-a^2) \\ (a^2-a) \end{bmatrix} \tag{5.95}$$

Equation (5.95) can be summarized as follows:

$$\begin{aligned} I_{fa0} &= 0 \\ I_{fa1} &= -I_{fa2} \end{aligned} \tag{5.96}$$

Therefore, no zero-sequence current is injected into the network at bus k, and hence, the zero-sequence network remains dead for an LL fault. Furthermore, the positive- and negative-sequence currents are negative of each other.

From Figure 5.34, the following expression for the voltage at the faulted point is written:

$$V_{kb} - V_{kc} = Z_f I_{fb} \tag{5.97}$$

Again, the line-to-line voltage is given by

$$\begin{aligned} V_{kb} - V_{kc} &= V_{kb0} + V_{kb1} + V_{kb2} - V_{kc0} - V_{kc1} - V_{kc2} \\ &= (V_{kb1} - V_{kc1}) + (V_{kb2} - V_{kc2}) \\ &= (a^2 - a)\, V_{ka1} + (a - a^2)\, V_{ka2} = (a^2 - a)\,(V_{ka1} - V_{ka2}) \end{aligned} \tag{5.98}$$

From (5.96), we have $I_{fb} = I_{fa} = 0$ and $I_{fa1} = -I_{fa2}$. Therefore, we can write the fault current as

$$I_{fb} = I_{fb0} + I_{fb1} + I_{fb2} = a^2 I_{fa1} + a I_{fb2} = (a^2 - a) I_{fa1} \tag{5.99}$$

Combining (5.97) to (5.99), the following equation is obtained:

$$V_{ka1} - V_{ka2} = Z_f I_{fa1} \tag{5.100}$$

Equations (5.97) and (5.100) indicate that the positive- and negative-sequence networks are in parallel. The sequence network is then as shown in Figure 5.35. From this network, the fault current is obtained as

$$I_{fa1} = -I_{fa2} = \frac{V_f}{Z_{kk1} + Z_{kk2} + Z_f} \tag{5.101}$$

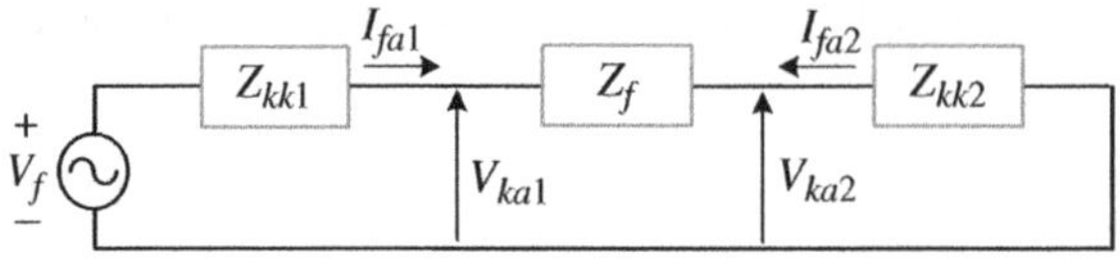

Figure 5.35 Thevenin equivalent circuit of a L–L fault.

Example 5.11 Let us consider the same generator as given in Example 5.10. Assume that the generator is unloaded when a bolted ($Z_f = 0$) short circuit occurs between phases b and c, that is, $I_{fb} = -I_{fc}$. Also, since the generator is unloaded, the current $I_{fa} = 0$. Therefore, from (5.46), we have

$$V_{an} = E_{an} = 1.0$$

$$V_{bn} = E_{bn} - j0.225 I_{fb} = 1\angle -120^\circ - j0.225 I_{fb}$$

$$V_{cn} = E_{cn} - j0.225 I_{fc} = 1\angle 120^\circ + j0.225 I_{fb}$$

Also, since $V_{bn} = V_{cn}$, the last two equations above are combined to give

$$I_{fb} = -I_{fc} = \frac{1\angle -120^\circ - 1\angle 120^\circ}{j0.45} = -3.849$$

Then

$$\mathbf{I_{fa012}} = \mathbf{C}\begin{bmatrix} 0 \\ -3.849 \\ 3.849 \end{bmatrix} = \begin{bmatrix} 0 \\ -j2.2222 \\ j2.2222 \end{bmatrix}$$

The above equation can also be obtained from (5.101) as

$$I_{fa1} = -I_{fa2} = \frac{1}{j0.225 + j0.225} = -j2.2222$$

Also, since the neutral current I_n is zero, we can write $V_a = 1.0$ and

$$V_b = V_c = V_{bn} = -0.5$$

Hence, the sequence components of the line voltages are

$$\mathbf{V_{a012}} = \mathbf{C}\begin{bmatrix} 1.0 \\ -0.5 \\ -0.5 \end{bmatrix} = \begin{bmatrix} 0 \\ 0.5 \\ 0.5 \end{bmatrix}$$

Also note that

$$V_{a1} = 1.0 - j0.225 I_{fa1} = 0.5$$

$$V_{a2} = -j0.225 I_{fa2} = 0.5$$

which are the same as obtained before.

5.6.3 Double-Line-to-Ground (2LG) Fault

The faulted segment for a 2LG fault is shown in Figure 5.36 where it is assumed that the fault has occurred at node k of the network. Here, the phases b and c got shorted through the impedance Z_f to the ground. Since the system is unloaded before the occurrence of the fault, the same condition as (5.93) is assumed for the phase-a current. Therefore,

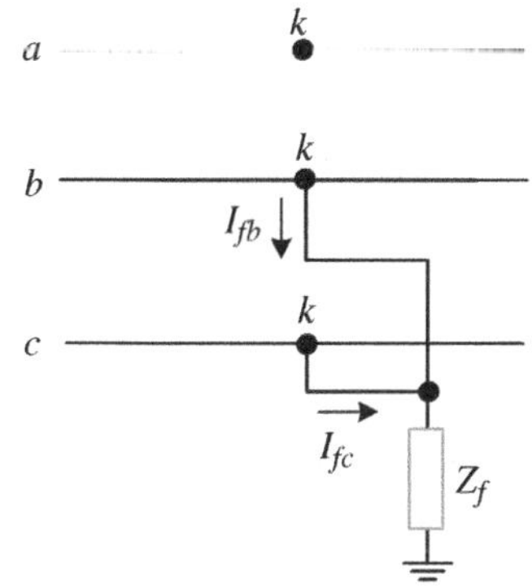

Figure 5.36 Representation of 2LG fault.

$$I_{fa0} = \frac{1}{3}\left(I_{fa} + I_{fb} + I_{fc}\right) = \frac{1}{3}\left(I_{fb} + I_{fc}\right)$$
$$\Rightarrow 3I_{fa0} = I_{fb} + I_{fc} \tag{5.102}$$

Also, the voltages of phases b and c are given by

$$V_{kb} = V_{kc} = Z_f(I_b + I_c) = 3Z_f I_{fa0} \tag{5.103}$$

Therefore,

$$\mathbf{V_{ka012}} = \mathbf{C}\begin{bmatrix} V_{ka} \\ V_{kb} \\ V_{kb} \end{bmatrix} = \frac{1}{3}\begin{bmatrix} V_{ka} + 2V_{kb} \\ V_{ka} + \left(a + a^2\right) V_{kb} \\ V_{ka} + \left(a + a^2\right) V_{kb} \end{bmatrix} \tag{5.104}$$

The following two equations are obtained from (5.104):

$$V_{ka1} = V_{ka2} \tag{5.105}$$

$$3V_{ka0} = V_{ka} + 2V_{kb} = V_{ka0} + V_{ka1} + V_{ka2} + 2V_{kb} \tag{5.106}$$

Substituting (5.103) and (5.105) in (5.106) and rearranging, we get

$$V_{ka1} = V_{ka2} = V_{ka0} - 3Z_f I_{fa0} \tag{5.107}$$

Also, since $I_{fa} = 0$, the sum of the symmetrical components of phase-a current will be zero, that is,

$$I_{fa0} + I_{fa1} + I_{fa2} = 0 \tag{5.108}$$

The Thevenin equivalent circuit for 2LG fault is shown in Figure 5.37. The following equation is obtained from this figure:

$$I_{fa1} = \frac{V_f}{Z_{kk1} + Z_{kk2} \|^{l} \left(Z_{kk0} + 3Z_f\right)} = \frac{V_f}{Z_{kk1} + \dfrac{Z_{kk2}\left(Z_{kk0} + 3Z_f\right)}{Z_{kk2} + Z_{kk0} + 3Z_f}} \tag{5.109}$$

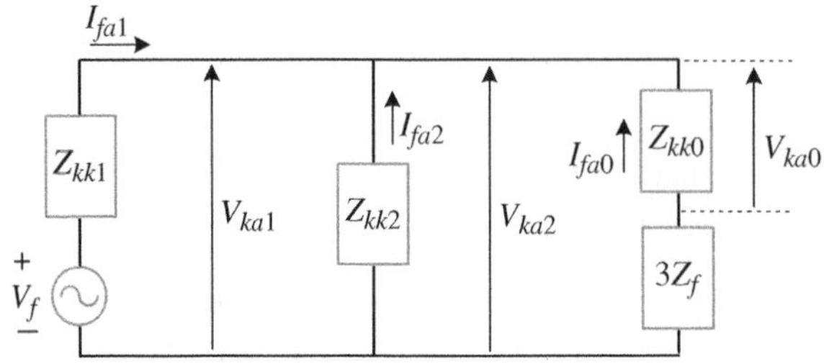

Figure 5.37 Thevenin equivalent circuit of a 2LG fault.

The zero- and negative-sequence currents can be obtained using the current divider principle as

$$I_{fa0} = -I_{fa1}\left(\frac{Z_{kk2}}{Z_{kk2} + Z_{kk0} + 3Z_f}\right) \tag{5.110}$$

$$I_{fa2} = -I_{fa1}\left(\frac{Z_{kk0} + 3Z_f}{Z_{kk2} + Z_{kk0} + 3Z_f}\right) \tag{5.111}$$

Example 5.12 Let us consider the same generator as given in Examples 5.10 and 5.11. Assume that the generator is operating without any load when a bolted 2LG fault occurs in phases b and c. The equivalent circuit for this fault is shown in Figure 5.38. From this figure, we can write

$$E_{bn} + V_n = 1\angle -120^\circ + V_n = j0.2I_{fb} - j0.025I_{fc}$$
$$E_{cn} + V_n = 1\angle 120^\circ + V_n = j0.2I_{fc} - j0.025I_{fb}$$
$$V_n = -j0.05\left(I_{fb} + I_{fc}\right)$$

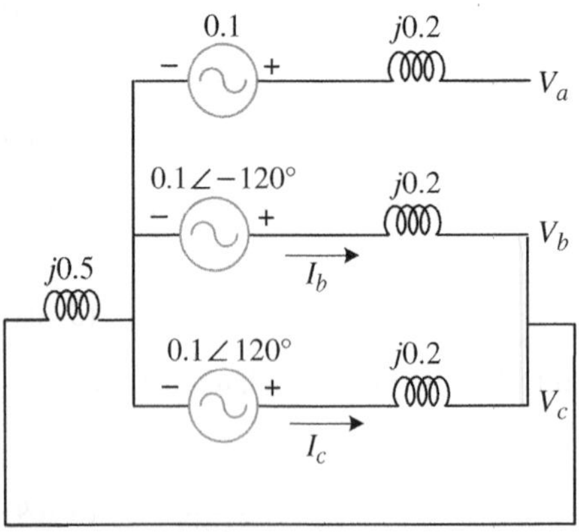

Figure 5.38 Equivalent circuit depicting the 2LG fault of Example 5.12.

The above three equations can be written in a combined matrix–vector form as

$$j\begin{bmatrix} 0.25 & 0.025 \\ 0.025 & 0.25 \end{bmatrix}\begin{bmatrix} I_{fb} \\ I_{fc} \end{bmatrix} = \begin{bmatrix} 1\angle -120^\circ \\ 1\angle 120^\circ \end{bmatrix}$$

The solution of the above equation yields

$$I_b = -3.849 + j1.8182$$
$$I_c = 3.849 + j1.8182$$

Hence

$$\mathbf{i_{fa012}} = \mathbf{C}\begin{bmatrix} 0 \\ -3.849 + j1.8182 \\ -3.849 + j1.8182 \end{bmatrix} = \begin{bmatrix} j1.2121 \\ -j2.8283 \\ j1.6162 \end{bmatrix}$$

Note that these values can also be obtained using (5.109) to (5.111). Also, from Example 5.10, we have

$$Z_1 = Z_2 = j0.225, \quad Z_0 = j(0.15 + 3 \times 0.05) = j0.3 \text{ and } Z_f = 0$$

Then

$$I_{fa1} = \frac{1.0}{j0.225 + \left(\dfrac{j0.225 \times j0.3}{j0.525}\right)} = -j2.8283$$

$$I_{fa2} = -I_{fa1}\frac{j0.3}{j0.525} = j1.6162$$

$$I_{fa0} = -I_{fa1}\frac{j0.225}{j0.525} = j1.2121$$

Then, the sequence components of the voltages are

$$V_{a1} = 1.0 - j0.225 \times I_{fa1} = 0.3636$$

$$V_{a2} = -j0.225 \times I_{fa2} = 0.3636$$

$$V_{a0} = -j0.3 \times I_{fa0} = 0.3636$$

Also note from Figure 5.38 that

$$V_a = E_{an} + V_n + j0.0225\left(I_{fb} + I_{fc}\right) = 1.0909$$

and $V_b = V_c = 0$. Therefore,

$$V_{a012} = C\begin{bmatrix} 1.0909 \\ 0 \\ 0 \end{bmatrix} = \begin{bmatrix} 0.3636 \\ 0.3636 \\ 0.3636 \end{bmatrix}$$

which are the same as obtained before.

5.6.4 Fault Current Computation Using Sequence Networks

The use of sequence networks in the calculation of fault currents will be demonstrated in this section through some examples.

Example 5.13 Consider the radial network shown in Figure 5.39. The system parameters are given below:

Generator G: 50 MVA, 20 kV, $X'' = X_1 = X_2 = 20\%$, $X_0 = 7.5\%$
Motor M: 40 MVA, 20 kV, $X'' = X_1 = X_2 = 20\%$, $X_0 = 10\%$, $X_n = 5\%$
Transformer T_1: 50 MVA, 20 kVΔ/110 kVY, $X = 10\%$
Transformer T_2: 50 MVA, 20 kVΔ/110 kVY, $X = 10\%$
Transmission line: $X_1 = X_2 = 24.2\,\Omega$, $X_0 = 60.5\,\Omega$

We shall find the fault current when (a) 1LG, (b) LL, and (c) 2LG fault occurs at Bus-2.

Let us choose a base in the circuit of the generator G, that is, 50 MVA and 20 kV. Then the per-unit impedances of the generator are:

$$X_{G1} = X_{G2} = 0.2, \quad X_{G0} = 0.075$$

Since the transformer ratings are the same as the base MVA and kV, the per-unit impedances of the two transformers are

$$X_{T1} = X_{T2} = 0.1$$

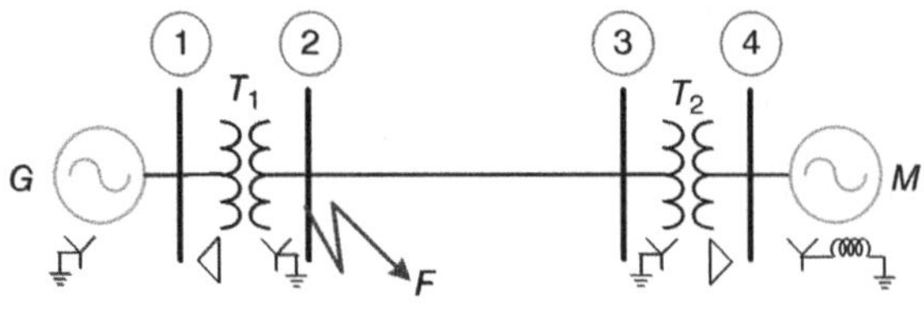

Figure 5.39 Radial power system of Example 5.13.

For the motor M, the kV base is 20; however, the MVA base of the motor is 40. The per-unit impedances of the motor are then converted into the common base as

$$X_{M1} = X_{M2} = 0.2 \times \frac{50}{40} = 0.25, \quad X_{M0} = 0.1 \times \frac{50}{40} = 0.125,$$

$$X_n = 0.05 \times \frac{50}{40} = 0.0625$$

For the transmission line

$$Z_{base} = \frac{110^2}{50} = 242\ \Omega$$

Therefore,

$$X_{L1} = X_{L2} = \frac{24.2}{242} = 0.1, \quad X_{L0} = \frac{60.5}{242} = 0.25$$

Let us neglect the phase shift associated with the Y/Δ transformers. Then the positive-, negative-, and zero-sequence networks are shown in Figures 5.40 to 5.42, respectively.

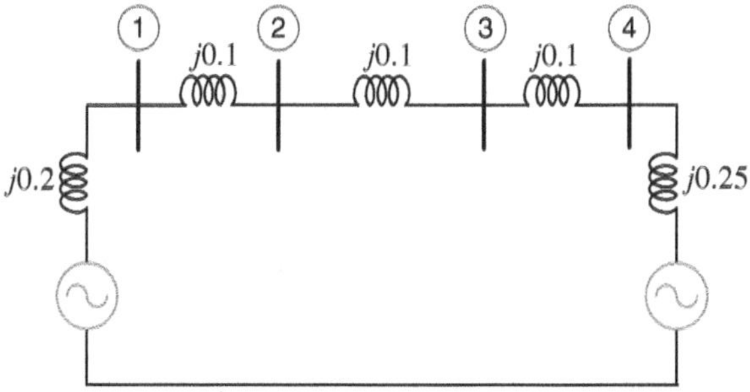

Figure 5.40 Positive-sequence network of the power system of Figure 5.39.

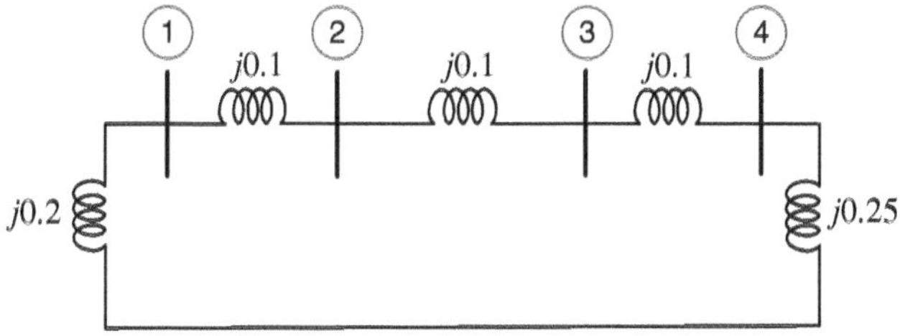

Figure 5.41 Negative-sequence network of the power system of Figure 5.39.

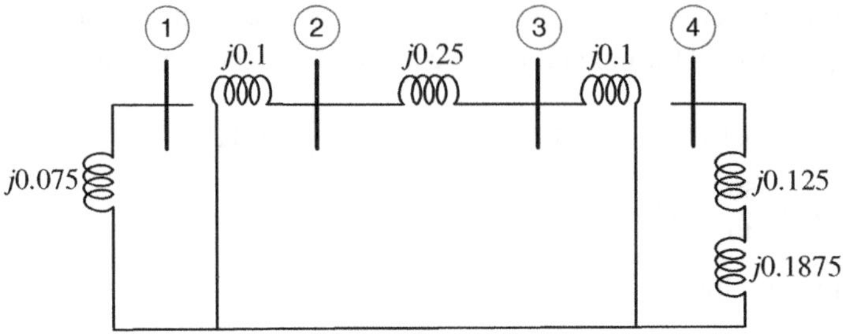

Figure 5.42 Zero-sequence network of the power system of Figure 5.39.

From Figures 5.40 and 5.41, the following $\mathbf{Y_{bus}}$ matrix is obtained for both positive and negative sequences:

$$\mathbf{Y_{bus1}} = \mathbf{Y_{bus2}} = j\begin{bmatrix} -15 & 10 & 0 & 0 \\ 10 & -20 & 10 & 0 \\ 0 & 10 & -20 & 10 \\ 0 & 0 & 10 & 14 \end{bmatrix}$$

Inverting the above matrix, the following $\mathbf{Z_{bus}}$ matrix is computed:

$$\mathbf{Z_{bus1}} = \mathbf{Z_{bus2}} = j\begin{bmatrix} 0.1467 & 0.1200 & 0.0933 & 0.0667 \\ 0.1200 & 0.1800 & 0.1400 & 0.1000 \\ 0.0933 & 0.1400 & 0.1867 & 0.1333 \\ 0.0667 & 0.1000 & 0.1333 & 0.1667 \end{bmatrix}$$

Similarly, from Figure 5.42, the following $\mathbf{Y_{bus}}$ and $\mathbf{Z_{bus}}$ matrices are obtained for the zero-sequence

$$\mathbf{Y_{bus0}} = j\begin{bmatrix} -13.3333 & 0 & 0 & 0 \\ 0 & -14 & 4 & 0 \\ 0 & 4 & -14 & 0 \\ 0 & 0 & 0 & -3.2 \end{bmatrix}$$

$$\mathbf{Z_{bus0}} = j\begin{bmatrix} 0.075 & 0 & 0 & 0 \\ 0 & 0.0778 & 0.0222 & 0 \\ 0 & 0.0222 & 0.0778 & 0 \\ 0 & 0 & 0 & 0.3125 \end{bmatrix}$$

Hence, for a fault in Bus-2, we have the following Thevenin impedances, which are (2,2) elements of the $\mathbf{Z_{bus}}$ matrices

$$Z_1 = Z_2 = j0.18, \quad Z_0 = j0.0778$$

Alternatively, the Thevenin equivalent can also be computed from the parallel combination of impedances from Figures 5.40 and 5.42 as

$$Z_1 = Z_2 = j0.3||^l j0.45 = j0.18$$

$$Z_0 = j0.1||^l j0.35 = j0.0778$$

<u>*(a) Single-Line-to-Ground Fault*</u>: The bolted 1LG fault occurs at Bus-2 when the system is unloaded with bus voltages being 1.0 per unit. Then from (5.88), we get

$$I_{fa0} = I_{fa1} = I_{fa2} = \frac{1}{j(2 \times 0.18 + 0.0778)} = -j2.2841 \text{ per unit}$$

Therefore, the fault current through phase-a is given as

$$I_{fa} = 3I_{fa0} = -j6.8524 \text{ per unit}$$

Furthermore, the currents of the other two phases are zero, that is, $I_{fb} = I_{fc} = 0$. The sequence components of the voltages are then computed from (5.89) as

$$V_{2a0} = -j0.0778 I_{fa0} = -0.1777$$
$$V_{2a1} = 1 - j0.18 I_{fa1} = 0.5889$$
$$V_{2a2} = -j0.18 I_{fa2} = -0.4111$$

Therefore, the voltages at the faulted bus are

$$\begin{bmatrix} V_a \\ V_b \\ V_c \end{bmatrix} = \mathbf{C}^{-1} \begin{bmatrix} V_{2a0} \\ V_{2a1} \\ V_{2a2} \end{bmatrix} = \begin{bmatrix} 0 \\ 0.9061\angle -107.11^\circ \\ 0.9061\angle 107.11^\circ \end{bmatrix}$$

<u>*(b) Line-to-Line Fault*</u>: For a bolted LL fault, we can write from (5.96)

$$I_{fa0} = 0, I_{fa1} = -I_{fa2} = \frac{1}{j2 \times 0.18} = -j2.7778 \text{ per unit}$$

Then the fault currents are

$$\begin{bmatrix} I_{fa} \\ I_{fb} \\ I_{fc} \end{bmatrix} = \mathbf{C}^{-1} \begin{bmatrix} 0 \\ I_{fa1} \\ I_{fa2} \end{bmatrix} = \begin{bmatrix} 0 \\ -4.8113 \\ 4.8113 \end{bmatrix}$$

The sequence components of Bus-2 voltages are computed from Figure 5.35 as

$$V_{2a0} = 0$$
$$V_{2a1} = 1 - j0.18 I_{fa1} = 0.5$$
$$V_{2a2} = -j0.18 I_{fa2} = 0.5$$

Hence, the faulted bus voltages are

$$\begin{bmatrix} V_a \\ V_b \\ V_c \end{bmatrix} = \mathbf{C}^{-1} \begin{bmatrix} V_{2a0} \\ V_{2a1} \\ V_{2a2} \end{bmatrix} = \begin{bmatrix} 1.0 \\ -0.5 \\ -0.5 \end{bmatrix}$$

<u>*(c) Double-Line-to-Ground Fault*</u>: For a bolted 2LG fault at Bus-2, the parallel combination of the negative- and zero-sequence components is computed as

$$Z_{eq} = j0.18 \| {}^{l} j0.0778 = j0.0543$$

Thus, the positive-sequence current is given from (5.109) as

$$I_{fa1} = \frac{1}{j0.18 + Z_{eq}} = -j4.2676 \text{ per unit}$$

The zero- and negative-sequence currents are then computed respectively from (5.110) and (5.111) as

$$I_{fa0} = -I_{fa1}\frac{j0.18}{j(0.18 + 0.0778)} = j2.9797 \text{ per unit}$$

$$I_{fa2} = -I_{fa1}\frac{j0.0778}{j(0.18 + 0.0778)} = j1.2879 \text{ per unit}$$

Therefore, the fault currents flowing in the line are

$$\begin{bmatrix} I_{fa} \\ I_{fb} \\ I_{fc} \end{bmatrix} = \mathbf{C}^{-1} \begin{bmatrix} I_{fa0} \\ I_{fa1} \\ I_{fa2} \end{bmatrix} = \begin{bmatrix} 0 \\ 6.657\angle 137.11^\circ \\ 6.657\angle 42.89^\circ \end{bmatrix}$$

Furthermore, the sequence components of Bus-2 voltages are computed from Figure 5.37 as

$$V_{2a0} = -j0.0778 I_{fa0} = 0.2318$$
$$V_{2a1} = 1 - j0.18 I_{fa1} = 0.2318$$
$$V_{2a2} = -j0.18 I_{fa2} = 0.2318$$

Therefore, voltages at the faulted bus are

$$\begin{bmatrix} V_a \\ V_b \\ V_c \end{bmatrix} = \mathbf{C}^{-1} \begin{bmatrix} V_{2a0} \\ V_{2a1} \\ V_{2a2} \end{bmatrix} = \begin{bmatrix} 0.6954 \\ 0 \\ 0 \end{bmatrix}$$

Example 5.14 Let us now assume that a 2LG fault has occurred in Bus-4 of the system of Example 5.13 instead of the one in Bus-2. Therefore,

$$X_1 = X_2 = j0.1667, \quad X_0 = j0.3125$$

Furthermore, the equivalent impedance is

$$Z_{eq} = j0.1667 \|{}^{l} j0.3125 = j0.1087$$

Hence

$$I_{fa1} = \frac{1}{j0.1667 + Z_{eq}} = -j3.631 \text{ per unit}$$

$$I_{fa0} = -I_{fa1}\frac{j0.1667}{j(0.1667 + 0.3125)} = j1.2631 \text{ per unit}$$

$$I_{fa2} = -I_{fa1}\frac{j0.3125}{j(0.1667 + 0.3125)} = j2.3678 \text{ per unit}$$

Therefore, the fault currents flowing in the line are

$$\begin{bmatrix} I_{fa} \\ I_{fb} \\ I_{fc} \end{bmatrix} = \mathbf{C}^{-1}\begin{bmatrix} I_{fa0} \\ I_{fa1} \\ I_{fa2} \end{bmatrix} = \begin{bmatrix} 0 \\ 5.5298\angle 159.96^\circ \\ 5.5298\angle 20.04^\circ \end{bmatrix}$$

We shall now compute the currents contributed by the generator and the motor to the fault. Let us denote the current flowing to the fault from the generator side by I_g, while that flowing from the motor by I_m. Then from Figure 5.40 using the current divider principle, the positive-sequence currents contributed by the two buses are

$$I_{ga1} = I_{fa1} \times \frac{j0.25}{j0.75} = -j1.2103 \text{ per unit}$$

$$I_{ma1} = I_{fa1} \times \frac{j0.5}{j0.75} = -j2.4206 \text{ per unit}$$

Similarly from Figure 5.41, the negative-sequence currents are given as

$$I_{ga2} = I_{fa2} \times \frac{j0.25}{j0.75} = j0.7893 \text{ per unit}$$

$$I_{ma2} = I_{fa2} \times \frac{j0.5}{j0.75} = j1.5786 \text{ per unit}$$

Finally, notice from Figure 5.42 that the zero-sequence current flowing from the generator to the fault is 0. Then we have

$$I_{ga0} = 0$$

$$I_{ma0} = I_{fa0} = j1.2631 \text{ per unit}$$

Therefore, the fault currents flowing from the generator side are

$$\begin{bmatrix} I_{ga} \\ I_{gb} \\ I_{gc} \end{bmatrix} = \mathbf{C}^{-1} \begin{bmatrix} I_{ga0} \\ I_{ga1} \\ I_{ga2} \end{bmatrix} = \begin{bmatrix} 0.4210\angle -90^\circ \\ 1.7445\angle 173.07^\circ \\ 1.7445\angle 6.93^\circ \end{bmatrix}$$

and those flowing from the motor are

$$\begin{bmatrix} I_{ma} \\ I_{mb} \\ I_{mc} \end{bmatrix} = \mathbf{C}^{-1} \begin{bmatrix} I_{ma0} \\ I_{ma1} \\ I_{ma2} \end{bmatrix} = \begin{bmatrix} 0.4210\angle 90^\circ \\ 3.8512\angle 154.07^\circ \\ 3.8512\angle 25.93^\circ \end{bmatrix}$$

Verify that adding I_g and I_m the fault current I_{fa}, I_{fb} and I_{fc} computed before are obtained.

In the above two examples, the phase shifts of the Y/Δ transformers are neglected. However, according to the American standard, the positive-sequence components of the high-tension side lead those of the low-tension side by 30°, while the negative-sequence behavior is the reverse of the positive-sequence behavior. Usually, the high-tension side of a Y/Δ transformer is Y-connected. Therefore, as evident from Figure 5.27, the positive-sequence component of the Y side leads the positive-sequence component of the Δ side by 30° while the negative-sequence component of the Y side lags that of the Δ side by 30°. The phase shift will now be used to compute the fault current for an unsymmetrical fault.

Example 5.15 Let us consider the same system as given in Example 5.14. Since the phase shift does not alter the zero sequence, the circuit of Figure 5.42 remains unchanged. The positive and the negative-sequence circuits must however include the respective phase shifts. These circuits are redrawn as shown in Figures 5.43 and 5.44. Note from these figures that the term $\sqrt{3}a$ is dropped vis-à-vis that of Figure 5.27. This is because the per-unit impedances remain unchanged when referred to the either high-tension or low-tension side of an ideal transformer. Therefore, the per-unit impedances will also not be altered.

Since the zero sequence remains unaltered, these currents will not change from those computed in Example 5.14. Thus,

$$I_{ga0} = 0 \text{ and } I_{ma0} = j1.2631 \text{ per unit}$$

Now the positive-sequence fault current from the generator I_{ga1}, being on the Y-side of the Y/Δ transformer will lead I_{ma1} by 30°. Therefore,

$$I_{ga1} = -j1.2103 \times 1\angle 30^\circ = 1.2103\angle -60^\circ \text{ per unit}$$

$$I_{ma1} = -j2.4206 \text{ per unit}$$

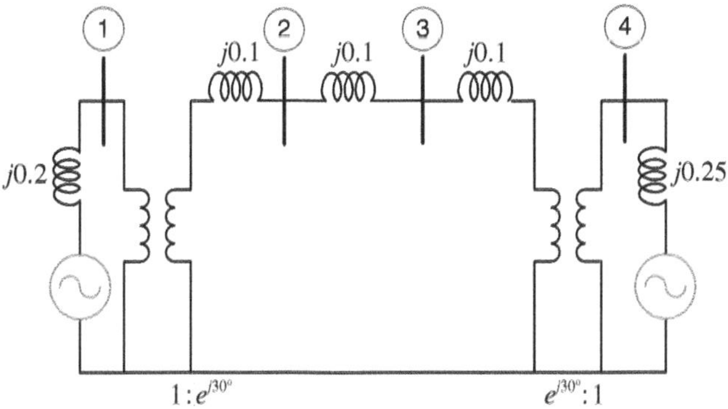

Figure 5.43 Positive-sequence network of the power system of Figure 5.39 including transformer phase shift.

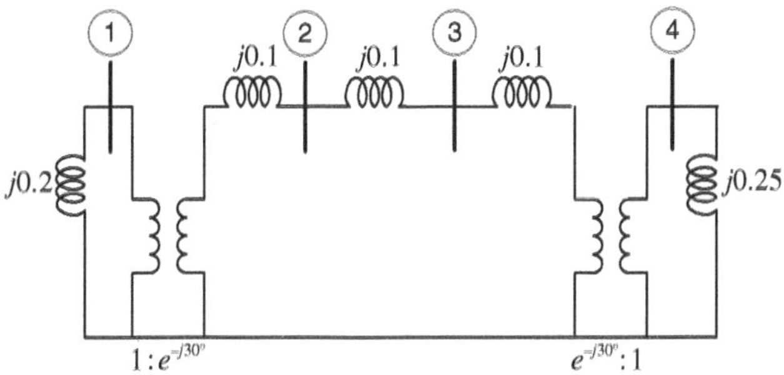

Figure 5.44 Negative-sequence network of the power system of Figure 5.39 including transformer phase shift.

Finally, the negative-sequence current I_{ga2} will lag I_{ma2} by 30° and hence

$$I_{ga2} = j0.7893 \times 1\angle -30^\circ = j0.7893\angle 60^\circ \text{ per unit}$$

$$I_{ma2} = j1.5786 \text{ per unit}$$

Therefore,

$$\begin{bmatrix} I_{ga} \\ I_{gb} \\ I_{gc} \end{bmatrix} = \mathbf{C}^{-1} \begin{bmatrix} I_{ga0} \\ I_{ga1} \\ I_{ga2} \end{bmatrix} = \begin{bmatrix} 1.0642\angle -20.04^\circ \\ 1.9996\angle -180^\circ \\ 1.0642\angle 20.04^\circ \end{bmatrix}$$

The fault currents flowing from the motor will also remain unaltered. Note that the currents flowing into the fault remain unchanged. This implies that the phase shift of the Y/Δ transformers does not affect the fault currents.

5.7 Concluding Remarks

This chapter presents a thorough treatment for the calculations of fault currents in a power system. Both symmetrical and unsymmetrical fault current calculations are presented. Usually, the fault currents are calculated in the phasor domain, where a synchronous generator is replaced by its subtransient voltage and reactances. Furthermore, the system is assumed to be unloaded before the fault. Since the magnitude of the fault current is significantly higher than that of the pre-fault current, this assumption does not affect the main purpose of fault current calculation, which is the selection of circuit breakers. The philosophy of circuit breaker selection is discussed in Chapter 6.

A circuit behaves differently when it is unbalanced from when it is balanced. In a balanced circuit, the magnitude of the phase voltages (and/or currents) will be the same and all the phases will be displaced by 120° from each other. A set of unbalanced voltages or currents can be resolved into three sets of balanced voltages or currents using the symmetrical component transformation. Moreover, different power components behave differently during unsymmetrical operations. Therefore, they are resolved into their respective sequence circuits. These are then carefully joined together to form sequence networks. These networks are then used for finding fault currents during unsymmetrical faults. A comprehensive and detailed analysis of faults and symmetrical components is given in [1].

Reference

1 J. J. Grainger and W. D. Stevenson, *Power System Analysis*, McGraw-Hill, New York, 1994.

Problems

P5.1 Consider the three-bus power system shown in Figure P5.1, where all the quantities are given in per unit. It is assumed that the generators are unloaded and operate at a voltage of 1 per unit, when a bolted three-phase fault occurs at Bus-2. Find,

(a) the fault current,

(b) the voltage of buses 1 and 3,

(c) the subtransient currents flowing from generators 1 and 2, and

(d) the current flowing between in the line connecting buses 1 and 2.

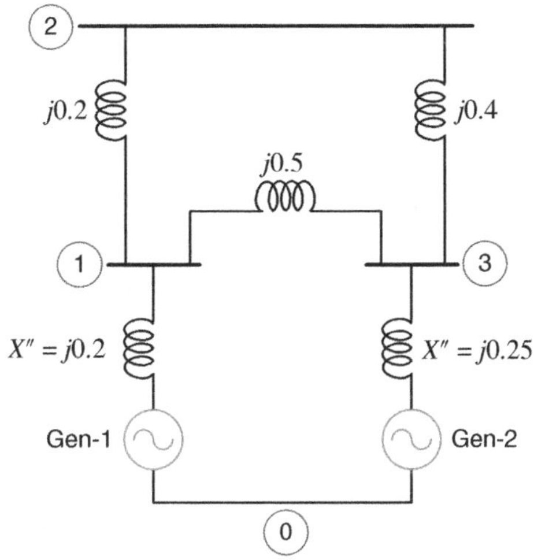

Figure P5.1 Three-bus power system of Problem 5.1.

P5.2 A 500 kVA, 20 kV, generator with $X''_d = 0.2$ per unit is connected to a bus through a circuit breaker as shown in Figure P5.2. Connected through circuit breakers to the same bus are three synchronous motors. The subtransient reactance (X''_m) of each motor, calculated on a base of 250 KVA, 24 kV, 1.0 power factor and 90% efficiency, is 0.2 per unit. The motors operate at full load at unity power factor and rated voltage with the load equally distributed amongst them. Assume that the reactance (X_m) of each motor for interrupting current is 1.5 times the motor subtransient reactance (X''_m).

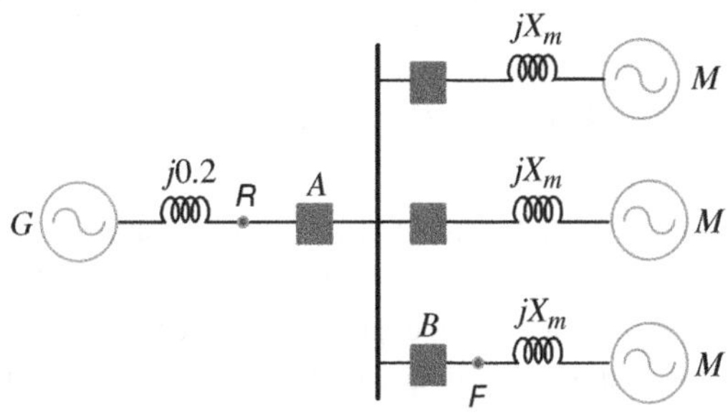

Figure P5.2 A synchronous generator is connected with three motor loads of Problem 5.2.

(a) Find the symmetrical short circuit current which must be interrupted by the breakers A and B for a three-phase fault at Point F. Simplify calculations by neglecting the pre-fault current.
(b) Repeat Part (a) for a three-phase fault at Point R.

P5.3 Consider the power system shown in Figure P5.3, where all quantities are given in per unit. A balanced three-phase occurs at Bus-3. Assuming that the system is unloaded before the fault and $V_f = 1.0$ per unit, find the fault current.

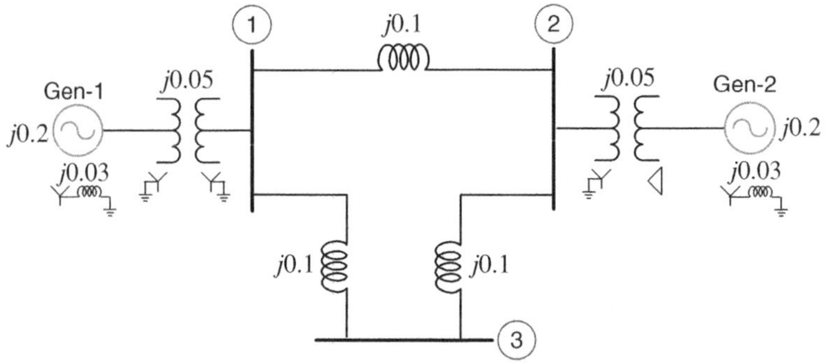

Figure P5.3 Three-bus power system of Problem 5.3.

P5.4 Let the abc to 012 transformation matrix be given by

$$\mathbf{C} = K \begin{bmatrix} 1 & 1 & 1 \\ 1 & a & a^2 \\ 1 & a^2 & a \end{bmatrix}$$

Find the constant K such that the transformation matrix is orthogonal, that is,

$$P_{abc} + jQ_{abc} = \mathbf{V}_{\mathbf{a012}}^T \mathbf{I}_{\mathbf{a012}}^* = V_{a0}I_{a0}^* + V_{a1}I_{a1}^* + V_{a2}I_{a2}^*$$

P5.5 The voltage at the terminals of a balanced load consisting of three 10 Ω resistors connected in Y are $V_{ab} = 100 \angle 0°$ V, $V_{bc} = 80.8 \angle -121.44°$ V and $V_{ca} = 90 \angle 130°$ V. Determine a general expression between the symmetrical components of the line and phase voltages, that is between (1) V_{ab1} and V_{an1} and (2) between V_{ab2} and V_{an2}. Assuming that the neutral of the load is not connected, find the current in phase-a from the symmetrical components of the given voltages.

P5.6 In the circuit shown in Figure P5.6, an open delta load is supplied by a balanced three-phase supply. The value of the load impedance $Z = 20 + j10\ \Omega$. Assume that the line-to-line supply voltage $V_{ab} = 400 \angle 0°$ V.

(a) Find the currents I_{ab} and I_{bc}.

(b) Using the value calculated in (a), find I_{ab0}, I_{ab1}, and I_{ab2}.

(c) Using the values of I_{ab0}, I_{ab1}, and I_{ab2} obtained in (b), find the sequence components of the line currents, that is, I_{a0}, I_{a1}, and I_{a2}.

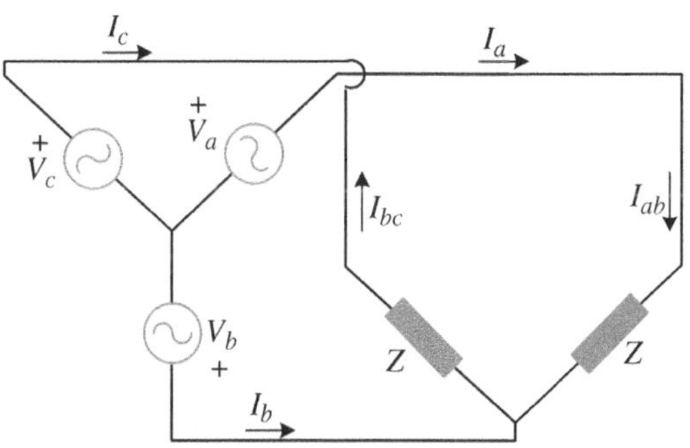

Figure P5.6 An open delta load supplied by a balanced Y-connected source of Problem 5.6.

P5.7 A Y-connected and solidly grounded generator is operating at rated voltage and no load. It is disconnected from the rest of the system. Given $X_m'' = X_1 = X_2 = 0.15$ per unit and $X_0 = 0.05$ per unit. Find the ratio of subtransient line current for a single-line-to-ground fault to the subtransient line current for a three-phase fault.

P5.8 A generator supplies a motor through a Y–Δ transformer. The generator is connected to the grounded Y-side of the transformer. A fault occurs between the motor terminals and the transformer. The per unit symmetrical components of the subtransient current in the motor flowing toward the fault is

$$I_{a1} = -0.8 - j2.6, \quad I_{a2} = -j2.0, \quad I_{a0} = -j3.0$$

and from the transformer toward the fault are

$$I_{a1} = 0.8 - j0.4, \quad I_{a2} = -j1.0, \quad I_{a0} = 0$$

Assume $X_d'' = X_1 = X_2$ both for the motor and the generator.

(a) Draw the sequence network and describe the type of fault.

(b) Find the subtransient fault current in per unit.

P5.9 Consider the power system shown in Figure 5.28, the sequence diagrams of which are shown in Figure 5.29. The system parameters are:

Generator G_1: 200 MVA, 20 kV, $X'' = 20\,\%$, $X_0 = 10\%$
Generator G_2: 300 MVA, 18 kV, $X'' = 20\,\%$, $X_0 = 10\%$
Generator G_3: 300 MVA, 20 kV, $X'' = 25\,\%$, $X_0 = 15\%$
Transformer T_1: 300 MVA, 220Y/22Δ kV, $X = 10\%$
Transformer T_2: Three single-phase units each rated 100 MVA, 130/25 kV, $X = 10\%$
Transformer T_3: 300 MVA, 220Y/22Y kV, $X = 10\%$
Line *B-C*: $X_1 = X_2 = 75\,\Omega$, $X_0 = 100\,\Omega$
Line *C-D*: $X_1 = X_2 = 75\,\Omega$, $X_0 = 100\,\Omega$
Line *C-F*: $X_1 = X_2 = 50\,\Omega$, $X_0 = 75\,\Omega$

Neglecting the phase shifts of Y/Δ-connected transformers, and assuming that the system is unloaded, find the fault current for a 1LG fault at Bus-*C*. Choose the circuit of Generator G_3 as the base.

P5.10 The reactances of a generator rated 500 MVA and 20 kV are $X_1 = X_2 = 20\%$ and $X_0 = 5\%$. The generator is connected to a transformer rated 500 MVA, 20Δ/400Y kV with the leakage reactance of 10%. The neutral point of the transformer is solidly grounded. The terminal voltage of the transformer is 20 kV when a double-line-to-ground fault occurs on the high-tension side of the transformer. Neglecting the phase shift, find the fault current in all the three phases.

P5.11 Two generators G_1 and G_2 are connected respectively through transformers T_1 and T_2 to a high-voltage bus as shown in Figure P5.11. A fault occurs at point *F* shown in the figure. The pre-fault voltage at point *F* is 515 kV. The specifications of the various apparatus are

G_1: 1,000 MVA, 20 kV, $X''_d = X_1 = X_2 = 10\%$, $X_0 = 5\%$

G_2: 800 MVA, 22 kV, $X''_d = X_1 = X_2 = 15\%$, $X_0 = 8\%$

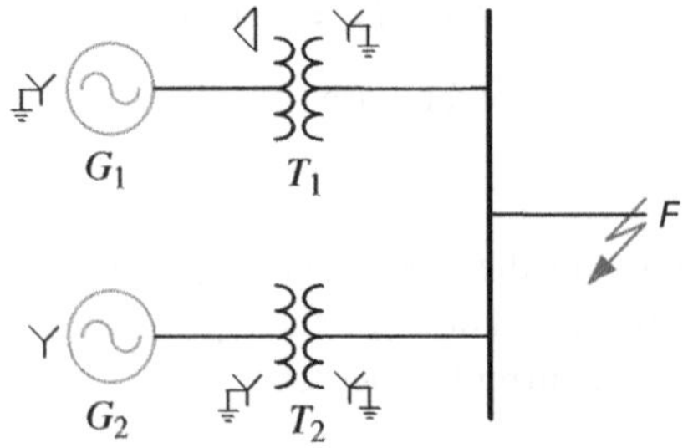

Figure P5.11 Two generators connected to a high-voltage bus of Problem 5.11.

T_1: 1,000 MVA, 500Y/20Δ kV, $X = 17.5\%$

T_2: 800 MVA, 500Y/22Y kV, $X = 16\%$

Line: $X_1 = X_2 = 15\%$, $X_0 = 40\%$ on a base of 1, 500 MVA, 500 kV

Neglecting the pre-fault current and the transformer phase shift, find the subtransient current for a (a) 1LG fault and (b) LL fault at point F. Choose a base of 1,000 MVA, 500 kV in the circuit of the transmission line.

P5.12 The single-line diagram of a power system is shown in Figure P5.12. The per unit system parameters that are represented on a common base are

Generators:
G_1: $X_1 = X_2 = 0.18$ and $X_0 = 0.07$
G_2: $X_1 = X_2 = 0.2$ and $X_0 = 0.1$
G_3: $X_1 = X_2 = 0.2539$, $X_0 = 0.08464$ and $X_n = 0.08464$
G_4: $X_1 = X_2 = 0.3386$ and $X_0 = 0.1129$

Transformers:
T_1: $X = 0.1$
T_2: $X = 0.1$
T_3: $X = 0.24$
T_4: $X = 0.1467$

Transmission lines:
Line joining buses 1 and 2: $X_1 = X_2 = 0.0854$ and $X_0 = 0.2563$
Line joining buses 1 and 3: $X_1 = X_2 = 0.0683$ and $X_0 = 0.1709$
Line joining buses 2 and 3: $X_1 = X_2 = 0.0683$ and $X_0 = 0.1709$

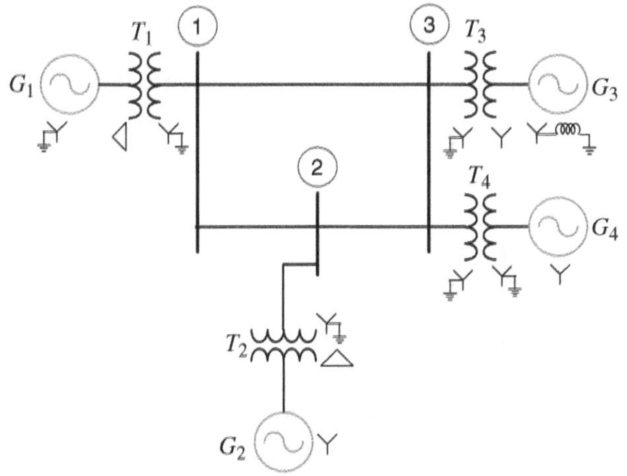

Figure P5.12 Three-bus, four-generator power system of Problem 5.12.

Draw the positive-, negative-, and zero-sequence impedance diagrams of the system. Neglecting the phase shifts of Y–Δ transformers, find the fault currents using the sequence networks for bolted 2LG fault Bus-3.

P5.13 Repeat Problem 5.13 using Thevenin impedance.

P5.14 Consider the single-line diagram of a power system shown in Figure P5.14, where all quantities are defined on a common per unit base. The system parameters are as follows:

$$G_1\text{: } X_1 = X_2 = 0.2,\ \ X_0 = 0.05$$

$$G_2\text{: } X_1 = X_2 = 0.2,\ \ X_0 = 0.05$$

$$T_1\text{: } X = 0.05$$

$$T_2\text{: } X = 0.05$$

$$TL_{12}\text{: } X_1 = X_2 = 0.1,\ \ X_0 = 0.3$$

$$TL_{13}\text{: } X_1 = X_2 = 0.1,\ \ X_0 = 0.3$$

$$TL_{23}\text{: } X_1 = X_2 = 0.1,\ \ X_0 = 0.3$$

(a) Draw the positive-, negative-, and zero-sequence impedance diagram for the circuit.
(b) Find the driving point impedances of positive, negative, and zero sequences at Bus-3 for the three networks drawn in Part-(a).
(c) Find the three-phase voltage and current for a
   A) balanced fault at Bus-3,
   B) 1LG fault at Bus-3,
   C) LL fault at Bus-3, and
   D) 2LG fault at Bus-3.

The fault impedance is assumed to be zero, that is, $Z_f = 0$. The generators are assumed to be unloaded with the voltage at the fault point being $1\angle 0°$.

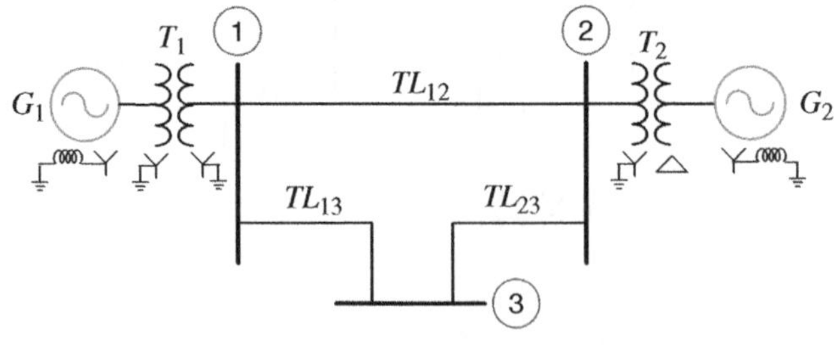

Figure P5.14 Three-bus power system of Problem 5.14.

6

Power System Protection

A power system, containing a vast spread-out area, is left to the vagaries of Mother Nature. There can be thunderstorms, cyclones, snowstorms, or many such natural occurrences that may result in the uprooting of transmission or distribution lines causing interruption in power transfer and delivery. During a thunderstorm, a branch of a tree can cause a short circuit between two lines. As we have seen earlier, the magnitude of a short-circuit current is several times the normal operating current. Therefore, the power apparatus (such as generators and transformers) must be protected from such a large current.

Furthermore, electrical systems can fail due to causes such as short circuits and overheating. The aim of the protection system is to protect the remaining equipment, consumers, and as well as, life, and property. Consider, for example, a household toaster. Due to its continuous use, a fault can develop, and it may cause a short circuit in its coils. The protection system should prevent the flow of continuous fault currents that may lead to fires. At the same time, the supply of power should not be disrupted to the entire street or even the entire house due to this fault. Some of the main reasons for protection are listed below.

- To prevent voltage dips for customers (quality of supply).
- To minimize loss of revenue for the supply industry and customers.
- To prevent and/or minimize plant damage.
- To maintain system stability (for transmission networks).
- To maintain public and personnel safety.
- To ensure discrimination and maximize reliability.

Modern Power System, First Edition. Arindam Ghosh.

Published 2026 by John Wiley & Sons, Inc.
Companion website: www.wiley.com/go/ghoshpowersystem

A protection circuit usually has three components:

- Protective devices such as circuit breakers and fuses.
- Instrument transformers as current transformers (CTs) or potential transformers (PTs), produce reduced magnitude currents or voltages for protective relays.
- Protective relays take inputs from instrument transformers and send trip signals to circuit breakers.

In this chapter, we shall discuss all the three aspects of protection. Furthermore, how protective relays can be coordinated will also be discussed.

One of the major challenges that currently the power industry is facing is the integration of power converter-interfaced renewable energy sources in the power distribution systems. These can be distributed throughout a system. Overcurrent relay coordination assumes unidirectional power flow from the substation down a feeder toward loads. However, with the connections of renewable energy sources anywhere in a feeder, the unidirectional nature of power flow is not valid anymore. Therefore, directional relays may not be effective. The other problem is that the maximum current that power converters can supply is limited, and thus, they may not be able to supply high-level fault currents, unlike synchronous generators. Consequently, the relay coordination becomes more challenging in the presence of distributed renewable generators (DRGs). This will also be discussed in this chapter.

Finally, modern substations are equipped with microprocessor-based digital relays. Again, all the relays need not be supplied by the same vendor. Therefore, it is imperative that a substation automation process be established such that all the relays can work effectively and collectively to protect the power system. IEC 81850 is a substation automation protocol that will be discussed briefly at the end of the chapter.

6.1 Protective Elements

Usually, there are two types of protecting elements – fuses and circuit breakers. These are briefly discussed in this section.

6.1.1 Fuses

This is the simplest form of protection. If the current through a small section of a thin conductor exceeds a rating, the metal will melt, and an open circuit will be created. There is a limited amount of heat that can be conducted from the narrow section of metal and if the dissipation is too high, the temperature will run away leading to melting. For low-voltage applications, fuses are (or used to be) very common. In its primitive form, fuses are wires that are placed in porcelain holders, which are placed in series between two segments of a line. The wires connect two conductors that are embedded in the porcelain holders and are easy to replace. The fuse functions by melting excess current. A wire can glow red hot at 5 A, but blow after two minutes on 5.5 A. Since it is rather difficult to get the calibration accurately, this form of protection cannot be precise.

The specifications of a fuse are normally based on the following four factors:

- *Voltage Rating*: A blown fuse should be able to withstand its voltage rating. This rating determines the ability of a fuse to suppress the internal arc that occurs after the fuse melts.
- *Continuous Current Rating*: This is the rating of the current that the fuse can carry without melting.
- *Interrupting Current Rating*: This is the largest current that the fuse can safely interrupt.
- *Time Response*: The melting and clearing time of a fuse depends on the magnitude of the overcurrent or fault current and is usually specified by a time–current curve. The curve has time along the y-axis and current along the x-axis, somewhat like an exponential curve.

The low cost and simplicity of operation are the benefits of the fuse. The limitations are that the precise current of melting will vary making it difficult to give a precise protection level. Another limitation is that the process of restoration of the circuit cannot be achieved remotely. There is also an additional time delay associated with the changing of the fuse element itself. These limitations have resulted in replacing them in most household circuits, which are now protected by miniature circuit breakers. However, the individual items of equipment are still protected by fuses.

The expulsion-type fuse is a protection method for medium-voltage levels that has lower costs than circuit breakers. They are typically mounted at the top of the pole in 11–33 kV systems. It consists of a straight tube that is open at both ends. The tube is usually made of fiber and the fuse is placed inside the tube. This fuse vents exhaust gases during the interruption process, which is usually boric acid. During a fault, arc heat decomposes the boric acid into water vapor, which deionizes the arc path preventing the arc reignition. Once the fuse blows, the fuse chamber, which is attached to the line, drops down and hangs from the line. This helps the workers to identify the failure point easily. Expulsion fuses release gases during fault interruptions and, thus, are not suitable for indoor use. Different aspects of fuses and their operating characteristics are discussed in [1, 2].

6.1.2 Circuit Breakers

The circuit breakers are the main protection element of a power system. These items need to open when a fault current flows through the circuit. A typical circuit breaker operating time is shown in Figure 6.1. Once a fault occurs, the protective devices get activated. A certain amount of time elapses before the protective relays can determine that there is an overcurrent in the circuit and initiate the trip command. This is called the *detection time*. The contacts of the circuit breakers are held together by the spring mechanism, and, with the trip command, the spring mechanism releases the contacts. When two current-carrying contacts separate from each other, a voltage instantly appears at the contacts and a large voltage gradient appears in the medium between the two contacts. This voltage gradient ionizes the medium thereby maintaining the flow of current. This current generates extreme heat and light, which is called the electric arc. Different mechanisms are used for elongating the arc such that it can be cooled and extinguished. Therefore, the circuit breaker must withstand fault current from the instant of initiation of the fault to the time the arc extinguishes.

Note that the contacts need to open quickly such that the arc that is created is elongated and cooled. The arc will extinguish at the current zero but if the arc remains hot enough or the contacts remain too close when the voltage begins to rise again, there will still be enough ionized gases for the reignition of the arc. A circuit breaker design must consider the following aspects:

- It must elongate the arc paths.
- It must limit the amount of ionized material and cool it rapidly.
- It must ensure that the contacts are sufficiently opened such that the arc will not reestablish.

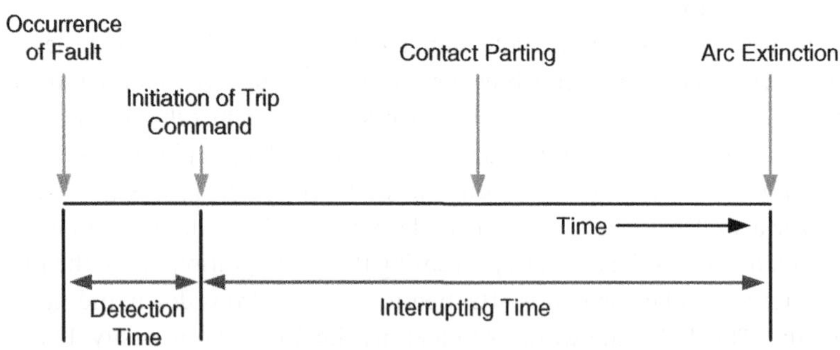

Figure 6.1 Typical circuit breaker operating cycle.

Once a circuit breaker moves the contacts apart, arcs will inevitably be formed. These arcs must be kept sufficiently cool to be able to extinguish them at the next current zero. Some of the prominent circuit breakers used are:

- Air blast circuit breakers that produce a blast of air to lengthen the arcs, thus cooling them.
- Magnetic expulsion circuit breakers that use the fault current to force the arcs into a path with insulating elements to cool them.
- Similar arc lengthening is used inside sulfur hexafluoride (SF6) breakers. SF6 provides good insulating and thermal conduction properties and can be used in magnetic lengthening of the arc.
- Vacuum circuit breakers are commonly used for low-voltage circuits between 11 kV and 33 kV. Once the contacts part, due to the lack of medium to sustain the arc, the arc particles condense on the contact surfaces. This creates a gap between the contacts, and the arc can be extinguished in the next zero crossing of the current.

Two factors are of utmost importance for the selection of circuit breakers. These are:

- The maximum instantaneous current that a breaker must withstand.
- The current when the breaker contacts part.

As we have shown in Chapter 5, the instantaneous current following a fault will also contain the DC component. In a high-power circuit breaker selection, the subtransient current is multiplied by a factor of 1.6 to determine the RSM value of the current the circuit breaker must withstand. This current is called the *momentary current*. The *interrupting current* of a circuit breaker is lower than the momentary current and will depend on the speed of the circuit breaker. The interrupting current may be asymmetrical because some DC components may still remain and continue to decay.

Breakers are usually classified by their nominal voltage, continuous current rating, rated maximum voltage, K-factor which is the voltage range factor, rated short circuit current at maximum voltage, and operating time. The K-factor is the ratio of the rated maximum voltage to the lower limit of the range of the operating voltage. The maximum symmetrical interrupting current of a circuit breaker is given by K times the rated short circuit current.

Reclosers or auto-reclosers are used mainly on power distribution systems. Many of the faults that occur in distribution systems are self-clearing types. For example, high wind can cause the wires of two phases to clash momentarily or can cause a tree branch to clash with the wires. These are not permanent faults but can cause momentary disruptions in the power supply. A recloser automatically tests to

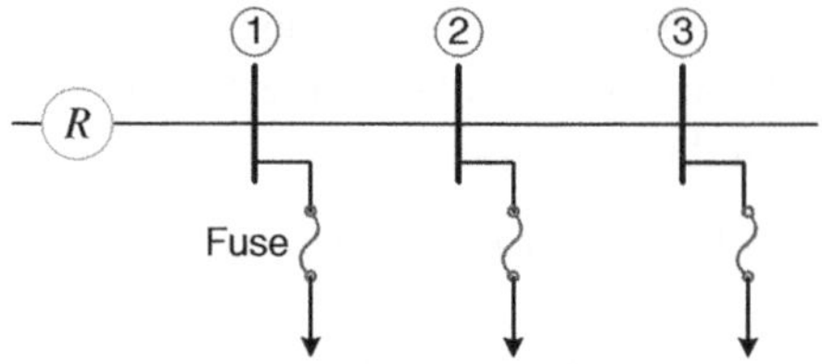

Figure 6.2 A circuit containing recloser and fuses.

determine whether the fault has been cleared. If the problem is only temporary, then the recloser automatically resets itself and restores the electric power. Usually, a recloser has an on–off duty cycle. When it recloses for the first time, if the fault persists, it will open the circuit again. After a certain interval, it will reclose the breaker. If the fault persists further, then it will open and will remain open as this might be a permanent type of fault that will require human intervention.

A radial circuit can have upstream reclosers, closer to the substation. The downstream loads can then be protected by fuses. Consider, for example, the system shown in Figure 6.2, where the recloser is indicated by R. For temporary faults, the recloser is set for one or more operations, by which it will try to clear temporary faults. Suppose a fault occurs between the recloser and bus-1. The recloser will try to clear the fault. If it persists, then after a few tries, it will lock out the circuit by opening permanently, and manual intervention will be required. If, on the other hand, the fault is downstream from say bus-1, the fuse connected to it will melt isolating the line connected to bus-1 from the rest of the network and the recloser can restore power. However, the fuse types and their operating time should be chosen carefully such that they can coordinate well with the recloser.

6.2 Instrument Transformers

In the previous section, we discussed fuses and circuit breakers. These components break a faulted circuit. However, a circuit breaker, is not capable of determining whether a fault has occurred. We need additional components to detect a fault and trip a circuit breaker to isolate the fault. These are usually called protective relays. However, the relays cannot withstand high voltage or current. Instrument transformers are used to reduce voltage and current levels significantly such that they can be safely given as input to the relays. The instrument transformers offer a twofold advantage:

a) to obtain input levels for relays and associated hardware containing electronic components and
b) to protect maintenance personnel from high-voltage and current risks.

In principle, these transformers are no different from ordinary transformers. However, their usage is highly specialized. For example, their power requirement is fairly low as they are used for driving relays.

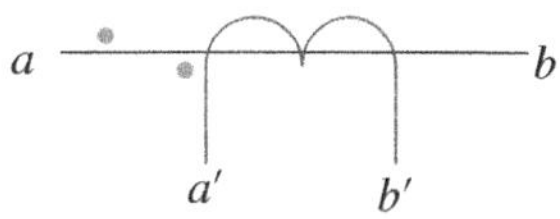

Figure 6.3 Schematic diagram of a CT.

6.2.1 Current Transformer (CT)

The schematic diagram of a current transformer (CT) is shown in Figure 6.3. The primary winding consists of a single turn (shown by the straight-line joining a and b). The single turn is obtained by threading the primary conductor through one or more toroidal steel cores. The secondary winding (terminals marked a' and b') are multiple turn windings wound on toroidal cores. The dots placed on the primary and secondary sides have the same connotation as a conventional transformer. CTs are available in various ranges and sizes. Usually, the secondary winding of a CT is rated 5 A in most countries. However, during short circuits, it is capable of withstanding 20 times that value.

As mentioned above, the most common CT secondary full-load current is 5 A, which matches the standard 5 A full-scale current rating of switchboard indicating devices, power metering equipment, and protective relays. However, CTs with a 1 A full-load value and matching instruments with a 1 A full-range value are also available. Many new protective relays are programmable for either value. CT ratios are expressed as a ratio of the rated primary current to the rated secondary current. For example, a 300 : 5 CT will produce 5 A of secondary current when 300 A flows through the primary. As the primary current changes, the secondary current will vary accordingly. With 150 A through the 300 A rated primary, the secondary current will be 2.5 A. When the rated primary current is exceeded, which is usually the case when a fault occurs in the system, the amount of secondary current will increase. However, depending on the magnetic saturation in the CT, the output may not be exactly proportional.

Loads on instrument transformers are usually referred to as burdens. Ideally, the CT current-sensing device connected to the secondary-side should have a miniscule impedance such that the total current flows through to the sensing device. In practice, however, the secondary current is divided into two paths – a low impedance path for the sensing device and a high impedance path for the CT shunt excitation impedance. A CT can have a ratio error which can either be calculated or determined by tests. The error can be quite high if the impedance burden is large. The error, however, can be kept at an acceptable level with the proper selection of CT. The CT performance analysis is discussed in [3].

6.2.2 Potential Transformer (PT)

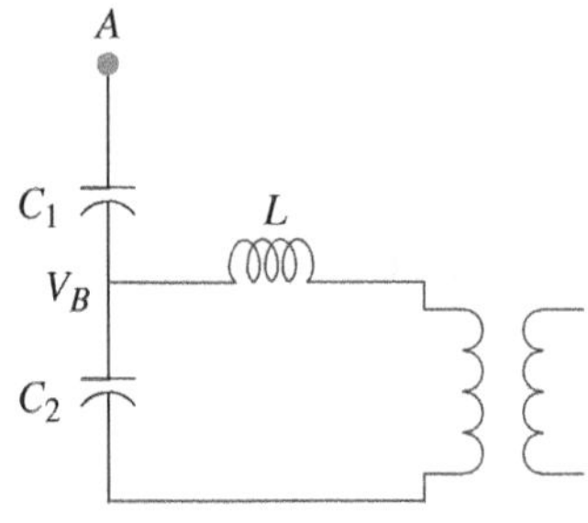

Figure 6.4 High-voltage PT connection.

A potential transformer (PT) is also referred to as a voltage transformer (VT) in the electric power industry. Two types of PTs are commonly used. For low-voltage applications, standard transformers are used, which typically use 115 V (L-L) for the secondary side (in the Australian power industry, the standard secondary rating is 110 V, L-L). For high-voltage applications, a PT is used in a potential divider circuit as shown in Figure 6.4. In this, the values of C_1 and C_2 are so adjusted that a voltage of few kV is obtained across C_2 when terminal A is at the system potential. This voltage is then further reduced by the PT.

Consider the circuit of Figure 6.4. The open circuit voltage across C_2 is given by

$$V_B = \frac{V_A(1/j\omega C_2)}{1/j\omega C_1 + 1/j\omega C_2} = V_A \frac{C_1}{C_1 + C_2} \tag{6.1}$$

The short circuit current is

$$I_{sc} = \frac{V_A}{1/j\omega C_1} = j\omega C_1 V_A \tag{6.2}$$

Therefore, the Thevenin impedance calculated from (6.1) and (6.2) is

$$Z_{TH} = \frac{V_B}{I_{sc}} = \frac{1}{j\omega (C_1 + C_2)} \tag{6.3}$$

Let the leakage impedance of the PT be L. Let the values of C_1 and C_2 be chosen such that

$$-\frac{1}{j\omega (C_1 + C_2)} = j\omega L \Rightarrow L = \frac{1}{\omega^2 (C_1 + C_2)} \tag{6.4}$$

This selection creates a series resonance circuit such that the output of the PT is in phase with the line voltage, and there is no phase angle error between the input voltage and the output voltage.

6.3 Protective Relays

There are a large number of protective relays that are available. Their use is usually application dependent. In this section, different types of relays and their characteristics will be discussed.

6.3.1 Overcurrent Relay

This relay responds when the magnitude of its input current exceeds a certain threshold to trip the faulted circuit. The threshold is adjustable. To discuss the relay, we assume that the fault current is referred to the secondary side of the CT and it is denoted by I_f. The current threshold level at which a fault is detected is called the pickup current I_p. The relay characteristic is given by

$$\begin{aligned} &\text{If } |I_f| > |I_p| \text{ then trip} \\ &\text{elseif } |I_f| < |I_p| \text{ then block} \end{aligned} \tag{6.5}$$

here trip indicates that the relay issues a command to a circuit breaker to trip, while block means that the relay does not take any action, i.e., do not send a trip command to a circuit breaker.

Traditionally, overcurrent relays are constructed using the magnetic induction principle. However, these days they are constructed using microprocessors. The typical schematic diagram of a traditional overcurrent relay is shown in Figure 6.5. This contains a magnetic core and an aluminum disk. An AC signal, when applied to the coil, will produce a magnetic field that is perpendicular to the conducting aluminum disk. These two will then interact to produce a torque on the disk. The rotation of the disk is restricted by a spiral spring by pulling the movable contact a apart from the fixed contact b. When the input current exceeds the pickup current, the spring is released and the movable contact must travel over an angle θ to meet the fixed contact b. The greater the coil current, the greater is the torque resulting in rapid disk rotation. The operating time of the relay is inversely proportional to the level of the fault current. Therefore, this relay is also referred to as an inverse definite minimum time (IDMT) relay.

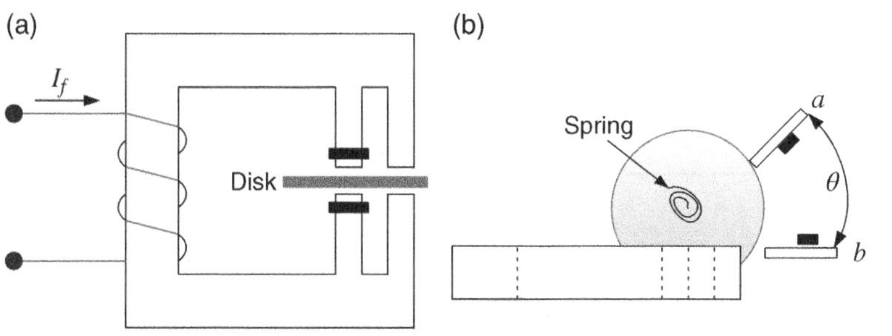

Figure 6.5 (a) Side view and (b) top view of an induction-type overcurrent relay.

The operating characteristics of the relay are defined in terms of the following two quantities:

- Current tap setting (CTS): The pickup current in A.
- Time dial setting (TDS): The adjustable time delay amount in s.

The pickup time of an overcurrent relay can be defined as [4]

$$t_p = \left(\frac{A}{M^p - 1} + B\right) \times \text{TDS} \tag{6.6}$$

where t_p is the pickup or operating time; M is the ratio of $|I_f|/|I_p|$; and A, B, and p are constants.

The constants vary slightly for induction relays but are constant for microprocessor-based relays.

The values of the constants A, B, and p vary with the nature of the relay curve. Usually, three types of curves are available – moderate inverse, very inverse, and extreme inverse. The values for different inverse types are listed in Table 6.1. The overcurrent relay characteristics are shown in Figure 6.6 for moderate inverse. Along the x-axis, the ratio $M = |I_f|/|I_p|$ is plotted, while the operating time t_p is plotted along the y-axis. The curves shown are for various values of TDS. These curves allow different time delay adjustments relay coordination. For example, let us assume that the fault current I_f is 2 A. Then if we choose CTS (I_p) of 1 A, then M will be 2. Then for a TDS of 0.5, the operating time will be 1.9 s, while for a TDS of 2, it will be 7.6 s. To reduce the operating time, we can increase the value of M by reducing the pickup current.

6.3.2 Directional Relay

When the system becomes complex or non-radial, time delay coordination of overcurrent relays may become too complicated and impossible to achieve. Consider

Table 6.1 Parameters of inverse-time overcurrent relays.

Characteristic	*A*	*B*	*p*
Moderate inverse	0.0515	0.1140	0.02
Very inverse	19.61	0.491	2.0
Extreme inverse	28.2	0.1217	2.0

Source: EEE Standard, Inverse-time characteristic equation for overcurrent relays, *IEEE Standard*, C37, 112, 1996 [4].

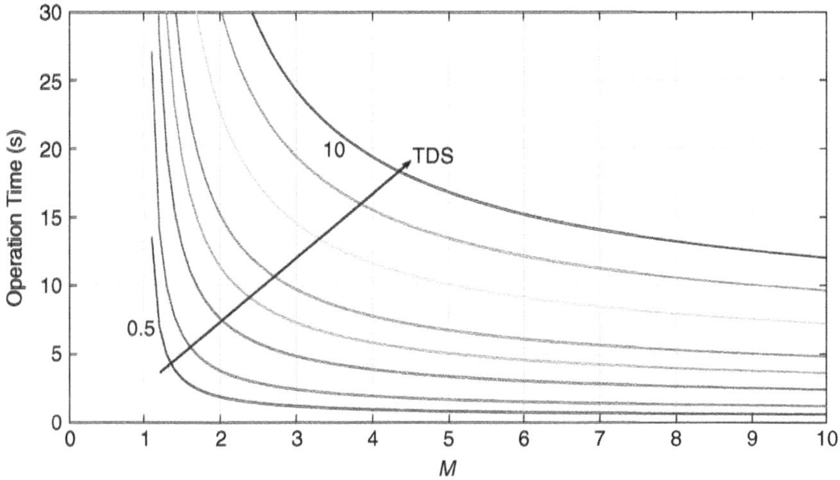

Figure 6.6 Inverse time overcurrent relay characteristics.

the mesh system shown in Figure 6.7 (a). Assume that all the breakers are operated through overcurrent relays. Let us assume that there is a fault at point X. We would like the circuit breakers B_{13} and B_{31} to clear the fault such that no bus gets affected. The overcurrent relays associated breakers B_{13}, B_{31}, B_{23}, and B_{32} will sense the fault current. However, the problem lies in the coordination between breakers B_{31} and B_{32}. If breaker B_{31} is faster than B_{32}, then breakers B_{13} and B_{31} will isolate the fault and only line segment 1-3 will be isolated, as desired. However, if B_{32} is faster than B_{31}, it will trip first. Breaker B_{31} may or may not trip under this condition. Assuming that the breaker B_{31} has tripped, the power supply to bus-3 will be lost. To avoid this problem, let us choose B_{32} to operate slower than B_{31}. However, unfortunately

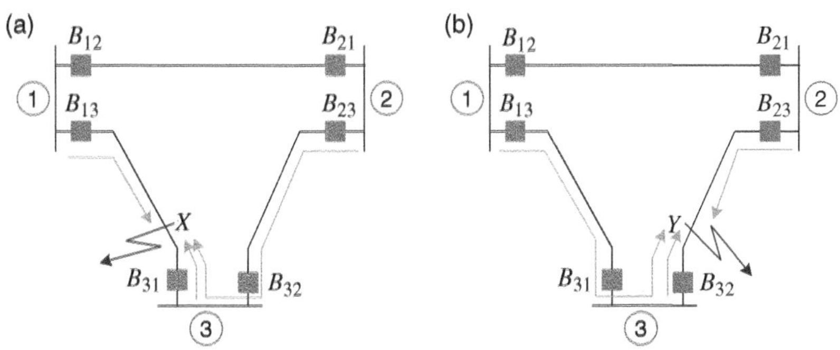

Figure 6.7 A typical mesh system with (a) fault in point X and (b) fault in point Y.

this will cause B_{31} to operate before B_{32} for a fault at point Y, as shown in Figure 6.7 (b). It is therefore impossible to coordinate overcurrent relays in such a system without losing a bus.

The directional relay uses the principle that a transmission line is mostly reactive. Thus, a fault at point X (Figure 6.7 a) will result in a fault current flowing from bus-3 to bus-1 that will lag the voltage at bus-3 by nearly 90°. On the other hand, the current will lead the bus-3 voltage by almost 90° for a fault at point Y (Figure 6.7 b). In this way, the breakers of the circuit of Figure 6.7 can be coordinated using the directional feature. The tripping and blocking zones of a directional relay (also called a reactive relay) are shown in Figure 6.8.

To illustrate the functioning of the directional relay, consider the radial system shown in Figure 6.9 in which the directions of the voltage V_T and current I are as shown. The directional relay is a 2-input relay as shown in the figure. Let us now define the voltage V_T to the reference voltage, that is, $V_T = |V_T| \angle 0^\circ$. Then noting that the line is mostly inductive, the current I is given by

$$I = \frac{|V|\angle 0^\circ}{R + jX} = \frac{|V|}{\sqrt{R^2 + X^2}} \angle - \tan^{-1}(X/R) = |I| \angle - \varphi \tag{6.7}$$

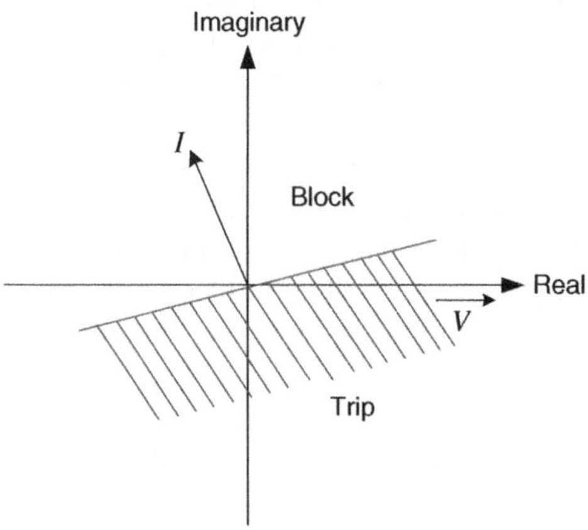

Figure 6.8 Trip and block regions of directional relays.

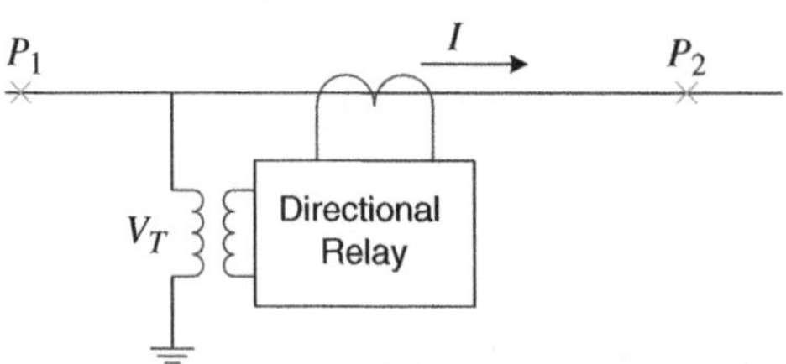

Figure 6.9 Schematic diagram of a directional relay.

where R and X are line resistance and inductance, respectively. It can be seen from the above equation that $\phi = \tan^{-1}(X/R)$. When $R = 0$, this angle will be 90°. For a finite X/R ratio, this angle will be around 90° provided that the direction of the current is as shown in Figure 6.9. Therefore, for a fault in point P_2, the current will lag the voltage. However, for a fault at point P_1, the direction of the current will be reversed, and hence, the current will lead the voltage by around 90°. Thus, from (6.7) and Figure 6.8, we can conclude that this relay will block a fault at point P_1 (upstream from the relay), while it will trip for a fault at point P_2.

6.3.3 Distance Protection

Suppose both CT and PT are located on the bus under protection, similar to that shown in Figure 6.9. Also, let us define the voltage across PT as V_T and the current through CT as I. Then the impedance seen at this point is given by

$$Z = \frac{V_T}{I}$$

Noting that during a fault, the voltage dips and the current increases, we can write the following relay operating relationship:

$$\begin{aligned} &\text{If } |Z| < |Z_r| \text{ then trip} \\ &\text{elseif } |Z| > |Z_r| \text{ then block} \end{aligned} \tag{6.8}$$

This implies that if Z becomes less than Z_r, then send a trip signal. Otherwise, keep the relay blocked. Noting that $Z = R + jX$, we find that the relay operating region is a circle as shown in Figure 6.10. Under normal operating conditions Z will have a very large magnitude and an arbitrary phase angle and will be outside the circle $|Z_r|$. Once the fault occurs resulting in an increase of current and decrease of voltage, the impedance will move inside the circle $|Z_r|$ and the relay will trip. However, unfortunately this relay does not have a direction component associated with it.

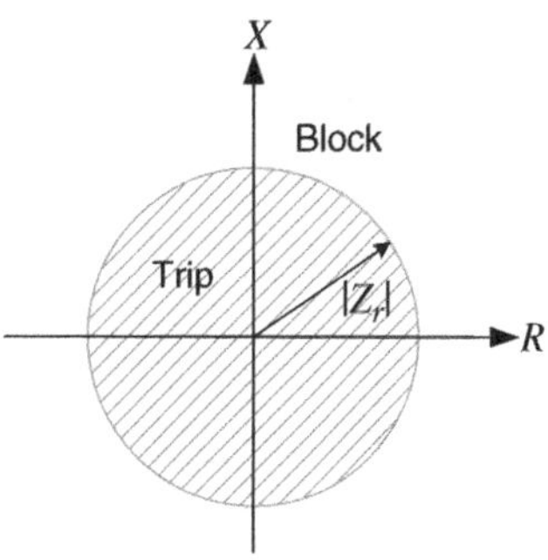

Figure 6.10 Schematic diagram of an impedance relay.

A simple modification to the impedance relay characteristics is often very useful as this enables us to add a directional component to the relay. The circle $|Z_r|$, without being centered at the origin, can be offset by an amount Z' producing the characteristics of an offset impedance relay as shown in Figure 6.11. This relay trip and block relations can then be given as

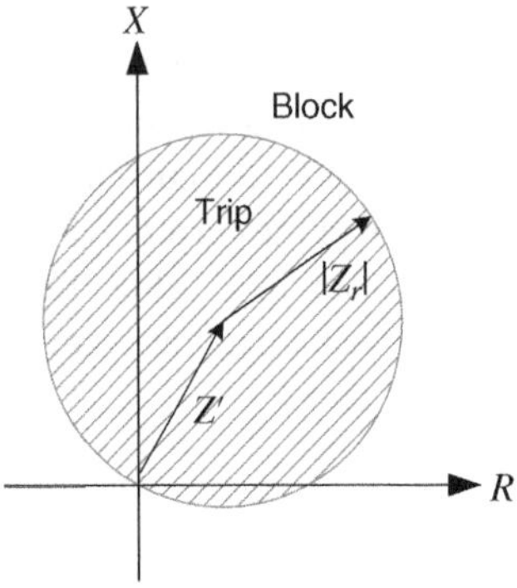

Figure 6.11 Characteristics of a modified impedance or mho relay.

$$\begin{aligned} &\text{If } |Z - Z'| < |Z_r| \text{ then trip} \\ &\text{elseif } |Z - Z'| > |Z_r| \text{ then block} \end{aligned} \tag{6.9}$$

The factor Z' introduces a direction component to this relay. For example, if the zonal protection chosen is such that an impedance lies in the third quadrant, the relay will block the fault. This was not possible by a pure impedance relay. We shall discuss the use of this relay in Section 6.5 for zonal protection.

6.3.4 Differential Protection

The operation of a differential relay is based on vector differences of two or more similar quantities. The most common application is a differential relay that considers the current entering into and leaving from a protected element as shown in Figure 6.12. If the difference exceeds the pickup value, the relay trips. Consider the differential protection scheme shown in Figure 6.12. Under normal operating conditions, the difference in current will be zero, that is,

$$I_1 - I_2 = 0 \tag{6.10}$$

However, for a fault inside the protected region (see Figure 6.12), the current difference will be

$$I_1 - I_2 = I_f \tag{6.11}$$

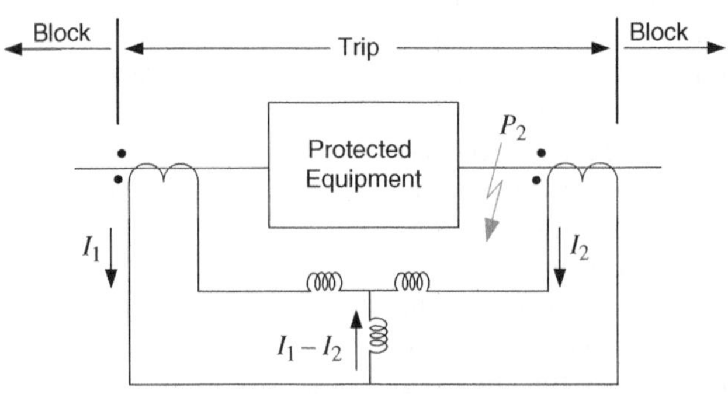

Figure 6.12 Schematic diagram of differential protection.

Note that (6.10) and (6.11) may not be exactly true in practice due to errors in measurements. We can therefore choose a small relay pickup current I_p such that the relay operating characteristic is given by

$$\begin{aligned} &\text{If } |I_1 - I_2| < |I_p| \text{ then block} \\ &\text{elseif } |I_1 - I_2| > |I_p| \text{ then trip} \end{aligned} \tag{6.12}$$

Often difficulties are encountered due to CT errors which increase with the increase in currents. To alleviate this problem, *percentage differential relays* are used. In such relays, a fixed pickup current is not chosen. Instead, the pickup current is chosen as an average value of I_1 and I_2. The operating characteristic of this relay is

$$\begin{aligned} &\text{If } |I_1 - I_2| < K\left|\frac{1}{2}(I_1 - I_2)\right| \text{ then block} \\ &\text{elseif } |I_1 - I_2| > K\left|\frac{1}{2}(I_1 - I_2)\right| \text{ then trip} \end{aligned} \tag{6.13}$$

where K is a positive constant.

6.3.5 Transformer Protection

Consider the single-phase transformer that is protected by a differential relay as shown in Figure 6.13. Let us denote the turns ratio of CT_1 as $1 : n_1$ and that of CT_2 as $1 : n_2$. We then have

$$I_1' = \frac{I_1}{n_1} \text{ and } I_2' = \frac{I_2}{n_2} \tag{6.14}$$

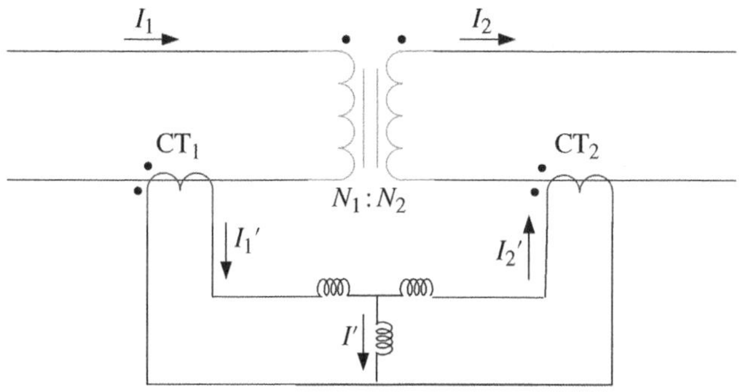

Figure 6.13 Differential protection of a single-phase transformer.

Using Kirchoff's current law (KCL), the current through the relay is given by

$$I' = I'_1 - I'_2 = \frac{I_1}{n_1} - \frac{I_2}{n_2} \tag{6.15}$$

Assuming that the transformer is ideal and there is no internal fault, its primary and secondary currents are related by

$$I_1N_1 = I_2N_2 \Rightarrow I_2 = \frac{I_1N_1}{N_2} \tag{6.16}$$

Combining (6.15) and (6.16), we get

$$I' = \frac{I_1}{n_1}\left(1 - \frac{N_1/N_2}{n_2/n_1}\right) \tag{6.17}$$

To prevent, the differential relay from tripping for external fault, we must choose the relay current I' to be zero, that is,

$$\frac{N_1/N_2}{n_2/n_1} = 1 \Rightarrow \frac{n_2}{n_1} = \frac{N_1}{N_2} \tag{6.18}$$

If there is a fault internal to the transformer, the relay current I' is nonzero and the relay will trip.

Example 6.1 Consider a 10 MVA, 80 kV/20 kV single-phase transformer that will be differentially protected. We must choose a CT ratio assuming that it is to be protected by a differential relay.

The transformers primary rated current is

$$I_{1rated} = \frac{10 \times 10^6}{80 \times 10^3} = 125 \text{ A}$$

We therefore choose a 150 : 5 A CT_1 that will give the rated current for unfaulted condition as

$$I'_{1rated} = 125 \times \frac{5}{150} = 4.17 \text{ A}$$

In a similar way, the secondary rated current I_{2rated} is

$$I_{2rated} = \frac{10 \times 10^6}{20 \times 10^3} = 500 \text{ A}$$

We therefore choose a 600 : 5 A CT_2 that will give the rated current for unfaulted condition as

$$I'_{2rated} = 500 \times \frac{5}{600} = 4.17\ \text{A}$$

We can therefore see that I' will be zero for unfaulted conditions.

One of the problems of using differential protection for transformers is a mismatch of the relay currents may occur when standard CT ratios are used. For example, if the voltage of the primary side in Example 6.1 was 90 kV instead of 80 kV, the rated current would have been $I_{1rated} = 111.11$ A. Choosing a 150 : 5 A CT would have given $I'_{1rated} = 3.7$ A. There will be a mismatch of about 12.7% if the secondary voltage is 20 KV and a 600 : 5 A CT is chosen. One way of avoiding this problem is to use auxiliary CTs which provide a wide-ranging turns ratio. However, unfortunately the auxiliary CTs add their own burden to the main CTs and also increase transformation errors. A better solution to the problem is to use a tap setting on the relays themselves.

It is to be noted that in modern numeric differential relays, the transformer configuration and CT ratios can be corrected within the relays. Additionally, it should be noted that oil-filled power transformers are also protected by the Buchholz relay (also known as gas or sudden pressure relay). This relay is used as a protective device sensitive to the effects of dielectric failure inside the equipment. This safety device is mounted on oil-filled power transformers, and equipped with an external overhead oil reservoir called a conservator. Due to a slight overload, gas can accumulate slowly due to the decomposition of the insulating oil. The gas accumulates at the top of the relay and forces the oil level down. A float-operated switch in the relay is used to initiate an alarm signal. This same switch will also operate at low oil levels, such as a slow oil leak. This relay was first developed by Max Buchholz in 1921.

A three-phase transformer protection scheme is shown in Figure 6.14. It is assumed that the transformer is Y-Δ connected. Then the CTs in the Y-connected side are connected in Δ, while those in the Δ-connected side are connected in Y.

6.3.6 Pilot Relays

A pilot relay is a form of differential relay that compares the quantities at its terminals using some form of communication channel. In general, generators or transformers do not need pilot protection. However, differential protection of

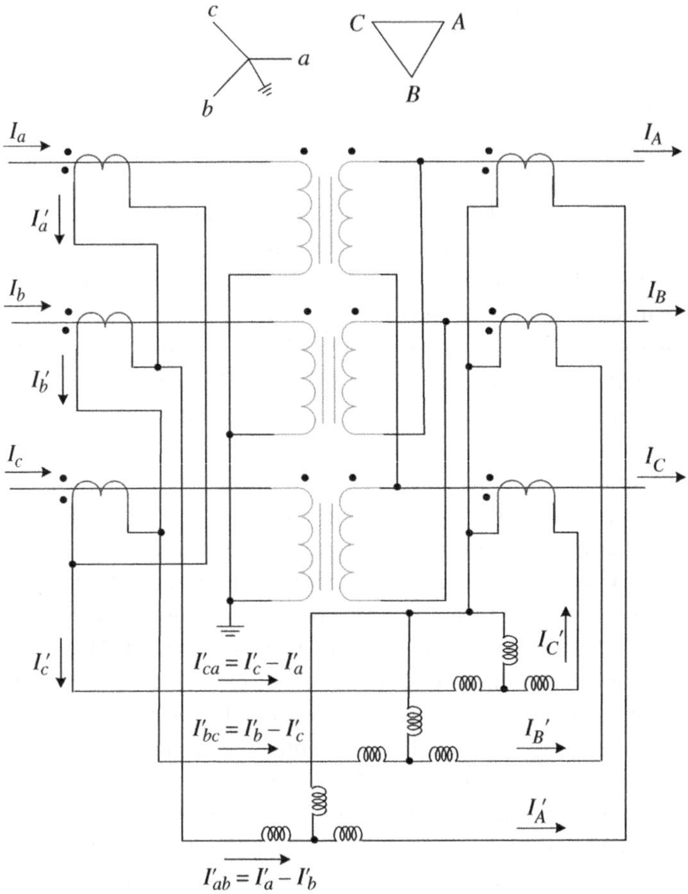

Figure 6.14 Differential protection of a three-phase Y-Δ connected transformer.

transmission lines indeed requires pilot wires as the two ends are very far apart. The following differential protections are available for transmission lines:

- *Pilot Wires*: These are separate electrical circuits operating at DC, 50 or 60 Hz, or audio frequencies.
- *Power-Line Carrier*: The transmission line itself is used as a communication circuit. In this, the signals can be transmitted at a frequency between 30 and 300 Hz.
- *Microwave*: A 2–12 GHz signal is transmitted by the line-of-sight paths between the terminals using dish antennas.
- *Fiber Optic Cables*: A point-to-point communication can be made by using very high-speed fiber optic cables. This is the safest and most reliable form of pilot communication.

6.4 Overcurrent Relay Coordination

The relays discussed in the previous section are responsible for removing a fault as soon as possible. The relay that is mainly responsible for clearing a particular fault is called the *primary protection*. However, it is possible that one of the subsystems of the primary protection may fail in times of contingency. In this case, *backup protection* is provided to save the system from catastrophic failure. It is to be noted that the backup protection is not necessarily a duplicate of the primary protection. A duplication of the primary not only adds to the cost but it may also not always be helpful because some of the components such as CTs, PTs, or circuit breakers are common with the primary. Thus, the duplicated primary may also fail when the primary fails. Therefore, it is desirable that the backup protection is remote from the primary protection such that the failure of the components of the primary protection does not affect it. In this section, we shall discuss the coordination of overcurrent relays in a radial system to provide primary and backup protection. We shall illustrate this with the help of the following example.

Example 6.2 Consider the radial system shown in Figure 6.15. The system contains five buses that are supplied from a substation through a 110/22 kV transformer. The system is protected by four circuit breakers B1 to B4 and the corresponding relays are denoted as R1 to R4, respectively. The system frequency is 50 Hz, and all the quantities are referred to the low-tension (22 kV) side of the transformer. The parameters are:

$$\text{Transformer}: Z = j0.5\,\Omega \text{ and } Z_n = j0.025\,\Omega$$

$$\text{Feeders}: Z_{12} = Z_{34} = j2.42\,\Omega, Z_{23} = j1.21\,\Omega, Z_{45} = j2.9\,\Omega$$

It is assumed that the feeders have no mutual couplings. Then the positive and zero-sequence impedances from the substation to bus-5 are

$$Z_1 = Z_2 = j0.5 + 2 \times j2.42 + j1.21 + j2.99 = j7.51\,\Omega$$

$$Z_0 = Z_1 + 3 \times j0.05 = j7.585\,\Omega$$

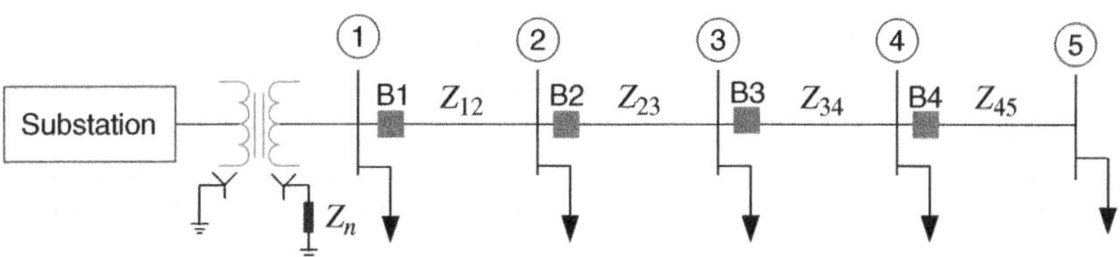

Figure 6.15 Radial system for coordination of overcurrent relays of Example 6.2.

For a symmetrical fault at bus-5, the fault current is

$$I_{f3-ph} = \frac{22 \times 10^3}{\sqrt{3} \times Z_1} = -j1{,}691.3\ \text{A}$$

For a LL fault at bus-5, the fault current is

$$I_{fabcLL} = P^{-1} \begin{bmatrix} 0 \\ I_{f3-ph} \\ -I_{f3-ph} \end{bmatrix} = \begin{bmatrix} 0 \\ -1{,}464.7 \\ 1{,}464.7 \end{bmatrix} \text{A}$$

We can also find that the magnitude of the fault current for 1LG and 2LG faults are 1,685.7 A and 1,688.5 A, respectively. Therefore, the maximum fault current occurs for a three-phase fault and the minimum fault current is obtained for an LL fault. The maximum and minimum fault currents for faults at different buses are listed in Table 6.2. For the relay coordination, the extreme inverse curve parameters of Table 6.1 will be used in a digital protection scheme.

Now consider the system of Figure 6.15. If for a fault at bus-5, all the circuit breakers trip, then the power supplied to all the buses is lost. Ideally, only breaker B4 should trip. Similarly, for a fault at bus-4, only breaker B3 will trip such that the power supplied to the upstream buses is not interrupted. For the coordination of overcurrent relays, first, the relay that is furthest from the substation bus is considered, that is, relay R4. This will protect the system from a fault in bus-5. In the event of relay R4 or the corresponding circuit breaker B4 failing to clear the fault, the backup is provided by R3 and B3. In a similar fashion, all the other relays are coordinated.

Setting for Relay R4: This relay must operate for the minimum fault current at bus-5, that is, 1,464.7 A. However, from the reliability point of view, the relay pickup current is set as one-third of this value, that is,

$$I'_P = \frac{1{,}464.7}{3} = 488.23\ \text{A}$$

Let us choose a CT ratio of 500 : 5. Then, the pickup current at the secondary of the CT is

$$I_P = I'_P \times \frac{5}{500} = 4.88\ \text{A}$$

Table 6.2 Minimum and maximum fault currents for faults at different buses.

Fault at bus-1	1	2	3	4	5
Maximum fault current (A)	25,403	4,379.9	3,090.4	2,755.3	1,691.3
Minimum fault current (A)	22,000	3,793.1	2,676.4	2,386.1	1,464.7

For simplicity, the pickup current (I_P) is chosen as 5 A. Also, the TDS is chosen as 0.2. It is to be noted that the TDS is restricted to a minimum of ½ for electromechanical relays. However, for microprocessor-based relays, low values can be chosen.

Setting for Relay R3: This relay must provide backup for R4, and hence, it must pick up the minimum current seen by relay R4. We therefore choose the same CT ratio of 500 : 5 and the pickup current of 5 A. To determine the TDS, we must provide a minimum *discrimination time* of 0.3 s. This time is provided such that R3 operates 0.3 s after the highest (not lowest) fault current seen by R4. Therefore, R3 operates in no less than 0.3 s after every possible fault seen by R4.

It is to be noted that the highest current seen by relay R4 is not 1,691.3 A as given in Table 6.2. For a fault immediately after B4, R4 will see a fault current that is equal to the fault current seen by bus-4. Therefore, the highest fault current seen by R4 is 2,755.3 A (Table 6.2). The current seen by both secondary CTS of relay R3 and R4 for this fault will be

$$I_f = 2{,}755.3 \times \frac{5}{500} = 27.553\ \text{A}$$

The ratio M given in (6.6) will then be

$$M = \frac{|I_f|}{|I_P|} = \frac{27.53}{5} = 5.51$$

The TDS for relay R4 is chosen as 0.2. Therefore, for a fault immediately after bus-4, the tripping time of R4 is calculated using (6.6) as

$$t_p = \left(\frac{28.2}{5.51^2 - 1} + 0.1217\right) \times 0.2 = 1.082 \times 0.2 = 0.2164\ \text{s}$$

As has been mentioned earlier, relay R3 must provide a discrimination time of 0.3 s for a failure of R4. Therefore, the operating time of the relay is 0.2164 + 0.3 = 0.5164 s. Since R3 also has an M of 5.51, the TDS is calculated from (6.6) as

$$TDS = \frac{t_p}{\dfrac{A}{M^p - 1} + B} = \frac{0.5164}{1.082} = 0.4773 \approx 0.5$$

The operating time is then $1.082 \times 0.5 = 0.541$ s, maintaining a discrimination time of 0.3 s.

Setting for Relay R2: This relay must provide a backup for relay R3. The smallest fault current seen by R2 to provide a backup for R3 is 2,386.1 A (Table 6.2). For a reliable operation, we choose one-third of this current, that is, 795.37 A. Choosing a CT ratio of 1,000 : 5, the pickup current is then

$$I_p = 795.37 \times \frac{5}{1{,}000} = 3.977\ \text{A}$$

Let us choose a CTS (i.e., pickup current) of 4.0 A. We now have to determine the TDS of R2.

The maximum current seen by R3 is 3,090.4 A. Then, at R3, for a CT ratio of 500 : 5 and a CTS of 5 A, we get an M of

$$M = 3{,}090.4 \times \frac{5}{500} \times \frac{1}{5} = 6.181$$

For the above values of M and a TDS of 0.5, the operating time of relay R3 is 0.4399 s. Thus, relay R2 should add a discrimination time of 0.3 s, that is, the operating time should be 0.7399 s.

Now relay R2 is a backup for relay R3 and therefore it will see the same fault current of 3,090.4 A. Then M for this fault is

$$M = 3{,}090.4 \times \frac{5}{1{,}000} \times \frac{1}{4} = 3.863$$

For this value of M, we get a TDS of $0.3446 \approx 0.35$. This gives an operating time of 0.7515 s.

Setting for Relay R1: This relay must provide a backup for relay R2. Thus, we choose the same CT ratio and CTS for this relay as well. The maximum fault current seen by R2 is 4,379.9 A. Then, at R2, for a CT ratio of 1,000 : 5 and a CTS of 4 A, we get an M of

$$M = 4{,}379.9 \times \frac{5}{1{,}000} \times \frac{1}{4} = 5.475$$

For the above M and a TDS of 0.35, the operating time of relay R2 is 0.3832 s. Thus, relay R1 should add a discrimination time of 0.3 s, that is, the operating time should be 0.6832 s.

Now relay R1 is a backup for relay R2 and therefore it will see the same fault current of 4,379.9 A. Then, for the same M, the TDS is $0.624 \approx 0.65$. This will give an operating time of 0.7117 s. The calculated relay settings are given in Table 6.3.

The coordination of overcurrent relays has been explained in [5], where the relays considered are of electrotechnical type. The characteristic curves of these

Table 6.3 Relay settings for the circuit of Example 6.2.

	R1	R2	R3	R4
CT ratio	1,000 : 5	1,000 : 5	500 : 5	500 : 5
CTS, A	4	4	5	5
TDS	0.65	0.35	0.5	0.2

relays are similar to those shown in Figure 6.6. However, an arbitrary choice of TDS is not feasible with these types of relays. Even though the flexibility of the choice is possible with microprocessor-based relays, the design principles are the same with both these types of relays. For example, consider any bus-k. For the primary protection of the relay connected to this bus, it must guarantee protection from the minimum current seen for any downstream fault, that is, faults at buses $(k+1)$, $(k+2)$, etc. However, the relay connected to bus-$(k-1)$, which provides secondary protection to the relay connected to bus-k, must provide a discrimination time (chosen as 0.3 s in the example) for the maximum possible current seen by the relay at bus-k. This simple principle has been used in Example 6.2.

6.5 Zones of Protection

The philosophy of protective zones is to assign responsibility to all the protective devices in a power system. The protective zones overlap with each other. All the circuit breakers are located in overlapping regions such that for a fault anywhere in the zone, all the circuit breakers open to isolate the fault. Consider the system shown in Figure 6.16. It has nine breakers and nine zones. It can be seen that all the zones are overlapping. Also, no part of the system is left unprotected.

For example, if a fault occurs at point P1, which is only in zone-3, the circuit breakers B3 and B4 will open to isolate the fault. However, if a fault occurs at point P2, then it falls under the overlapping zones 7 and 8. Thus the circuit breakers B8, B9, and B10 must open. Therefore, there is no possibility of any of the sources supplying the fault current. Again, if there is a fault at point P3, both B5 and B6 must

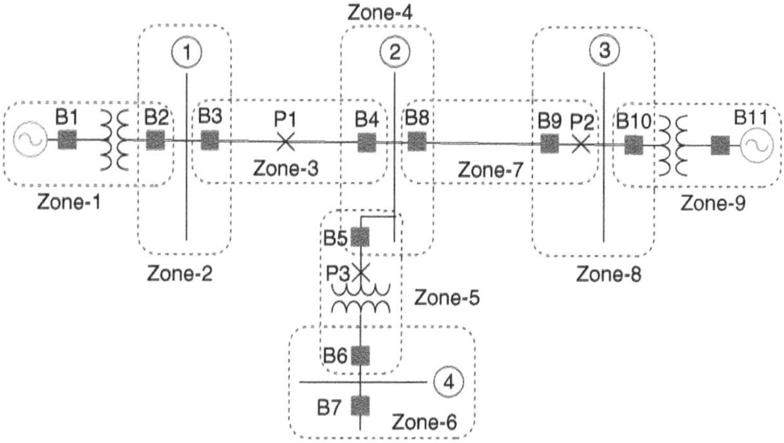

Figure 6.16 Typical zones of protection of a four-bus power system.

open. This is irrespective of the fact that bus-4 may not have a source or active load and there is no possibility of the fault current being fed through bus-4.

Let us now consider the system shown in Figure 6.17. We have seen in Section 6.3.2 that it is rather difficult to coordinate an overcurrent relay to protect such a system. Distance relays will now be used to protect this system. To discuss this, let us denote the bus voltages as V_i, where i is the bus number. A current will be denoted by I_{ik} when it flows from the ith to the kth bus. Consider now the relay that opens the breaker B12. The impedance is then given by $Z = V_1/I_{12}$. During normal operation, the current is small, and the relay will be blocked as has been discussed with respect to Figures. 6.10 and 6.11. However, for a fault at point P1, the current I_{12} will become large, while the voltage V_1 will become small and hence the relay will trip. Now consider a fault at point P3. The impedance Z appears to the relay as negative from the relay to the fault. If the impedance is less than the pickup value, the relay will send a trip signal to B12. Thus, the impedance relay does not have a directional feature, and it can trip for a fault to the left or to the right.

We shall now use a modified impedance relay or mho relay to protect the system. The *reach* of a mho relay specifies how far down the line the mho relay protects a fault. For example, a 70% reach means that the relay will detect any three-phase

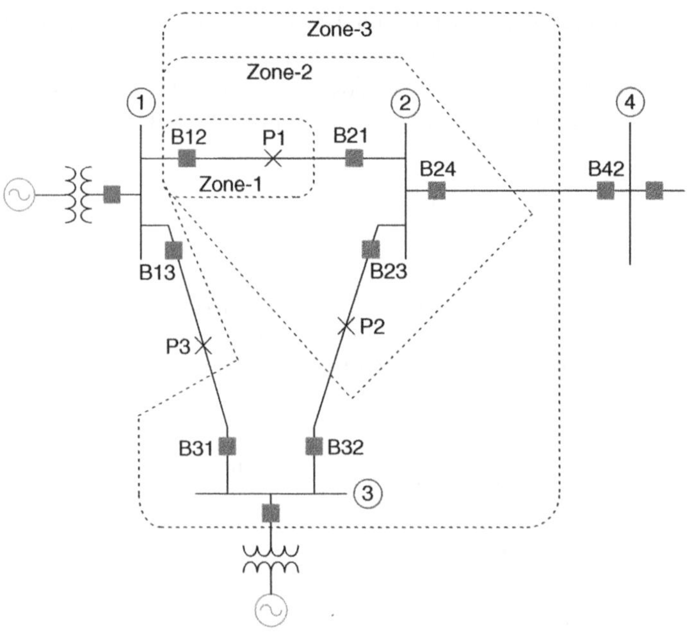

Figure 6.17 Protection of a loop system using distance protection.

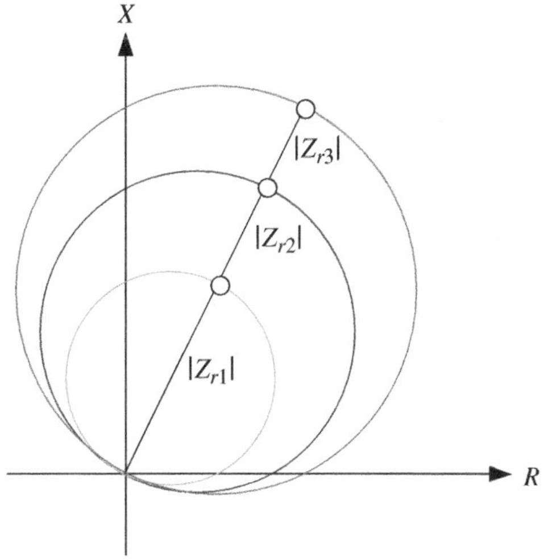

Figure 6.18 Mho relay characteristics for the three zones.

fault between the relay location and 70% of the line length. This is why such a relay is also called a *distance relay*.

It is a common practice to use three mho relays per phase with increasing reaches and longer time delays, as shown in Figure 6.18. Consider, for example, Figure 6.18, where the three-zone protection scheme is also depicted for B12. Zone-1 is set for a reach of 80% and for instantaneous operation in order to provide primary protection for relay B12. Zone-2 relay is set for 120% reach and obviously, this extends beyond bus-2. The typical time delay is set for 0.2 or 0.3 s. Zone-2 relay provides backup protection for faults in line 1-2 as well as remote backup for line 2-3 or 2-4 in zone-2.

Zone-3 relay reach is set for 100% of line 1-2, plus 120% of either line 2-3 or line 2-4, whichever is longer. The time delay for this relay can be set as high as 1 s. The mho relay characteristics for the different zones are shown in Figure 6.18. For a fault in line 2-3 at point P2, it is desirable that relay B23 trips. The impedance seen by B12 is almost the same for a fault near bus-2 either in line 1-2 or line 2-3. We have set an 80% reach for the relay B12. Therefore, for a fault at P2, B23 should trip instantaneously. If it fails to trip, then B12 will trip after a time delay. In this way, both primary and backup protection can be provided using mho relays.

Example 6.3 The line data for the system of Figure 6.17 are given in Table 6.4. Let us assume that the CT and PT ratios of B12 are 1,500 : 5 and 3,000 : 1, respectively. The impedance seen by B12 is

$$Z = \frac{V_{1(L-N)}}{I_{12}}$$

Table 6.4 Data for the system of Figure 6.17.

Line	Positive-sequence impedance (Ω)
1-2	$8 + j50$
2-3	$8 + j50$
2-4	$5.3 + j33$
1-3	$4.3 + j27$

Using the CT and PT ratios mentioned above we have

$$Z' = \frac{V_{1(L-N)}/\left(\frac{3{,}000}{1}\right)}{I_{12}/\left(\frac{1{,}500}{5}\right)} = \frac{Z}{10}$$

Now zone-1 of the B12 relay is set for 80% reach, that is, 80% of line 1-2 (secondary) impedance. Therefore,

$$Z_{r1} = 0.80 \times \frac{8 + j50}{10} = 0.64 + j4 = 4.05\angle 80.9^\circ\ \Omega$$

The setting for zone-2 for the B12 relay, with a reach of 120%, is

$$Z_{r2} = 1.2 \times \frac{8 + j50}{10} = 0.96 + j6 = 6.08\angle 80.9^\circ\ \Omega$$

It can be seen from Table 6.4 that line 2-4 have a larger impedance than line 2-3. Therefore, we set B12 for zone-3 as 100% of line 1-2 and 120% of line 2-4, such that

$$Z_{r3} = 1 \times \frac{8 + j50}{10} + 1.2 \times \frac{8 + j50}{10} = 1.55 + j8.96 = 9.07\angle 80.9^\circ\ \Omega$$

Suppose now the voltage at bus-1 is 345 kV and the maximum current for an emergency loading condition. 1,500 A. Then

$$Z' = \frac{Z}{10} = \frac{1}{10} \times \frac{345 \times 10^3/\sqrt{3}}{1{,}500\angle -18.19^\circ} = 13.28\angle 18.19^\circ\ \Omega$$

Because this impedance exceeds the zone-3 trip setting, the impedance during the emergency loading condition is outside the trip settings of any of the zones. Therefore, none of the relays will trip. Moreover, the impedance during a normal loading condition will be even less and hence it will be further away from the trip regions.

6.6 Protection in the Presence of Distributed Renewable Generators

Consider the radial system of Figure 6.15. When a fault occurs in the network, the over-current relays respond to isolate the portion of the network resulting in power interruption to the customers downstream from the fault location. For example, a fault occurs in the line segment joining buses 3 and 4 is cleared by the opening of the breaker B3. This means that the power supplied to buses 4 and 5 will be lost till the fault is cleared. However, when distributed renewable generators (DRGs or DGs) are connected to different buses, they must be allowed to supply power to the unfaulted portion of the network such that the smallest portion of the network is isolated.

The impact of DGs on overcurrent relays has been discussed in [6]. Consider, for example, the two systems that have a DG connected to distribution feeders, as shown in Figure 6.19. In the system shown in Figure 6.19 (a), without the DG, the fault current will flow from the upstream network, that is, $I_f = I_s$. However, with the DG, the fault current now will be $I_f = I_s + I_{dg}$. This will cause a reduction in the source current I_s that has been used for setting the relay for breaker B1, and, as a result, it might not operate to isolate the fault. Now consider the system of Figure 6.19 (b) in which the fault occurs in the feeder where the DG is not connected. However, the DG will supply part of the fault current in response to the fault in the upstream direction, as shown in the figure. This might trip breaker B2 and/or interfere with the operation of B1.

The DG in Figure 6.19 has been assumed to supply enough fault current to upset the relay settings. Indeed, the presence of biodiesel-based synchronous generators placed downstream from the substation can alter or affect the relay settings. However, converter-interfaced renewable generators will have their current limits based on the converter ratings. In general, converters cannot provide fault currents that are much higher than their rated currents, unlike synchronous generators. A voltage source converter (VSC) can work as a voltage source or a current source.

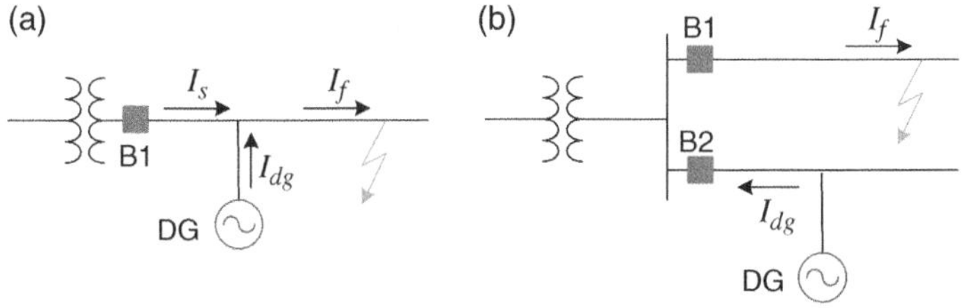

Figure 6.19 DG connected distribution network: (a) single feeder and (b) double feeders.

In the voltage control mode, the VSC produces a set of balanced voltage at the system frequency while supplying real and reactive power to the grid. The VSC will experience an increase in the output current flow when a fault occurs. There are the following two options for the VSC.

- *Option-A*: The VSC can switch off. However, this may have an impact on the loads that it is supplying, and they may have to be curtailed as well.
- *Option-B*: The VSC switches from voltage control to current control mode where it supplies a finite amount of balanced current to the grid. However, this may cause a delay in overcurrent relays to respond. This problem can be avoided by choosing bigger converter sizes than are required for the power level that the renewable generator is supplying. This will increase the installation cost and may result in an increase in converter losses.

6.6.1 Protection Using Directional Overcurrent Relays

An example of protection of a distribution system containing DGs has been presented in [7]. Consider the system shown in Figure 6.20. It contains four buses and three DGs with loads connected to each bus (L-1 to L-4). There are three breakers protecting the buses (B1 to B3), and another three at the outputs of each DG (BD1 to BD3), which are used for isolating the DGs from the grid. Each DG can also have its local load (LL). It has been assumed that the DGs operate in voltage control mode during normal operations but switch over to current control mode once a fault is detected. A communication link coordinates the relays. The link checks for the status of the DGs, that is, if they are online or offline. Accordingly, the protection in the reverse direction will be coordinated. The advantage of this scheme is that breakers B1 to B3 can be reclosers.

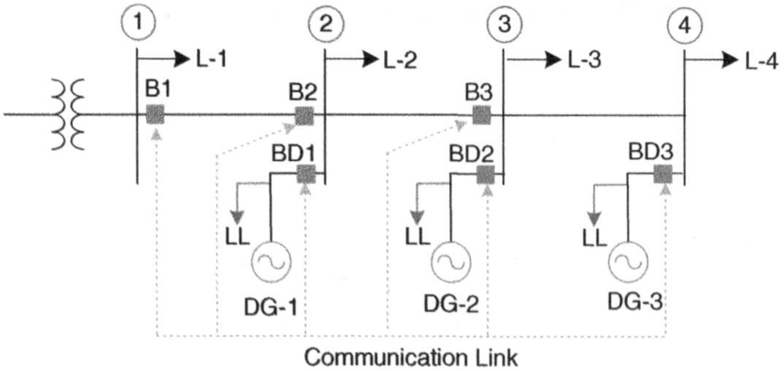

Figure 6.20 Directional overcurrent scheme in DG connected distribution network.

Of the three overcurrent relays R1 to R3 (for B1 to B3, respectively), R2 and R3 are directional. In the forward direction, all three relays are coordinated using the same principle as outlined in Example 6.2. However, in the reverse direction, relays R2 and R3 have definite time characteristics. This implies that these two relays will trip after a fixed predetermined time period once the current exceeds a threshold, irrespective of the peak magnitude of the current above the threshold. Therefore, to coordinate the directional relays, the current ratings of the DGs must be considered.

It has been assumed that the DGs nominally operate in voltage control mode but switch over to current control mode once the rated current is exceeded. Let the current ratings of the DGs be as follows:

$$\text{DG-1} : 25\text{ A},\ \text{DG-2} : 35\text{ A},\ \text{and DG-3} : 20\text{ A}$$

The maximum withstanding currents of the DGs are assumed to be 1.5 times the rated current. Based on these, the calculated relay setting in the reverse direction for relays R2 and R3 are listed in Table 6.5, where DG status 0 indicates that it is disconnected, while 1 indicates that it is connected. As the table shows, the relay R3 sees reverse currents from DGs 2 and 3, while the relay R2 faces reverse currents from all three DGs. Moreover, the DG protection systems BD1 to BD3 should also have a predetermined time setting such that they do not supply the maximum withstanding current for a long duration. However, the duration should be more than what is required for trip breakers B2 and B3.

Assume that a fault has occurred between buses 1 and 2. It is expected that B1 will isolate the fault from the substation side. However, all three DGs can also clear the fault by tripping B2. This will depend on how many DGs are in operation at

Table 6.5 Overcurrent relay setting in the reverse direction.

DG status DG1–DG2–DG3	R2 setting		R3 setting	
	Maximum load current (A)	Pickup current (A)	Maximum load current (A)	Pickup current (A)
000	0	Blocked	0	Blocked
001	20	30	20	30
010	35	52.5	35	52.5
011	65	97.5	65	97.5
100	25	49.2	0	Blocked
101	45	67.5	20	30
110	60	90	35	52.5
111	80	120	65	97.5

that time. For example, if all DGs are in operation, then a total of 120 A current will flow towards the fault as the DGs will supply the maximum current available to them. If, on the other hand, only DG1 and DG2 are operational, then the maximum current available will be 90 A. Therefore, the current threshold for the reverse direction of R2 can be dynamically set using a microprocessor. It is interesting to note that if B1 does not reclose due to a permanent fault, the DGs can supply loads L-2 to L-4, as well as their local loads if they have sufficient capacity. If any of the DGs exceed their capacity, the respective breaker BD will open, and the DG will supply only its local load.

6.6.2 Inverse Time Admittance (ITA) Relay

Inverse time admittance (ITA) relay [8–10] does not need communication channels but relies on the calculation of normalized admittance. Consider the radial feeder shown in Figure 6.21 (a), where the relay is placed at node R, while the total admittance of the feeder is Y_f. Point K is an arbitrary point on the feeder. The admittance between R and M is Y_m. Then the normalized admittance is given by

$$Y_r = \left|\frac{Y_m}{Y_f}\right| \tag{6.19}$$

The impedance of the fault point M vis-à-vis the total length of the line is given by $|Z_m| < |Z_f|$. Therefore, for a fault at point M, the normalized admittance Y_r must be greater than 1.

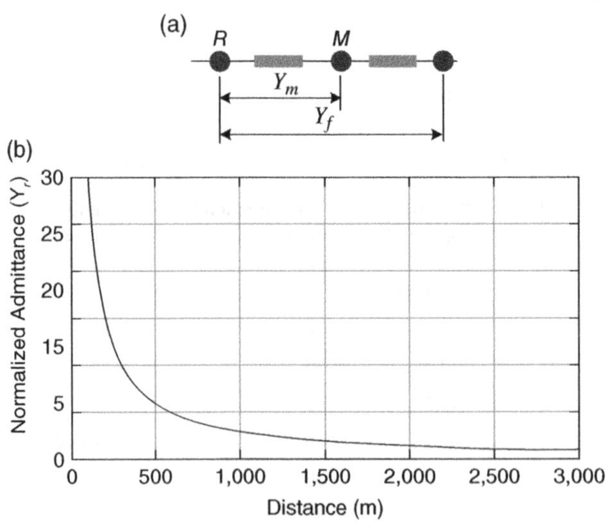

Figure 6.21 (a) A radial distribution feeder and (b) its ITA relay characteristic.

Consider, for example, a 3,000 m long feeder with an impedance of $(0.65 + j4.817) \times 10^{-4}$ Ω per m. The normalized admittance is shown in Figure 6.21 (b). The normalized admittance decreases with distance. The change in the normalized admittance along a feeder is used for inverse time-tripping characteristics of the ITA relays. This is given by

$$t_p = \frac{A}{Y_r^\rho - 1} + k \tag{6.20}$$

where A, ρ, and k are constants. The relay characteristics for $A = 0.0047$, $\rho = 0.08$, and two different values of k are shown in Figure 6.22.

The connection of an ITA relay at a bus is shown in Figure 6.23 (a), while its relay tripping process is depicted in Figure 6.23 (b). The relay needs both measurements of bus voltage and current injected into the feeder. The fundamental components of these two quantities are then extracted. These are then used for admittance calculation. This admittance is then used for the calculation of normalized admittance calculation. Based on the relay reach setting, the trip signal is generated in the event of a fault. Because the relay is based on the fundamental component of current, it can determine whether the fault is in the forward or in the reverse direction, that is, the relay has directional capacity. The operation of the relay is dependent on the calculation of the fundamental components, even in the presence of harmonics, DC offset, and unbalance. A fast Fourier transform (FFT) can be used for fundamental calculation. There are other suitable FFT-based methods that can also be employed for this process.

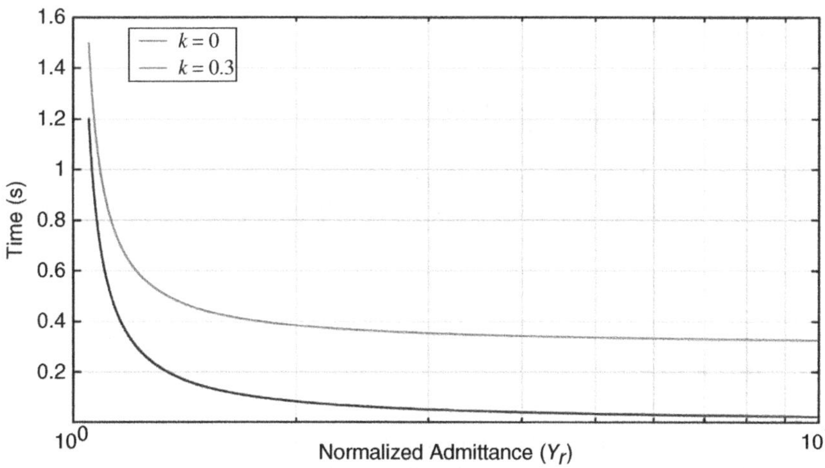

Figure 6.22 ITA relay characteristics for two different values of k.

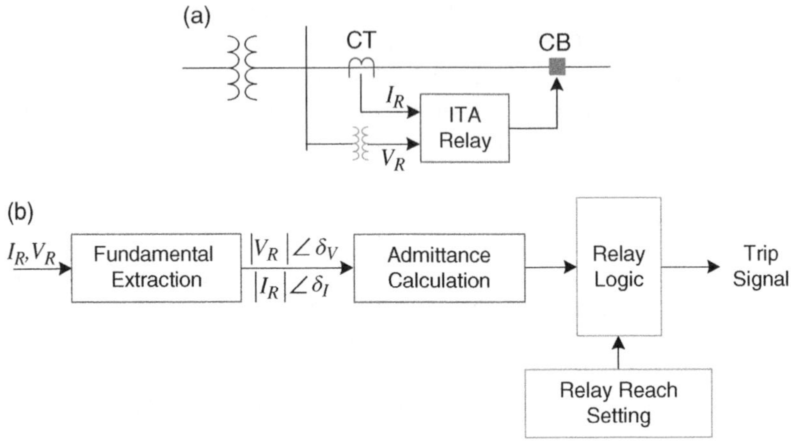

Figure 6.23 ITA relay (a) connection diagram and (b) relay decision-making process.

An ITA relay can be used for zonal protection as well. For example, consider the system of Figure 6.20. To set the protection setting for CB2, the relay R2 will cover 120% of the line 1-2 and 200% of the line. The protection scheme for the two zones is then calculated as follows [10]:

$$\text{Zone}-1: Y_f = \frac{1}{1.2 \times Z_{12}}$$

$$t_p = \frac{0.0037}{Y_r^{0.08} - 1} + 0.05$$

$$\text{Zone}-2: Y_f = \frac{1}{2 \times Z_{12}}$$

$$t_p = \frac{0.0037}{Y_r^{0.1} - 1} + 0.15$$

Note that the values of A, ρ, and k are chosen to maintain the discrimination between the zonal operations. A detailed description of the ITA relay and its operation including hardware results are presented in [10].

6.7 IEC 61850

International Electrotechnical Commission (IEC) 61850 is a standard for *Communication Networks and Systems in Substations* and a substation automation protocol. This has been increasingly deployed in power systems. This protocol is a prime example of interoperability as different intelligent electronic devices (IEDs) from different vendors can be seamlessly integrated together for substation automation. This reduces the dependence on a single vendor. This standard uses a common data model such that integration, maintenance, and expansion of protection systems become much more simplified.

IEC 61850 consists of three levels, as shown in Figure 6.24. These are as follows:

- *Process Level*: This is the lowest level in which measurements are collected from CTs, PTs, other sensors, and circuit breakers (CBs), and trip signals are issued to CBs.
- *Bay Level*: This level contains IEDs that are used for protection and control.
- *Station Level*: This level provides a data gateway to operators through human–machine interface (HMI) that can be used for network management.

There are several advantages of IEC 61850. Some of these are listed below.

- *Interoperability*: It uses a standard protocol that allows protection and control devices from different manufacturers to communicate with each other. This removes the dependency on a single manufacturer.
- *Standardization*: A common data model is used where unique naming conventions make the maintenance and expansion of the protection system simple.

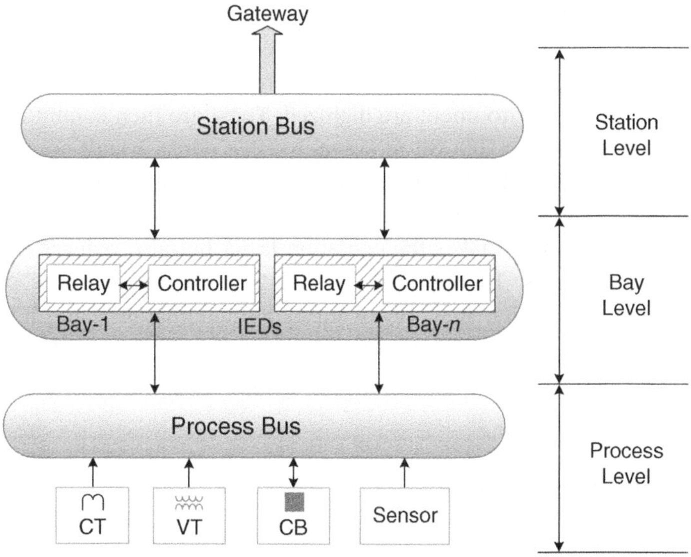

Figure 6.24 IEC 61850 architecture.

- *High-speed Communication*: IEC 61850 uses ethernet for communication for data exchange between different devices.
- *Flexibility and Scalability*: IEC 61850 allows easy reconfiguration of protection schemes and allows the addition of more protective devices as the power system expands.
- *System Integration*: Each IED used in IEC 61850 integrates the functions of protection, control, and monitoring, thereby reducing the cost and complexity of having separate systems for each function.
- *Event Logging*: IEC 61850 records each event and disturbance for post-event analysis and to improve the performance of protective devices over time.

The fast and reliable communication and data exchange in IEC 61850 is done through the communication protocol called *Generic Object-Oriented Substation Event*, which is simply known as GOOSE. Real-time GOOSE messages about trips and alarms are broadcasted over the ethernet network with low latency such that several IEDs receive the signals simultaneously. Moreover, these messages are event-driven, that is, the messages are sent only when an event occurs. This reduces network traffic congestion. GOOSE messages also have built-in redundancies, that is, they are sent several times, albeit with reduced frequency. Importantly, GOOSE facilitates interlocking functions between different devices enhancing safety. It can prevent conflicting actions such as preventing the closing of a circuit breaker unless it is safe to do so.

IEC 61850 simplifies the protection system. Consider, for example, the differential protection of a transformer. For three-phase transformer protection, several CTs will be required to cover both the primary and secondary sides. Needless to say that this will be a complex and time-consuming process. The measurements from CTs in an IEC 61850 environment are digitized. These are then communicated over the ethernet through GOOSE messages to IEDs. Once a fault occurs, a GOOSE message is delivered to IEDs instantaneously. This will allow the protection relays to detect a fault and send necessary trip commands to relays quickly. The use of IEC 61850 not only makes the operation faster but also reduces the physical wiring requirements. Therefore, the installation and maintenance of relays can become much simpler. A survey of substation automation using IEC 61850 is presented in [11], while the performance evaluation of the IEC 61850 communication system has been modeled and evaluated in [12].

6.8 Concluding Remarks

Power system protection is an important area for power utility companies. The relays and the associated circuit breakers not only protect expensive power apparatus from damage but also protect personnel from electrocution. Protective devices evolve with the introduction of different types of power equipment. In Example 6.2, an overcurrent protection coordination has been discussed assuming power flows from upstream to downstream. However, due to the inclusion of renewable sources such as solar photovoltaic (PV), the radial nature of the power flow is disrupted. Practicing engineers and researchers are working on a solution to this problem. Moreover, most of the relays these days are digital, which have much better functionality and faster response time. These, combined with solid-state circuit breakers, will change the protection strategies in the near future. This chapter introduces the subject and gives an overall picture of the available protective devices. The main aim of this chapter is to give the readers a flavor of the topic.

References

1 P. M. Anderson, C. Henville, R. Rifaat et al., *Power System Protection*, IEEE Press-Wiley, Hoboken, NJ, 2022.

2 L. Hewitson, M. Brown and B. Ramesh, *Practical Power Systems Protection*, Elsevier, Oxford, 2004.

3 J. D. Glover, M. S. Sarma and T. J. Overbye, *Power Systems Analysis and Design*, 5th Ed., Cengage Learning, Stamford, CT, 2012.

4 IEEE Standard, "Inverse-time characteristic equation for overcurrent relays", *IEEE Standard* C37, 112, 1996.

5 W. D. Stevenson, *Power System Analysis*, 4th Ed., McGraw-Hill, New York, 1982.

6 J. A. Martinez and J. Martin-Arnedo, "Impact of distributed generation on distribution protection and power quality," *IEEE Power and Energy Society General Meeting, Calgary*, 2009.

7 A. Ghosh and M. Dewadasa, "Operation, control and energy management of distributed generation," *CSIRO Intelligent Grid Research Cluster – Project* 7, 2011. Available. https://igrid.net.au/resources/index.html#project7

8 M. Dewdasa, A. Ghosh and G. Ledwich, "Fold back current control and admittance protection scheme for a distribution network containing distributed generators," *IET Generation, Transmission and Distribution*, Vol. 4, No. 8, pp. 952–962, 2010.

9 M. Dewdasa, A. Ghosh, G. Ledwich and M. Wishart, "Fault isolation in distributed generation connected distribution networks," *IET Generation, Transmission and Distribution*, Vol. 5, No. 10, pp. 1053–1061, 2011.

10 M. Dewdasa, *Protection of Distributed Generation Interfaced Networks*, Ph.D. Thesis, Queensland University of Technology, Brisbane, Australia, 2010.

11 M. A. Aftaba, S. M. S. Hussainb, I. Alic and T. S. Ustunb, "IEC 61850 based substation automation system: a survey," *Electric Power and Energy Systems*, Vol. 120, 2020. https://doi.org/10.1016/j.ijepes.2020.106008.

12 T. S. Siddhu and Y. Yin, "Modelling and simulation for performance evaluation of IEC61850-based substation communication systems," *IEEE Transactions on Power Systems*, Vol. 22, No. 7, pp. 1482–1489, 2007.

Problems

P6.1 The conductor of one phase of a three-phase transmission line operating at 400 kV, 500 MW, 0.95 power factor lagging has a CT and a PT connected to it. The PT is connected between the line and ground. The CT ratio is 1,000 : 5 and the PT ratio is 5,000 : 1. Determine the CT secondary current and PT secondary voltage.

P6.2 The input current to an overcurrent relay is 10 A. Using the overcurrent relay characteristics of (6.6) and using the extreme inverse curve of Table 6.1, determine the relay operating time for the following current tap setting (CTS) and time dial setting (CTS):
(a) CTS = 1.0 and TDS = ½;
(b) CTS = 2.0 and TDS = 1.5;
(c) CTS = 2.0 and TDS = 7;
(d) CTS = 3.0 and TDS = 7, and
(e) CTS = 12.0 and TDS = 1.

P6.3 A single-phase 5 MVA, 20/8.66 kV transformer is protected by a differential relay with taps. Available relay tap settings are 5 : 5, 5 : 5.5, 5 : 6.6, 5 : 7.3, 5 : 8, 5 : 9, and 5 : 10. Select CT ratios and relay tap settings. Also determine the percentage mismatch for the selected tap settings.

P6.4 An 11 kV radial system is shown in Figure P6.4. The system has the following impedance parameters:

Line 1-2: Positive-sequences: j0.8 Ω, zero-sequence: j2.5 Ω
Line 2-3: Positive-sequences: j2.4 Ω, zero-sequence: j7.5 Ω
Transformers: Positive-sequences: j2 Ω, zero-sequence: j3.5 Ω

Under emergency conditions, the system may operate with only one transformer. The system has to be protected for a 1LG fault. The CT ratio chosen for R2 (associated with Breaker B2) is 250 : 5 and its TDS is 0.5 s. Find the fastest operating time for this relay using the (a) moderate inverse curve, (b) very inverse curve and (c) extreme inverse curve of Table 6.1.

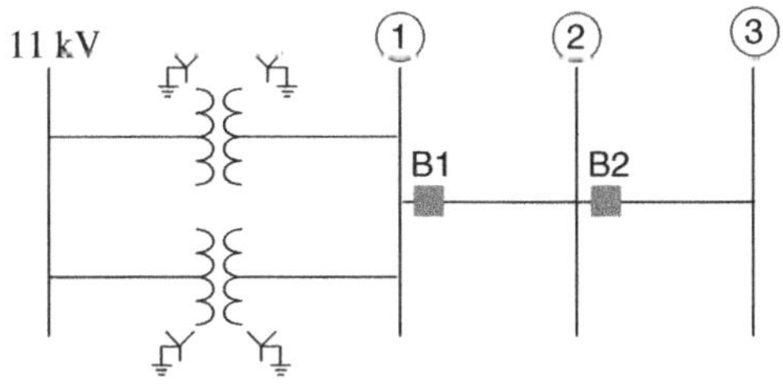

Figure P6.4 Radial system of Problem 6.4.

P6.5 Consider the single-phase capacitive voltage transformer (CVT) shown in Figure P6.5. The open circuit voltage requirement of the CVT is 100 V, while the line voltage connected across terminal A is 100 kV. Find the values of C1 and C2 such that there is no phase displacement between the line voltage and the output of the CVT. The leakage inductance of the transformer is 1 mH and the supply frequency is 50 Hz.

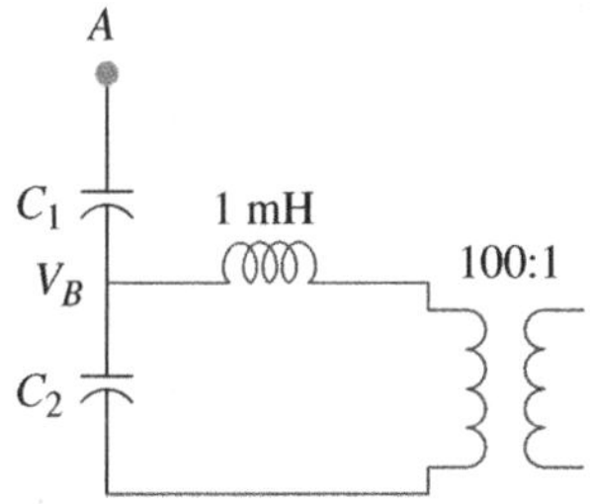

Figure P6.5 CVT configuration of Problem 6.5.

P6.6 Three-zone mho relays are used for the protection of the power system shown in Figure P6.6. The positive-sequence impedances are given in Table P6.6. The rated voltage at bus-1 is 500 kV and the CT and PT ratios are 1,500 : 5 and 4,500 : 1, respectively. In a three-zone scheme for B3, zone-1 protects 80% of line 1-2, zone-2 protects 120% of line 1-2 and zone-3 protects 100% of line 1-2 and 120% of line 2-3.

(a) Determine the settings of Z_{r1}, Z_{r2}, and Z_{r3}, the blocking impedances for the three zones.

(b) Determine in which zone the relay will trip for a bolted three-phase at bus-3.

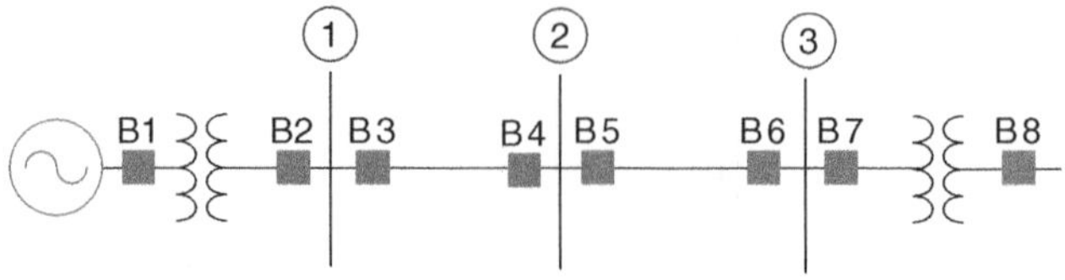

Figure P6.6 Three-zone protection scheme of Problem 6.6.

Table P6.6 Positive-sequence impedances of Figure P6.6.

Line	Positive-sequence impedance (Ω)
1-2	$4.5 + j45$
2-3	$9 + j90$

P6.7 Consider the system shown in Figure P6.7. Directional overcurrent relays are used for breakers B12, B21, B23, B32, B34, and B43. Only overcurrent relays are used for breakers B1 and B4.

(a) Which breakers will not operate for a fault at P1? Which breakers should be coordinated?

(b) Repeat part (a) for a fault at P2.

(c) Repeat part (a) for a fault at P3.

(d) Explain how the system is protected against bus faults.

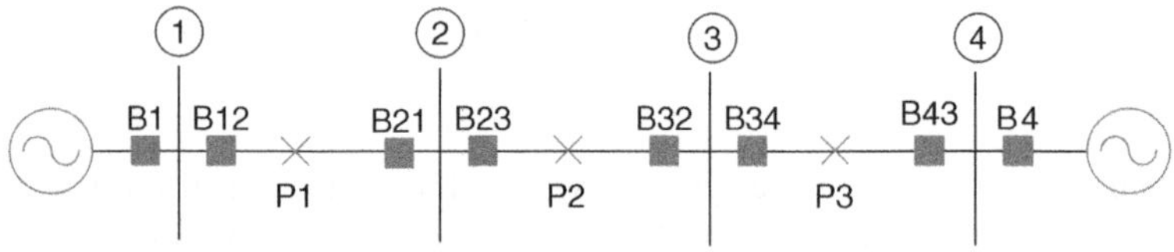

Figure P6.7 Protection system of Problem 6.7.

P6.8 Draw the zones of protection for the power system shown in Figure P6.8, assuming that each relay is protected in two zones. Which circuit breakers will open for a fault at (a) P1, (b) P2, and (c) P3?

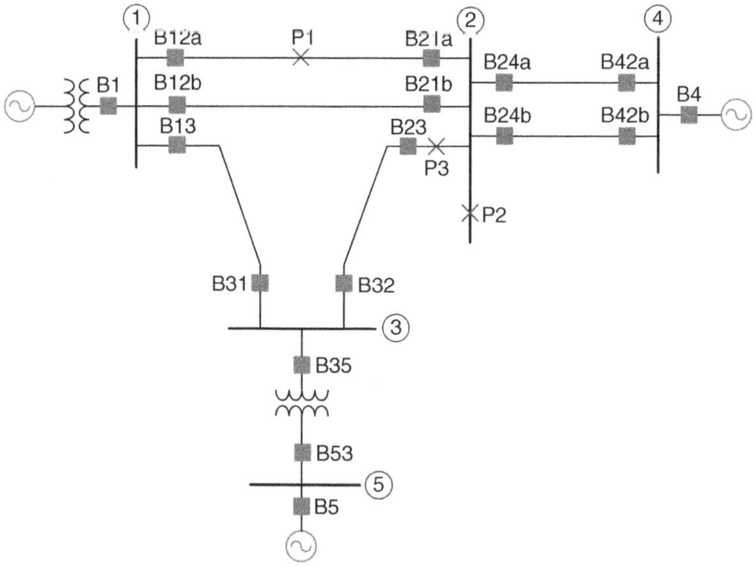

Figure P6.8 Zonal protection system of Problem 6.8.

P6.9 Three-zone mho relays are used for the protection of the power system shown in Figure P6.9. Positive-sequence impedances are listed in Table P6.9. The rated voltage at bus-1 is 500 kV and the CT and PT ratios are 1,500 : 5 and 4,500 : 1, respectively. In a three-zone scheme for B3, zone-1 protects 80% of line 1-2, zone-2 protects 120% of line 1-2, and zone-3 protects 100% of line 1-2 and 120% of line 2-3.

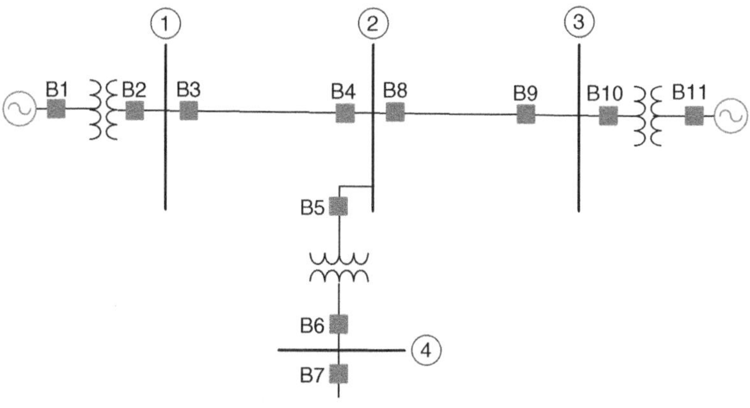

Figure P6.9 Protection system of Problem 6.9.

Table P6.9 Positive-sequence impedances of Figure P6.9.

Line	Positive-sequence impedance (Ω)
1-2	$6+j60$
2-3	$5+j50$

(a) Determine the settings of Z_{r1}, Z_{r2}, and Z_{r3}.

(b) Find the maximum current through line 1-2 during an emergency loading condition of 1,400 A at 0.9 power factor lagging. Will any of the relays trip during this condition?

7

Power System Stability and Control

The first electric power system was a DC system built by Edison in 1882. The subsequent power systems that were constructed in the late 19th century were all DC systems. However, despite the initial popularity of DC systems, by the turn of the 20th century, AC systems started to outnumber them. The AC systems were thought to be superior as AC machines were cheaper than their DC counterparts and more importantly AC voltages were easily transformable from one level to the other using transformers. The early stability problems of AC systems were experienced in 1920 when insufficient damping caused spontaneous oscillations or hunting. These problems were solved using generator damper winding and the use of turbine-type prime movers [1]. Damper windings are short-circuited copper bars that are placed on the rotor like a squirrel cage induction motor. When the rotor oscillates against the rotating magnetic field, eddy currents are induced in the damper bars. Following Lenz's law, the eddy currents oppose the magnetic field that produced them, thereby providing damping. Usually, this is called a damping torque.

Moreover, improvement was also made to the prime movers that drove the generators, where early steam engines were replaced by turbines. These turbines could handle fluctuations in power more efficiently. Following this, hydraulic or mechanical governors were introduced to control the turbine speed. The governors adjust the flow of steam on the turbines based on the changes in generator speed. Additionally, the *fast valving* technique was also employed in large turboalternators. Through fast valving, the steam flow in the turbine is fully or partially closed rapidly to reduce the steam flow, thereby reducing the mechanical power output. Despite all these improvements, a power system is prone to stability problems, some of which will be discussed in this chapter. Additionally, some countermeasures to these stability problems will also be discussed in the chapter.

A system is stable when it returns to a steady state after being subjected to an external disturbance. For a linear dynamical system to be stable, all the poles or

Modern Power System, First Edition. Arindam Ghosh.

Published 2026 by John Wiley & Sons, Inc.
Companion website: www.wiley.com/go/ghoshpowersystem

eigenvalues must have negative real parts or must be at the origin in the Laplace domain (s-plane). However, a power system is a large nonlinear system and there are various aspects of stability that have been studied extensively.

In a traditional power system, the active power is generated by synchronous generators that operate in synchronism with the rest of the system. Two systems are said to be synchronized when both have the same frequency, voltage, and phase sequence. A synchronous generator losing synchronism with the power system to which it is connected will cause large oscillations in the power flow thereby causing damage to its mechanical components and can also induce oscillations in the power system. From this perspective, power system stability is defined as the ability of a generator to return to a steady state without losing synchronism. Usually, power system stability is categorized into the following types:

- *Steady State Stability*: These studies are restricted to small and gradual changes in the operating conditions of a power system. Here, the bus voltages are restricted close to their nominal values. Furthermore, it is ensured that the phase angle difference between the two buses is not too large. These studies are performed to check for the overloading of the power equipment and transmission lines. These checks are usually done using power flow studies.
- *Transient Stability*: This involves the study of the power system following a major disturbance. Following the disturbance, a synchronous alternator output voltage angle changes due to the sudden acceleration of the rotor shaft. The objective of the transient stability study is to ascertain whether the load angle returns to a steady value following the clearance of the disturbance.
- *Dynamic Stability*: In this, the ability of a power system to maintain stability under continuous small disturbances is investigated. This is also known as small-signal stability. These small disturbances occur due to random fluctuations in loads and generation levels. In an interconnected power system, these random variations can lead to catastrophic failure as this may force the rotor angle to increase steadily.

In this chapter, we shall start our discussion with transient stability. The machine rotor is represented by the so-called swing equation, which dictates how the machine will accelerate or decelerate depending on the input and output power balance. The swing equation is used to determine the transient stability limits based on the calculation of the critical clearing angle, which is the angle up to which the machine can swing without losing its stability. Furthermore, the critical clearing time is computed from the critical clearing angle based on machine inertia.

The transient stability analysis of multimachine systems is presented next. The classical method involves the determination of the pre-fault system operating conditions based on load flow studies and then the use of a swing equation to determine the stability condition. In the final section of this chapter, dynamic stability is discussed, specifically the analysis of the effects of an automatic voltage regulator

(AVR) on the system stability and how the stability can be enhanced using power system stabilizers (PSS).

7.1 Transient Stability

The response of a synchronous generator to a fault in the system is investigated in transient stability studies. Large turboalternators are tied to a chain of shafts forming a large rotating mass. As a result, these generators are not able to rapidly change their output power in response to a fault. On the other hand, the rotating mass stores rotational kinetic energy, which can be released temporarily to meet a change in the load demand.

Modern power transmission systems are predominantly inductive. Therefore, a fault is akin to creating a short circuit through an inductor, where only reactive power flows without any real power. Therefore, depending on the transmission system structure, the real power flowing out of a generator is either zero or largely reduced during a three-phase bolted fault. If the fault lasts more than a few cycles, the machine might lose synchronism. during the time the fault persists, it is assumed that the mechanical power input to the generator remains constant because it cannot change instantaneously. Therefore, there will be a mismatch between the input power and the output power. This will cause the machine to accelerate. Synchronous generators, as the name suggests, operate at the rated frequency when tied to a power system. If the machine acceleration is allowed to persist, it will deviate from the synchronous frequency and will eventually be disconnected from the power system, losing synchronism. The aim of transient stability studies is to determine when a machine will lose synchronization.

7.1.1 Power–Angle Curve

Modern transmission lines have negligible resistance compared to the line reactance. Thus, the line resistance can be neglected. Such systems are called lossless because there are no I^2R losses in the line. A lossless system is shown in Figure 7.1.

Figure 7.1 A lossless SMIB system.

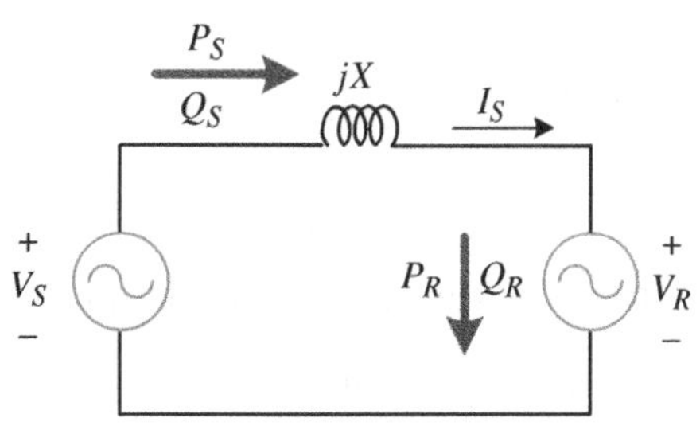

In this, two voltage sources are connected through an external reactance X. Such a system is often called a single-machine, infinite-bus (SMIB) system. Here, V_S is the sending-end internal emf of the machine (synchronous generator) and V_R is the infinite bus voltage. Reactance X includes the reactance of the transmission line and the synchronous reactance or the transient reactance of the generator.

The generator and the infinite bus voltages are given respectively by

$$V_S = V_1\angle\delta_1 \text{ and } V_R = V_2\angle\delta_2$$

The current flowing through the line is denoted by I_S. The active and reactive power of the sending and the receiving ends are denoted by P_S, Q_S, P_R, and Q_R, as shown in Figure 7.1. The current flowing through the line is given by

$$I_S = \frac{V_1\angle\delta_1 - V_2\angle\delta_2}{jX} \tag{7.1}$$

The complex power flowing from the source is then

$$P_S + jQ_S = V_S \times I_S^* = V_1\angle\delta_1 \times \frac{V_1\angle-\delta_1 - V_2\angle-\delta_2}{-jX} = \frac{V_1^2 - V_1V_2\angle(\delta_1-\delta_2)}{-jX} \tag{7.2}$$

Expanding (7.2), we get

$$P_S + jQ_S = \frac{1}{-jX} \times \left[V_1^2 - V_1V_2\cos(\delta_1-\delta_2) - jV_1V_2\sin(\delta_1-\delta_2)\right]$$

Separating the real and imaginary components of the above equation, the real and reactive power dispatched by the source are

$$P_S = \frac{V_1V_2\sin(\delta_1-\delta_2)}{X} \tag{7.3}$$

$$Q_S = \frac{V_1^2 - V_1V_2\cos(\delta_1-\delta_2)}{X} \tag{7.4}$$

The complex power received by the infinite bus is

$$P_R + jQ_R = V_R \times I_S^* = V_2\angle\delta_2 \times \frac{V_1\angle-\delta_1 - V_2\angle-\delta_2}{-jX} = \frac{V_1V_2\angle(\delta_2-\delta_1) - V_2^2}{-jX} \tag{7.5}$$

Equation (7.5) is expanded as

$$P_R + jQ_R = \frac{1}{-jX}\left[V_1V_2\cos(\delta_1-\delta_2) - jV_1V_2\sin(\delta_1-\delta_2) - V_2^2\right]$$

Then the real and reactive power delivered at the receiving end are

$$P_R = \frac{V_1V_2\sin(\delta_1-\delta_2)}{X} \tag{7.6}$$

$$Q_R = \frac{V_1 V_2 \cos(\delta_1 - \delta_2) - V_2^2}{X} \tag{7.7}$$

Comparing (7.3) with (7.6), it can be seen that $P_S = P_R$, which is not surprising since the line is lossless. We define the electric power output generator as $P_e = P_S = P_R$.

Equation (7.1) is expanded as

$$I_S = \frac{1}{X}[-jV_1 \cos(\delta_1) + jV_2 \cos(\delta_2) + V_1 \sin(\delta_1) - V_2 \sin(\delta_2)]$$

Therefore

$$|I_S|^2 = \frac{1}{X^2}\{[-V_1 \cos(\delta_1) + V_2 \cos(\delta_2)]^2 + [V_1 \sin(\delta_1) - V_2 \sin(\delta_2)]^2\}$$

The solution to the equation is

$$|I_S|^2 = \frac{1}{X^2}\left[V_1^2 + V_2^2 - 2V_1V_2 \cos(\delta_1 - \delta_2)\right]$$

Therefore, the reactive drop in the line is given by

$$Q_L = |I_S|^2 X = \frac{1}{X}\left[V_1^2 + V_2^2 - 2V_1V_2 \cos(\delta_1 - \delta_2)\right] \tag{7.8}$$

This can also be computed from the relation $Q_L = Q_S - Q_R$ using (7.4) and (7.7).

Note from (7.3) and (7.6) that the power transfer depends on the relative angle difference $(\delta_1 - \delta_2)$. If $\delta_1 > \delta_2$, power flows from the sending to the receiving end. On the other hand, if $\delta_1 < \delta_2$, the power flow direction will be reversed, that is, from the receiving end to the sending end. Because the power flow depends on the angle difference and not on the actual values of the angles, usually the infinite bus voltage angle is chosen as 0°, and the angle difference $(\delta_1 - \delta_2)$ is chosen as δ, which is called the *load angle*.

Example 7.1 Consider the system of Figure **7.1**, where $V_R = 1\angle 0°$ per unit, $X = 0.25$ per unit, and $V_S = 1.1$ per unit. Then, the power transfer is given by

$$P_S = P_R = P_e = \frac{1 \times 1.1}{0.25} \sin(\delta) = 4.4 \sin(\delta)$$

From the above expression, we find that the maximum power (P_{max}) that can be transferred through the line is 4.4 per unit, and it occurs when $\delta = 90°$, that is,

$$P_{max} = \frac{V_1 V_2}{X} \sin(90°) = 4.4$$

The power–angle curve is shown in Figure 7.2, where the angle δ varies from 0° to 180°. The horizontal lines show the constant power transfer. At P_{max}, the

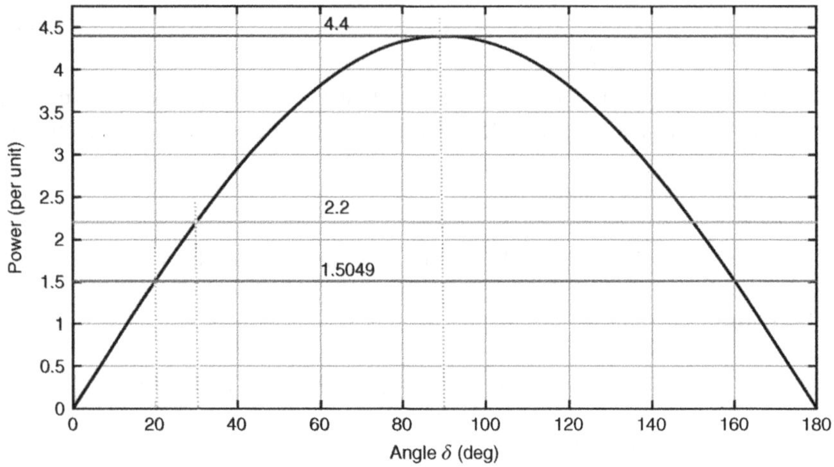

Figure 7.2 The power–angle curve of the system of Example 7.1.

straight line for a power transfer of 4.4 per unit is a tangent to the power–angle curve, parallel to the x-axis. Again, when $\delta = 20°$, the power output that is obtained from (7.3) is

$$P_e = 4.4\sin(20°) = 1.5049 \text{ pu}$$

Similarly, when $\delta = 30°$, $P_e = 2.2$ per unit. Therefore, the interactions of the constant power lines with load-angles define the load angles. However, note that each line intersects the curve at two points. For example, when $P_e = 2.2$ per unit, the line intersects the curve at 30° and $180° - 30° = 150°$. These angles are usually referred to as δ_0 and δ_{max}, respectively. As will be evident later in this section, we shall choose the operating angle of $\delta = 30°$.

The main lessons that are learned from the power–angle relationships are summarized below.

$$P_{max} = \frac{V_1 V_2}{X} \tag{7.9}$$

$$\delta_0 = \sin^{-1}\left(\frac{P_0}{P_{max}}\right) \tag{7.10}$$

$$\delta_{max} = 180° - \delta_0 \tag{7.11}$$

7.1.2 Swing Equation

Consider a three-phase synchronous alternator that is driven by a prime mover. In a rotational system, the equation of motion relates to the torque and angular acceleration by $T = J\alpha$, where, T is the applied torque, J is the moment of inertia, and α is the angular acceleration. The net torque in a synchronous generator is the difference between the applied mechanical torque and the output electrical torque plus losses in the energy conversion process. Neglecting losses, the equation of motion of the generator rotor is given by

$$J\frac{d^2\theta}{dt^2} = T_m - T_e = T_a \tag{7.12}$$

where J is the total moment of inertia of the rotor mass in kg m^2; T_m is the mechanical power supplied by the prime mover in N-m; T_e is the electrical torque output of the alternator in N-m; T_a is the accelerating torque in N-m, and θ is the angular position of the rotor in radians.

It is evident from (7.12) that the accelerating torque is the difference between the mechanical input torque and the electrical output torque. During the steady state, the electrical torque is equal to the mechanical torque and the accelerating torque is zero. At that time, the alternator rotor moves at *synchronous speed* ω_s rad/s.

The angle θ is measured with respect to a stationary reference frame and it continuously increases with time. However, the rotor speed that is relative to the synchronous speed is of interest to us as it provides the measure of the rotor angular position with respect to a reference axis that rotates at synchronous speed. Therefore, the angle θ is defined as

$$\theta = \omega_{sm}t + \delta_m \tag{7.13}$$

where ω_{ms} is the synchronous speed in mechanical rad/s, while δ_m is the displacement of the rotor from the reference axis in mechanical radians. Taking the first and second derivatives of (7.13), we get

$$\frac{d\theta}{dt} = \omega_{sm} + \frac{d\delta_m}{dt} \tag{7.14}$$

$$\frac{d^2\theta}{dt^2} = \frac{d^2\delta_m}{dt^2} \tag{7.15}$$

Equation (7.15) is substituted in (7.12) to obtain

$$J\frac{d^2\delta_m}{dt^2} = T_m - T_e = T_a \tag{7.16}$$

The multiplication of both sides of (7.16) by rotor mechanical speed ω_m results in

$$J\omega_m\frac{d^2\delta_m}{dt^2} = P_m - P_e = P_a \text{ W} \tag{7.17}$$

An inertial constant H is defined as

$$H = \frac{\text{KE in mega Joules}}{\text{Generator MVA rating } (S_{rated})} \tag{7.18}$$

where KE is the stored kinetic energy in the rotor shaft at synchronous speed, which is given by

$$\text{KE} = \frac{1}{2} J\omega_{sm}^2 \tag{7.19}$$

Comparing (7.19) with (7.18), we get

$$J\omega_{sm}^2 = 2H \times S_{rated} \tag{7.20}$$

Note that a synchronous machine operates very close to the synchronous speed, and therefore, we can assume $\omega_{sm} = \omega_m$. Then, the substitution of (7.20) in (7.17) results in

$$\frac{2H}{\omega_m}\frac{d^2\delta_m}{dt^2} = \frac{P_m - P_e}{S_{rated}} = \frac{P_a}{S_{rated}} \tag{7.21}$$

Note that both δ_m and ω_m are expressed in mechanical radians. Therefore, both of them can be replaced by electrical radians. Furthermore, power in Watts divided by the machines MVA rating is the power in per unit. Hence, (7.21) is rewritten as

$$\frac{2H}{\omega_s}\frac{d^2\delta}{dt^2} = P_m - P_e = P_a \text{ per unit} \tag{7.22}$$

Equation (7.14) can be written in terms of electrical radians as

$$\frac{d\theta}{dt} = \omega_s + \frac{d\delta}{dt} \tag{7.23}$$

The angular speed of the rotor is defined as

$$\omega_r = \frac{d\theta}{dt} \tag{7.24}$$

Comparing (7.23) with (7.24), the following equation is obtained

$$\omega_r - \omega_s = \frac{d\delta}{dt} \tag{7.25}$$

We can therefore conclude that the rotor angular speed is equal to the synchronous speed only when $d\delta/dt$ is equal to zero. We can therefore term $d\delta/dt$ as the error in speed.

Example 7.2 Consider a 4-pole, 50 Hz turboalternator rated 500 MVA and 20 kV with the inertia constant (*H*) of 5 MJ/MVA. We want to find out (a) the kinetic energy stored in the rotor shaft, and (b) the angular acceleration in rpm/s when the electrical power output is 450 MW for an input of 740 000 hp. (1 hp. = 746 W roughly). Neglect the rotational losses.

a) The kinetic energy is equal to H times the machine rating in MVA = $6 \times 500 = 3000$ MJ.
b) The mechanical input power is

$$P_m = 740{,}000\ \text{hp} = 740 \times 746 \times 10^3 = 552.04\ \text{MW}$$

The accelerating power then is

$$P_a = P_m - P_e = 552.04 - 450\ \text{MW} = \frac{102.04}{500} = 0.2041\ \text{pu}$$

From (7.22), we have

$$\frac{2H}{\omega_s}\frac{d^2\delta}{dt^2} = 0.2041$$

$$\Rightarrow \frac{d^2\delta}{dt^2} = 0.2041 \times \frac{\omega_s}{2H} = 0.2041 \times \frac{100\pi}{10} = 5.343\ \text{elec rad/s}^2$$

This is then converted in electrical degree/s^2 using

$$\frac{d^2\delta}{dt^2} = 5.343 \times \frac{180}{\pi} = 306.12\ \text{elec deg/s}^2$$

In a 4-pole machine, the mechanical system rotates by 180° when the electrical system rotates by 360°, that is,

$$\frac{d^2\delta}{dt^2} = \frac{306.12}{2} = 153.06\ \text{mech deg/s}^2$$

Now, one revolution rotor shaft translates into a mechanical rotation of 360°. Therefore,

$$\frac{d^2\delta}{dt^2} = \frac{153.06}{360}\ \text{rev/s}^2 = 0.4252\ \text{rps/s} = 60 \times 0.4252 = 25.51\ \text{rpm/s}$$

The angular acceleration in rpm/s as H changes from 2 to 10 MJ/MVA is shown in Figure 7.3. It can be seen that as H increases, the rotor acceleration gets slower. This result is intuitive, as larger the machine is, the slower it moves. This concept will be illustrated while discussing the critical clearing time of Section 7.1.4.

7.1.3 Critical Clearing Angle

Consider the situation in which the synchronous machine is operating in a steady state delivering a power equal to P_m when a fault occurs in the system. The opening of the circuit breakers in the faulted section subsequently clears the fault. The circuit breakers take a few cycles to open and the subsequent post-fault transients last

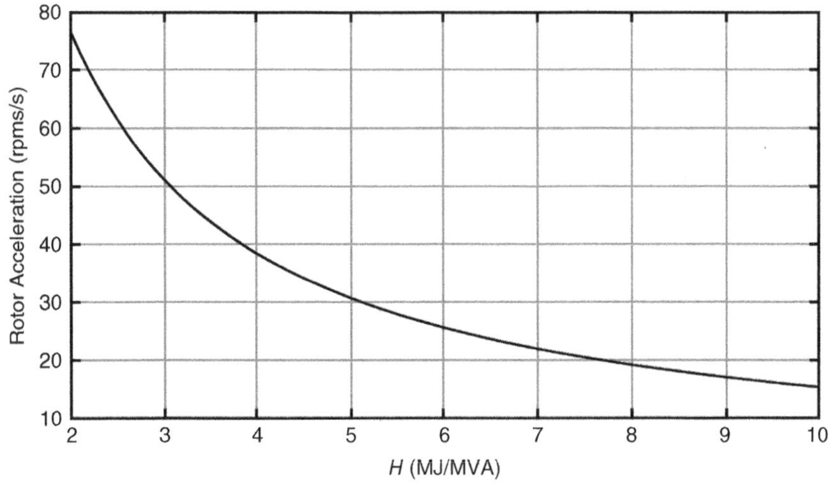

Figure 7.3 Change in the angular acceleration with the changes in inertia.

for another few cycles. The input power, on the other hand, is supplied by a prime mover that is usually driven by a steam turbine. The time constant of the turbine mass system is of the order of a few seconds, while the electrical system time constant is in milliseconds. Therefore, for all practical purposes, mechanical power remains constant during the electrical transients. The transient stability study therefore concentrates on the ability of the power system to recover from the fault and deliver the constant power P_m with a possible new load angle δ.

Consider the power–angle curve of an SMIB system shown in Figure 7.4. In this, the generator is steady state delivering a power of P_m at an angle of δ_0, when a fault occurs such that the real power transfer becomes zero. Because P_m remains constant, the accelerating power P_a becomes equal to P_m. The difference in the power gives rise to the rate of change of stored kinetic energy in the rotor masses. Thus, the rotor will accelerate under the constant influence of nonzero accelerating power, and hence, the load angle will increase. Now suppose the fault is cleared at an angle δ_c. The power will then revert to the normal operating curve. At that point, the electrical power will be more than the mechanical power and the accelerating power will be negative. This will cause the machine to decelerate. However, due to the inertia of the rotor masses, the load angle will keep on increasing. The increase in this angle may eventually stop and the rotor may start decelerating, otherwise, the system will lose synchronism.

Note that for an arbitrary variable u, we have

$$\frac{d}{dt}u^2 = 2u\frac{du}{dt}$$

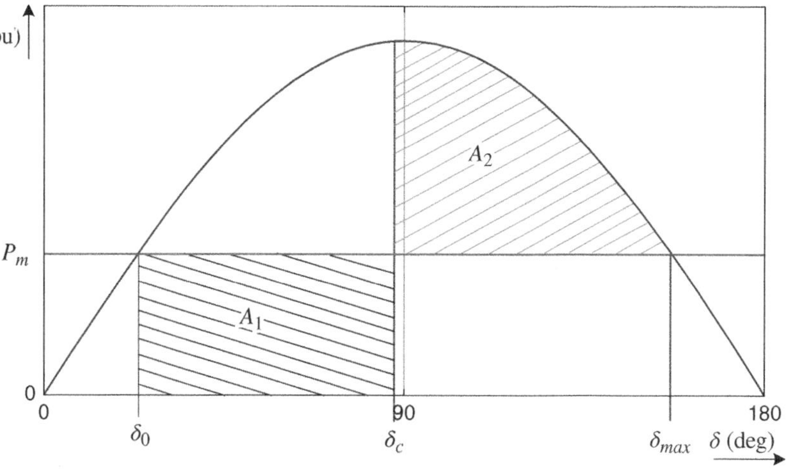

Figure 7.4 The power–angle curve of an SMIB system.

Then substituting $u = d\delta/dt$, we get

$$\frac{d}{dt}\left(\frac{d\delta}{dt}\right)^2 = 2\left(\frac{d\delta}{dt}\right)\left(\frac{d^2\delta}{dt^2}\right) \tag{7.26}$$

Multiplying both sides of (7.22) by $d\delta/dt$, the following equation is obtained:

$$\frac{2H}{\omega_s}\left(\frac{d\delta}{dt}\right)\left(\frac{d^2\delta}{dt^2}\right) = (P_m - P_e)\frac{d\delta}{dt} \tag{7.27}$$

Using (7.26), (7.27) is modified as

$$\frac{H}{\omega_s}\frac{d}{dt}\left(\frac{d\delta}{dt}\right)^2 = (P_m - P_e)\frac{d\delta}{dt}$$

Multiplying both sides of the above equation by dt and then integrating between two arbitrary angles δ_0 and δ_c, the following expression is obtained:

$$\frac{H}{\omega_s}\left(\frac{d\delta}{dt}\right)^2\Bigg|_{\delta_0}^{\delta_c} = \int_{\delta_0}^{\delta_c}(P_m - P_e)d\delta \tag{7.28}$$

The left-hand side of the above equation is the departure of the machine from the synchronous speed as per (7.25). Defining $d\delta/dt = \Delta\omega$, (7.28) is rewritten as

$$\frac{H}{\omega_s}\left(\Delta\omega_c^2 - \Delta\omega_0^2\right) = \int_{\delta_0}^{\delta_c}(P_m - P_e)d\delta \tag{7.29}$$

Now suppose the generator is at rest at δ_0, such that $d\delta/dt = 0$. Once a fault occurs, the machine starts accelerating. Once the fault is cleared, the machine keeps on accelerating before it reaches its peak at δ_c, at which point we again have $d\delta/dt = 0$. Thus, the area under accelerating is given from (7.29) as

$$A_1 = \int_{\delta_0}^{\delta_c} (P_m - P_e)d\delta \tag{7.30}$$

In a similar way, the area under deceleration is defined as

$$A_2 = \int_{\delta_c}^{\delta_{max}} (P_e - P_m)d\delta \tag{7.31}$$

The accelerating and decelerating areas are indicated in Figure 7.4.

Now consider the case when the line is reclosed at an arbitrary angle δ_c such that the area of acceleration is larger than the area of deceleration, that is, $A_1 > A_2$. The generator load angle will then cross the point δ_{max}, beyond which the electrical power will be less than the mechanical power forcing the accelerating power to be positive. The generator will therefore start accelerating before it slows down completely and will eventually become unstable. If, on the other hand, $A_1 < A_2$, that is, the decelerating area is larger than the accelerating area, the machine will decelerate completely before accelerating again. The rotor inertia will force the subsequent acceleration and deceleration areas to be smaller than the first ones and the machine will eventually attain the steady state. If the two areas are equal, that is, $A_1 = A_2$, then the accelerating area is equal to the decelerating area, and this defines the boundary of the stability limit. The clearing angle δ_c in this case is called the *critical clearing angle* and is denoted by δ_{cr}. We then get the following equation from Figure 7.4 by substituting $\delta_c = \delta_{cr}$.

$$\int_{\delta_0}^{\delta_{cr}} (P_m - P_e)d\delta = \int_{\delta_{cr}}^{\delta_{max}} (P_e - P_m)d\delta \tag{7.32}$$

The critical clearing angle is derived from the above equation. Because the critical clearing angle depends on the equality of the areas, this is called the *equal area criterion*.

Example 7.3 A generator is connected to an infinite bus through a lossless transmission line with a reactance of 0.3 per unit (see Figure **7.5**). The synchronous reactance of the generator is 0.2 per unit and the voltage magnitude of the infinite bus is 1.0 per unit with its angle being 0°. The generator delivers 0.9 per unit of power to the constant voltage bus when the angle δ_t of the terminal voltage is 15°. The generator is operating in the steady state when a fault occurs causing the real power transfer to be zero. The fault is cleared by circuit breakers, thereby

Figure 7.5 SMIB system of Problem 7.3.

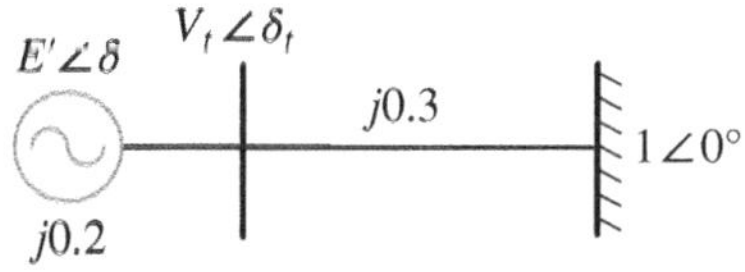

restoring the power supply. We have to find the critical clearing angle δ_{cr}, that is, the maximum angle the generator can swing to before losing synchronism.

Let us denote the magnitude of the terminal voltage as V_t. Because its angle is 15°, the power flow through the line is given from (7.3) by

$$P_e = 0.9 = \frac{V_t \times 1}{0.3}\sin(15°)$$

The solution to the above equation gives $V_t = 1.0432$ per unit. Therefore, the current flowing through the line is

$$I = \frac{1.0432\angle 15° - 1}{j0.3} = 0.9 - j0.0255 \text{ per unit}$$

The internal emf of the generator then is

$$E' = V_t\angle 15° + j0.2I = 1.0128 + j0.45 = 1.1082\angle 23.96° \text{ per unit}$$

Therefore, $\delta_0 = 23.96° = 0.4142$ rad and the maximum power transferred is

$$P_{max} = \frac{1.1082 \times 1}{0.5} = 2.2164 \text{ per unit}$$

It is important to note that, even though we define the angles in degrees for easy reference, the computation is always performed in terms of radians, where 1 rad = $180/\pi = 57.2958°$. Now to determine the critical clearing angle δ_{cr}, the accelerating and decelerating areas need to be calculated. To compute the accelerating area, it is to be noted that P_e in (7.30) is zero. Therefore, this area is

$$A_1 = \int_{0.4182}^{\delta_{cr}} 0.9\, d\delta = 0.9(\delta_{cr} - 0.4182) = 0.9\delta_{cr} - 0.3764$$

To determine the decelerating area, the maximum angle that the rotor is allowed to swing is given by

$$\delta_{max} = 180° - 23.96° = 156.04° = 2.7234 \text{ rad}$$

Then from (7.31) we have

$$A_2 = \int_{\delta_{cr}}^{2.7234} (2.2164\sin\delta - 0.9)\, d\delta = [-2.2164\cos\delta - 0.9\delta]_{\delta_{cr}}^{2.7234}$$

The solution of the above equation is

$$A_2 = 2.2164\cos(\delta_{cr}) + 0.9\delta_{cr} - 0.4257$$

Equating the two areas, that is, $A_1 = A_2$, we have

$$0.9\delta_{cr} - 0.3764 = 2.2164\cos(\delta_{cr}) + 0.9\delta_{cr} - 0.4257$$

$$\Rightarrow \delta_{cr} = \cos^{-1}\left(\frac{0.4257 - 0.3764}{2.2164}\right) = 1.5486 \text{ rad} = 88.73^\circ$$

7.1.4 Critical Clearing Time

A frequently asked question is "what does the critical clearing angle mean?" Because we are interested in finding out the maximum time that any circuit breaker may be allowed before opening, we should be more concerned about the critical clearing time rather than the critical clearing angle. Furthermore, notice that the clearing angle is independent of the generalized inertia constant H. Hence, we can comment that the critical clearing angle in this case is true for any generator that has a d-axis transient reactance of 0.20 per unit and a transmission line reactance of 0.3 per unit while transmitting 0.9 per unit of power.

During the fault, the electrical power is $P_e = 0$. Substituting this in (7.22), the following equation is obtained:

$$\frac{d^2\delta}{dt^2} = \frac{\omega_s}{2H}P_m \tag{7.33}$$

Because the machine starts accelerating only when the fault occurs, that is, the initial acceleration being zero, the integration of the above equation with respect to t gives

$$\frac{d\delta}{dt} = \int_0^t \frac{\omega_s}{2H}P_m dt = \frac{\omega_s}{2H}P_m t \tag{7.34}$$

Again, the integration of (7.34) with respect to t will lead to

$$\delta = \int_0^t \frac{\omega_s}{2H}P_m t\, dt = \frac{\omega_s}{4H}P_m t^2 + \delta_0 \tag{7.35}$$

Replacing δ by δ_{cr} and t by t_{cr} in (7.35), the critical clearing time is given by

$$t_{cr} = \sqrt{\frac{4H}{\omega_s P_m}(\delta_{cr} - \delta_0)} \tag{7.36}$$

Example 7.4 Consider the system of Example 7.3, where the system frequency is chosen as 50 Hz, such that $\omega_s = 100\pi$ rad/s. Let us choose H as 4 MJ/MVA. From Example 7.3, the following quantities are obtained:

$$P_m = 0.9 \text{ per unit}, \delta_0 = 0.4182 \text{ rad and } \delta_{cr} = 1.5486 \text{ rad}$$

Therefore, the critical clearing time is computed from (7.36) as

$$t_{cr} = \sqrt{\frac{16}{90\pi}(1.5486 - 0.4182)} = 0.2529 \text{ s}$$

Table 7.1 lists the values of the critical clearing angle as H changes. It can be seen that the critical clearing angle increases as H increases. This is in accordance with that shown in Figure 7.3.

To illustrate the response of the load angle δ, the swing equation is simulated in MATLAB. The swing equation of (7.22) is expressed as [2]

$$\begin{aligned} \frac{d\Delta\omega_r}{dt} &= \frac{1}{2H}(P_m - P_e) \\ \frac{d\delta}{dt} &= \omega_s \times \Delta\omega_r \end{aligned} \tag{7.37}$$

where $\Delta\omega_r$ is the deviation for the rotor speed from the synchronous speed ω_s. It is to be noted that the swing equation of (7.37) does not contain any damping. Usually a damping term, that is proportional to the machine speed $\Delta\omega_r$, is added with the accelerating power. Without any damping, the roots of (7.22) on the complex s-plane will be at

$$\delta = \pm j\sqrt{\frac{\omega_s P_a}{2H}}$$

Table 7.1 The effect of inertia on critical clearing angle of Example 7.4.

H (MJ/MVA)	t_{cr} (s)
2	0.1788
4	0.2529
6	0.3098
8	0.3577
10	0.3959

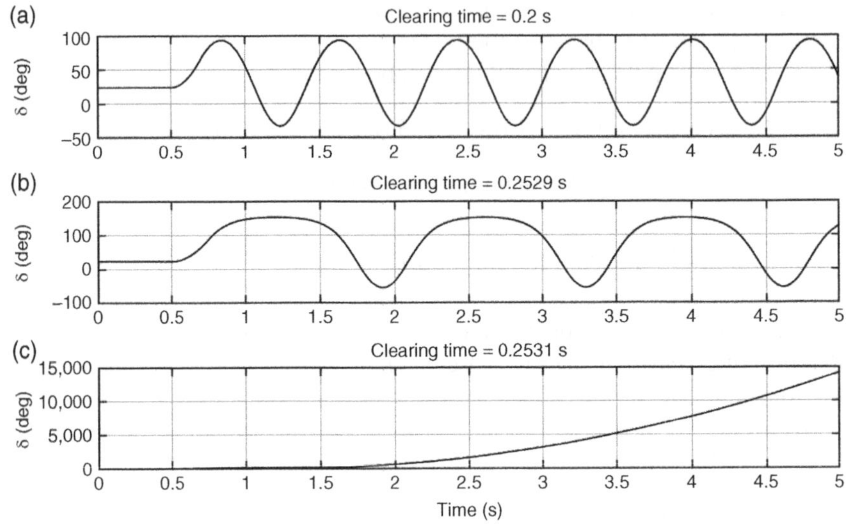

Figure 7.6 System response as a function of clearing time: (a) stable, (b) critically stable, and (c) unstable.

Therefore, the load angle will exhibit a sustained oscillation even when the system remains stable with the fault being cleared within the critical clearing time.

Figure 7.6 depicts the response of the load angle δ for three different values of clearing time t_{cl}. It is assumed that the fault occurs at 0.5 s when the system is operating in a steady state delivering 0.9 per unit of power. The load angle during this time is constant at 24.96°. When the clearing time is chosen as 0.2 s, the system remains stable, albeit with sustained oscillations, as shown in Figure 7.6 (a). The critically stable response is shown in Figure 7.6 (b). However, the system becomes unstable when the clearing time is 0.2531 s with a clearing angle of 88.77° and the load angle increases asymptotically, as shown in Figure 7.6 (c). This is called the loss of synchronism. It is to be noted that such an increase in the load angle is not permissible, and the protection device will isolate the generator from the system.

Equation (7.35) is derived assuming that $P_e = 0$. However, this may not always be true. The following example highlights this.

Example 7.5 Consider the system shown in Figure **7.7** in which a generator is connected to an infinite bus through a double circuit transmission. All quantities in this figure are defined in per unit. The infinite bus voltage is $1 \angle 0°$ per unit. The generator delivers 1.0 per unit of power to the infinite bus with the voltage of bus-1 being 1.05 per unit. While the generator is operating in a steady state, a three-phase

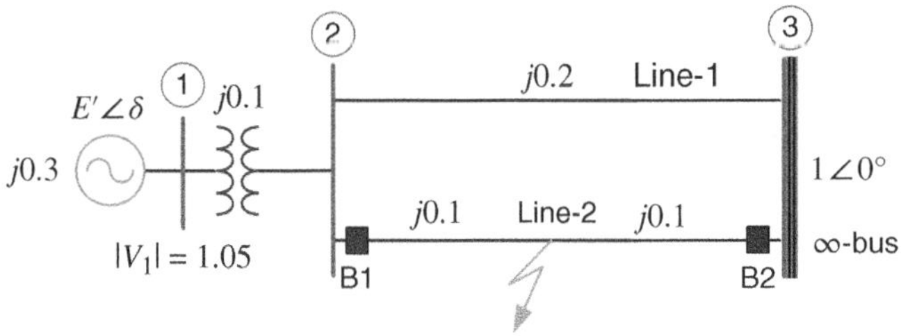

Figure 7.7 An SMIB system connected through a double circuit line.

bolted short circuit occurs at the midpoint of line-2, as shown in Figure **7.7**. The fault is cleared by opening the circuit breakers B1 and B2.

Before the fault, the equivalent impedance between bus-1 and the infinite bus is

$$X_{eq} = 0.1 + \frac{0.2 \times 0.2}{0.4} = 0.2 \text{ per unit}$$

Then the power from bus-1 to the infinite bus is given by

$$P_e = 1 = \frac{1.05 \times 1}{X_{eq}} \sin(\delta_1) \Rightarrow \delta_1 = \sin^{-1}\left(\frac{0.2}{1.05}\right) = 10.9806°$$

where δ_1 is the angle of bus-1 voltage. Therefore, the voltage of bus-1 is $V_1 = 1.05 \angle 10.98°$ per unit. Now, the current through the equivalent impedance is

$$I = \frac{1.05\angle 10.98° - 1}{j0.2} = 1.0118\angle - 8.75° \text{ per unit}$$

The reactance between the machines internal voltage and the bus-1 voltage is $j0.3$ per unit. The machines internal voltage is

$$E'\angle\delta_0 = V_1 + j0.3I = 1.1874\angle 24.904° \text{ per unit}$$

Therefore, the machines internal voltage is $E' = 1.1874$ per unit and its angle (δ_0) is 24.904° or 0.4347 rad. As the next step, we shall compute the maximum power that can be transferred (a) before the fault, (b) after the fault is cleared, and (c) during the fault, which is not zero for the circuit of Figure 7.7.

Pre-fault: The reactance between the machine voltage and the infinite bus is $j(0.3 + X_{eq}) = j0.5$ per unit. Therefore, the maximum power is given by

$$P_{m1} = \frac{1.1874}{0.5} = 2.3747 \text{ per unit}$$

Post-fault: After the fault is cleared by opening the circuit breakers B1 and B2, the power transfer through line-2 is zero, and therefore, the line joining bus-4 and bus-3 is also zero. Then, the impedance between the generator and the infinite bus is $j(0.3+0.1+0.2) = j0.6$ per unit. The post-fault power transfer equation is given by

$$P_{m3} = \frac{1.1874}{0.6} = 1.9789 \text{ per unit}$$

Furthermore, the maximum allowable angle to which the generator rotor can swing is

$$\delta_{max} = \pi - \sin^{-1}\left(\frac{1}{2.2688}\right) = 2.6118 \text{ rad}$$

During Fault: The power transfer during fault is not zero for the circuit of Figure 7.7. The equivalent circuit during fault is shown in Figure 7.8 (a). Once Bus-4 is short circuited, the impedances of the three lines form a Δ-connection, as shown in the figure. These are then converted into an equivalent Y-connection. With respect to the circuit of Figure 7.8 (a), let us define the following:

$$X_{ab} = j0.2, X_{bc} = j0.1 \text{ and } X_{ca} = j0.1$$

The Δ-Y transform of these quantities are

$$X_a = \frac{X_{ab} \times X_{ca}}{X_{ab} + X_{bc} + X_{ca}} = j\frac{0.2 \times 0.1}{0.5} = j0.05$$

$$X_b = \frac{X_{ab} \times X_{bc}}{X_{ab} + X_{bc} + X_{ca}} = j\frac{0.2 \times 0.1}{0.5} = j0.05$$

$$X_c = \frac{X_{bc} \times X_{ca}}{X_{ab} + X_{bc} + X_{ca}} = j\frac{0.1 \times 0.1}{0.5} = j0.025$$

The equivalent circuit is shown in Figure 7.8 (b).

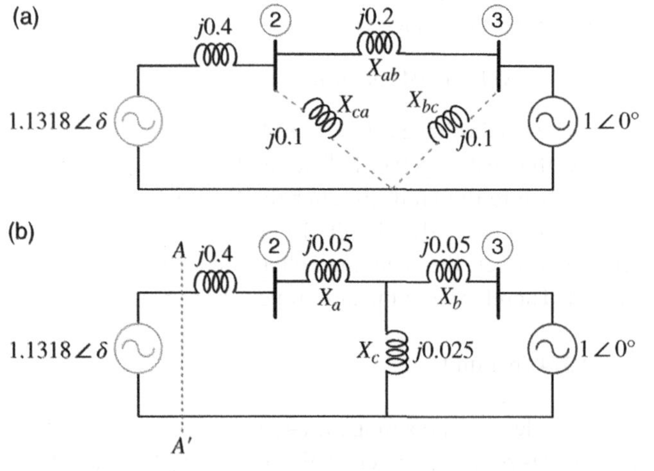

Figure 7.8 Equivalent circuit during fault: (a) Δ formation and (b) after Δ-Y transformation.

The Thevenin impedance to the right of line joining points A and A' of Figure 7.8 (b) is

$$X_{th} = j0.4 + X_a + \frac{X_b \times X_c}{X_b + X_c} = j0.4667 \text{ per unit}$$

Furthermore, the open circuit Thevenin voltage is

$$V_{th} = 1\angle 0^\circ \times \frac{X_c}{X_b + X_c} = 0.3333\angle 0^\circ \text{ per unit}$$

Therefore, the maximum power during fault is

$$P_{m2} = \frac{1.1874 \times 0.3333}{0.4667} = 0.8481 \text{ per unit}$$

The power–angle curves are shown in Figure 7.9.

From Figure 7.9, the accelerating area is given by

$$A_1 = \int_{\delta_0}^{\delta_{cr}} (1 - P_{m2} \sin \delta)\, d\delta$$

Expanding the above equation, we get

$$A_1 = \delta_{cr} - 0.4347 + 0.8481 \cos \delta_{cr} - 0.8481 \cos(0.4347)$$
$$= \delta_{cr} + 0.8481 \cos \delta_{cr} - 1.2039$$

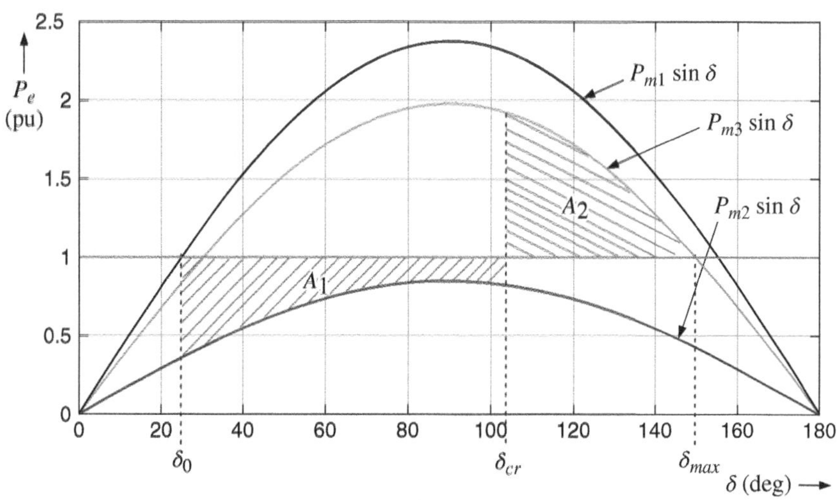

Figure 7.9 The power–angle curve of the system of Example 7.5.

Again, the decelerating area is

$$A_2 = \int_{\delta_{cr}}^{\delta_{max}} (P_{m3}\sin\delta - 1)\, d\delta$$

Expanding the above equation, we get

$$A_2 = -1.9789\cos(2.6118) + 1.9789\cos\delta_{cr} - 2.6118 + \delta_{cr}$$

$$= \delta_{cr} + 1.9789\cos\delta_{cr} - 0.9042$$

For finding the critical clearing angle, the two areas are equated to give

$$\delta_{cr} + 0.8481\cos\delta_{cr} - 1.2039 = \delta_{cr} + 1.9789\cos\delta_{cr} - 0.9042$$

$$\Rightarrow \delta_{cr} = \cos^{-1}\left(\frac{-0.2997}{1.1308}\right) = 1.8391 \text{ rad} = 105.37°$$

In the above example, we have calculated δ_0 and δ_{cr}. Then, (7.22) can be written during the fault as

$$\frac{2H}{\omega_s}\frac{d^2\delta}{dt^2} = P_m - P_{m2}\sin\delta$$

This equation needs to be integrated twice within the intervals of δ_0 and δ_{cr}, which is rather difficult. To solve this problem, a step-by-step integration procedure using Euler's method is employed to determine the clearing time in [3–6]. Instead, a simple MATLAB-based method is used to determine the clearing time here.

Equation (7.37) is written in a vector–matrix form as

$$\frac{d}{dt}\mathbf{x} = \mathbf{A}\mathbf{x} + \mathbf{B}(P_m - P_e) \tag{7.38}$$

where

$$P_e = P_{m2}\sin\delta, \mathbf{x} = \begin{bmatrix} \Delta\omega_s \\ \delta \end{bmatrix}, \mathbf{A} = \begin{bmatrix} 0 & 0 \\ \omega_s & 0 \end{bmatrix} \text{ and } \mathbf{B} = \begin{bmatrix} \frac{1}{2H} \\ 0 \end{bmatrix}$$

Choosing an integration interval of Δt, (7.38) can be discretized as [7]

$$\mathbf{x}(k+1) = \mathbf{F}\mathbf{x}(k) + \mathbf{G}P_m - \mathbf{G}P_e(k) \tag{7.39}$$

where

$$\mathbf{F} = e^{\mathbf{A}\Delta T} \text{ and } \mathbf{G} = \left(\int_0^{\Delta T} e^{\mathbf{A}t}dt\right)\mathbf{B}$$

Note that the discretization process of (7.39) assumes that the forcing function remains constant between the instants k and $k+1$. The term P_m is constant.

However, P_e varies with sin δ. If Δt is chosen small, then the variation of P_e between k and $k+1$ will be small. Let us now demonstrate this with the help of the following example.

Example 7.6 This is a continuation of Example 7.5, where we have $P_{m3} = 0.8083$ per unit and $\delta_0 = 0.3802$ rad. Let us choose $H = 2$ MJ/MVA and a frequency of 50 Hz, that is, $\omega_s = 100\pi$ rad/s. Choosing $\Delta t = 10^{-6}$ s, the matrices **F** and **G** are computed as

$$\mathbf{F} = \begin{bmatrix} 1 & 0 \\ 0.0003 & 1 \end{bmatrix} \text{ and } \mathbf{G} = \begin{bmatrix} 0.25 \times 10^{-6} \\ 0 \end{bmatrix}$$

The initial state vector is $\mathbf{x} = [\delta_0 \;\; 0]^T$. The MATLAB program is given below.

```
del = del0;

while (del < = dcr)

  x = f*x + g*(1-Pm2*sin[del]);
  t = t + td;
  indx = indx+1;
  del = x(2);

end

The answers are del and t
```

The clearing time with different values of H is listed in Table 7.2. The response of the system for five values of H is shown in Figure 7.10, where the clearing times given in Table 7.2 are chosen as the fault clearing times. From this figure, it is evident that the system is on the verge of instability for all the cases. Moreover, as the system inertia increases, the frequency of oscillation decreases.

Table 7.2 The effect of inertia on the critical clearing angle of Example 7.6.

H (MJ/MVA)	t_{cr} (s)	Clearing time (s)
2	0.2745	105.3339
4	0.3882	105.3530
6	0.4754	105.3456
8	0.5490	105.3601
10	0.6138	105.3609

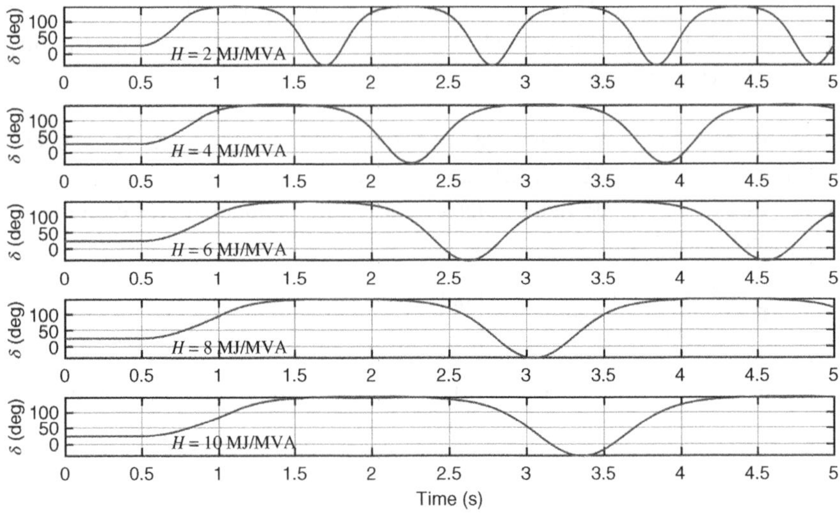

Figure 7.10 System response for five values of H for Example 7.6.

7.1.5 Simplified Calculation of Critical Clearing Angle

There is a simplified formula that can be used to calculate the critical clearing angle, as has been given in [4, 5]. Let us define the following:

Before fault: $P_e = P_{max} \sin \delta$

During fault: $P_e = r_1 P_{max} \sin \delta$

Post-fault: $P_e = r_2 P_{max} \sin \delta$

From (7.30), the accelerating area is given by

$$A_1 = \int_{\delta_0}^{\delta_{cc}} (P_m - r_1 P_{max} \sin \delta) d\delta = [P_m \, \delta + r_1 P_{max} \cos \delta]_{\delta_0}^{\delta_{cr}} \\ = P_m(\delta_{cr} - \delta_0) + r_1 P_{max}(\cos \delta_{cr} - \cos \delta_0) \tag{7.40}$$

Similarly, the deceleration is given from (7.31) as

$$A_2 = \int_{\delta_{cr}}^{\delta_{max}} (r_2 P_{max} \sin \delta - P_m) \, d\delta = [-r_2 P_{max} \cos \delta - P_m \, \delta]_{\delta_{cr}}^{\delta_{max}} \\ = r_2 P_{max}(\cos \delta_{cr} - \cos \delta_{max}) + P_m(\delta_{cr} - \delta_{max}) \tag{7.41}$$

Equating the two areas, we have

$$P_m(\delta_{cr} - \delta_0) + r_1 P_{max}(\cos \delta_{cr} - \cos \delta_0) \\ = r_2 P_{max}(\cos \delta_{cr} - \cos \delta_{max}) + P_m(\delta_{cr} - \delta_{max})$$

The above equation is rearranged as

$$P_{max}(r_2 - r_1)\cos\delta_{cr} = P_m(\delta_{max} - \delta_0) + r_2 P_{max}\cos\delta_{max} - r_1 P_{max}\cos\delta_0 \tag{7.42}$$

The critical clearing angle is then given by

$$\delta_{cr} = \cos^{-1}\left[\frac{\left(P_m/P_{max}\right)(\delta_{max} - \delta_0) + r_2\cos\delta_{max} - r_1\cos\delta_0}{r_2 - r_1}\right] \tag{7.43}$$

Example 7.7 In this example, we shall verify the answers obtained in Examples 7.3 and 7.4.

a) From Example 7.3, we have the following:

Before fault: $P_e = P_{max}\sin\delta = 2.2164\sin\delta$

During fault: $P_e = r_1 P_{max}\sin\delta = 0$

Post-fault: $P_e = r_2 P_{max}\sin\delta = 2.2164\sin\delta$

Therefore,

$$r_1 = 0 \text{ and } r_2 = 1$$

Furthermore, $P_m = 0.9$ per unit, $\delta_0 = 0.4182$ rad, and $\delta_{max} = 2.7234$ rad. Substituting these values in (7.43), we have

$$\delta_{cr} = \cos^{-1}[(0.9/2.2164)(2.7234 - 0.4182) + \cos(2.7234)] = 1.5486 \text{ rad} = 88.73°$$

b) From Example 7.5, we have the following:

Before fault: $P_e = P_{max}\sin\delta = 2.3747\sin\delta$

During fault: $P_e = r_1 P_{max}\sin\delta = 0.8481\sin\delta$

Post-fault: $P_e = r_2 P_{max}\sin\delta = 1.9789\sin\delta$

$P_m = 1$ per unit, $\delta_0 = 0.3802$ rad and $\delta_{max} = 2.684$ rad

Therefore,

$$r_1 = \frac{0.8481}{2.3747} = 0.3571 \text{ and } r_2 = \frac{1.9789}{2.3747} = 0.8333$$

Substituting these values in (7.43), we have

$$\delta_{cr} = \cos^{-1}(-0.2651) = 1.8391 \text{ rad} = 105.37°$$

7.2 Multimachine System Stability

A helpful mechanical analog of multimachine stability has been given by Elgerd [8]. Consider Figure 7.11, which depicts several weights that are suspended by elastic strings. The weights represent generators, with the electric transmission lines being represented by the strings. Note that in a transmission system, each transmission line is loaded below its static stability limit. Similarly, when the mechanical system is in a static steady state, each string is loaded below its break point. Let us suppose that the strings are at their stretching points when one of the strings is suddenly cut. This will result in transient oscillations in the coupled strings and all the weights will wobble. In the best possible case, the coupled system may settle down to a new steady state. On the other hand, in the worst possible scenario, the disturbance may lead to the breaking of one more additional string, resulting in a chain reaction in which more strings may break forcing the system to collapse. In a similar way, in an interconnected electric power network, the tripping of a transmission line may cause a catastrophic failure in which many generators are lost forcing a blackout in a large area.

Modern power systems are interconnected and operate close to their transient and steady-state stability limits. In large interconnected systems, it is common to find a natural response of a group of closely coupled machines oscillating against other groups of machines. These oscillations have a frequency range of 0.1 Hz–0.8 Hz. The lowest-frequency mode involves all generators of the system [2]. This oscillation groups the system into two parts with generators in one part oscillating against the other part. The higher-frequency modes are usually localized with small groups oscillating against each other. Unfortunately, the inter-area oscillation can be initiated by a small disturbance in any part of the system. These small frequency oscillations fall under the category of dynamic stability and

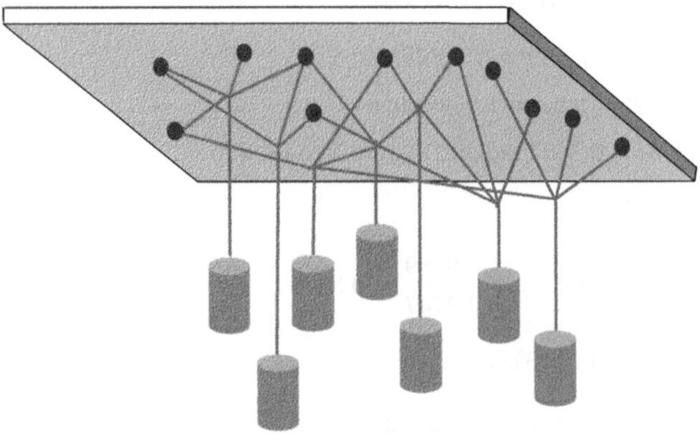

Figure 7.11 Mechanical analog of an interconnected electric power network.

are analyzed in the small signal linear domain through the linearization of the entire interconnected systems model.

Inter-area oscillations manifest wherever the power system is heavily interconnected. The oscillations, unless damped, can lead to grid failure and total system collapse. Low-frequency oscillations in the range of 0.04 Hz–0.06 Hz were observed in the Pacific Northwest region as early as 1950 [9]. Improper speed governor control of hydro units created these oscillations. The Northern and Southern regions of WSCC (Western Systems Coordinating Council) were interconnected by a 230 kV line in 1964. Immediately the system experienced a 0.1 Hz oscillation resulting in over 100 instances of opening of the tie line in the first nine months of operation. Some system damping was provided through the modification of the hydro turbine governors.

A 500 kV Pacific intertie and another ±400 HVDC (High voltage DC) system were commissioned in 1968. This raised the frequency of oscillation from 0.1 Hz to 0.33 Hz and these oscillations could no longer be controlled through governor action alone. In late the 1980s, a new intertie joined the WSCC system in Alberta and British Columbia in Canada. As a result of this interconnection, the two different oscillation frequencies manifested – one at 0.29 Hz and the other at 0.45 Hz.

Ontario Hydro is one of the largest utilities in North America. Due to the vast and sparsely populated topology of Canada, the operating span of Ontario Hydro is over 1000 km from East to West and from North to South. The Ontario Hydro system is connected to the neighboring Canadian provinces and the Northwestern region of the United States. In 1959, Ontario Hydro was connected to Michigan in the South and Quebec Hydro in the East. As a result of this connection, a 0.25 Hz oscillation was observed and as a result of this, it was decided to remove the tie with Quebec and retain the tie to Michigan [9]. The Western portion of Ontario was connected to neighboring Manitoba in 1956 and then Manitoba was connected to its neighbor Saskatchewan in 1960. This resulted in oscillation in the frequency range 0.35 Hz–0.45 Hz often tripping the tie. As a result of this, Ontario Hydro decided to commission PSS for all their generating units since the early 1960s. It has also sponsored extensive research in this area.

Through research, it was established that the action of AVRs caused these oscillations. An AVR regulates the generator terminal voltage and also helps in the enhancement of transient stability by reducing the peak of the first swing following any disturbance. However, its high gain contributed to the negative damping of the system [10]. The knowledge of this relation resulted in the commissioning of PSS. It was observed that these oscillations were the results of the periodic interchange of kinetic energy between the generator rotors. A PSS provides negative feedback in response to the changes in rotor kinetic energy when it is connected to the excitation system thereby providing damping to these small oscillations. The PSS has been the subject of extensive research. Its characteristics, tuning, and choice of feedback signal have thoroughly been discussed in a three-part paper [11]. The team of Dr. P. Kundur, then with Ontario Hydro, and his coworkers have done extensive research on PSS tuning and its characteristics. Through their vast experience and extensive research, they reported the enhancement of inter-area and local modes through PSS in [12]. Because a power system is piece-wise linear, its system characteristics change with an operating point. Therefore, an adaptive controller that can tune with the changes in the system has been developed and reported in [13, 14]. It was shown that the adaptive PSS is effective in damping large as well as small disturbances.

The power flow between generators was seen in Section 7.1 to be dependent on the angle between those generators. The stable operating point of the power system is where the generated power at each station is matched by the electrical power sent out from that station. When there is a mismatch between electrical power output and the generated mechanical shaft power, the generator accelerates or decelerates at a rate determined by the power mismatch and the machine inertia as given in (7.22).

7.2.1 Classical Method

The classical method of transient multimachine stability studies assumes that each individual generator oscillates at a frequency between 0.5 and 2 Hz. This is then superimposed on the nominal frequency of 50 or 60 Hz. Thus, the transmission system frequency is not perturbed from the nominal frequency by a large amount, and the frequency-dependent system reactances do not change. There are a few further assumptions required. These are:

- The mechanical power input to each generator remains constant.
- As in the swing equation of (7.22), the damping is neglected.
- Each machine is represented by its constant internal voltage and constant transient reactance.
- All loads are Y-connected, and their values do not change during the transients.

Consider the power system that is used in load flow studies, as shown in Figure 7.12 (a). Bus-1 is assumed to be the slack bus, which is considered the infinite bus. The PV buses are numbered M_1 to M_n and PQ buses are numbered L_1 to L_k. As mentioned in Chapter 3, the generators are not included in the load flow procedure and the loads are represented by their real and reactive power drawn. The first step in the classical multimachine stability analysis is to perform load flow with the system data given. The next step is to convert the PQ loads by their equivalent impedances and include the generators, as shown in Figure 7.12 (b).

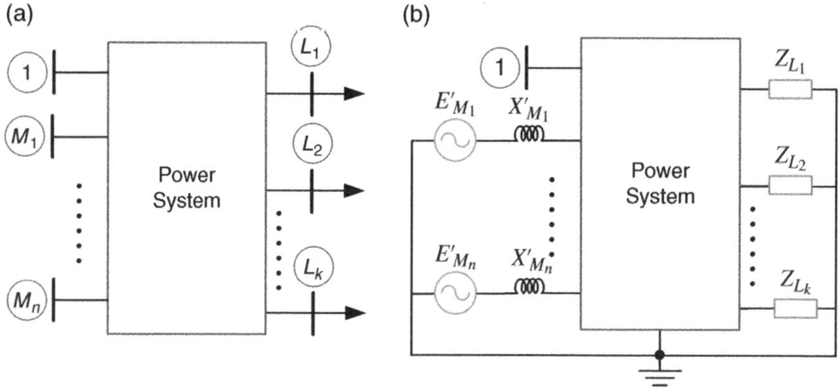

Figure 7.12 (a) Original power network and (b) its augmented version.

7.2.2 Pre-fault Bus Admittance Matrix

Let us consider bus-M_1 first. Load flow results for this bus are as follows:

$$V_{M_1} = |V_{M_1}|\angle\delta_{M_1} \text{ and } P_{M_1} + jQ_{M_1} = V_{M_1}I^*_{M_1}$$

Therefore,

$$I_{M_1} = \left(\frac{P_{M_1} + jQ_{M_1}}{V_{M_1}}\right)^* \tag{7.44}$$

Given the generator transient reactance, the generator transient internal voltage is computed as

$$E'_{M_1} = V_{M_1} + jX'_{M_1}I_{M_1} \tag{7.45}$$

The internal voltage of all the other PV buses can also be computed similarly. The loads are converted into constant RL loads, from the load flow data as well. The admittance for bus-L_1 is given by

$$Y_{L_1} = \frac{P_{L_1} - jQ_{L_1}}{|V_{L_1}|^2} \tag{7.46}$$

Converting the remaining load real and reactive power in admittances, the $\mathbf{Y_{bus}}$ matrix is modified to represent the pre-fault system.

Example 7.8 Consider the power system shown in Figure 7.13, where the infinite bus is assumed to be the slack bus. There are two generators connected to buses 2 and 3 and two loads connected to buses 4 and 5. The base voltage is chosen

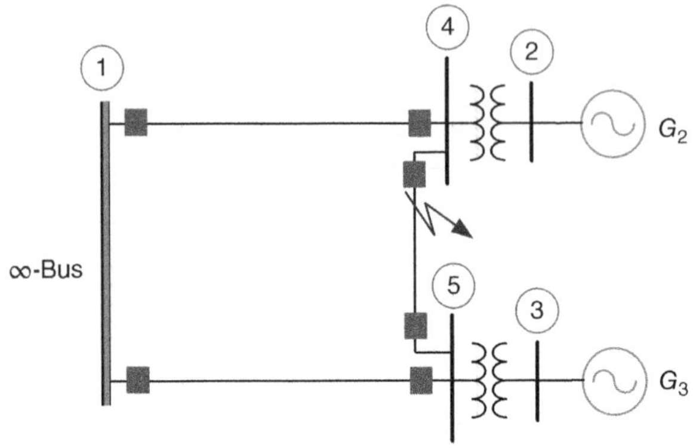

Figure 7.13 Five-bus power system of Example 7.8.

Table 7.3 Impedance data for the system of Figure 7.13.

Bus to bus	Resistance (pu)	Reactance (pu)
1-4	0.005	0.05
1-5	0.005	0.05
2-3	0	0.025
3-5	0	0.015
4-5	0.003	0.015

as 20 kV in the circuit of generator G_2. The base MV is chosen as 100. The impedance data are listed in Table 7.3. The generator parameters are selected as follows:

Generator G_2: $X'_2 = 0.075$ pu and $H = 10$ MJ/MVA
Generator G_3: $X'_3 = 0.1$ pu and $H = 8$ MJ/MVA

The bus admittance matrix is given by

$$\mathbf{Y_{bus}} = \begin{bmatrix} 3.96 - j39.6 & 0 & 0 & -1.98 + j19.8 & -1.98 + j19.8 \\ 0 & -j40 & 0 & j40 & 0 \\ 0 & 0 & -j66.67 & 0 & j66.67 \\ -1.98 + j19.8 & j40 & 0 & 14.8 - j123.9 & -12.82 + j64.1 \\ -1.98 + j19.8 & 0 & j66.67 & -12.82 + j64.1 & 14.8 - j150.57 \end{bmatrix}$$

The load flow results are shown in Table 7.4.

We shall now augment the system, as shown in Figure 7.12 (b) using (7.44) to (7.46). Each of the buses is considered separately.

Table 7.4 Load flow results for Example 7.8.

		Generation		Load	
Bus number	**Voltage**	*P*	*Q*	*P*	*Q*
1	$1.0\angle 0°$	−2.3352	−0.4511	0	0
2	$1.05\angle 6.695°$	2.5	1.2835	0	0
3	$1.02\angle 4.084°$	1.2	0.2633	0	0
4	$1.0212\angle 3.354°$	0	0	−0.9	−0.5
5	$1.0163\angle 3.089°$	0	0	−0.45	−0.25

Bus-1: Slack bus.

$$E_1' = 1\angle 0°$$

Bus-2: From (7.44),

$$I_2 = \left(\frac{2.5 + j1.2845}{1.05\angle 6.695}\right)^* = 2.5073 - j0.9663$$

From (7.45),

$$E_2' = V_2 + jX_2' I_2 = 1.05\angle 6.695° + j0.075 I_2 = 1.113 + j0.31 = 1.1556\angle 15.588°$$

Bus-3:

$$I_3 = \left(\frac{1.2 + j0.2633}{1.02\angle 4.084}\right)^* = 1.1919 - j0.1737$$

$$E_3' = V_3 + jX_3' I_3 = 1.03\angle 4.084° + j0.1 I_2 = 1.035 + j0.192 = 1.0524\angle 10.504°$$

Bus-4: Using (7.46),

$$Y_4 = \frac{P_4 - jQ_4}{|V_4|^2} = \frac{0.9 - j0.5}{1.0212^2} = 0.8630 - j0.4795$$

Bus-5:

$$Y_5 = \frac{P_5 - jQ_5}{|V_5|^2} = \frac{0.45 - j0.25}{1.0163^2} = 0.4357 - j0.2420$$

The augmented pre-fault $\mathbf{Y_{bus}}$ matrix is then formed. These are given by

$$Y_{b22} = \frac{1}{j0.025 + j0.075} = -j10$$

$$Y_{b24} = Y_{b42} = -Y_{b22}$$

$$Y_{b33} = \frac{1}{j0.025 + j0.1} = -j8.696$$

$$Y_{b35} = Y_{b53} = -Y_{b33}$$

The $\mathbf{Y_{bus}}$ matrix is formed using the following two steps. First, modify the matrix using the parameters obtained above. This will produce the following matrix:

$$\mathbf{Y_{bus}} = \begin{bmatrix} 3.96-j39.6 & 0 & 0 & -1.98+j19.8 & -1.98+j19.8 \\ 0 & -j10 & 0 & j10 & 0 \\ 0 & 0 & -j8.698 & 0 & j8.698 \\ -1.98+j19.8 & j10 & 0 & 14.8-j123.9 & -12.82+j64.1 \\ -1.98+j19.8 & 0 & j8.698 & -12.82+j64.1 & 14.8-j150.57 \end{bmatrix}$$

Because of these changes, in the second and third rows of the $\mathbf{Y_{bus}}$ matrix, the diagonal elements of the last two rows will also change. Moreover, the admittances Y_4 and Y_5 also need to be added with the diagonal elements. Therefore, these elements are obtained as follows:

$$Y_{b44} = \sum\nolimits_{i=1}^{5} Y_{b4i} - Y_{b44} + Y_4 = 15.664 - j95.384$$

$$Y_{b55} = \sum\nolimits_{i=1}^{5} Y_{b5i} - Y_{b55} + Y_5 = 15.236 - j92.842$$

The pre-fault $\mathbf{Y_{bus}}$ matrix is then

$$Y_{bus}^{pre-fault} = \begin{bmatrix} 3.96-j39.6 & 0 & 0 & -1.98+j19.8 & -1.98+j19.8 \\ 0 & -j10 & 0 & j10 & 0 \\ 0 & 0 & -j8.698 & 0 & j8.698 \\ -1.98+j19.8 & j10 & 0 & 15.664-j95.384 & -12.82+j64.1 \\ -1.98+j19.8 & 0 & j8.698 & -12.82+j64.1 & 15.236-j92.842 \end{bmatrix}$$

7.2.3 Reduction of Bus Admittance Matrix

Sometimes it is desirable to reduce the network by eliminating the nodes in which the current does not enter or leave. The bus admittance is then written in the form

$$\begin{bmatrix} \mathbf{I_A} \\ \mathbf{0} \end{bmatrix} = \begin{bmatrix} \mathbf{K} & \mathbf{L} \\ \mathbf{L^T} & \mathbf{M} \end{bmatrix} \begin{bmatrix} \mathbf{V_A} \\ \mathbf{V_x} \end{bmatrix} \tag{7.47}$$

In the above equation, $\mathbf{I_A}$ is a vector containing the currents that are injected, and the $\mathbf{Y_{bus}}$ matrix is portioned with the matrices **K**, **L**, and **M**. Note that the $\mathbf{Y_{bus}}$ matrix contains both **L** and $\mathbf{L}^T$ due to its symmetric nature.

We get the following two sets of equations from (7.47)

$$\mathbf{I_A} = \mathbf{KV_A} + \mathbf{LV_x} \tag{7.48}$$

$$\mathbf{0} = \mathbf{L^T V_A} + \mathbf{MV_x} \Rightarrow \mathbf{V_x} = -\mathbf{M^{-1}L^T V_A} \tag{7.49}$$

Substituting (7.49) in (7.48), we get

$$\mathbf{I_A} = \left(\mathbf{K} - \mathbf{LM^{-1}L^T}\right)\mathbf{V_A} \tag{7.50}$$

From (7.50), the reduced bus admittance matrix is obtained as

$$\mathbf{Y_{bus}^{reduced}} = \mathbf{K} - \mathbf{LM^{-1}L^T} \tag{7.51}$$

7.2.4 Bus Admittance Matrices During Fault and Post-Fault

The $\mathbf{Y_{bus}}$ reduction is now used to calculate the bus admittance matrices during faults and post-fault.

Example 7.9 This is a continuation of Example 7.8. Equation (7.51) will be used to obtain the two bus admittance matrices. For this, the fault is assumed to have occurred on line 4-5, immediately adjacent to bus-4. Because the fault has occurred on bus-4, the 4th row, and the 4th column of the admittance of the admittance matrix are removed. Furthermore, because current is injected at bus-5, this bus can be removed from the computation using (7.51). We then have

$$\boldsymbol{K} = \begin{bmatrix} 3.96 - j39.6 & 0 & 0 \\ 0 & -j10 & 0 \\ 0 & 0 & -j8.698 \end{bmatrix}, \boldsymbol{L} = \begin{bmatrix} -1.98 + j19.8 \\ 0 \\ j8.698 \end{bmatrix} \text{ and}$$

$$M = 15.236 - j92.842$$

The bus admittance matrix during fault will then contain buses 1–3, where the impedance matrix is computed using (7.51) as

$$Y_{bus}^{df} = \begin{bmatrix} 3.806 - j35.397 & 0 & 0.116 + j1.836 \\ 0 & -j10 & 0 \\ 0.116 + j1.836 & 0 & 0.13 - j7.903 \end{bmatrix}$$

The fault is removed by the opening of the two breakers at the two ends of line 4-5. This modifies the diagonal elements of the bus admittance matrix as

$$Y_{b44} = Y_{b44} - Y_{b45} \text{ and } Y_{b55} = Y_{b55} - Y_{b54}$$

The resulting bus admittance matrix is then

$$\mathbf{Y_{bus}} = \begin{bmatrix} 3.96 - j39.6 & 0 & 0 & -1.98 + j19.8 & -1.98 + j19.8 \\ 0 & -j10 & 0 & j10 & 0 \\ 0 & 0 & -j8.698 & 0 & j8.698 \\ -1.98 + j19.8 & j10 & 0 & 2.843 - j30.281 & 0 \\ -1.98 + j19.8 & 0 & j8.698 & 0 & 2.416 - j28.74 \end{bmatrix}$$

The last two rows and columns are eliminated using (7.51) to produce the following post-fault bus admittance matrix.

$$\mathbf{Y_{bus}^{pf}} = \begin{bmatrix} 1.004 - j13.015 & -0.04 + j6.543 & -0.095 + j5.999 \\ -0.04 + j6.543 & 0.307 - j6.727 & 0 \\ -0.095 + j5.999 & 0 & 0.22 - j6.083 \end{bmatrix}$$

7.2.5 Multimachine Swing Equation

For a multimachine system, the swing equation for machine-i is given by

$$\frac{2H_i}{\omega_s}\frac{d^2\delta_i}{dt^2} = P_{mi} - P_{ei} \tag{7.52}$$

In (7.52), P_{mi} is the mechanical power input to machine-i, and its electrical power output is given by

$$P_{ei} = \left|E_i'\right|^2 G_{ii} + \sum_{k=1,k\neq i}^{n} \left|E_i'\right|\left|E_k'\right|\left|Y_{bik}\right|\cos(\delta_{ik} - \theta_{ik}) \tag{7.53}$$

where $\delta_{ik} = \delta_i - \delta_k$. The elements of the $\mathbf{Y_{bus}}$ matrix will depend on the state of operation. For example, for the faulted operation, elements of $\mathbf{Y_{bus}^{df}}$ will be chosen and the elements of $\mathbf{Y_{bus}^{pf}}$ will be chosen for the post-fault operation.

Example 7.10 This is the continuation of Examples 7.8 and 7.9. We shall consider the faulted system first.

During fault: Since bus-4 is short circuited, we have

$$P_{e2} = 0$$

Noting from Table 7.4 that $P_{m2} = 2.5$ per unit, the swing equation for G_2 is then

$$\frac{d^2\delta_2}{dt^2} = \frac{\omega_s}{2\times 10}(2.5 - 0) = 0.125\omega_s$$

Using (7.53) and $\mathbf{Y}_{\mathbf{bus}}^{\mathbf{df}}$ computed in Example 7.9, the electric power supplied by G_3 is

$$P_{e3} = |E_3'|^2 G_{33}^{df} + |E_1'||E_3'|\left|Y_{b13}^{df}\right| \cos\left(\delta_{31} - \theta_{31}^{df}\right)$$

$$= (1.0524)^2 \times 0.13 + 1 \times 1.0524 \times 1.8393 \times \cos(\delta_3 - 86.39°)$$

$$= 0.1442 + 1.9357 \cos(\delta_3 - 86.39°)$$

It is to be noted that $Y_{b12}^{df} = 0$ and therefore this term does not appear in the above equation. Note from Table 7.4 that $P_{m3} = 1.2$ per unit. The swing equation is then given from (7.52) as

$$\frac{d^2\delta_3}{dt^2} = \frac{\omega_s}{16}[1.2 - \{0.1442 + 1.9357 \cos(\delta_3 - 86.39°)\}]$$

Post-fault: For this case, line 4-5 has been removed and therefore $Y_{b23} = 0$. Then, for bus-2 we have

$$P_{e2} = |E_2'|^2 G_{22}^{pf} + |E_1'||E_2'|\left|Y_{b21}^{pf}\right| \cos\left(\delta_2 - \theta_{21}^{pf}\right)$$

$$= (1.1556)^2 \times 0.3074 + 1 \times 1.1556 \times 6.5432(\delta_3 - 90.35°)$$

$$= 0.4104 + 7.561 \cos(\delta_3 - 90.91°)$$

The swing equation is then

$$\frac{d^2\delta_2}{dt^2} = \frac{\omega_s}{20}[2.5 - \{0.4104 + 7.561 \cos(\delta_2 - 90.35°)\}]$$

Similarly, for bus-3, the power equation is given by

$$P_{e3} = |E_3'|^2 G_{33}^{pf} + |E_1'||E_3'|\left|Y_{b13}^{pf}\right| \cos\left(\delta_{31} - \theta_{13}^{pf}\right)$$

$$= (1.0524)^2 \times 0.2196 + 1 \times 1.0524 \times 6 \times \cos(\delta_3 - 90.901°)$$

$$= 0.2432 + 6.3146 \cos(\delta_3 - 90.901°)$$

The swing equation is then

$$\frac{d^2\delta_3}{dt^2} = \frac{\omega_s}{16}[1.2 - \{0.2432 + 6.3146 \cos(\delta_3 - 90.901°)\}]$$

The system is now simulated using (7.37), where $\omega = 100\pi$ (50 Hz). The angle response of the two generators is shown in Figure 7.14, when the clearing time is 3 cycles (0.06 s). It can be seen that both generators are stable. However, when the clearing time is 0.26 s, generator G_2 becomes unstable, as shown in Figure 7.15. However, generator G_2 remains stable. Several methods for solving the multimachine problem are given in [15].

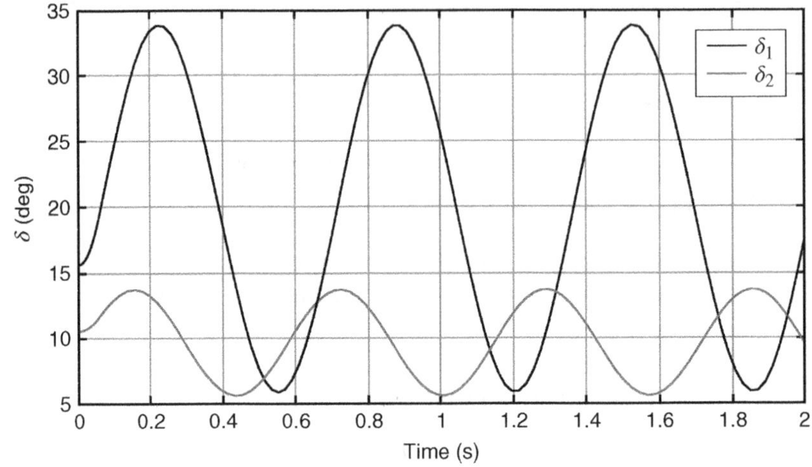

Figure 7.14 Response of generator angles when the clearing time is 0.06 s.

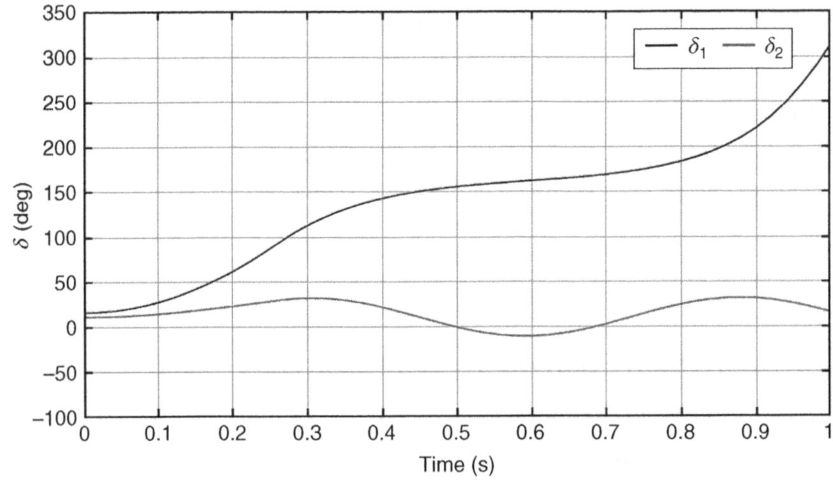

Figure 7.15 Response of generator angles when the clearing time is 0.26 s.

7.2.6 Oscillations in a Two-Area System

Consider the simple power system shown in Figure 7.16 in which two areas are represented by two machines that are connected through a tie-line. Let us assume that starting with the initial angles δ_1 and δ_2 with respect to some reference at

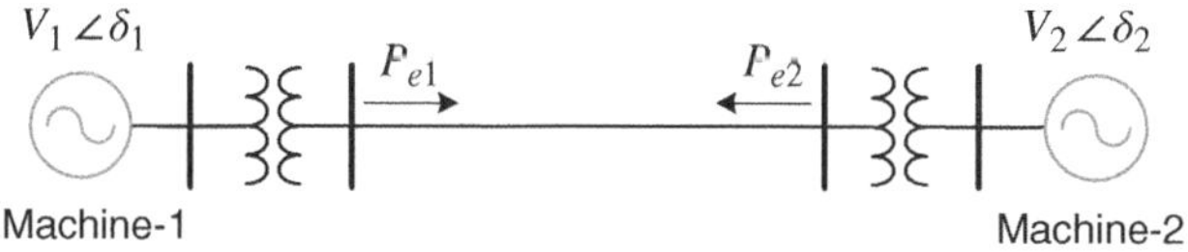

Figure 7.16 A simple two-area power system.

nominal frequency, Machine-1 accelerates while Machine-2 decelerates from this nominal frequency. We then have

$$\begin{aligned}\frac{2H_1}{\omega_s}\ddot{\delta}_1 &= P_{m1} - P_{e1}\\ \frac{2H_2}{\omega_s}\ddot{\delta}_2 &= P_{m2} - P_{e2}\end{aligned} \tag{7.54}$$

where the subscripts 1 and 2 refer to machines 1 and 2, respectively. Let us assume that the transmission line is lossless. Then, in the simple case where the power from Machine-1 flows to Machine-2, we get

$$P_{e1} = -P_{e2} = \frac{V_1V_2}{X}\sin(\delta_1 - \delta_2) = \frac{V_1V_2}{X}\sin\delta_{12} \tag{7.55}$$

where $\delta_{12} = \delta_1 - \delta_2$.

Now because the system is lossless, (7.55) will also imply that $P_{m1} = -P_{m2}$. This means that in the steady state, the power generated at Machine-1 is absorbed by Machine-2. Combining (7.54) with (7.55), the following equation is obtained:

$$\frac{2H_1}{\omega_s}\ddot{\delta}_1 - \frac{2H_2}{\omega_s}\ddot{\delta}_2 = 2P_{m1} - P_{e1} + P_{e2} = 2P_{m1} - \frac{2V_1V_2}{X}\sin\delta_{12} \tag{7.56}$$

Let us now assume the following:

$$H_1 = H_2 = H, V_1 = V_2 = 1.0 \text{ per unit and } P_{m1} = 0$$

Then (7.56) is rewritten as

$$\ddot{\delta}_{12} = -\frac{\omega_s}{HX}\sin\delta_{12} = -\omega^2\sin\delta_{12} \tag{7.57}$$

where the oscillation frequency ω is given by

$$\omega = \sqrt{\omega_s/HX} \tag{7.58}$$

Thus, the weighted difference of angles approximate a simple harmonic motion for small changes in δ_{12} and the frequency will decrease for an increase in inertia H

or impedance X. Another aspect can be seen by adding the two equations in (7.54), which gives

$$\begin{aligned} &\frac{2H_1}{\omega_s}\ddot{\delta}_1 + \frac{2H_2}{\omega_s}\ddot{\delta}_2 = P_{m1} + P_{m2} = 0 \\ &\Rightarrow H_1\ddot{\delta}_1 + H_2\ddot{\delta}_2 = 0 \end{aligned} \tag{7.59}$$

This implies that the overall acceleration of the machine group will depend on the overall balance between the power generated and consumed. Usually, there are governors on the generators to reduce generated power if the system frequency increases.

7.3 Excitation Control

The power transfer equation in an SMIB system is given in (7.3) and (7.6) as

$$P_e = \frac{V_1 V_2}{X}\sin\delta$$

It can be seen from this equation that P_e will drop if V_1 drops. To prevent this drop, a very fast-acting voltage regulator is required to hold the generator voltage to a reasonable level. In the transient stability problem studied in Section 7.1, it has been mentioned that the transient can last only for a short interval. During this time, a turbine-governor system cannot take necessary action due to its slow speed of response. Similarly, for a sudden increase in load, the slow response of the turbine-governor system will prevent it from supplying the load demand immediately. During this time, the stored kinetic energy in the rotor shaft is released first, followed by rotating energy till the governor action takes place.

The dynamic stability problem differs from the transient stability where the stability of a machine under small changes or low oscillation frequency is studied. Assume that several machines that are connected together through transmission lines are supplying a group of loads. At any given instant, one of the loads changes. The machines that are closest to the load will see large changes compared to the machines that are further from the load. As we have discussed in the previous paragraph, the governor's action will be inadequate to cater to the load change immediately. Then the machines that are closer to the load will contribute to the increase in the load. As a result, each machine will oscillate at its own natural frequency and thus, these machines will "ring" at different frequencies till the damping actions suppress the oscillations.

All the modern turboalternators are equipped with an AVR that regulates the output voltage through the control of the excitation system of the machine. During the first swing following a transient, the AVR rapidly alters the excitation input towards its ceiling to hold the output voltage constant. Therefore, AVRs usually have very high gains. This action may have deleterious effects on the system damping. This will be discussed next.

7.3.1 Linearized Swing Equation

The small signal linearized model of an SMIB system is discussed in [2, 10, 16]. The linearized generator swing equation is written as

$$\Delta\omega_r = \frac{1}{2Hs + K_D}(\Delta T_m - \Delta T_e) \tag{7.60}$$

$$\Delta\delta = \frac{\omega_s}{s}\Delta\omega_r \tag{7.61}$$

Where Δ indicates a small variation from the steady-state value and K_D is the damping constant, which has been neglected in (7.22). This quantity is multiplied by the machine speed ω to add damping to the swing equation. Equations (7.60) and (7.61) are represented by the block diagram of Figure 7.17 (a). Let us now

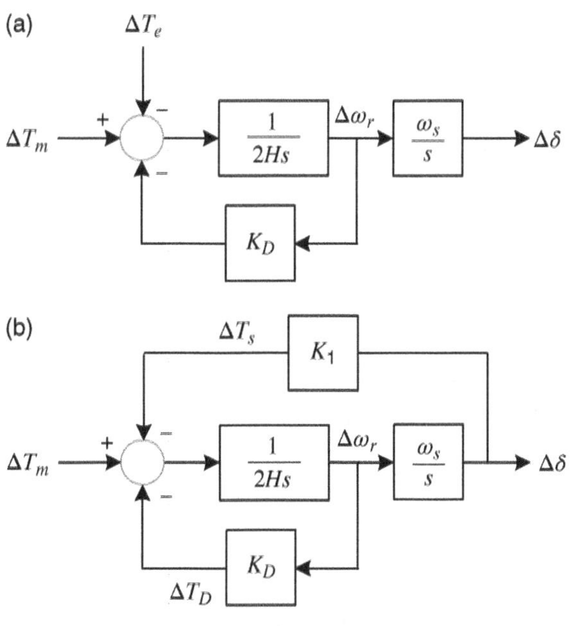

Figure 7.17 Linearized swing equation of an SMIB system: (a) swing equation and (b) with constant flux.

define a constant K_1 as the change in electrical torque for a change in rotor angle for a constant flux linkage in the d-axis, which is defined for constant excitation voltage $\Delta E_{fd} = 0$. The constant is then written as

$$K_1 = \left.\frac{\Delta T_e}{\Delta \delta}\right|_{\Delta E_{fd} = 0}$$

Then Figure 7.17 (a) is modified to Figure 7.17 (b).

Comparing Figure 7.17 (a) with Figure 7.17 (b), we can write the following expression:

$$\Delta T_e = \Delta T_S + \Delta T_D = K_1 \Delta \delta + K_D \Delta \omega_r \tag{7.62}$$

Two terms added to Figure 7.17 (b) are:

- Synchronizing torque (T_S), which is proportional to the changes in the machine angle.
- Damping torque (T_D), which is proportional to the changes in the machine speed.

From Figure 7.17 (b), we can write

$$\frac{\Delta \delta(s)}{\Delta T_m(s)} = \frac{G(s)}{1 + G(s)H(s)}$$

where

$$G(s) = \frac{\omega_s}{s} \frac{\frac{1}{2Hs}}{1 + \frac{K_D}{2Hs}} = \frac{\omega_s}{s} \frac{1}{2Hs + K_D} = \frac{\omega_s}{2Hs^2 + K_D s}$$

$$H(s) = K_1$$

Then the closed-loop transfer function is given as

$$\frac{\Delta \delta(s)}{\Delta T_m(s)} = \frac{\omega_s}{2Hs^2 + K_D s + K_1 \omega_s} = \frac{\omega_s/2H}{s^2 + (K_D/2H)s + K_1 \omega_s/2H} \tag{7.63}$$

From (7.63), the natural frequency of oscillation is given as

$$\omega_n = \sqrt{K_1 \omega_s/2H}$$

The natural frequency is inversely proportional to the under the root of the inertia constant, which is expected. The damping ratio is computed from

$$2\xi\omega_n = \frac{K_D}{2H} \Rightarrow \xi = \frac{K_D}{4H\sqrt{K_1 \omega_s/2H}} = \frac{K_D}{2\sqrt{2HK_1 \omega_s}}$$

The damping ratio is directly proportional to the damping constant K_D, and it decreases as the inertia constant increases. It can therefore be surmised that the system's stability can be endangered by the lack of either the synchronizing torque or the damping torque.

Example 7.11 Consider an SMIB system with the following parameters:

$$K_1 = 0.8, H = 4 \text{ and } \omega = 100\pi \text{ rad/s}.$$

From Figure 7.17 (b), we get the following equations:

$$\Delta\dot{\delta} = \omega_s \Delta\omega_r$$

$$\Delta\dot{\omega}_r = \frac{1}{2H}(\Delta T_m - K_1\Delta\delta - K_D\Delta\omega)$$

Choosing $\Delta P_m = 0.005$ per unit, the system response when $K_D = 0$ is shown in Figure 7.18. Because the damping torque is zero when $K_D = 0$, the system oscillates at the undamped natural frequency given by

$$\omega_n = \sqrt{K_1\omega_s/2H} = 5.605 \text{ rad/s} \Rightarrow f_n = 0.8921 \text{ Hz}$$

This implies that the oscillation period is 1.121 s. Obviously, the damping torque is zero.

The characteristic equation of the closed-loop system is given from (7.63) as

$$2Hs^2 + K_D s + K_1\omega_s = 0$$

Rearranging this, we have the following equation:

$$1 + \frac{K_D s}{2Hs^2 + K_1\omega_s} = 0$$

The root locus plot of the system is shown in Figure 7.19. It is obvious that all the roots are stable. Note that the undamped natural frequency (ω_n) depends on the system frequency, the constant K_1, and the inertia constant H and is independent of K_D. However, damping ratio and damped natural frequency are dependent on K_D. The damped natural frequency is given by $\omega_d = \omega_n\sqrt{1-\xi^2}$.

For $K_1 = 0.8$, the undamped natural frequency, the damped natural frequency, and the damped ratio are listed in Table 7.5. It can be seen that system damping increases with the increase in K_D.

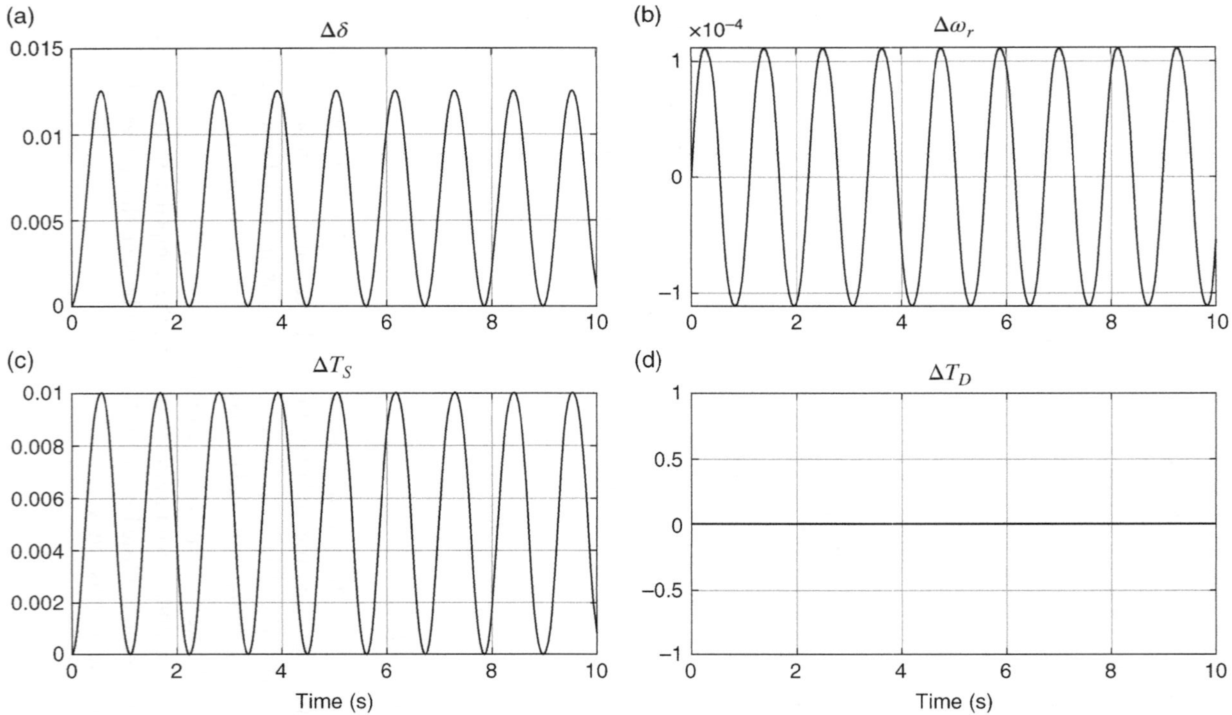

Figure 7.18 Linearized swing equation when $K_D = 0$. Changes in (a) machine angle, (b) machine speed, (c) synchronizing torque and (d) damping torque.

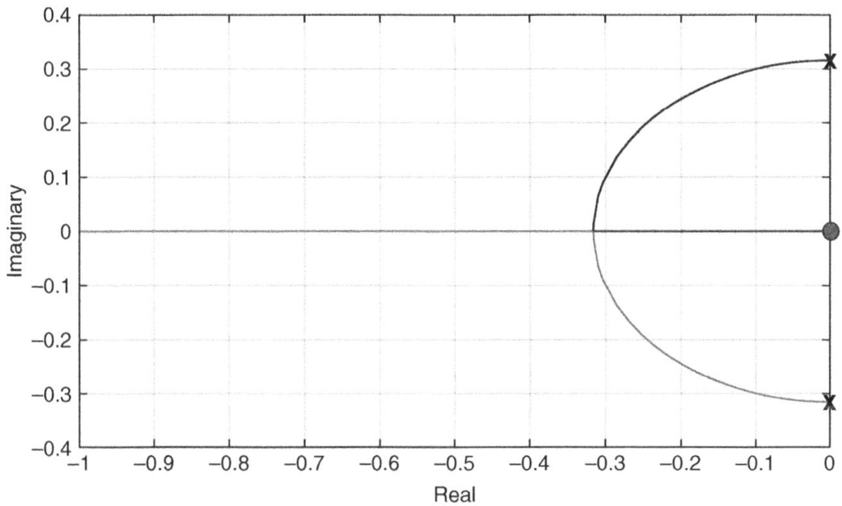

Figure 7.19 Root locus plot of the system of closed-loop system.

Table 7.5 Variation in system damping with H and K_D.

	$H = 4$ (MJ/MVA)			$H = 8$ (MJ/MVA)		
K_D	ω_n (rad/s)	ξ	ω_d (rad/s)	ω_n (rad/s)	ξ	ω_d (rad/s)
0	5.605	0	5.605	3.9633	0	3.9633
2	5.605	0.0223	5.6036	3.9633	0.0158	3.9628
4	5.605	0.446	5.5994	3.9633	0.0315	3.9614
6	5.605	0.0669	5.5924	3.9633	0.0473	3.9589
8	5.605	0.0892	5.5826	3.9633	0.0631	3.9554
10	5.605	0.1115	5.57	3.9633	0.0788	3.951

7.3.2 Excitation System

The block diagram of the SMIB system with constant excitation voltage (i.e., $\Delta e_{fd} = 0$) is shown in Figure 7.20. In this figure,

ψ_{fd} is the d-axis flux linkage,
K_3 and T_3 respectively are the gain and time constant of the field circuit, and
K_4 is the demagnetizing effect of the rotor angle on the field excitation.

From Figure 7.20, the electrical torque equation is given by

$$\Delta T_e(s) = \left[K_1 - \frac{K_2 K_3 K_4}{1 + T_3 s}\right] \Delta\delta(s) \tag{7.64}$$

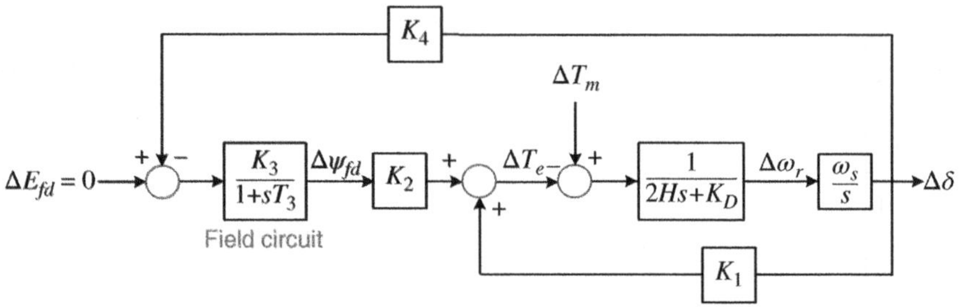

Figure 7.20 Block diagram SMIB system with constant excitation voltage.

The torque equation has two components – one purely due to the swing equation through K_1, and the other due to the field circuit, that is, due to $\Delta\psi_{fd}$. We can express the latter as follows:

$$\Delta T_e(s)|_{\Delta\psi_{fd}} = -\frac{K_2K_3K_4}{1+T_3s}\Delta\delta(s) \tag{7.65}$$

Replacing s in (7.65) by $j\omega$, the following equation is obtained:

$$\Delta T_e(j\omega)|_{\Delta\psi_{fd}} = -\frac{K_2K_3K_4}{1+jT_3\omega}\Delta\delta(j\omega) \tag{7.66}$$

At very low frequencies, the above torque equation is written as

$$\Delta T_e(j\omega)|_{\Delta\psi_{fd}} = -K_2K_3K_4\Delta\delta(j\omega)$$

From (7.64), the above equation implies that

$$\Delta T_e(j\omega) = (K_1 - K_2K_3K_4)\Delta\delta(j\omega)$$

Since K_2, K_3, and K_4 are positive, this torque component produces a negative synchronizing torque due to armature reaction. The synchronizing torque will become negative when $K_1 < K_2K_3K_4$. Therefore, the limiting value for the synchronizing torque is

$$K_1 = -K_2K_3K_4 \tag{7.67}$$

At higher frequencies, the torque component of (7.66) is given by

$$\Delta T_e(s)|_{\Delta\psi_{fd}} = -\frac{K_2K_3K_4}{jT_3\omega}\Delta\delta(s) = j\frac{K_2K_3K_4}{T_3\omega}\Delta\delta(s) \tag{7.68}$$

This implies that, at higher frequencies, the torque component due to $\Delta\psi_{fd}$ is phase quadrature, leading $\Delta\delta$ by 90°. This implies that this component is in phase with $\Delta\omega_r$ and, therefore, is the positive damping torque.

Example 7.12 Consider a system with the following parameters:

$$K_1 = 0.8, K_D = 2.5, K_2 = 0.89, K_3 = 0.35, K_4 = 1.5 \text{ and } T_3 = 2.35$$

Then, the steady-state synchronizing torque component is given by

$$K_S = K_1 - K_2K_3K_4 = 0.8 - 0.4672 = 0.3328$$

From (7.65), we have

$$\begin{aligned} \Delta T_e|_{\Delta\psi_{fd}} &= -\frac{K_2K_3K_4}{1-T_3^2s^2}\Delta\delta + \frac{K_2K_3K_4T_3}{1-T_3^2s^2}s\Delta\delta \\ &= -\frac{K_2K_3K_4}{1-T_3^2s^2}\Delta\delta + \frac{K_2K_3K_4T_3}{1-T_3^2s^2}\omega_s\Delta\omega_r \end{aligned}$$

The synchronizing and the damping torque are then given from the above equation as

$$\text{Synchronizing torque: } \Delta T_S|_{\Delta\psi_{fd}} = -\frac{K_2K_3K_4}{1-T_3^2s^2}$$

$$\text{Damping torque: } \Delta T_D|_{\Delta\psi_{fd}} = \frac{K_2K_3K_4T_3}{1-T_3^2s^2}\omega_s$$

The variations in these torques as the frequency changes from 1 rad/s to 10 rad/s are shown in Figure 7.21. For a frequency of 1 Hz, that is, 6.2832 rad/s, the two torques are given by

$$\Delta T_S|_{\Delta\psi_{fd}} = -0.0021 \text{ pu and } \Delta T_D|_{\Delta\psi_{fd}} = 1.575 \text{ pu}$$

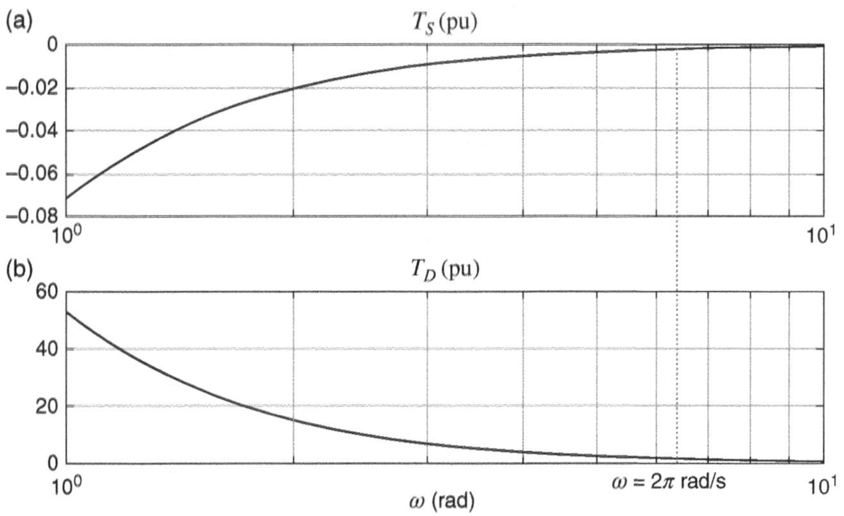

Figure 7.21 Variations in (a) synchronizing and (b) damping torques with frequency.

These points are also shown in Figure 7.21. The total synchronization coefficient is however equal to

$$\Delta T_S|_{\Delta\psi_{fd}} = 0.8 - 0.0021 = 0.7979 \text{ pu}$$

7.3.3 Automatic Voltage Regulator (AVR)

The excitation system with a static exciter of gain K_A is shown in Figure 7.22. The machine terminal voltage is measured through a transducer. The output of the transducer is compared with the reference voltage V_{ref}. We define two more constants as given below:

$$K_5 = \left.\frac{\Delta e_{fd}}{\Delta\delta}\right|_{\Delta\psi_{fd}=0} \quad \text{and} \quad K_6 = \left.\frac{\Delta e_{fd}}{\Delta\psi_{fd}}\right|_{\Delta\delta=0}$$

Then, the machine terminal voltage is written as

$$\Delta E_t = K_5\Delta\delta + K_6\Delta\psi_{fd} \tag{7.69}$$

The SMIB system of Figure 7.20 is then modified using (7.69) and Figure 7.22 as shown in Figure 7.23.

From Figure 7.23, we can write

$$\Delta T_e = K_1\Delta\delta + K_2\Delta\psi_{fd} \tag{7.70}$$

This implies that there are two torque components – one due to $\Delta\delta$, and the other due to $\Delta\psi_{fd}$. Again, note that $\Delta V_{ref} = 0$ in Figure 7.23 we have

$$\Delta\psi_{fd} = \frac{K_3}{1+sT_3}\left[-\frac{K_A}{1+sT_R}\left(K_5\Delta\delta + K_6\Delta\psi_{fd}\right) - K_4\Delta\delta\right] \tag{7.71}$$

Rearranging (7.70), the following expression is obtained:

$$\Delta\psi_{fd} = -\frac{K_AK_3K_5 + K_3K_4(1+sT_R)}{K_AK_3K_6 + (1+sT_3)(1+sT_R)}\Delta\delta \tag{7.72}$$

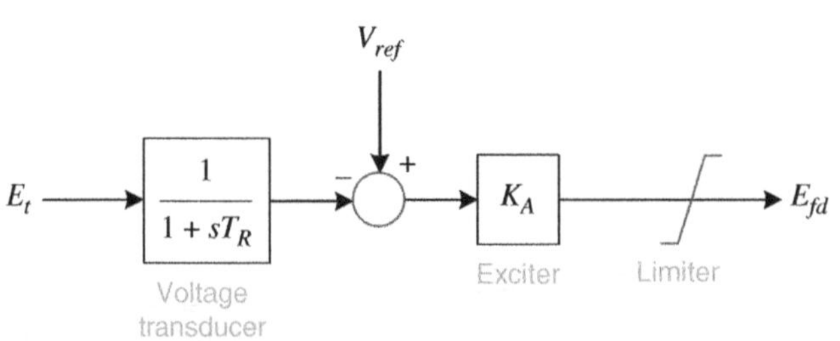

Figure 7.22 Excitation system with automatic voltage regulator.

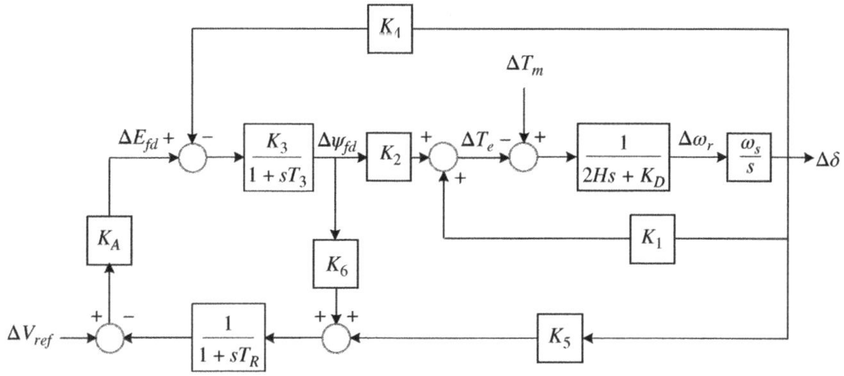

Figure 7.23 SMIB system with AVR and exciter.

Again, from Figure 7.23, it can be seen that the change in T_e due to a change in ψ_{fd} is given by

$$\Delta T_e|_{\Delta\psi_{fd}} = K_2 \Delta\psi_{fd} \tag{7.73}$$

Using (7.72), (7.73) is modified as follows:

$$\Delta T_e|_{\Delta\psi_{fd}} = -K_2 \frac{K_A K_3 K_5 + K_3 K_4 (1 + sT_R)}{K_A K_3 K_6 + (1 + sT_3)(1 + sT_R)} \Delta\delta \tag{7.74}$$

Equation (7.74) is expanded as follows:

$$\Delta T_e|_{\Delta\psi_{fd}} = -K_2 \frac{K_3 K_4 T_R s + K_A K_3 K_5 + K_3 K_4}{T_3 T_R s^2 + (T_3 + T_R)s + K_A K_3 K_6} \Delta\delta \tag{7.75}$$

Replacing s by $j\omega$ in the above equation, we have

$$\Delta T_e|_{\Delta\psi_{fd}} = -K_2 \frac{jK_3 K_4 T_R \omega + K_A K_3 K_5 + K_3 K_4}{-T_3 T_R \omega^2 + j(T_3 + T_R)\omega + K_A K_3 K_6} \Delta\delta = \frac{a_{11} + ja_{12}}{b} \Delta\delta \tag{7.76}$$

where

$$a_{11} = -K_2\left[(K_A K_3 K_5 + K_3 K_4)(K_A K_3 K_6 - T_3 T_R \omega^2) + K_3 K_4 T_R (T_3 + T_R)\omega^2\right]$$
$$a_{12} = -K_2\left[K_3 K_4 T_R \omega (K_A K_3 K_6 - T_3 T_R \omega^2) - (T_3 + T_R)\omega (K_A K_3 K_5 + K_3 K_4)\right]$$
$$b = (K_A K_3 K_6 - T_3 T_R \omega^2)^2 + (T_3 + T_R)^2 \omega^2$$

The synchronizing and damping components of torque are then

$$\Delta T_S = \Delta T_e|_{\Delta\delta} + \Delta T_e|_{\Delta\psi_{fd}} = K_1 \Delta\delta + \frac{a_{11}}{b} \Delta\delta \tag{7.77}$$

$$\Delta T_D = \frac{a_{12}}{b} (j\Delta\delta) \tag{7.78}$$

Now from (7.61), the following expression can be written:

$$s\Delta\delta = \omega_s \Delta\omega_r \Rightarrow j\omega\Delta\delta = \omega_s \Delta\omega_r \Rightarrow j\Delta\delta = \frac{\omega_S}{\omega}\Delta\omega_r$$

Therefore, the damping torque equation of (7.78) is rewritten as

$$\Delta T_D = \frac{a_{12}}{b}\frac{\omega_S}{\omega}\Delta\omega_r \tag{7.79}$$

The sign of the constant K_5 depends on the operating conditions and the external network impedance [2]. When it is negative, it can affect the system stability depending on the exciter gain K_A. The following example illustrates this.

Example 7.13 Consider the same system parameters as given in Example 7.12. Additionally, we choose the following parameters:

$$K_5 = -0.2, K_6 = 0.35 \text{ and } T_R = 0.025$$

Generally, the small signal frequency range of 0.5–2.5 Hz is of interest. For this example, we choose $\omega = 9.45$ rad/s (i.e., 1.512 Hz).

The gain K_A is varied from 0 to 100. The synchronizing and damping torque are shown in Figure 7.24. Figure 7.24 (a) shows the component of synchronizing torque that is due to a_{11}/b. It is negative below $K_A = 50$. After that, the demagnetizing

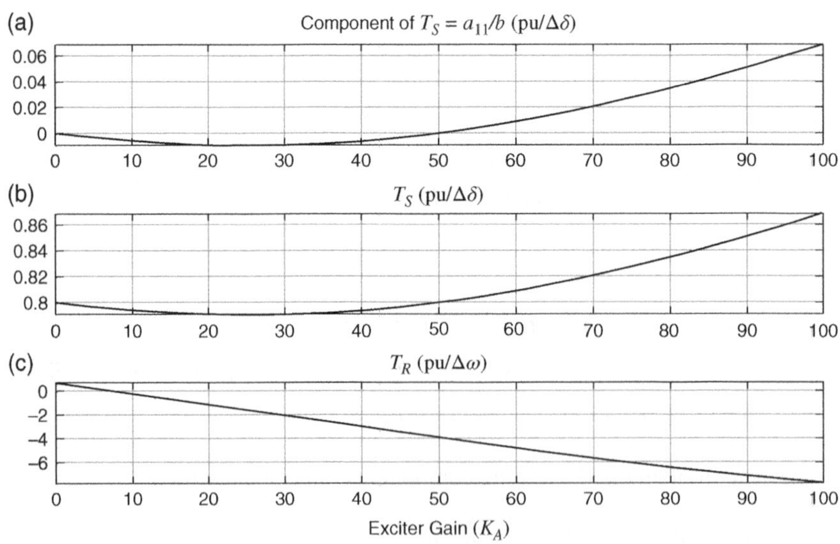

Figure 7.24 Variations in synchronizing and damping torques with exciter gain. (a) component of synchronizing torque (a_{11}/b) in math mode, (b) synchrnizing torque and (c) damping torque.

effect due to the armature reaction is compensated by the AVR, as shown in Figure 7.24 (b). However, the damping torque becomes negative when $K_A = 8$, as shown in Figure 7.24 (c). When $K_A = 100$, the synchronizing torque is 0.8685 per unit, while the damping torque is –7.825 per unit.

7.3.4 Power System Stabilizer (PSS)

The best way of providing additional damping is to use an auxiliary stabilizing signal. Figure 7.25 shows the partial block diagram of the system where G_{PSS} is the transfer function of the PSS. The main aim of the PSS is to extend the stability limits by modulating the generator excitation to provide damping in the frequency range of 0.25–2.5 Hz.

From Figure 7.25, the following expression is written:

$$\Delta\psi_{fd} = \frac{K_A K_3}{1 + sT_3}\left[\Delta v_{PSS} - \frac{K_6}{1 + sT_R}\Delta\psi_{fd}\right] \tag{7.80}$$

Equation (7.79) is rearranged as

$$\frac{\Delta\psi_{fd}}{\Delta v_{PSS}} = \frac{K_A K_3(1 + sT_R)}{T_3 T_R s^2 + (T_3 + T_R)s + K_A K_3 K_6 + 1} \tag{7.81}$$

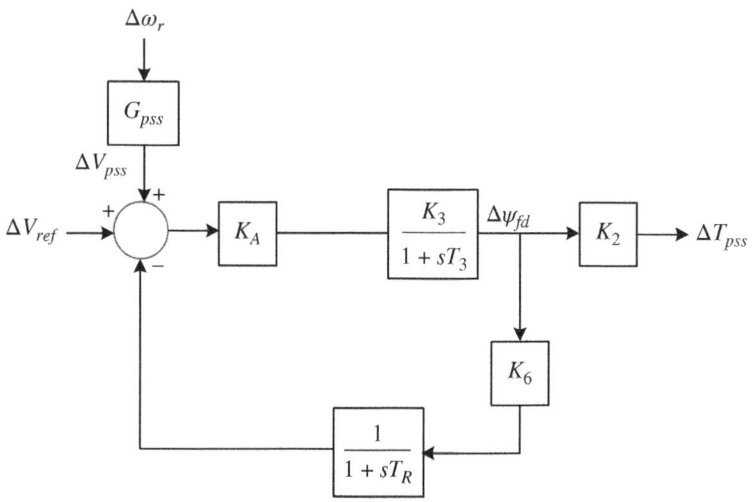

Figure 7.25 Block diagram showing the effect of PSS on ψ_{fd}.

Example 7.14 The parameters of Example 7.13 are also considered for this example. For $K_A = 100$, the transfer function of (7.80) is obtained as

$$\frac{\Delta\psi_{fd}}{\Delta v_{PSS}} = \frac{0.755s + 35}{0.0588s^2 + 2.375s + 13.25}$$

At frequency $s = j\omega = j9.5$ rad/s, the above expression is converted to

$$\frac{\Delta\psi_{fd}}{\Delta v_{PSS}} = \frac{35 + j8.269}{8 + j22.44} = 1.509\angle -57.08^\circ$$

From Figure 7.25, we write the following expression for the torque produced due to PSS:

$$\frac{\Delta T_{PSS}}{\Delta v_{PSS}} = \frac{\Delta T_{PSS}}{G_{PSS}\Delta\omega_r} = K_2\frac{\Delta\psi_{fd}}{\Delta v_{PSS}} = 1.3433\angle -57.08^\circ$$

$$\Rightarrow \frac{\Delta T_{PSS}}{\Delta\omega_r} = G_{PSS}(1.3433\angle -57.08^\circ)$$

Now, the damping torque is in phase with the rotor speed. The PSS transfer function should be so chosen that it provides a phase lead of 57.08° at 9.5 rad/s. Assuming that the PSS is able to provide the exact amount of phase lead, the damping torque is computed using the result of Example 7.13 as

$$K_D = -7.825 + 1.3433|G_{PSS}|_{\omega = j9.5}$$

Therefore, the PSS gain at 9.5 rad/s should be

$$|G_{PSS}|_{\omega = j9.5} \geq \frac{7.825}{1.3433} = 5.8253$$

Usually, the PSS gain must be large enough to provide sufficient damping torque. However, it is to be noted that the PSS can reduce the synchronizing torque component [2].

The basic block diagram of a PSS is shown in Figure 7.26. It contains a washout block, the PSS gain, and phase compensator blocks. The purpose of the washout

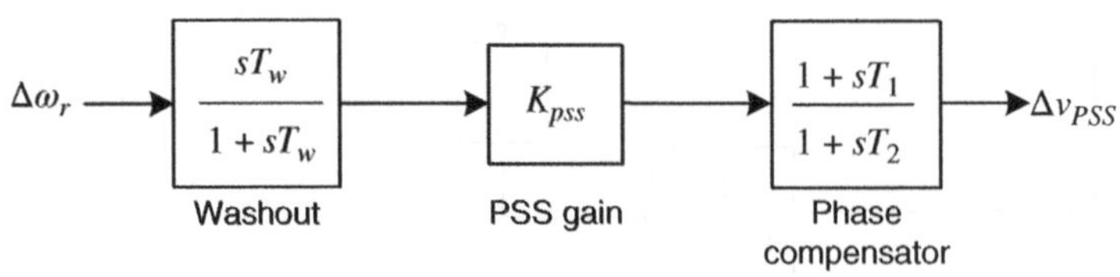

Figure 7.26 Block diagram of a typical PSS.

block is to remove any DC offset such that the output of the PSS is not biased. The PSS gain is chosen such that the required amount of damping torque is obtained. The phase compensator block adds the amount of phase lead needed to cancel out the phase bias at the operating frequency. From the discussions presented in this section, the most natural choice for the PSS input is speed deviation. However, there are other signals that also have been considered, namely, the frequency input and power input. A detailed analysis regarding the choice of PSS input is presented in [11].

Example 7.14 shows how a PSS can be tuned for a particular frequency. However, the PSS must cover the entire frequency range of 0.2–2.5 Hz. Therefore, careful consideration must be given to the PSS tuning [11]. The main problem is that a stabilizer with excess damping often detracts from the effect of AVR. Therefore, its control efforts should be limited during the period immediately following a large transient. However, the PSS must damp out the subsequent oscillations irrespective of the operating conditions or the frequency of oscillation. An adaptive PSS can perform these tasks seamlessly. The typical block diagram of an adaptive PSS system is shown in Figure 7.27. It contains a system identification block that estimates the system transfer function parameters θ based on the system output y and the system input u. The control system then computes the control action u based on a suitable control action [13, 14]. Some typical adaptive controllers are discussed in [7].

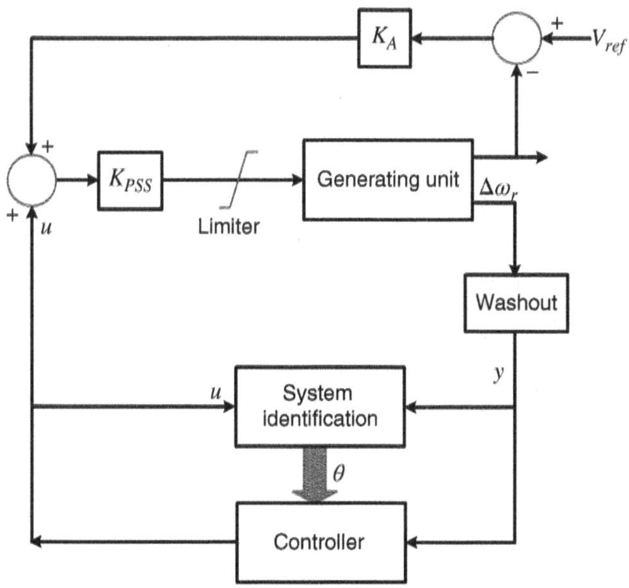

Figure 7.27 Schematic diagram of adaptive PSS.

7.4 Concluding Remarks

The main aim of power system stability is to evaluate the ability of the system to recover following a large transient. Two different kinds of stability have been discussed in this chapter. The transient stability problem determines the critical clearing angle beyond which the system becomes unstable. Both single-machine, infinite bus and multimachine system stability have been discussed. The instant of time at which the critical clearing angle occurs usually cannot be computed in most cases. Usually, a step-by-step integration method is employed for the calculation of the critical clearing time. Instead, a simple MATLAB-based method has been presented that can be used for the calculation of this time.

The dynamic stability of a power system deals with the system oscillations in a very low-frequency range. This can become very acute in large, interconnected power systems. In Section 7.3, the impact of AVR gain on the system stability has been discussed. It has been shown that an AVR with a high gain can have detrimental effects on the system damping torque. To enhance the damping torque, PSS are used. Note that in Section 7.3, the system analysis is presented using the K constants. These constants are derived based on the d-q axis model of generators, which has not been discussed in this book. See [2, 10] for more information. However, these constants are defined as follows:

K_1 is the change in the electrical torque for a change in the rotor angle with constant flux.
K_2 is the change in the electrical torque for a change in flux linkage with a constant angle.
K_3 is an impedance factor.
K_4 defines the impedance ratio between subtransient and transient reactances.
K_5 is the change in the terminal voltage for a change in rotor angle for constant flux.
K_6 is the change in the terminal voltage for a change in flux for rotor angle.

Of these constants, only K_5 can become negative due to the demagnetizing effect due to the armature reaction.

Power system control is a very wide area of study. In the next two chapters, we shall study how power flow through a transmission line can be increased, while maintaining the system stability at the same time.

References

1 P. Kundur, "Introduction to the special publication on inter-area oscillations," *IEEE Power Engineering Society Publication on Inter-Area Oscillations in Power Systems*, No. 95 TP 101, pp. 7–9, 1994.

2 P. Kundur, *Power System Stability and Control*, McGraw-Hill, New York, 1994.

3 J. D. Glover, M. S. Sarma and T. J. Overbye, *Power Systems Analysis and Design*, 5th Ed., Cengage Learning, Stamford, CT, 2012.

4 W. D. Stevenson, *Power System Analysis*, 4th Ed., McGraw-Hill, New York, 1982.

5 J. J. Grainger and W. D. Stevenson, *Power System Analysis*, McGraw-Hill, New York, 1994.

6 D. P. Kothari and I. J. Nagrath, Power System Engineering, 2nd Ed., McGraw Hill Education (India), New Delhi, 2008.

7 A. Ghosh and F. Zare, *Control of Power Electronic Converters with Microgrid Applications*, IEEE Press-Wiley, Hoboken, New Jersey, 2023.

8 O. I. Elgerd, *Electric Energy Systems Theory*, McGraw-Hill, New York, 1971.

9 P. M. Anderson, "Power system oscillations, summary of utility experience," *IEEE Power Engineering Society Publication on Inter-Area Oscillations in Power Systems*, No. 95 TP 101, pp. 43–54, 1994.

10 P. DeMello and C. Concordia, "Concepts of synchronous machine stability as affected by excitation control," *IEEE Transactions on Power Apparatus & Systems*, Vol. PAS-88, pp. 316–329, 1969.

11 E. V. Larsen and D. A. Swan, "Applying power system stabilizers, Part I, II and III," *IEEE Transactions on Power Apparatus & Systems*, Vol. PAS-100, pp. 3017–3043, 1981.

12 P. Kundur, M. Klein, G. J. Rogers, and M. S. Zywno, "Application of power system stabilizers for enhancement of overall system stability," *IEEE Transactions on Power Systems*, Vol. 4, No. 3, pp. 614–626, 1989.

13 A. Ghosh, G. Ledwich, O. P. Malik and G. S. Hope, "Power system stabilizers based on adaptive control techniques," *IEEE Transactions on Power Applications & Systems*, Vol. PAS-103, pp. 1983–1989, 1984.

14 A. Ghosh, G. Ledwich, G. S. Hope and O. P. Malik, "Power systems stabilizers for large disturbances," *Proceedings of IEE*, Vol. 132, Pt. C, pp. 14–25, 1985.

15 M. A. Pai and D. Chatterjee, *Computer Techniques in Power System Analysis*, 3rd Ed., McGraw Hill Education (India), New Delhi, 2014.

16 Hefron, W.G. and Phillips, R.A. (1952). Effect of a modern amplidyne voltage regulator on underexcited operation of large turbine generators. *American Institute of Electrical Engineers Transactions on Power Apparatus and Systems* **71** (3): 692–697.

Problems

P7.1 Consider a 2-pole, 50 Hz turboalternator rated 500 MVA and 20 KV with the inertia constant (H) of 10 MJ/MVA. Neglecting the rotational losses, find the angular acceleration in rpm/s when the electrical power output is 400 MW for an input of 700 000 hp.

P7.2 The generator in Problem P7.1 is synchronized with a large power system. Starting from $P_a = 0$, the generator accelerates for 10 cycles. Find the speed in rpm at the end of this period.

P7.3 A 100 MVA, 50 Hz synchronous generator has an internal emf of 1.2 per unit. It is connected to an infinite bus of voltage 1.0 per unit through a lossless transmission line. The combined reactance of the generator and the line is 0.3 per unit. The generator is transmitting 1.0 per unit of power to the infinite bus at unity power factor when a three-phase short circuit occurs to force the power transfer to zero. Subsequently, circuit breakers operate to isolate the fault such that the reactance between the generator and the infinite bus becomes 0.4 per unit.

(a) Calculate the critical clearing angle.
(b) If the generator inertia constant is 2.5 MJ/MVA, calculate the critical clearing time.

P7.4 Consider the power system shown in Figure P7.4. Rating of the different components are as follows:

Generator: 25 kV, 200 MV, Y-connected, $X' = 0.15$ pu
Transformer (T-1): 25 Y/400 Y kV, 200 MVA, $X = 0.05$ pu
Transformer (T-2): 100 Y/400 Y kV, 200 MVA, $X = 0.05$ pu

Convert all the quantities to a common per unit base in terms of 100 kVA and 200 MVA in the circuit of the infinite bus. The generator is delivering 200 MW and 40 MVAr to the infinite bus when the circuit breaker B-1 opens inadvertently. Assuming the infinite bus voltage angle to be the reference, find the maximum angle to which the rotor will swing to.

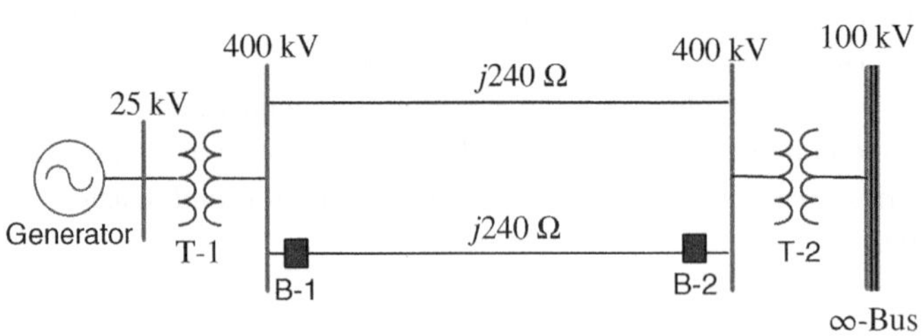

Figure P7.4 SMIB system of Problem P7.4.

P7.5 A 50-Hz generator is supplying 60% of its maximum power (P_{max}) to an infinite bus through a reactive network. A fault occurs which increases the reactance of the network between the generator and the infinite bus to

400%. When the fault is cleared, the maximum power that can be delivered is 80% of the pre-faulted maximum power level. Determine the critical clearing angle.

P7.6 A 50 Hz synchronous generator is connected to an infinite bus through a double circuit transmission line, as shown in Figure P7.6, where all quantities are given is a common per unit base. The infinite bus voltage is assumed to be $1\angle 0°$ per unit. The generator is transmitting 1.0 per unit of power when a bolted three-phase fault occurs at point P indicated in the figure. The fault is cleared by the opening of breakers B-1 and B-2. Determine the critical clearing time for the opening of the breakers when the inertia constant $H = 10$ MJ/MVA.

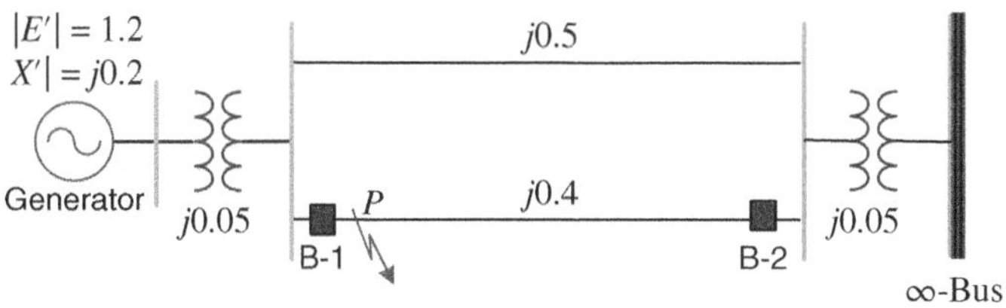

Figure P7.6 A generator connected to infinite bus through double circuit line of Problem P7.6.

P7.7 A 50 Hz synchronous generator is connected to an infinite bus through a double circuit transmission line, as shown in Figure P7.7, where all quantities are given is a common per unit base. The inertia constant (H) is 2.0 MJ/MVA. With the system operating in the steady state delivering 1.0 per unit of power to the infinite bus, a bolted three-phase occurs at point P shown in the figure. If the fault is cleared by the opening of the breakers B-1 and B-2 after 8 cycles, state, with justification, whether the system will remain stable or not.

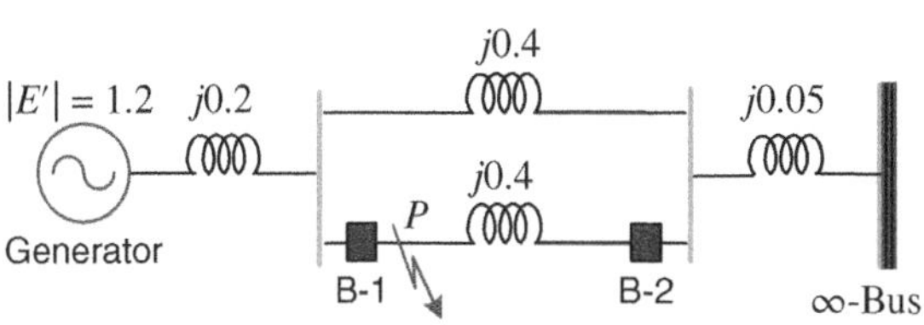

Figure P7.7 SMIB system of Problem P7.7.

P7.8 Consider the double-circuit SMIB system shown in Figure P7.8. The system is operating in the steady state delivering 1.5 per unit of power to the infinite bus when a bolted symmetrical fault occurs at the point shown in the figure.

(a) Will the system remain stable if the fault is not removed?

(b) Suppose the fault is cleared at angle of 60°, will the system remain stable?

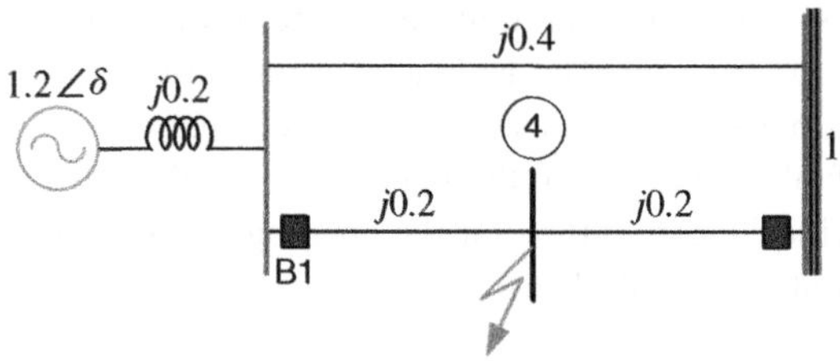

Figure P7.8 SMIB system of Problem P7.8.

P7.9 Consider a 50 Hz SMIB system that transfers 1.0 per unit of power to the infinite bus. The maximum power during three events is given by

$$P_{max} = 2.4 \text{ pu}$$

$$P_{max\_fault} = 0.857 \text{ pu}$$

$$P_{max\_post\_fault} = 2.0 \text{ pu}$$

Find the critical clearing angle and hence find when H in MJ/MVA is (a) 2, (b) 4, (c) 6, (d) 8, and (e) 10, using the MATLAB program given in Example 7.6.

P7.10 Consider the SMIB system shown in Figure P7.10, which has a braking resistor connected to the machine terminal. The machine is delivering a complex power $1 + j0.4$ per unit to the infinite bus when the circuit breaker CB1 opens inadvertently. Assume that the behavior of the braking resistor is ideal, that is, switch S is closed as soon as B1 opens. Then find the value of the critical clearing angle if the value of resistor R is 2.3 per unit.

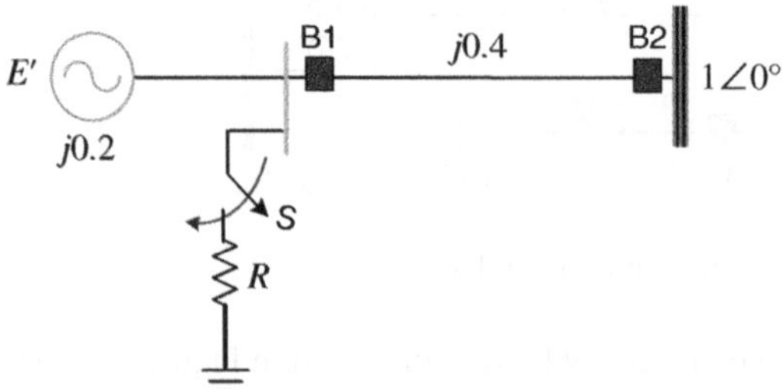

Figure P7.10 SMIB system with a braking resistor of Problem P7.8.

P7.11 A lossless 50 Hz SMIB system has the following parameters:

Generator internal voltage: 1.1 per unit
Infinite bus voltage: 1.0 per unit
Line reactance: 0.8 per unit
Inertia constant (H): 4.5 MJ/MVA
Power transmitted: 1.0 per unit

Determine the damping ratio and the undamped natural frequency of oscillation.

8

Reactive Power Compensation

The interconnected western power systems of the United States include the provinces of Alberta and British Columbia in Canada. It covers an area that is more than half the land area of the United States and provides power to 59 million people. The operation and the planning of this western interconnection that includes some 88 member systems are coordinated by the Western Systems Coordinating Council (WSCC). In July and August 1996, two large-scale power outages occurred in this system. July 2 outage was reported in [1, 2].

On August 10, it was very hot and humid with temperature soaring to 40 °C. Recent excellent rainfall made hydroelectric generation readily available. Furthermore, the combined effect of rain and hot weather also caused the trees to grow faster. Three 500 kV lines went out of service as they sagged and came in contact with trees. Even though these lines were lightly loaded at that time, their capacitance provided a much-needed reactive power support in the high-power transfer corridor between Canada and California [2].

The serious trouble started when a 500 kV line connecting Seattle and Portland carrying a load of over 1,300 MW sagged into a tree and went out of service. This resulted in a loss of voltage support, particularly in North-Central Oregon, at the head of the intertie to Canada. Initially, generation at the McNary hydroelectric plant sought to offset this problem by increasing voltage support. But within 5 minutes, all generating units at the McNary Dam shut down, shedding nearly 700 MW in 70 seconds. Growing oscillations then caused the tripping of the AC intertie causing severe generation loss.

The outages of 1996 were due to a lack of reactive power in the system. Reactive power is consumed or absorbed in an inductive circuit, while it is produced or generated in a capacitive circuit. The consumption of reactive power depresses transmission voltage. Conversely, the injection of reactive power tends to support voltage of a transmission line. As a transmission line is predominantly reactive,

Modern Power System, First Edition. Arindam Ghosh.

Published 2026 by John Wiley & Sons, Inc.
Companion website: www.wiley.com/go/ghoshpowersystem

if the voltages at its two ends are held constant, its production of reactive power is nearly constant, while its consumption of reactive power is inversely proportional to loading. Thus, generators must absorb reactive power during light load and generate it during heavy load. This reactive balance is however lost during a severe transient.

The term *voltage stability* defines the ability of the power system to return to nominal voltages of all buses following a disturbance in the system. In addition, the system will also be able to maintain the nominal voltage at buses in a steady state. Conversely, voltage instability occurs when the voltages at different buses drop continuously following a disturbance or load change. We start our discussions of this chapter with an introduction to voltage stability. This will be followed by discussions on reactive power compensation, where both ideal shunt and series compensation will be discussed.

8.1 Voltage Stability

Consider the radial system that is feeding a constant impedance load as shown in Figure 8.1. The line current is given by

$$I_S = \frac{V_S}{R_S + R_R + j(X_S + X_R)} \tag{8.1}$$

The receiving-end voltage and power are given by

$$V_R = (R_R + jX_R)I_R = \frac{V_S(R_R + jX_R)}{R_S + R_R + j(X_S + X_R)} = \frac{V_S}{1 + \dfrac{R_S + jX_S}{R_R + jX_R}} \tag{8.2}$$

$$P_R = \mathrm{Re}\left(V_R I_S^*\right) = \mathrm{Re}\left[\frac{V_R V_S^*}{R_S + R_R - j(X_S + X_R)}\right] = \frac{|V_S|^2 R_R}{(R_S + R_R)^2 + (X_S + X_R)^2} \tag{8.3}$$

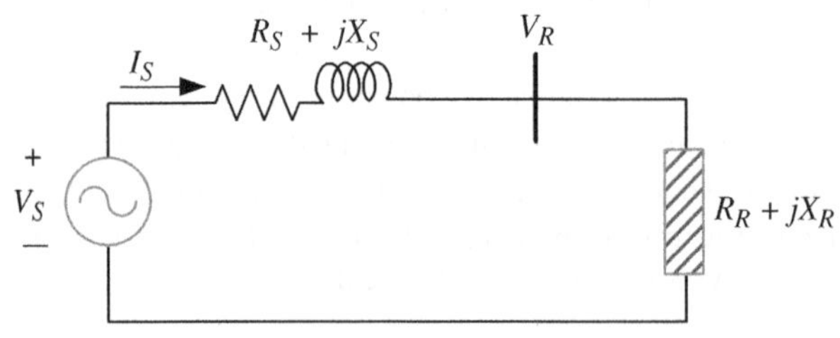

Figure 8.1 Schematic diagram of a radial system feeding a constant impedance load.

Let us define the following:

$$R_S + jX_S = |Z_S|\angle\theta_S = |Z_S|(\cos\theta_S + j\sin\theta_S) \tag{8.4}$$

$$R_R + jX_R = |Z_R|\angle\theta_R = |Z_R|(\cos\theta_R + j\sin\theta_R) \tag{8.5}$$

The ratio of the impedances is given by

$$\frac{|Z_S|\angle\theta_S}{|Z_R|\angle\theta_R} = |Z_{ratio}|\angle(\theta_S - \theta_R) \tag{8.6}$$

The receiving-end voltage of (8.2) is then rewritten as follows:

$$V_R = \frac{V_S}{1 + |Z_{ratio}|\angle(\theta_S - \theta_R)} \tag{8.7}$$

Using (8.4) and (8.5), the following expression is obtained:

$$\begin{aligned}(R_S+R_R)^2+(X_S+X_R)^2 &= (|Z_R|\cos\theta_R+|Z_S|\cos\theta_S)^2+(|Z_R|\sin\theta_R+|Z_S|\sin\theta_S)^2\\ &= |Z_R|^2+|Z_S|^2+|Z_R||Z_S|\cos(\theta_S-\theta_R)\end{aligned}$$

Then, the receiving-end power of (8.3) is modified as follows:

$$P_R = \frac{|V_S|^2|Z_R|\cos\theta_R}{|Z_R|^2 + |Z_S|^2 + 2|Z_R||Z_S|\cos(\theta_S - \theta_R)} \tag{8.8}$$

To determine the maximum power that can be transmitted to a constant power factor load, the derivative of (8.8) is taken with respect to $|Z_R|$ and the resultant is equated to zero. This gives

$$\begin{aligned}&\{2|Z_R|+2|Z_S|\cos(\theta_S-\theta_R)\}|Z_R|-\{|Z_R|^2+|Z_S|^2+2|Z_R||Z_S|\cos(\theta_S-\theta_R)\}=0\\ &\Rightarrow\{2|Z_R|^2+2|Z_S||Z_R|\cos(\theta_S-\theta_R)\}-\{|Z_R|^2+|Z_S|^2+2|Z_R||Z_S|\cos(\theta_S-\theta_R)\}=0\\ &\Rightarrow|Z_R|=|Z_S|\end{aligned}$$

Thus, the maximum power that can be transferred to a constant power factor load when the magnitdue of the load impedance is equal to the magnitude of the source impedance. The maximum power supplied to a constant power factor load is then written from (8.8) as

$$P_{Rmax} = \frac{|V_S|^2\cos\theta_R}{2|Z_S|\{1 + \cos(\theta_S - \theta_R)\}} \tag{8.9}$$

Defining P_{Rmax} as the base power, the per unit power is obtained by combining (8.8) and (8.9) as

$$\frac{P_R}{P_{Rmax}} = \frac{2|Z_R||Z_S|\{1 + \cos(\theta_S - \theta_R)\}}{|Z_R|^2 + |Z_S|^2 + 2|Z_R||Z_S|\cos(\theta_S - \theta_R)} \tag{8.10}$$

Dividing the numerator and the denominator of the right-hand side of (8.10) by $|Z_R|^2$, the following equation is obtained:

$$\frac{P_R}{P_{Rmax}} = \frac{2|Z_{ratio}|\{1 + \cos(\theta_S - \theta_R)\}}{1 + |Z_{ratio}|^2 + 2|Z_{ratio}|\cos(\theta_S - \theta_R)} \tag{8.11}$$

Example 8.1 Let us now consider a transmission line that has a line resistance that is 10% of the line reactance. This implies that $\theta_S = \tan^{-1}(10)$. Also, the line is feeding an inductive load with a power factor of 0.9 such that $\theta_R = \cos^{-1}(0.9)$. Per unit receiving-end power and the magnitude of the receiving-end voltage in per unit are plotted in Figure 8.2. Note that since $|Z_S|$ is fixed for a given transmission system, an increase in Z_{ratio} implies a decrease in $|Z_R|$. Usually, a reduction in the load impedance is associated with an increase in power supplied to the load. Thus, as the load demand increases, P_R increases to a maximum of P_{Rmax}. This occurs when $Z_{ratio} = 1$, that is, $|Z_S| = |Z_R|$. After this point, P_R decreases monotonically with the load demand. Also evident from this figure is that the receiving-end voltage drops monotonically due to the absence of a voltage regulator at this terminal.

The point at which the power is maximum is called the *critical operating point*. If the load demand increases beyond this point, the ability of the system to meet the demand reduces rapidly. Thus, a decrease in the load impedance will decrease both power and voltage. They will settle points that are much below satisfactory operating points.

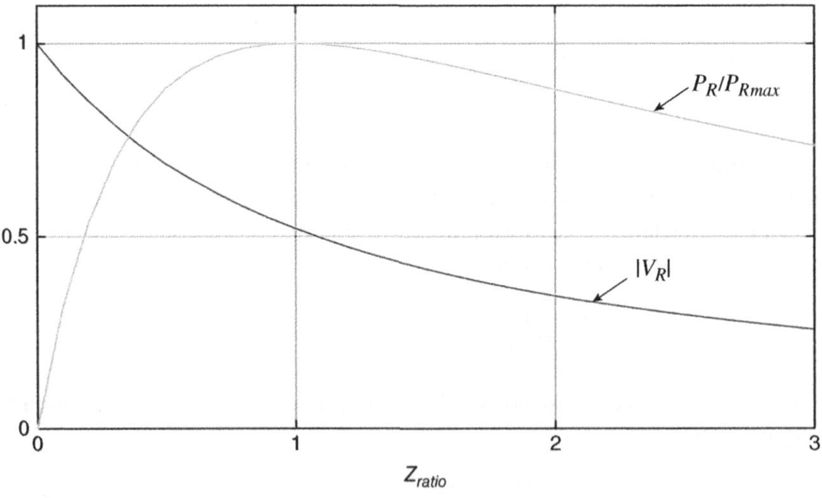

Figure 8.2 Variations in per unit receiving-end voltage and power with load demand.

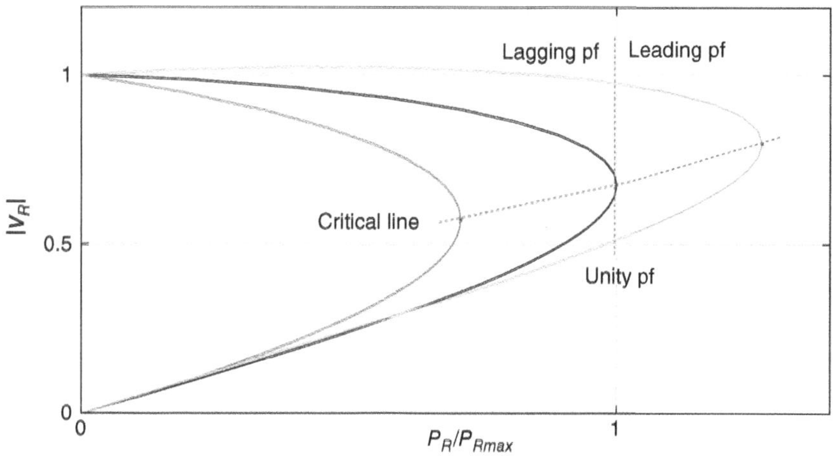

Figure 8.3 Power versus voltage curve of a radial power system feeding a static load.

Another important aspect is the power versus voltage curve. This is shown in Figure 8.3 for a range of values of load power factor. The load power is normalized with respect to the maximum power at unity power factor load, that is, $P_{max} = P_{Rmax}$ for $\theta_R = 0$. From the curves shown in Figure 8.3, the following can be concluded:

- For lagging power factor loads, the maximum power is less than P_{max}.
- For leading power factor loads, the maximum power is more than P_{max}.
- For unity factor loads, the maximum power is equal to P_{max}.

The locus of the critical operating point as the load power factor changes is shown by a dotted line in this figure. The maximum power that can be delivered to the load is a strong function of the load power factor. The more the load power factor is leading, the greater the power that can be transferred.

From Figure 8.3, it is evident that for a particular load power factor angle, there are two operating points of $|V_R|$ for a particular value of P_R. One of these points is above the critical line, which must be chosen for satisfactory operation. The point below the critical line should not be chosen as the voltage can rapidly decrease following a change in output power leading to a voltage collapse.

In the end, we can conclude that the maximum power that can be transferred can be raised by raising the source voltage or bringing the power factor close to unity for an inductive load. Since raising the source voltage is not always a viable alternative, the load power factor must be corrected. This can be achieved by supplying reactive power to compensate for the demands of reactive power of the load.

Example 8.2 Consider again, the system of Figure 8.1. The system parameters are given in per unit as

$$V_S = 1.0, Z_S = 0.05 + j0.3 \text{ and } Z_R = 1 + j1$$

Then

$$I_S = \frac{V_S}{Z_S + Z_R} = 0.376 - j0.4655 = 0.5984\angle -51.07^\circ$$

$$V_R = Z_R I_S = V_S - Z_S I_S = 0.8415 - j0.0895 = 0.8463\angle -6.07^\circ$$

The voltage drop is 0.1537 per unit. The phasor diagram is shown in Figure 8.4 (a), where I_R and I_Q, respectively, are the real and reactive components of the source current, that is,

$$I_R = 0.376 \text{ and } I_Q = -j0.4655$$

The power loss in the line is $|I_S|^2 R_S = 0.0179$ per unit. The complex power supplied to the receiving-end is

$$P_R + jQ_R = V_R I_S^* = 0.3581 + j0.3581 \text{ per unit}$$

Let us now introduce a capacitor with a reactance of $-jX_C$ in parallel with Z_R. The equivalent receiving-end impedance is then

$$Z_{eq} = \frac{-jX_C Z_R}{-jX_C + Z_R} = \frac{-jX_C \times (1 + j1)}{-jX_C + 1 + j1} = \frac{(-jX_C + X_C)(1 + jX_C - j1)}{1 + (1 - X_C)^2}$$
$$= \frac{X_C^2 + j(X_C^2 - 2X_C)}{1 + (1 - X_C)^2}$$

Now if we place the capacitor to make Z_{eq} purely resistive, then equating the imaginary component of the above equation to zero, we have

$$X_C^2 - 2X_C = 0 \Rightarrow X_C = 2$$

Then

$$Z_{eq} = \frac{-jX_C Z_R}{-jX_C + Z_R} = \frac{-j2(1 + j1)}{-j2 + 1 + j1} = \frac{2 - j2}{1 - j1} = 2$$

Therefore, the sending-end current and the receiving-end voltage are

$$I_S = \frac{V_S}{Z_S + Z_{eq}} = 0.4776 - j0.0699 = 0.4827\angle -8.33^\circ$$

$$V_R = Z_R I_S = 0.9552 - j0.1398 = 0.9653\angle -8.33^\circ$$

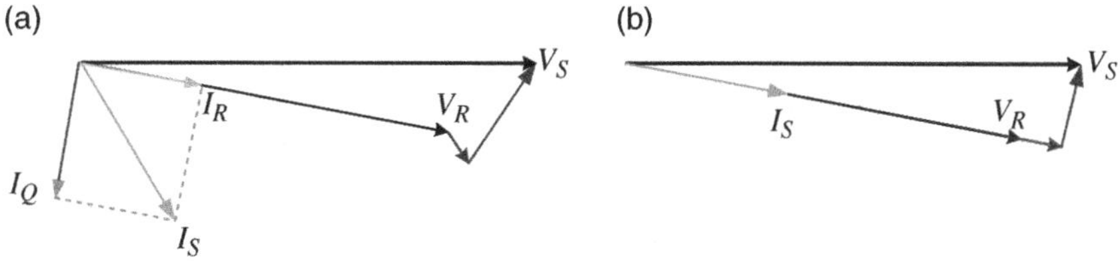

Figure 8.4 Phasor diagram of the system of Example 8.2 for (a) uncompensated system and (b) compensated system.

The power loss in the line is 0.0116 per unit and the power supplied to the load is equal to

$$P_R + jQ_R = V_R I_S^* = 0.4659 + j0 \text{ per unit}$$

Therefore, the power supplied by the load has also increased since the load voltage has increased. Furthermore, comparing the current I_S before and after the capacitor connection, we can conclude that the reactive power supplied by the source has reduced since the capacitor supplies the reactive power requirement of the load. Hence, the capacitor-connected system is called a (reactive power) compensated system. The phasor diagram of the capacitor-compensated system is shown in Figure 8.4 (b).

8.2 Ideal Reactive Compensation

The two major problems that modern power systems are facing are voltage and angle stability. There are various approaches to overcome the problem of stability arising due to small signal oscillations in an interconnected power system. As mentioned in Chapter 7, installing power system stabilizers with generator excitation control system provides damping to these oscillations. However, with the advancement in power electronic technology, various reactive power control equipment are increasingly being used in power transmission systems.

Power networks are mostly reactive. A synchronous generator usually generates active power that is specified by the mechanical power input. The reactive power supplied by the generator is dictated by the network and load requirements. A generator usually does not have any control over it. However, the lack of reactive power can cause voltage collapse in a system. It is therefore important to supply/absorb excess reactive power to/from the network. Shunt and series compensation are approaches that can be used to provide reactive power support.

A device that is connected in parallel with a transmission line is called a *shunt compensator*, while a device that is connected in series with the transmission line is called a *series compensator*. Usually, the maximum benefit can be obtained from a shunt compensator if it is always connected at the midpoint of the transmission system. On the other hand, a series compensator can be connected at any point in the line. We shall demonstrate that such connections in an single-machine, infinite-bus (SMIB) system improve the following:

- voltage profile
- power-angle characteristics
- stability margin
- damping of power oscillations

In this chapter, we shall discuss the ideal behavior of these compensating devices to broadly discuss the advantages of shunt and series compensation.

8.3 Ideal Shunt Compensation

An ideal shunt compensator is an ideal current source. We call this an ideal shunt compensator because we assume that it only supplies reactive power and no real power to the system. This assumption is not valid for practical systems. However, for an introduction, the assumption is more than adequate. We shall investigate the behavior of the compensator when connected in the middle of a transmission line, as shown in Figure 8.5. In Example 8.3, it will be demonstrated that this will cause the least voltage drop. In Figure 8.5, the shunt compensator, represented by an ideal current source. We shall demonstrate that such a configuration improves the four aspects that are mentioned above.

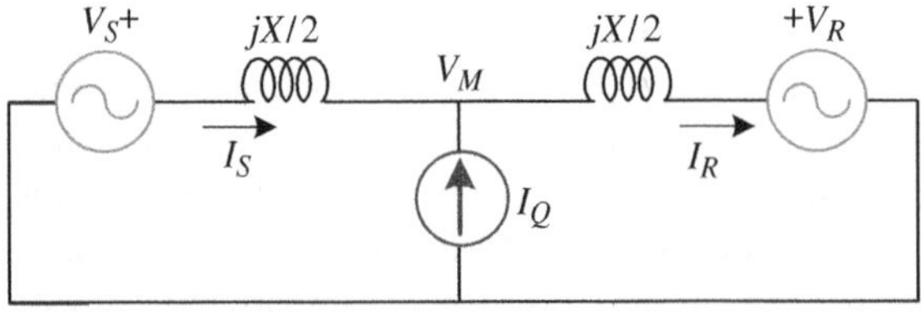

Figure 8.5 Schematic diagram of an ideal midpoint shunt compensation.

8.3.1 Improving Voltage Profile

For the analysis presented in this section, we define the following voltages:

$$V_S = V\angle\delta, V_R = V\angle 0^\circ \text{ and } V_M = |V_M|\angle\delta_M$$

From Figure 8.4, the following equations for the sending- and receiving-end currents are written:

$$I_S = \frac{V\angle\delta - |V_M|\angle\delta_M}{jX/2} \tag{8.12}$$

$$I_R = \frac{|V_M|\angle\delta_M - V}{jX/2} \tag{8.13}$$

Using Kirchoff's current law at the midpoint, we get

$$I_S + I_Q = I_R \tag{8.14}$$

Substituting (8.12) and (8.13) in (8.14), the following expression for the current injected by the shunt compensator is obtained:

$$I_Q = \frac{|V_M|\angle\delta_M - V}{jX/2} - \frac{V\angle\delta - |V_M|\angle\delta_M}{jX/2} = -j\frac{2}{X}[2|V_M|\angle\delta_M - V - V\angle\delta] \tag{8.15}$$

The complex power injected by the shunt compensator is then

$$\begin{aligned} P_Q + jQ_Q &= |V_M|\angle\delta_M \times I_Q^* = j\frac{2}{X}|V_M|\angle\delta_M[2|V_M|\angle(-\delta_M) - V - V\angle-\delta] \\ &= j\frac{2}{X}\left[2|V_M|^2 - V|V_M|\angle(\delta_M) - V|V_M|\angle(\delta_M - \delta)\right] \end{aligned} \tag{8.16}$$

Expanding the right-hand side of (8.16), the injected real and reactive power are written as

$$\begin{aligned} P_Q &= 2\frac{V|V_M|}{X}\{\sin(\delta_M) + \sin(\delta_M - \delta)\} \\ Q_Q &= \frac{2}{X}\left[2|V_M|^2 - V|V_M|\{\cos(\delta_M) + \cos(\delta_M - \delta)\}\right] \end{aligned} \tag{8.17}$$

It is desired that the shunt compensator provides pure reactive compensation, that is, it does not inject any active power. Therefore, from (8.17), we have

$$\begin{aligned} P_Q &= V|V_M|\{\sin(\delta_M) + \sin(\delta_M - \delta)\} = 0 \\ &\Rightarrow \sin(\delta_M) + \sin(\delta_M - \delta) = 0 \end{aligned} \tag{8.18}$$

To solve (8.18), the following identity will be used:

$$\sin(x) + \sin(y) = 2\sin\left(\frac{x+y}{2}\right)\cos\left(\frac{x-y}{2}\right)$$

Then, (8.18) is rewritten as

$$\sin(\delta_M) + \sin(\delta_M - \delta) = 2\sin\left(\frac{2\delta_M - \delta}{2}\right)\cos\left(\frac{\delta}{2}\right) = 0$$

Since $\delta/2$ cannot be equal to zero, the solution of (8.18) is

$$\sin\left(\delta_M - \frac{\delta}{2}\right) = 0 \Rightarrow \left(\delta_M - \frac{\delta}{2}\right) = n\pi, \quad n = 0, 1, 2, \ldots$$

Since δ_M must be less than δ for a power flow from the sending-end, for the solution of the above equation, n is chosen as 0 to produce

$$\delta_M = \frac{\delta}{2} \tag{8.19}$$

Let the midpoint voltage be chosen to be the same as the sending- and receiving-end voltages, that is, $|V_M| = V$. Then, the injected current in (8.15) is

$$\begin{aligned} I_Q &= -j\frac{2V}{X}[2\angle(\delta/2) - 1 - 1\angle\delta] \\ &= -j\frac{2V}{X}[2 - 1\angle(-\delta/2)\angle - 1\angle(\delta/2)]\angle(\delta/2) \end{aligned} \tag{8.20}$$

Now

$$\begin{aligned} 2 - 1\angle(-\delta/2)\angle - 1\angle(\delta/2) &= 2 - \cos(\delta/2) + j\sin(\delta/2) - \cos(\delta/2) - j\sin(\delta/2) \\ &= 2\{1 - \cos(\delta/2)\} \end{aligned}$$

Therefore, the injected current is

$$I_Q = -j\frac{4V}{X}\{1 - \cos(\delta/2)\}\angle(\delta/2) \tag{8.21}$$

The injected current is in phase quadrature with the midpoint voltage, indicating the real power injection to be zero. The reactive power injected by the ideal current source is computed from (8.17) as

$$Q_Q = \frac{2V^2}{X}[2 - 2\cos(\delta/2)] = \frac{4V^2}{X}[1 - \cos(\delta/2)] \tag{8.22}$$

Example 8.3 Consider the shunt compensated system shown in Figure 8.6. In this, two voltage sources are joined together by a transmission line with reactance of $X = X_1 + X_2$. In this, an ideal shunt compensator is placed at an arbitrary location Q, that is, X_1 is not necessarily equal to X_2. The following data are given in per unit basis:

Sending-end voltage: $V_S = 1\angle 30°$
Receiving-end voltage: $V_R = 1\angle 0°$
Bus-Q voltage: $V_Q = 1\angle \delta_Q$
Ideal shunt compensator: $I_Q = |I_Q|\angle(\delta_Q - 90°)$
Total line reactance: $X = X_1 + X_2 = 0.4$

The real and reactive power flowing at the different parts of the network are shown in Figure 8.6 as well.

Note that the line is lossless, and the ideal shunt compensator injects a current that is quadrature with the bus-Q voltage. Therefore, the active power dispatched from the sending-end will be received at the receiving-end. However, depending on the $X_1 : X_2$ ratio, the voltage profile, angle δ_Q and the active power flow over the line will vary.

Since this is a lossless line, the power transferred to bus-Q from Bus-1 will be the same as power transferred from bus-Q to Bus-2. Therefore, we have

$$\frac{1}{X_1}\sin(30° - \delta_Q) = \frac{1}{X_2}\sin\delta_Q$$

$$\Rightarrow \frac{X_2}{X_1}(\sin 30°\cos\delta_Q - \cos 30°\sin\delta_Q) = \sin\delta_Q$$

$$\Rightarrow \left(\frac{X_2}{X_1}\cos 30° + 1\right)\sin\delta_Q = \frac{X_2}{2X_1}\cos\delta_Q \Rightarrow \left(\frac{2X_1}{X_2} + 2\cos 30°\right)\sin\delta_Q = \cos\delta_Q$$

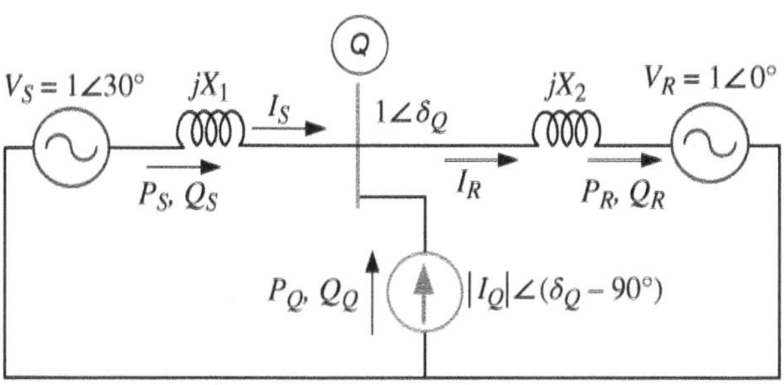

Figure 8.6 A transmission line is compensated by an ideal shunt compensator.

Let us define

$$\beta = \frac{2X_1}{X_2} + 2\cos 30^\circ$$

Then

$$\delta_Q = \tan^{-1}\left(\frac{1}{\beta}\right)$$

The currents flowing through the circuit are then given by

$$I_S = \frac{1\angle 30^\circ - 1\angle\delta_Q}{jX_1}, I_R = \frac{1\angle\delta_Q - V}{jX_2} \text{ and } I_Q = I_R - I_S$$

The real and reactive power flowing through the circuit are computed accordingly. The compensator is placed at three locations along the line length: (a) one-third, (b) midpoint, and (c) four-fifth. The results obtained are listed in Table 8.1. The voltage profiles along the line when there is no compensator and for the three cases are shown in Figure 8.7. It can be seen that the minimum voltage drop occurs when the compensator is placed at the midpoint. Without any compensation, the maximum voltage drop of 0.966 per unit occurs at the midpoint of the circuit. This is lower than the three compensated cases discussed here. This shows that the shunt compensator improves the voltage profile.

Table 8.1 System quantities for different compensator placement locations.

Line length	25%	50%	80%
Line reactance	$X_1 = 0.1,\ X_2 = 0.3$	$X_1 = 0.2,\ X_2 = 0.2$	$X_1 = 0.3,\ X_2 = 0.1$
Angle δ_Q	22.63°	15°	5.87°
Current $\lvert I_Q \rvert$	0.3393 (pu)	0.3407 (pu)	0.3386 (pu)
Sending-end real power P_S	1.2826 (pu)	1.2941 (pu)	1.2777 (pu)
Sending-end reactive power Q_S	0.0826 (pu)	0.1704 (pu)	0.2731 (pu)
Receiving-end real power P_R	1.2826 (pu)	1.2941 (pu)	1.2777 (pu)
Receiving-end reactive power Q_R	−0.2567 (pu)	−0.1704 (pu)	−0.0655 (pu)
Reactive power injected Q_Q	0.3393	0.3407	0.3386
Minimum voltage magnitude	0.9806 (pu)	0.9914 (pu)	0.9779 (pu)

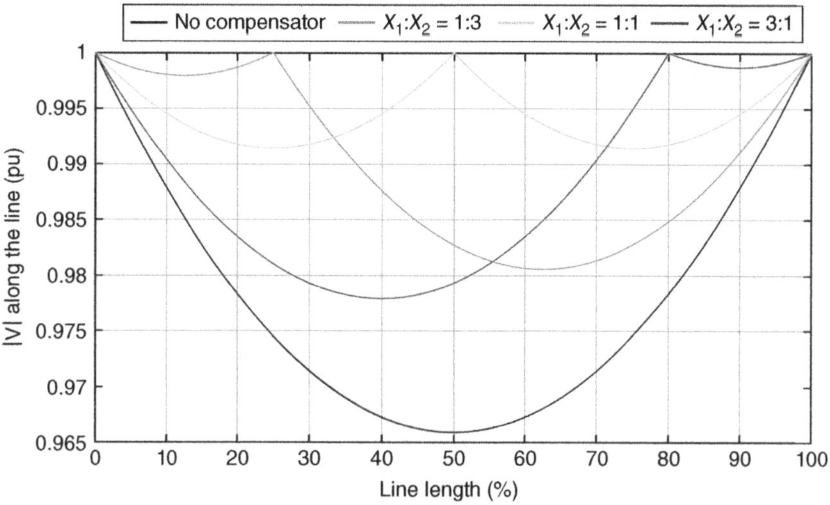

Figure 8.7 Voltage profile along the line for different locations of compensator placement.

In (8.20), we have assumed that the injected current I_Q lags the midpoint voltage V_M by 90°. However, we can modify this such that the injected current such that it leads the midpoint voltage by 90°. Then modifying (8.20), we have

$$I_Q = j\frac{2V}{X}[2\angle(\delta/2) - 1 - 1\angle\delta] = j\frac{4V}{X}\{1 - \cos(\delta/2)\}\angle(\delta/2) \tag{8.23}$$

The reactive power injected by the current source is then

$$Q_Q = -\frac{4V^2}{X}[1 - \cos(\delta/2)] \tag{8.24}$$

The voltage–current characteristic of the compensator is shown in Figure 8.8. This ideal behavior, however, is not feasible in practical systems where we get a slight drop in the voltage characteristic. This will be discussed in Chapter 9.

Figure 8.8 Voltage–current relationship of an ideal shunt compensator.

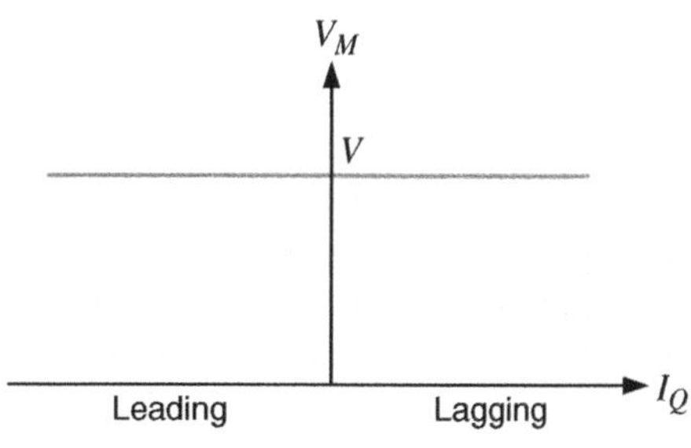

8.3.2 Improving Power-Angle Characteristics

Referring to Figure 8.5, the complex power supplied by source is

$$\begin{aligned} P_S + jQ_S = V_S I_S^* = V\angle\delta\left[\frac{V\angle-\delta - V\angle-(\delta/2)}{-jX/2}\right] = \frac{V^2 - V\angle(\delta/2)}{-jX/2} \\ = \frac{2V^2\sin(\delta/2)}{X} + j\frac{2V^2\{1-\cos(\delta/2)\}}{X} \end{aligned} \tag{8.25}$$

On the other hand, the complex power delivered to the receiving-end is

$$\begin{aligned} P_R + jQ_R = V_R I_R^* = V\left[\frac{V\angle-(\delta/2) - V}{-jX/2}\right] \\ = \frac{2V^2\sin(\delta/2)}{X} + j\frac{2V^2\{\cos(\delta/2)-1\}}{X} \end{aligned} \tag{8.26}$$

Therefore, the real power transmitted over the line is

$$P_e = P_S = P_R = \frac{2V^2}{X}\sin(\delta/2) \tag{8.27}$$

From (8.22), (8.25), and (8.26), we find the reactive power consumed by the line is

$$Q_e = Q_S + Q_Q - Q_R = \frac{8V^2}{X}\{1-\cos(\delta/2)\} \tag{8.28}$$

Example 8.4 Consider the midpoint shunt compensated line with $V_S = 1\angle\delta$ per unit, $V_R = 1\angle 0°$ per unit, $V_M = 1\angle(\delta/2)$ per unit, and $X = 1.0$ per unit.

From the data, it can be seen that $P_{max} = 1.0$ per unit, and this is chosen as the base power. Figure 8.9 shows the active and reactive power variations with the load angle. Since the line is lossless, the maximum power of 2.0 per unit occurs at $\delta = 90°$ as per (8.27). This is shown in Figure 8.9 (a). The reactive power at the various parts of the circuit is shown in Figure 8.9 (b). At $\delta = 90°$, the shunt compensator injects 0.5359 per unit of reactive power. On the other hand, if the real power transmitted is increased to 2 per unit, the shunt compensator must inject 4 per unit of reactive power. In practice, however, the amount of reactive power that can be injected will be restricted by the device rating.

Let us now relax the condition that the midpoint voltage is regulated to 1.0 per unit. This results in some very interesting plots as shown in Figure 8.10. In this figure, the x-axis shows the reactive power available from the shunt device, while the y-axis shows the maximum power that can be transferred over the line without

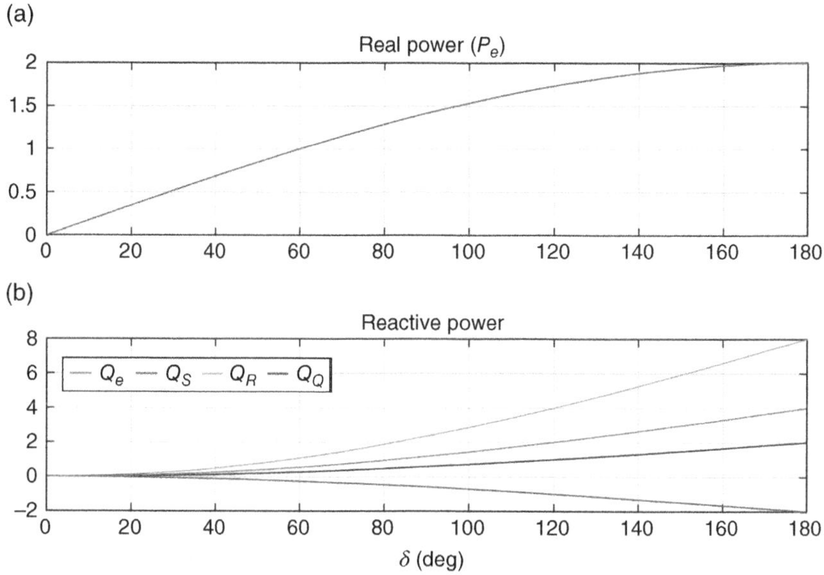

Figure 8.9 Power and reactive power variations with changes in the load angle.

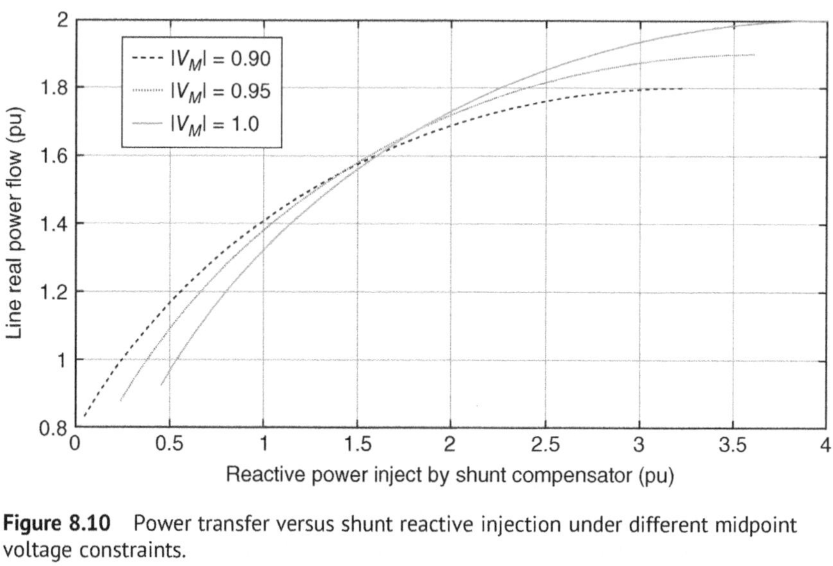

Figure 8.10 Power transfer versus shunt reactive injection under different midpoint voltage constraints.

violating the voltage constraint. There are three different P–Q relationships given for three midpoint voltage constraints. For a reactive power injection of 0.5 per unit, the power transfer can be increased from about 0.97 per unit to 1.17 per unit by lowering the midpoint voltage to 0.9 per unit. For a reactive power injection greater than 2.0 per unit, the best power transfer capability is obtained for $|V_M| = 1.0$ per unit. Thus, there will be no benefit in reducing the voltage constraint when the shunt device can inject a large amount of reactive power.

8.3.3 Improving Stability Margin

This is a consequence of the improvement in the power-angle characteristics and is one of the major benefits of using midpoint shunt compensation. As mentioned before, the stability margin of the system pertains to the regions of acceleration and deceleration in the power-angle curve. We shall use this concept to delineate the advantage of midpoint shunt compensation.

Consider the power-angle curves shown in Figure 8.11. The curve of Figure 8.11 (a) is for an uncompensated system, while that of Figure 8.11 (b) for the compensated system. Both these curves are drawn assuming that the base power is $P_{max} = V^2/X = 1.0$ per unit. Let us assume that the uncompensated system is operating on steady state delivering an electrical power equal to P_m with a load angle of δ_0 when a three-phase fault occurs that forces the real power to zero. To obtain the critical clearing angle δ_{cr} for the uncompensated system, we equate the accelerating area A_1 with the decelerating area A_2, where

$$A_1 = \int_{\delta_0}^{\delta_{cr}} P_m dt = P_m(\delta_{cr} - \delta_0)$$

$$A_2 = \int_{\delta_{cr}}^{\delta_{max}} (\sin\delta - P_m)\, dt = (\cos\delta_{cr} - \cos\delta_{max}) - P_m(\delta_{max} - \delta_{cr})$$

$$\delta_{max} = \pi - \delta$$

Equating the two areas, the critical clearing angle is obtained as

$$\delta_{cr} = \cos^{-1}[P_m(\delta_{max} - \delta_0) + \cos\delta_{max}] \tag{8.29}$$

Let us now consider that the midpoint shunt compensated system is working with the same mechanical power input P_m. The operating angle in this case is δ_1 and the maximum power that can be transferred in this case is 2 per unit. Let the fault be cleared at the same clearing angle δ_{cr} as given by (8.29). Then equating areas A_3 and A_4 in Figure 8.11 (b) we get δ_2, where

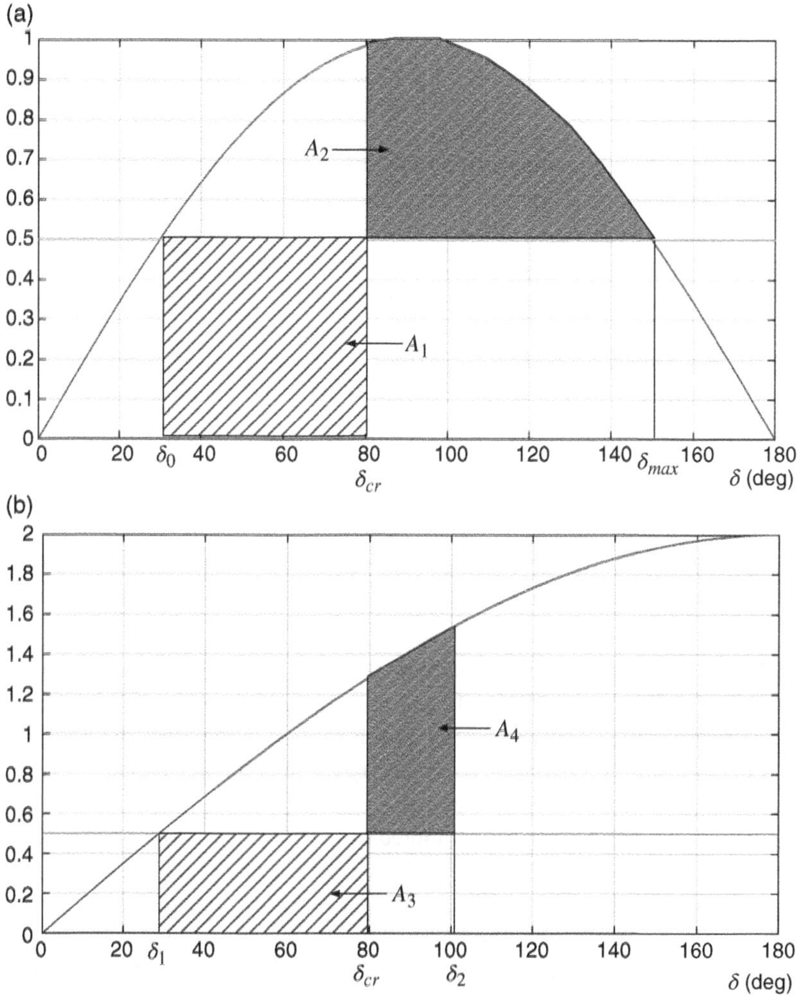

Figure 8.11 Power-angle curve showing clearing angles: (a) for uncompensated system and (b) for compensated system.

$$A_3 = \int_{\delta_1}^{\delta_{cr}} P_m \, dt = P_m(\delta_{cr} - \delta_1)$$

$$A_4 = \int_{\delta_{cr}}^{\delta_2} [2\sin(\delta/2) - P_m] \, dt = 4[\cos(\delta_{cr}/2) - \cos(\delta_2/2)] - P_m(\delta_2 - \delta_{cr}) \tag{8.30}$$

Equating the two areas given in (8.30), we have

$$\delta_{cr} = 2\cos^{-1}\left[\cos(\delta_2/2) + \frac{\delta_2 P_m - \delta_1 P_m}{4}\right] \tag{8.31}$$

Example 8.5 Let an uncompensated SMIB power system operates in a steady state with a mechanical power input P_m equal to 0.5 per unit. Then $\delta_0 = 30° = 0.5236$ rad and $\delta_{max} = 150° = 2.618$ rad. Consequently, the critical clearing angle is calculated from (8.29) as $\delta_{cr} = 79.56° = 1.3886$ rad.

Let us now put an ideal shunt compensator at the midpoint. The pre-fault steady-state operating angle of the compensated system can be obtained by solving $2\sin(\delta/2) = 0.5$, which produces $\delta_1 = 28.96° = 0.5054$ rad. Let us assume that we use the same critical clearing angle as obtained above for clearing a fault in the compensated system as well. The accelerating area is then given by $A_3 = 0.4416$. Equating this with area A_4 we get a nonlinear equation of the form

$$0.4416 = 3.0740 - 4\cos(\delta_2/2) - 0.5\delta_2 + 0.6943$$

Solving the above equation, we get $\delta_2 = 105.34° = 1.856$ rad. Needless to say that the stability margin has increased significantly in the compensated system.

Example 8.6 Consider an SMIB system with $P_{max} = V^2/X = 1.0$ per unit, $\omega_s = 100\pi$ rad/s, and $H = 2$ MJ/MVA. It has been assumed that when a fault occurs, the power transfer becomes zero till the fault is cleared. Thereafter, the nominal power transfer resumes. Therefore, $r_1 = 0$ and $r_2 = 1$ in (7.43) of Chapter 7. For an uncompensated system, the critical clearing angle (δ_{cr}) is computed from (7.43) and the critical clearing time (t_{cr}) is computed from (7.36). The variation of t_{cr} with P_m is shown in Figure 8.12 (a). If the fault is cleared in 5 cycles, that is, 0.1 s, the maximum power that can be transferred without losing synchronism is 0.745 per unit.

Let us now assume that the system is compensated by a midpoint shunt compensator. Then the maximum power that can be transferred is 2.0 per unit in this case. Then the initial angle is computed from

$$\delta_1 = \sin^{-1}\left(\frac{P_m}{2}\right)$$

Furthermore, from (8.27), we can surmise that the maximum power happens when $\delta = 180°$ and the power transfer becomes zero at $\delta = 0°$ and $\delta = 360°$. Therefore, the maximum angle the machine is allowed to swing without losing synchronism is

$$\delta_{max} = 2\pi - \delta_1$$

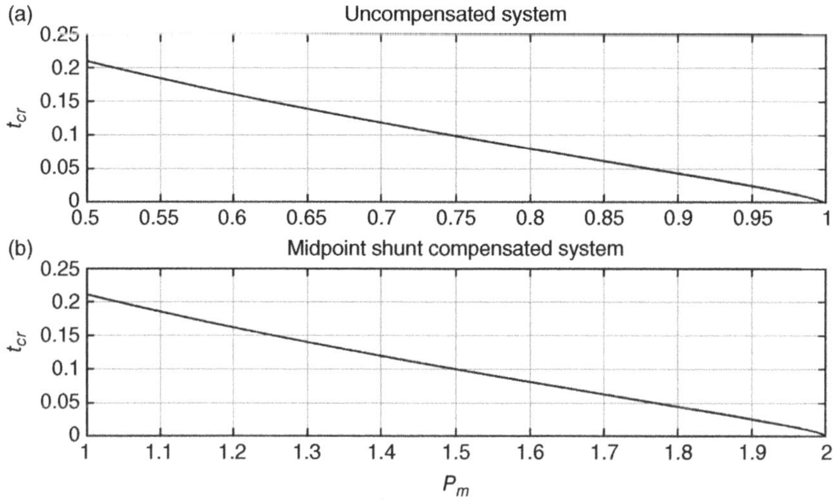

Figure 8.12 Critical clearing time: (a) for uncompensated system and (b) for compensated system.

The critical clearing angle is computed from (8.31) by replacing δ_2 by δ_{max} and critical clearing time is computed from (7.36). The maximum power that can be transmitted without losing synchronism for a fault clearing time of 0.1 s is 1.49, which is twice that of the uncompensated case, as shown in Figure 8.12 (b). Therefore, in general, it can be concluded that the line can be safely loaded to its thermal limit using the shunt compensator.

8.3.4 Power Swing Damping

The swing equation for a synchronous machine is given in (7.22). For any variation in the electrical quantities, the mechanical power input remains constant. Assuming that the magnitude of the midpoint voltage of the system is controlled by a shunt compensating device, the accelerating power in (7.22) becomes a function of two independent variables, magnitude of the midpoint voltage ($|V_M|$) and load angle (δ). Again, since the mechanical power is constant, its perturbation with the independent variables is zero. We then get the following small perturbation expression of the swing equation:

$$\frac{2H}{\omega_s}\frac{d^2\Delta\delta}{dt^2} + \frac{\partial P_e}{\partial |V_M|}\Delta|V_M| + \frac{\partial P_e}{\partial \delta}\Delta\delta = 0 \tag{8.32}$$

where Δ indicates a perturbation around the nominal values.

If the midpoint voltage is regulated at a constant magnitude, $\Delta|V_M|$ will be equal to zero. Hence, (8.32) will reduce to

$$\frac{2H}{\omega_s}\frac{d^2\Delta\delta}{dt^2} + \frac{\partial P_e}{\partial \delta}\Delta\delta = 0 \tag{8.33}$$

The second-order differential equation given in (8.33) can be written in the Laplace domain by neglecting the initial conditions as

$$\left(\frac{2H}{\omega_s}s^2 + \frac{\partial P_e}{\partial \delta}\right)\Delta\delta(s) = 0 \tag{8.34}$$

The roots of the above equation are located on the imaginary axis of the s-plane at locations $\pm j\omega_m$, where

$$\omega_m = \sqrt{\frac{\omega_s(\partial P_e/\partial \delta)}{2H}}$$

This implies that the load angle will have a sustained oscillation at a constant frequency of ω_m. Obviously, this solution is not acceptable. Thus, in order to provide damping, the midpoint voltage must be varied in sympathy with the rate of change in $\Delta\delta$. We can then write

$$\Delta|V_M| = K_M\frac{d\Delta\delta}{dt} \tag{8.35}$$

where K_M is a proportional gain. Substituting (8.35) in (8.32), we get

$$\frac{2H}{\omega}\frac{d^2\Delta\delta}{dt^2} + \frac{\partial P_e}{\partial |V_M|}K_M\frac{d\Delta\delta}{dt} + \frac{\partial P_e}{\partial \delta}\Delta\delta = 0 \tag{8.36}$$

For any $K_M > 0$, the introduction of the control action of (8.35) ensures that the roots of the second-order equation will have negative real parts. Therefore, through the feedback, damping to power swings can be provided by placing the poles of the above equation to provide the necessary damping ratio and undamped natural frequency of oscillations.

8.3.5 Shunt Compensator Representation

In Figure 8.5, the shunt compensator is realized using an ideal current source. However, a current source can be represented by a voltage source and an inductor. This is shown in Figure 8.13, where the shunt compensator is realized by the voltage source V_F that is connected to the midpoint bus through a pure inductor X_F, also known as an *interface inductor*. The voltage source V_F is driven such that it is always in phase with the midpoint voltage V_M. In this configuration, the shunt compensator can both supply and absorb reactive power. Since the current I_Q is

Figure 8.13 Shunt compensator realization using an ideal voltage source.

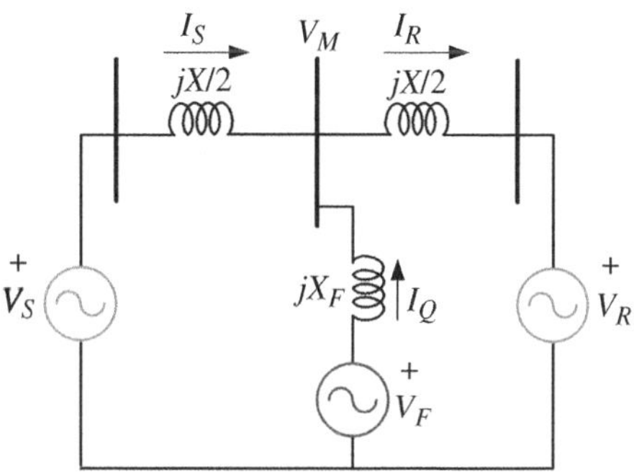

then purely inductive, its direction is dependent on the relative magnitudes of the two voltages:

- If $|V_M| > |V_F|$, that is, the magnitude of the midpoint voltage is higher than the voltage source, inductive current will flow from the ac system to the voltage source. This implies that the voltage source is absorbing var in this configuration.
- On the other hand, if $|V_M| < |V_F|$, the voltage source will generate var as its magnitude is higher than that of the midpoint voltage.

Example 8.7 Consider the system of Example 8.3 with $X = X_1 = X_2 = 0.2$ per unit and $X_f = 0.1$ per unit. We must regulate the midpoint voltage as $V_M = 1\angle 15°$ per unit. Then from Figure 8.13, we can write

$$I_S = \frac{1\angle 30° - 1\angle 15°}{j0.2} = 1.2059 + j0.4995 \text{ per unit}$$

$$I_R = \frac{1\angle 15° - 1\angle 0°}{j0.2} = 1.2941 + j0.1704 \text{ per unit}$$

Then

$$I_Q = I_R - I_S = 0.0882 + j0.3291 \text{ per unit}$$

Let us assume that $V_F = |V_F| \angle 15°$ per unit. Then from Figure 8.13, we have

$$|V_F|\angle 15° - 1\angle 15° = j0.1 I_Q = 0.0329 + j0.0088$$

We can then calculate $|V_F|$ from

$$|V_F| = \frac{0.0329}{\cos(15°)} + 1 = \frac{0.0088}{\sin(15°)} + 1 = 1.0341 \text{ per unit}$$

Alternatively, we can separate the real and imaginary component of

$$|V_F|\angle 15^\circ - 1\angle 15^\circ = 0.0329 + j0.0088, \textit{we have}$$
$$(|V_F| - 1)\cos(15^\circ) = 0.0329$$
$$(|V_F| - 1)\sin(15^\circ) = 0.0088$$

Squaring and adding the two equations, we have the following quadratic equation:

$$|V_F|^2 - 2|V_F| + 0.9988 = 0$$

The roots of this equation are $|V_F| = 1.0341$ and 0.9659. It can be verified that both the solutions are valid, depending on the direction of the current I_Q.

8.4 Ideal Series Compensation

Let us assume that the series compensator is represented by an ideal voltage source. This is shown in Figure 8.14. Let us further assume that the series compensator is an ideal voltage source, that is, it only supplies reactive power and no real power to the system. This assumption, although not valid for practical systems, is more than adequate for an introduction. It is to be noted that, unlike the shunt compensator, the location of the series compensator is not crucial, and it can be placed anywhere along the transmission line.

8.4.1 Impact of Series Compensator on Voltage Profile

In the single-line diagram of a series compensated power system shown in Figure 8.14, the receiving-end current is equal to the sending-end current, that is, $I_S = I_R$. The series voltage V_Q is injected in such a way that the magnitude of

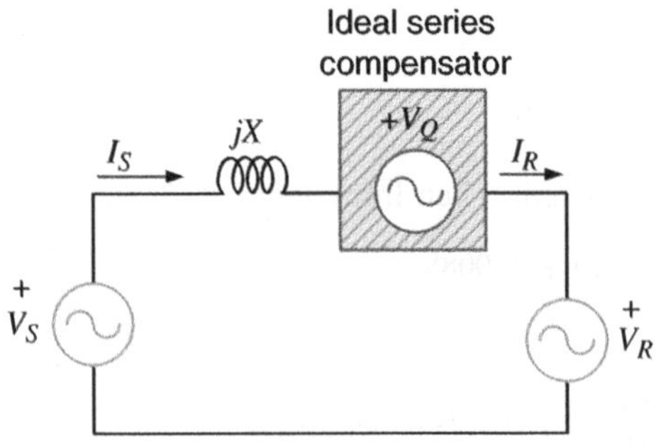

Figure 8.14 Single-line diagram of an ideal series compensated line.

the injected voltage is proportional to that of the line current. Furthermore, the phase of the voltage of V_Q is forced to be in quadrature with the line current. We then have

$$V_Q = \lambda I_S e^{\mp j90^\circ} \tag{8.37}$$

where λ is a proportionality constant. The ratio λ/X is called the *compensation level* and is often expressed in percentage. This compensation level is usually measured with respect to the transmission line reactance. For example, the compensation level as 50% when $\lambda = X/2$.

Applying KVL in the circuit of Figure 8.14, we get

$$V_S - V_Q - V_R = jXI_S \quad \Rightarrow \quad V_S - V_R = \mp j\lambda I_S + jXI_S$$

Assuming $V_S = V\angle\delta$ and $V_R = V\angle 0^\circ$, we get the following expression for the line current:

$$I_S = \frac{V\angle\delta - V}{j(X \mp \lambda)} \tag{8.38}$$

When we choose $V_Q = \lambda I_S e^{-j90^\circ}$, the line current equation becomes

$$I_S = \frac{V\angle\delta - V}{j(X - \lambda)} \tag{8.39}$$

Thus, we see that λ is subtracted from X. This choice of the sign corresponds to the voltage source acting as a pure capacitor. Hence, we call this the *capacitive mode of operation*. In contrast, if we choose $V_Q = \lambda I_S e^{+j90^\circ}$, λ is added to X, and this mode is referred to as the *inductive mode of operation*. We shall consider the implication of the series voltage injection on the transmission line voltage through the following example.

Example 8.8 Consider a lossless transmission line that has a line reactance of 0.5 per unit. The sending-end and receiving-end voltages are given by

$$V_S = 1\angle\delta = 1\angle 30^\circ \text{ and } V_R = 1\angle 0^\circ$$

Let us choose a capacitive mode of operation of the series compensator with a 30% compensation level. This implies that $\lambda = 0.15$. The line current is then given from (8.39) as $I_S = 1.479\angle 15^\circ$ per unit and the injected voltage calculated from (8.37) is $V_Q = 0.2218\angle -75^\circ$ per unit. The phasor diagrams of the sending- and receiving-end voltages, line current, and injected voltage are shown in Figure 8.15 (a), where V_R is chosen as the reference axis. Let us define the voltage on the left of the series compensator as V_{QL} and the voltage on the right of the series compensator as V_{QR}. We shall now consider a few different cases.

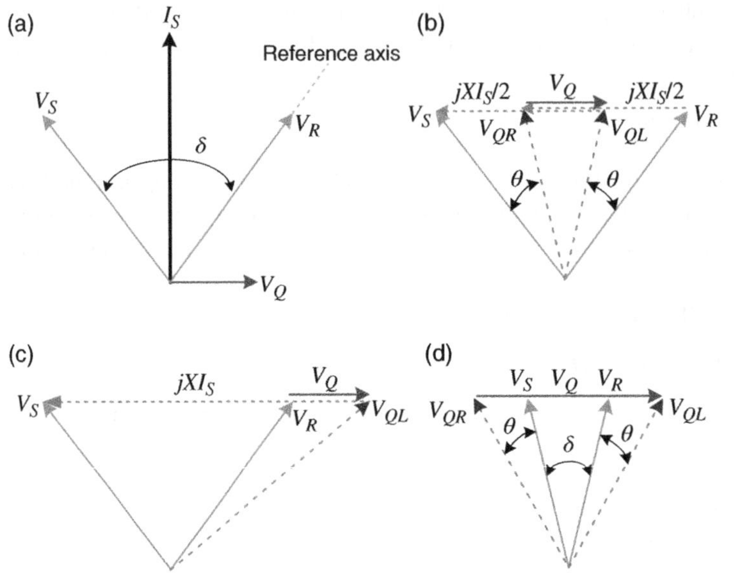

Figure 8.15 Phasor diagrams for (a) nominal case, (b) compensator in the middle, (c) end of the transmission line, and (d) 70% compensation.

Case-1: Let us assume that the compensator is placed in the middle of the line. Then, we have the following:

$$V_{QL} = V_S - j\left(\frac{X}{2}\right)I_S = 0.9723\angle 8.45^\circ \text{ pu}$$

$$V_{QR} = V_R + j\left(\frac{X}{2}\right)I_S = 0.9723\angle 21.55^\circ \text{ pu}$$

The difference between these two voltages is the injected voltage. This is shown in Figure 8.15 (b), where the angle $\theta = 8.45^\circ$. The worst-case voltage along the line will then be at the two points on either side of the series compensator where the voltage phasors are aligned with the line current phasor. These two points are equidistant from the series compensator. However, their particular locations will be dependent on the system parameters.

Case-2: Let the compensator be placed at the end of the transmission line, just before the infinite bus. We then have the following:

$$V_{QL} = V_R + V_Q = V_S - jXI_S = 1.079\angle -11.46^\circ \text{ pu}$$

$$V_{QR} = V_R = 1.0\angle 0^\circ \text{ pu}$$

This is shown in Figure 8.15 (c). The maximum voltage rise in the line is then to the immediate left of the compensator, that is, at V_{QL}. The maximum voltage drop, however, still occurs at the point where the voltage phasor is aligned with the line current phasor.

Case-3: When the compensator is placed immediately after the sending-end source, we have the following:

$$V_{QL} = V_S = 1.0\angle 30^\circ \text{ pu}$$
$$V_{QR} = V_R - V_Q = V_R + jXI_S = 1.079\angle 41.46^\circ \text{ pu}$$

The maximum voltage rise will be immediately to the right of the compensator.

Case-4: In this case, the level of compensation is increased from 30% to 70% (i.e., λ is changed from 0.15 to 0.35). However, the level of steady-state power transfer remains unchanged. The relation between power transfer and compensation level will be discussed in the next subsection. It will be suffice to say that increasing the compensation level, while retaining the power transfer level, can be accomplished by lowering the value of the angle δ of the sending-end voltage to 12.37°. Let us further assume that the series compensator is placed in the middle of the transmission line. We then have the following:

$$I_S = \frac{1\angle 12.37 - 1}{j(0.5 - 0.35)} = 1.436\angle 6.19^\circ \text{ pu}$$
$$V_Q = -j\lambda I_S = 0.003\angle -83.81^\circ \text{ pu}$$

Then

$$V_{QL} = V_S - j\left(\frac{X}{2}\right) I_S = 1.026\angle -8.01^\circ \text{ pu}$$
$$V_{QR} = V_R + j\left(\frac{X}{2}\right) I_S = 1.026\angle 20.38^\circ \text{ pu}$$

This is shown in Figure 8.15 (d), where $\theta = 8.01°$. It is obvious that the voltage along the line rises to a maximum level at either side of the series compensator.

8.4.2 Improving Power-Angle Characteristics

The sending-end complex power is given by

$$\begin{aligned} P_S + jQ_S = V_S I_S^* = V\angle\delta\left[\frac{V\angle -\delta - V}{-j(X \mp \lambda)}\right] = \frac{V^2 - V^2\angle\delta}{-j(X \mp \lambda)} \\ = \frac{V^2 \sin\delta}{X \mp \lambda} + j\frac{V^2(1 - \cos\delta)}{X \mp \lambda} \end{aligned} \tag{8.40}$$

Similarly, the receiving-end complex power is given by

$$\begin{aligned} P_R + jQ_R = V_R I_S^* &= V\left[\frac{V\angle -\delta - V}{-j(X \mp \lambda)}\right] \\ &= \frac{V^2 \sin\delta}{X \mp \lambda} + j\frac{V^2(\cos\delta - 1)}{X \mp \lambda} \end{aligned} \tag{8.41}$$

Therefore, the power transmitted over the line is

$$P_S = P_R = P_e = \frac{V^2}{X \mp \lambda}\sin\delta \tag{8.42}$$

The power-angle characteristics of a series compensated power system are given in Figure 8.16. In this figure, the base power is chosen as V^2/X. Three curves are shown. These are:

- For $\lambda = 0$. This is the uncompensated characteristic.
- For $\lambda = +0.5$. This is for capacitive compensation.
- For $\lambda = -0.5$. This is for inductive compensation.

The curves which have maximum power greater than the base power pertain to capacitive mode of operation. On the other hand, all curves the inductive mode of operation will have maximum values less than 1. These two regions are also included in Figure 8.16.

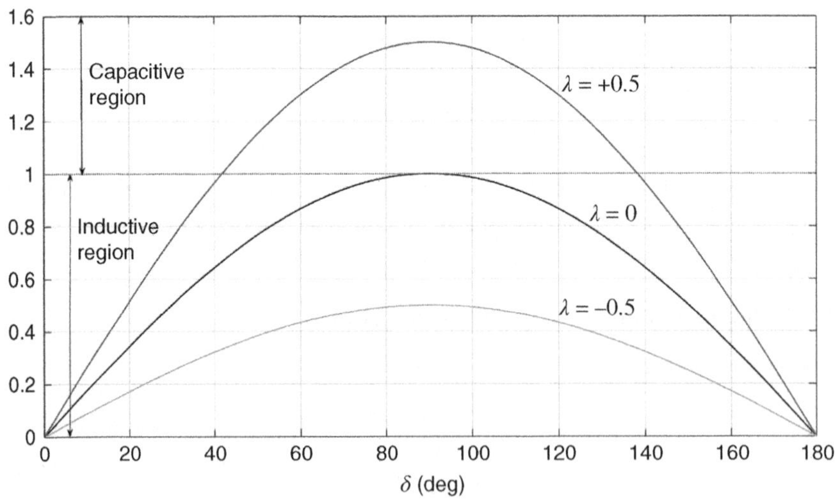

Figure 8.16 Power-angle characteristics for different modes of operation.

Let us now have a look at the reactive power. Since the capacitive mode of operation enhances the power transfer over a line, we shall restrict our attention to this mode only. From (8.37) and (8.39), the complex power transfer by the compensator is

$$\begin{aligned} P_Q + jQ_Q = V_Q I_S^* &= -j\lambda \frac{V\angle\delta - V}{j(X-\lambda)} \times \frac{V\angle(-\delta) - V}{-j(X-\lambda)} \\ &= -j\lambda \frac{V^2(2 - 1\angle\delta - 1\angle(-\delta))}{(X-\lambda)^2} \end{aligned} \tag{8.43}$$

Now

$$1\angle\delta + 1\angle(-\delta) = \cos\delta + j\sin\delta + \cos\delta - j\sin\delta = 2\cos\delta$$

Therefore, from (8.43), we have $P_Q = 0$, as expected and

$$Q_Q = -\frac{2\lambda V^2}{(X-\lambda)^2}(1 - \cos\delta) \tag{8.44}$$

Consider the system of Figure 8.14, where $V = 1$ per unit, $X = 0.5$ per unit, and $\delta = 30°$. In Figure 8.17, the reactive power injected by the series compensator is plotted against the maximum power transfer as the compensation level changes from 10% to 60%. As the compensation level increases, the maximum power transfer also increases. However, at the same time, the reactive injection requirement from the series compensator also increases.

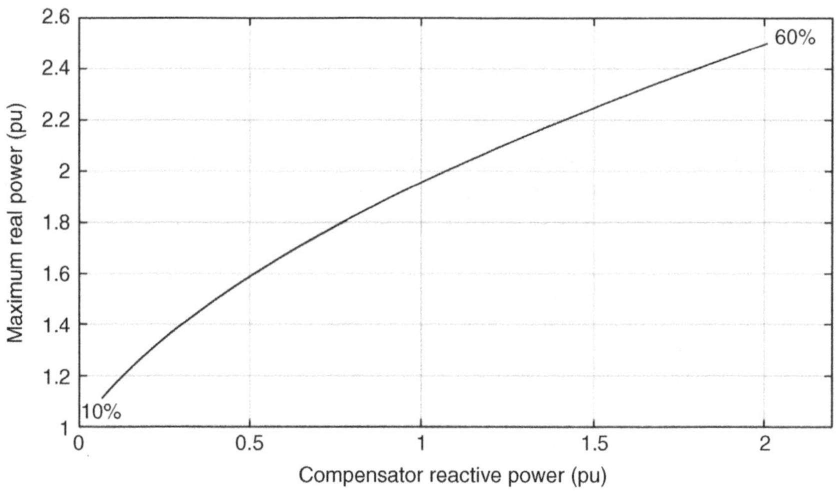

Figure 8.17 Reactive power injection by a series compensator versus maximum power transfer as the level of compensation changes.

8.4.3 Improving Stability Margin

As a direct consequence of the improved power-angle characteristics, the stability margin of the system improves. It can be seen from Figure 8.16 that the 50% series compensated system can transfer 50% more real power than an uncompensated system. Furthermore, an increase in the height in the power-angle curve means that a larger amount of decelerating area is available for a compensated system compared to an uncompensated system.

Example 8.9 In Example 8.5, we have seen that when $P_m = 0.5$ per unit, the uncompensated system has $\delta_0 = 30° = 0.5236$ rad, $\delta_{max} = 150° = 2.618$ rad, and the critical clearing angle is $\delta_{cr} = 79.56° = 1.3886$ rad. Consider a 50% series compensated system where for $\delta_0 = 30°$, the mechanical power input is $P_m = 0.75$ per unit, while $P_{max} = 1.5$ per unit. Then, the critical clearing angle, computed using (7.43) is $\delta_{cr} = 79.56°$, that is, the same as before. This means that 1.5 times power can be supplied through the line when the series compensator is employed. On the other hand, if P_m is chosen as 0.5 for a 50% series compensated system, then

$$\delta_0 = \sin^{-1}\left(\frac{0.5}{1.5}\right) = 19.47° = 0.34\,rad$$

$$\delta_{cr} = 97° = 1.6933\,rad$$

This implies that a much larger decelerating area is available, and hence the critical clearing angle can be delayed compared to the uncompensated system. This means that the stability margin has improved significantly.

8.4.4 Power Flow Control and Power Swing Damping

One of the major advantages of series compensation is that through its use real power flow over transmission corridors can be effectively controlled. Consider, for example, the SMIB system shown in Figure 8.18 in which the generator and infinite bus are connected through a double circuit transmission line, labeled Line-1 and Line-2. Of the two transmission lines, Line-2 is compensated by a series compensator. The compensator then can be utilized to regulate power flow over the entire system. For example, consider the following scenario:

- The system is operating in the steady state delivering a power of P_{m0} at a load angle of δ_0.
- Lines 1 and 2 are then sending power P_{e1} and P_{e2} respectively, such that $P_{m0} = P_{e1} + P_{e2}$.
- The mechanical power input suddenly goes up to P_{m1}, such that $P_{m1} = P_{m0} + \Delta P_m$.

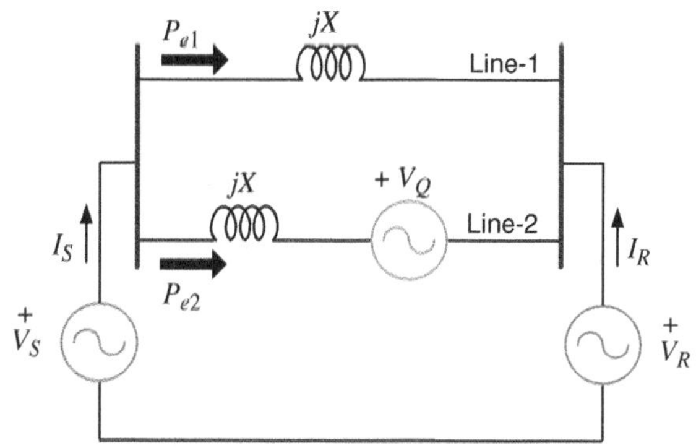

Figure 8.18 Double circuit line for power flow control by a series compensator.

There are two ways of controlling the power flow in this situation:

- *Regulating Control*: Sending the increased power ΔP_m through Line-1. This means that:
  - the series compensator maintains the power flow over Line-2 at P_{e2}, and
  - since the power will increase in Line-1, P_{e1} and the load angle will also increase in sympathy with the increase in power.
- *Tracking Control*: Sending the increased power ΔP_m through Line-2. This means that:
  - the power through Line-1 will remain unchanged, and hence the load angle will also remain unchanged at δ_0,
  - the increased amount power ΔP_m will flow through Line-2, which will adjust the compensation level.

Let us illustrate these two aspects with the help of the following numerical example.

Example 8.10 Consider the system of Figure 8.18 with the following parameters:

System Frequency = 50 Hz
$V_S = 1\angle 30°$ per unit
$V_R = 1\angle 0°$ per unit
$X = 0.5$ per unit

Nominal State: The power flow through Line-1 is then given by:

$$P_{e1} = \frac{\sin(30°)}{0.5} = 1.0 \text{ per unit}$$

Let us assume that the series compensator operates at 30% compensation level, that is, $\lambda = 0.15$. Note that the load angle must remain constant irrespective of the power flow through the two lines. Then, the power flow through Line-2 is

$$P_{e2} = \frac{\sin(30°)}{0.5 - 0.15} = 1.429 \text{ per unit}$$

Therefore,

$$P_{m0} = P_{e1} + P_{e2} = 2.429 \text{ per unit}$$

Control Action: The objective of the series compensator control for both schemes is to control the power through Line-2, where the reference power is denoted by P_{ref}. To accomplish this, a proportional-plus-integral (PI) controller is placed in the feedback loop of P_{e2}. In addition, to improve damping a term that is proportional to the deviation of machine speed is introduced in the feedback loop. The control law is then given by

$$C_L = K_P(P_{ref} - P_{e2}) + K_I \int (P_{ref} - P_{e2})\, dt + C_P \frac{d\Delta\delta}{dt}$$

where $C_L = \lambda/X$ is the compensation level. For the simulation studies performed, the following controller parameters are chosen:

$$K_P = 0.1, K_I = 1.0, \text{and } C_P = 75$$

Regulating Control: With the system operating in the nominal state, the mechanical power is increased by 10%, that is, $\Delta P_m = 0.2429$ per unit. Since the series compensator maintains the power flow through Line-2 at 1.429 per unit, the power flow through Line-1 is increased to 1.2429 per unit. Then

$$P_{e1} = \frac{\sin\delta}{0.5} = 1.2429 \Rightarrow \delta = \sin^{-1}(0.5 \times 1.2429) = 38.42°$$

To maintain the power flow constant through Line-2, the compensation level must change. This is given by

$$P_{e2} = \frac{\sin(38.42°)}{0.5 - \lambda} = 1.429 \Rightarrow \lambda = 0.5 - \frac{\sin(38.42°)}{1.429} = 0.065 \text{ per unit}$$

The compensation level will then change to $(0.0065 \div 0.5) \times 100 = 13\%$.

The system is simulated and the response with regulating control is shown in Figure 8.19. The figure shows that P_{e1}, P_{e2}, δ and percentage compensation settles at their calculated values.

Tracking Control: With the system operating in the nominal state, the mechanical power is increased by 25%, that is, $\Delta P_m = 0.607$ per unit. Since the entire power must flow through Line-2, P_{e1} will remain constant at 1.0 per unit and δ will

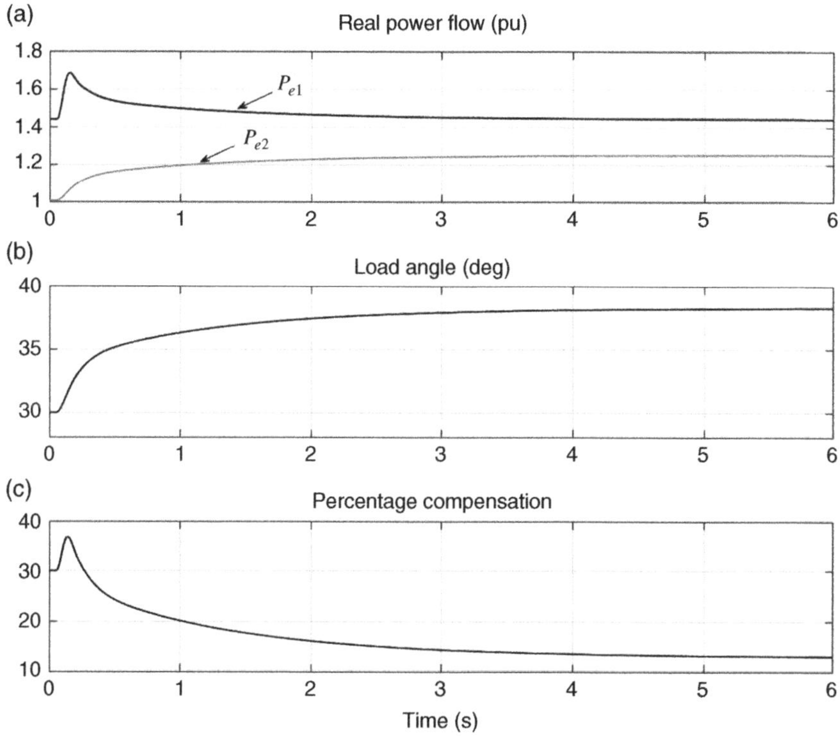

Figure 8.19 The system response with regulating control. (a) Real power flow, (b) load angle and (c) percentage compensation.

remain constant at 30°, power flow through Line-2, P_{e2} will change to $1.429 + 0.607 = 2.036$ per unit. Therefore,

$$P_{e2} = \frac{\sin(30°)}{0.5 - \lambda} = 2.036 \Rightarrow \lambda = 0.5 - \frac{\sin(30°)}{2.036} = 0.2544 \text{ per unit}$$

The compensation level will then change to $(0.2544 \div 0.5) \times 100 = 50.09\%$. The system response with tracking control is shown in Figure 8.20. They match the calculations presented above.

8.4.5 An Alternate Method of Series Compensation

A very interesting approach of generating series voltage is reported in [3]. In this, the voltage that is to be injected is independent of the line current, such that

$$V_Q = \lambda \frac{I_S}{|I_S|} e^{-j90°} \tag{8.45}$$

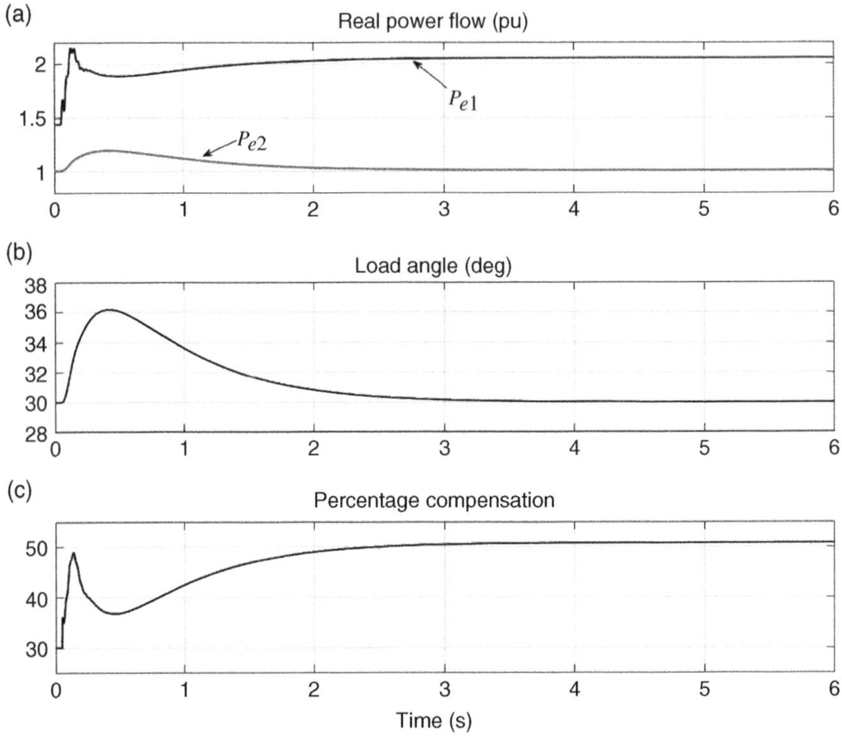

Figure 8.20 The system response with tracking control. (a) Real power flow, (b) load angle and (c) percentage compensation.

Consider for example that $I_S = |I_S| \angle \phi$. Then $I_S/|I_S| = 1 \angle \phi$, that is, it is a unit vector with an angle that is the same that of the current vector I_S. Thus, (8.45) becomes

$$V_Q = \lambda e^{(\phi - j90^\circ)}$$

Let us define

$$X_Q = \frac{V_Q}{I_S} = \frac{\lambda}{|I_S|} e^{-j90^\circ} = -j\frac{\lambda}{|I_S|} \tag{8.46}$$

The total line reactance in Figure 8.14 is then

$$X_{eq} = X - X_Q = X - \frac{\lambda}{|I_S|} \tag{8.47}$$

Kirchoff's voltage law of Figure 8.14 gives us the following relationship:

$$\begin{aligned} &V_S - V_R = jXI_S + V_Q \\ &\Rightarrow V_S - V_R = jXI_S - j\lambda\frac{I_S}{|I_S|} = j\left(X - \frac{\lambda}{|I_S|}\right)I_S \end{aligned} \tag{8.48}$$

Defining

$$V_S = V\angle\delta \text{ and } V_R = V\angle 0°$$

we have

$$\begin{aligned} |V_S - V_R| &= V|\cos\delta + j\sin\delta - 1| \\ &= V\sqrt{\cos^2\delta - 2\cos\delta + 1 + \sin^2\delta} = V\sqrt{2 - 2\cos\delta} \end{aligned} \tag{8.49}$$

Now

$$2 - 2\cos\delta = 2 - \left(1 - 2\sin^2(\delta/2)\right) = 4\sin^2(\delta/2)$$

Equation (8.49) is then modified as follows:

$$|V_S - V_R| = 2V\sin(\delta/2) \tag{8.50}$$

Substituting (8.50) in (8.48), the following equation is obtained:

$$\begin{aligned} |V_S - V_R| &= 2V\sin(\delta/2) = \left(X - \frac{\lambda}{|I_S|}\right)|I_S| = X|I_S| - \lambda \\ \Rightarrow |I_S| &= \frac{2V\sin(\delta/2) + \lambda}{X} \end{aligned} \tag{8.51}$$

Combining (8.51) with (8.47), the following equation for the total reactance of the line is obtained:

$$X_{eq} = X - \frac{\lambda}{|I_S|} = X - \frac{\lambda X}{2V\sin(\delta/2) + \lambda} = X\left[\frac{2V\sin(\delta/2)}{2V\sin(\delta/2) + \lambda}\right] \tag{8.52}$$

Then

$$\frac{1}{X_{eq}} = \frac{1}{X}\left[1 + \frac{\lambda}{2V\sin(\delta/2)}\right]$$

The power transfer equation is then given by

$$\begin{aligned} P_e &= \frac{V^2}{X_{eq}}\sin\delta = \frac{V^2}{X}\sin\delta + \frac{\lambda V}{2X}\frac{\sin\delta}{\sin(\delta/2)} \\ &= \frac{V^2}{X}\sin\delta + \frac{\lambda V}{X}\cos(\delta/2) \end{aligned} \tag{8.53}$$

Example 8.11 Let us consider the system of Figure 8.14 with $V = 1$ per unit and $X = 0.5$ per unit. This means the load angle $\delta_0 = 30°$ when there is no series compensation. The power-angle curves for three different values of λ are shown in Figure 8.21. Assume that the mechanical power in $P_m = 1.0$ per unit. Then, as evident from Figure 8.21, the choice of $\lambda = 0.75$ is not feasible. On the other hand, for $\lambda = 0.5$, 1.0 per unit power can be transferred even when the load angle $\delta_0 = 0°$. This is a unique property of this formulation. The required amount of power can be transferred for a load angle of $\delta_0 = 15°$ when $\lambda = 0.25$.

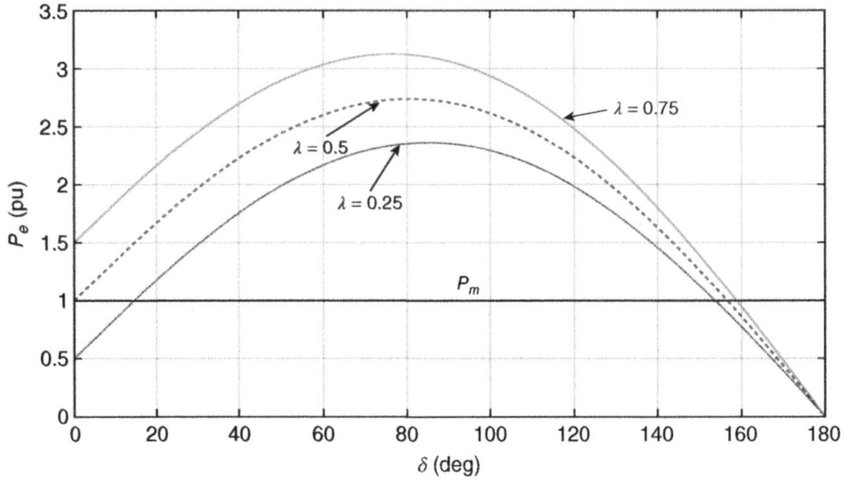

Figure 8.21 Power-angle curves for different values of λ for alternate formulation.

8.5 Concluding Remarks

The term flexible ac transmission systems (FACTS) was proposed by N. G. Hingorani in 1993 [4]. This pertains to the use of power electronics technology to overcome the difficulties of mechanically controlling electrical power systems. This chapter presents the basic concepts of shunt and series compensation of power systems using ideal current and voltage sources. It has been shown that the lack of reactive power can cause voltage stability problems in a power system. However, shunt or series compensators can provide much more than alleviating voltage problems – they can enhance power flow, increase stability limits, and damp power system oscillations.

Detailed descriptions of FACTS technology are presented in [3]. In the next chapter, we shall present some of the important components of this technology, which are much more than shunt and series compensation.

References

1 C. Alsberg, "WSCC unfold causes of the July 2 disturbance," *IEEE Power Engineering Review*, Vol. 16, No. 9, pp. 5–6, 1996.
2 C. W. Taylor, "Improving grid behavior," *IEEE Spectrum*, Vol. 36, No. 6, 1999.
3 N. G. Hingorani and L. Gyugyi, *Understanding FACTS: Concepts and Technology of Flexible AC Transmission Systems*, Wiley-IEEE Press, New Jersey, 2000.
4 N. G. Hingorani, "Flexible AC transmission," *IEEE Spectrum*, Vol. 30, No. 4, pp. 40–45, 1993.

Problems

P8.1 Consider the radial system shown in Figure 8.1, where

$$|Z_S| \angle \theta = R_S + jX_S, \quad |Z_R| \angle \phi = R_R + jX_R$$

The system parameters are

$$V_S = 1 \angle 0°, \quad X_S = 0.5, \quad R_S = 0.025, \quad \phi = \cos^{-1}(0.95)$$

Vary Z_{ratio} from 0.01 to 3 with steps of 0.01 for the fixed values of θ and ϕ. Then plot the curves P_R/P_{Rmax} versus Z_{ratio} and $|V_R|$ versus Z_{ratio}.

P8.2 In Example 8.2, the receiving-end power is calculated as 0.4659 per unit, while the power calculated using $|I_S|^2 R_R = 0.233$ per unit. What is the cause of this discrepancy?

P8.3 Consider the SMIB system shown in Figure P8.3, which is compensated by a midpoint shunt compensator. The system parameters, defined in per unit, are:

$$X = 0.25, \quad X_c = 0.1, \quad V_S = 1 \angle 30°, \quad V_R = 1 \angle 0°$$

Determine the compensator voltage (V_C) such that the midpoint voltage $V_M = 1 \angle 15°$.

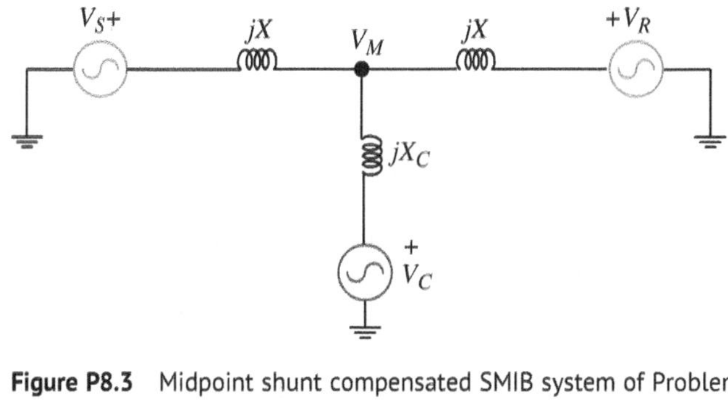

Figure P8.3 Midpoint shunt compensated SMIB system of Problem P8.2.

P8.4 Consider the system shown in Figure P8.4, in which a generator is connected with an infinite bus through a double circuit transmission line. One of the two lines is compensated by a variable series compensator with a capacitive reactance of XC. The generator voltage is $V_S = |V| \angle \delta$ and the infinite bus voltage is $V_\infty = |V| \angle 0°$. Find the value of X_C such that the maximum power that can be transferred by the compensated system is 1.4 times that of the uncompensated system.

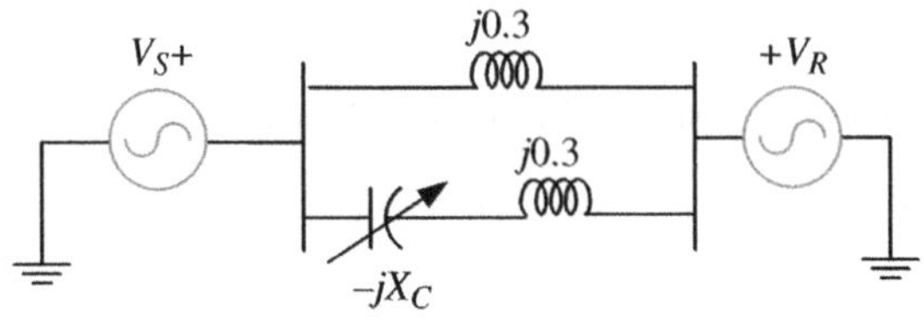

Figure P8.4 Series compensated transmission line of Problem P8.3.

P8.5 The system shown in Figure P8.5 is that of an SMIB system that is compensated by a shunt compensator, which is placed in the middle of the transmission line. Given $R = 0.005$ per unit, $X = 0.25$ per unit, $R_C = 0.002$ per unit, and $X_C = 0.1$ per unit, find the voltage V_C such that the midpoint voltage $V_M = 1\angle 15°$ per unit. Given $V_S = \angle 30°$ per unit and $V_R = 1\angle 0°$ per unit.

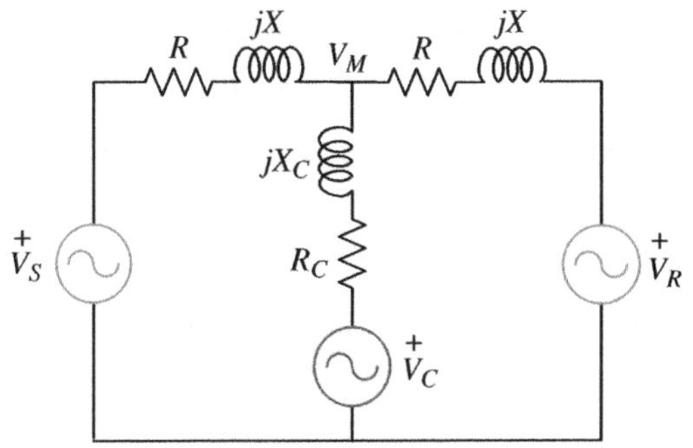

Figure P8.5 Shunt compensated transmission line of Problem P8.4.

P8.6 In the circuit of Figure P8.5, assume that the STATCOM is supplied by a dc capacitor, that is, the STATCOM cannot supply or absorb any real power.

Will the voltage V_C calculated in Problem 8.5 be able to sustain the operation of the STATCOM?

Now choose the midpoint voltage as $V_M = 1\angle \delta_M$. Then vary δ_M in small steps till an acceptable result is obtained.

What is the angle obtained (b) and what is complex power at the STATCOM terminal?

P8.7 Consider the lossless power system shown in Figure P8.7, which is compensated by a series compensator. The voltage injected by the series compensator is given by

$$V_Q = \lambda I_S e^{-j90°}$$

Find the value of λ, Q_S and Q_R if $P_S = P_R = 1.5$ per unit. Also, find the reactive drop across the line.

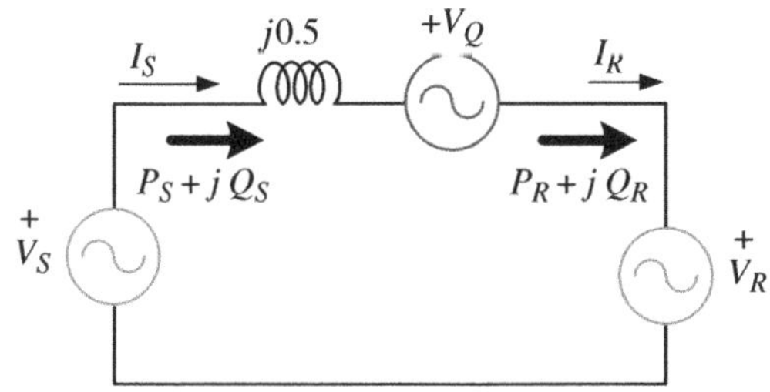

Figure P8.7 Series compensated transmission line of Problem P8.6.

P8.8 Consider an ideal shunt and series compensator. Suppose they are connected to two SMIB systems with the following characteristics:

$$|V_S| = |V_R| = 1.0 \text{ per unit and } X = 0.5 \text{ per unit}$$

The series compensation level is 20% in the capacitive mode. Suppose the per unit power transfer level in both the systems is the same, that is,

$$P_m = \frac{2V^2}{X}\sin(\delta/2) = \frac{V^2}{X-\lambda}\sin\delta$$

then find δ and P_m.

P8.9 Suppose the series compensator is working with the same δ and P_m obtained from Problem 8.8, when a fault occurs that forces the real power transfer to zero. Then, comment if the system can operate reliably under this condition.

P8.10 Consider the double-circuit transmission line shown in Figure P8.10. In this, Line-2 contains an ideal shunt compensator, as shown. Given that $V^2/X = 1.0$ per unit and $\delta = 30°$, find the total power transfer over the line.

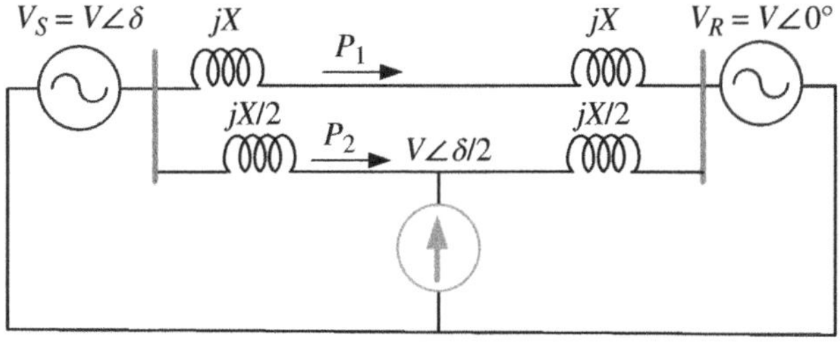

Figure P8.10 Double-circuit, shunt compensated transmission line of Problem P8.10.

9

Flexible AC Transmission Systems (FACTS)

The flexible AC transmission systems (FACTS) technology pertains to the application of power electronics in power transmission systems. In Chapter 8, the shunt or series compensation of a transmission system using ideal compensators has been discussed. In this chapter, both the physical structures of shunt and series compensators, along with several FACTS devices, will be discussed. The main purpose of these devices is to enhance power flow over transmission lines while improving system stability. Two different types of FACTS controllers are available. These are:

- Thyristor-based: These devices employ thyristor technology. Some of these devices are:
  - Static Var Compensator (SVC)
  - Thyristor-Controlled Series Compensator (TCSC)
  - Thyristor-Controlled Voltage Regulator (TCVR)
  - Thyristor-Controlled Phase Angle Regulator (TCPAR)
  - Thyristor-Controlled Braking Resistor (TCBR)
- Voltage source converter (VSC)-based: These devices employ synchronous voltage source (SVS) based on VSC technology. These include:
  - Static Compensator (STATCOM)
  - Static Synchronous Series Compensator (SSSC or S^3C)
  - Unified Power Flow Compensator (UPFC)

A thyristor is a type of controlled rectifier, also known as a silicon-controlled rectifier (SCR). Unlike an uncontrolled diode rectifier, which begins conducting immediately when the voltage in the forward direction becomes positive and stops when the current reverses polarity, a thyristor allows for delayed conduction. Instead of conducting as soon as the voltage turns positive, the thyristor's

Modern Power System, First Edition. Arindam Ghosh.

Published 2026 by John Wiley & Sons, Inc.
Companion website: www.wiley.com/go/ghoshpowersystem

conduction can be delayed by a specific angle. However, like a diode, the thyristor stops conducting (commutates) when the current reverses polarity. VSC-based devices usually use insulated gate bipolar transistors (IGBTs), which can be turned on or off through external commands. Using this principle, these devices can be used to construct an SVS, which can generate a voltage that is synchronized in phase and frequency with a reference system. Usually, these are supplied by a DC capacitor, and the output AC voltage is synchronized with the grid. However, for high-power use, both thyristor-based and IGBT-based power converters need several stages, as will be discussed in this chapter as well.

9.1 Static Var Compensator (SVC)

An SVC is the first-generation shunt compensator. It has been around since the 1960s. In the beginning, it was used for load compensation such as to provide var support to large industrial loads, and for flicker mitigation. However, with the advancement of semiconductor technology, the SVC started appearing in transmission systems in the 1970s. Today a large number of SVCs are connected to many transmission systems all over the world. An SVC is constructed using thyristor technology and therefore does not have a gate turn-off capability. There are two basic submodules of SVCs, which are discussed below.

9.1.1 Thyristor-Switched Capacitor (TSC)

In a thyristor-switched capacitor (TSC), a capacitor is connected in series with two opposite-poled thyristors, as shown in Figure 9.1 (a). An alternating current flows through the capacitor when the opposite-poled thyristors are gated. The current through the device can be stopped by blocking the thyristors. A TSC always comes in a pack, as shown in Figure 9.1 (b), which contains a total number of n TSC models. The effective reactance of the pack can be changed by switching a TSC on or off. Assuming that all the TSCs are identical, with each having a capacitance value C, the equivalent reactance of the TSC is given by

$$-jX_{TSC} = -j\frac{1}{k\omega C}, k = 1, \ldots, n \tag{9.1}$$

One of the main issues while using a TSC is switching transients. Since a TSC blocks current through it when the thyristors are blocked and allows it to pass when the thyristors are gated, it is obvious that severe switching transients will occur if a TSC is switched off while the current through it is not zero. Therefore, the device must be switched on at a particular instant of the voltage cycle. Consider Figure 9.1 (a), where the TSC is supplied by a voltage source v_S. The capacitor

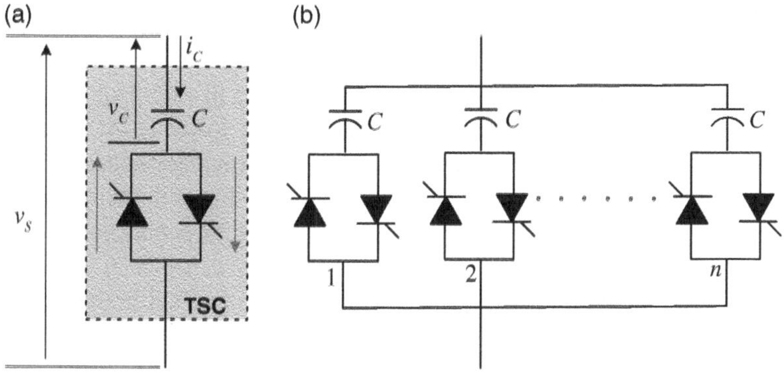

Figure 9.1 Schematic diagram of (a) single TSC module and (b) multiple TSC connections.

voltage across and current through the capacitor are denoted by v_C and i_C, respectively. Then the capacitor current is given by $i_C = C(dv_C/dt)$. Thus, the capacitor current is zero when $dv_C/dt = 0$, that is, when the capacitor voltage reaches its peak or trough. Thus, for transient-free switching, it must be ensured that the capacitor voltage is in either its positive peak or negative peak for either turn on or turn off. The transient-free switching is shown in Figure 9.2, in which the instant of switching on and off is also indicated.

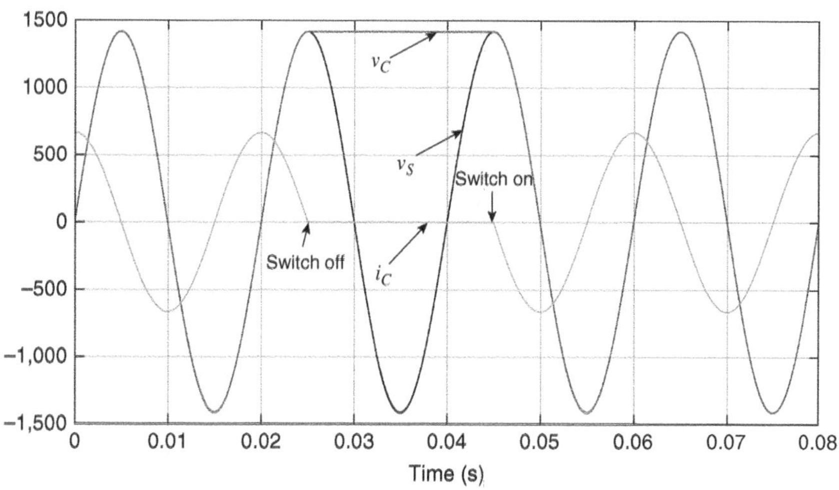

Figure 9.2 Transient-free switching of a TSC.

9.1.2 Thyristor-Controlled Reactor (TCR)

In a thyristor-controlled reactor (TCR), a reactor is connected in series with two opposite-poled thyristors. One of these thyristors conducts in each half cycle of supply frequency. The schematic diagram of a TCR connected to an AC voltage source is shown in Figure 9.3. The gating signal to each thyristor is delayed by an angle α (often called the *firing angle*) from the zero crossing of the supply voltage. The period during which the thyristors are on is called the *conduction period* and is denoted by σ.

Figure 9.4 shows the typical voltage–current waveforms of a TCR in the steady state. The firing angle must be in the range of $90° \leq \alpha \leq 180°$. When $\alpha = 90°$, the current conduction is continuous, as shown in Figure 9.5 (a). On the other hand,

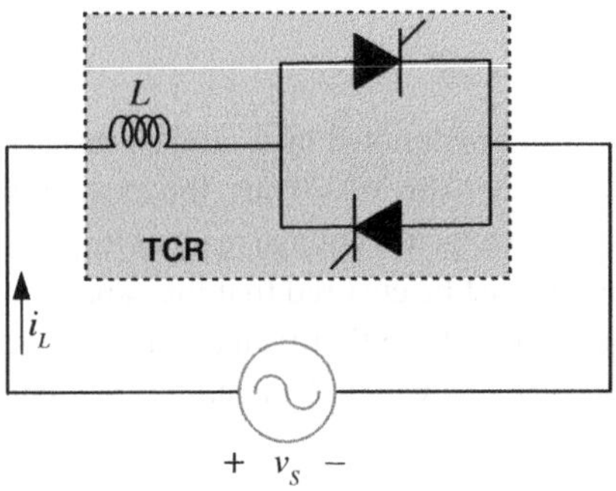

Figure 9.3 Schematic diagram of a TCR.

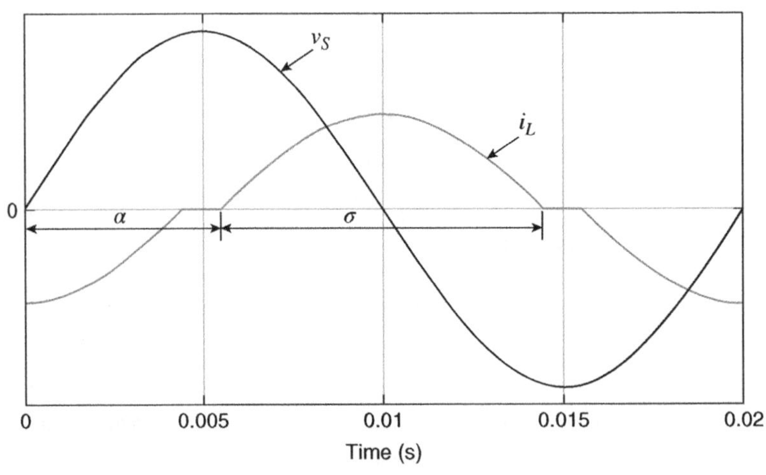

Figure 9.4 Steady-state voltage–current waveforms of a TCR.

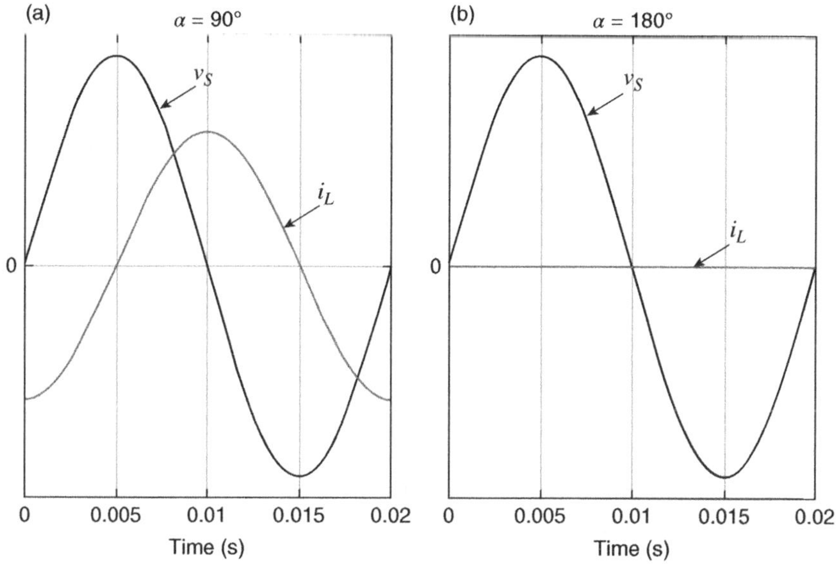

Figure 9.5 Steady-state voltage–current waveforms of a TCR for (a) α = 90° and (b) α = 180°.

when $\alpha = 180°$, the current will be zero, as shown in Figure 9.5 (b). In this case, the voltage across each thyristor will become negative as soon as the thyristor is turned on, which will force the thyristor to commutate. Therefore, no current will flow.

Let the supply voltage be given by

$$v_S = V_m \sin \omega t$$

From Figure 9.4, it can be seen that the current i_L is conducting between α and $\alpha + \sigma$. Therefore, the current is given by

$$i_L = \frac{1}{L}\int_{\alpha/\omega}^{t} V_m \sin(\omega\tau)d\tau \text{ for } \alpha \leq \omega t < \alpha + \sigma \tag{9.2}$$

The solution of (9.2) is given as follows:

$$i_L = \frac{V_m}{X_L}(\cos\alpha - \cos\omega t) \text{ for } \alpha \leq \omega t < \alpha + \sigma \tag{9.3}$$

where $X_L = \omega L$. From (9.3) and Figure 9.4, it is obvious that the current i_L is discontinuous and will therefore contain harmonics. Furthermore, from Figs. 9.4 and (9.3), it is also clear that the shape of the current i_L will vary with the firing angle α. Therefore, it is of interest to find the fundamental component of i_L as the firing angle α changes, so that we can define the fundamental susceptance of the TCR in terms of the firing angle α.

The fundamental current is computed when the system has reached its steady state. The current in (9.3) does not contain any DC. Thus, its fundamental component can be expressed as

$$i_{Lf} = a_1 \cos \omega t + b_1 \sin \omega t \tag{9.4}$$

Since the current i_L is half-cycle symmetric, the Fourier coefficients of (9.4) are given as follows:

$$a_1 = \frac{4}{T}\int_t^{t+T} i_L \cos(\omega\tau)d\tau \text{ and } b_1 = \frac{4}{T}\int_t^{t+T} i_L \sin(\omega\tau)d\tau \tag{9.5}$$

Let us assume that the conduction period of the thyristors is symmetric about the zero crossing of the source voltage in the steady state. Then, from Figure 9.4, we can write.

$$2\alpha + \sigma = 2\pi \Rightarrow \alpha = \pi - \frac{\sigma}{2} \tag{9.6}$$

Integrating i_L from (9.3) between 0 and 2π, we can find the constants of (9.5) as

$$a_1 = \frac{2\omega V_m}{\pi X_L}\int_{\alpha/\omega}^{(\alpha+\sigma)/\omega} (\cos\alpha - \cos\omega t)\cos\omega t\, dt = \frac{2\omega V_m}{\pi X_L}\left[\frac{\cos\alpha \sin\omega t}{\omega} - \frac{t}{2} - \frac{\sin(2\omega t)}{4\omega}\right]_{\alpha/\omega}^{(2\pi-\alpha)/\omega}$$
$$= \frac{2V_m}{\pi X_L}\left[-\frac{\sin(2\alpha)}{2} - \pi + \alpha\right]$$

$$b_1 = \frac{2\omega V_m}{\pi X_L}\int_{\alpha/\omega}^{(2\pi-\alpha)/\omega} (\cos\alpha - \cos\omega t)\sin\omega t\, dt = 0$$

Using (9.6), a_1 is rewritten as

$$a_1 = \frac{V_m}{\pi X_L}[\sin\sigma - \sigma] \tag{9.7}$$

Substituting (9.7) into (9.4), the fundamental current (i_{Lf}) is obtained as

$$i_{Lf} = \frac{V_m}{\pi X_L}[\sin\sigma - \sigma]\cos\omega t = \frac{V_m}{\pi X_L}[\sigma - \sin\sigma]\sin(\omega t - 90°) \tag{9.8}$$

It can be seen that the fundamental current lags the impressed voltage by 90°. The fundamental frequency susceptance of the TCR is then

$$B_L(\sigma) = \frac{I_{Lf}}{V_S} = \frac{\sigma - \sin\sigma}{\pi X_L} \tag{9.9}$$

This susceptance is a function of σ. The variation in the susceptance as a function of σ is shown in Figure 9.6 when X_L is chosen as 0.1 per unit. It can be seen that the susceptance is zero for $\sigma = 0°$, that is, $\alpha = 180°$. This means that the system impedance will become infinite, and this will block any current flow, as evident from Figure 9.5 (b). On the other hand, the conductance is equal to 10 (i.e., reciprocal of the chosen value of X_L) for $\sigma = 180°$, that is, $\alpha = 90°$. Hence, the current is continuous for this value of firing angle, as evident from Figure 9.5 (a).

The current shown in Figure 9.4 will only contain odd harmonics. The magnitudes of the harmonic currents are given by [1]

$$I_{Ln} = \frac{4V_m}{\pi X_L}\left[\frac{\sin\{(n+1)\alpha\}}{2(n+1)} + \frac{\sin\{(n-1)\alpha\}}{2(n-1)} - \cos\alpha\frac{\sin n\alpha}{n}\right] \text{ for } n = 3, 5, \ldots$$

Let us define magnitude ratios as

$$\text{Magnitude Ratio} = \frac{I_{Ln}}{I_{Lf}} \text{ for } n = 3, 5, \ldots$$

These are plotted for four different values of the firing angle, as shown in Figure 9.7. It is obvious from this figure that as the firing angle increases, the

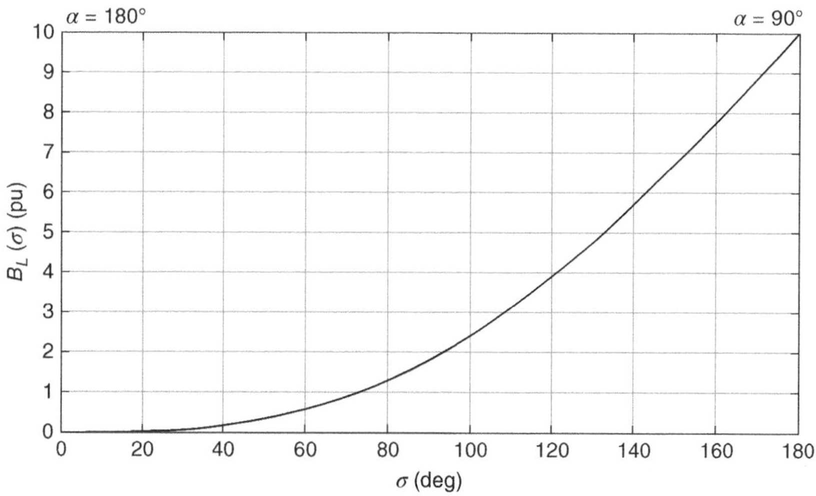

Figure 9.6 The variation in the fundamental frequency susceptance with σ.

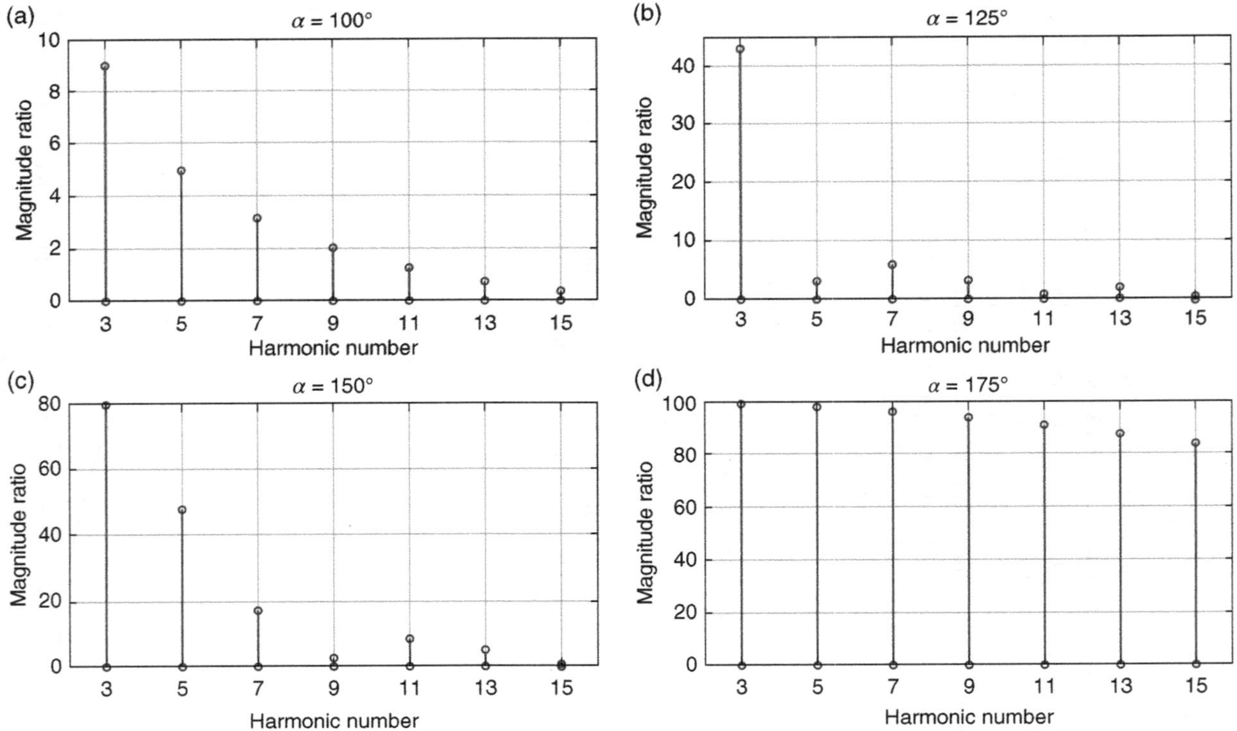

Figure 9.7 The variation in the harmonic content with changes in (a) $\alpha = 100°$, (b) $\alpha = 125°$, (c) $\alpha = 150°$ and (d) $\alpha = 175°$.

harmonic components increase. This is obvious since the peak of the current waveform and the conduction period decrease as the firing angle increases.

9.1.3 Composition of SVC

An SVC contains a TCR and a bank of TSCs with some associated tuned filters, as shown in Figure 9.8. The tuned filters usually target lower-order harmonics, for example, third, fifth, and seventh. An SVC is usually connected at a medium voltage (MV), and hence it is connected to the high voltage (HV) bus through a transformer. Also, there are protection and measuring transformers (CTs and PTs) and controllers, which are not shown in the figure. Because an SVC contains a TCR, it can only be represented by its fundamental reactance. This is discussed below.

The fundamental impedance of an SVC is a parallel combination of the fundamental component of the TCR and the effective reactance of the TSC bank. Let the fundamental frequency impedance of the TCR be given by

$$X_{TCR} = \frac{1}{B_L(\sigma)}$$

Then, the fundamental frequency impedance of an SVC is the parallel combination of TCR reactance jX_{TCR} and TCS reactance $-jX_{TCS}$, that is,

$$X_{SVC} = \frac{X_{TCR}X_{TSC}}{j(X_{TCR} - X_{TSC})} \tag{9.10}$$

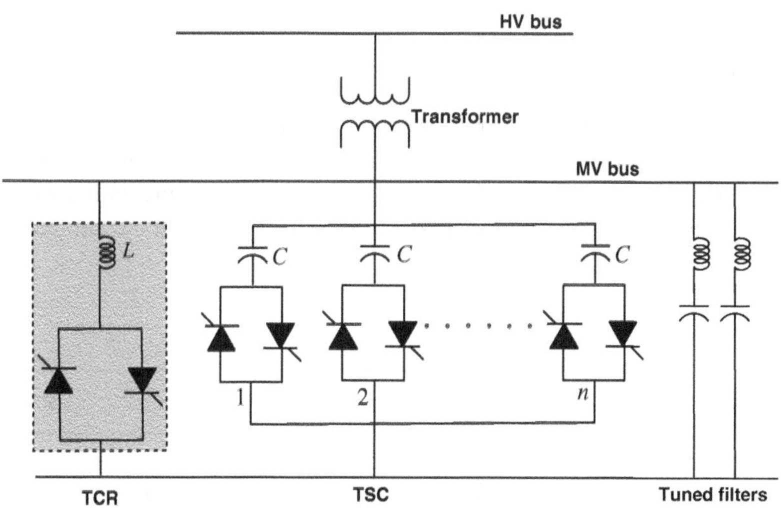

Figure 9.8 Schematic diagram of an SVC.

The susceptance of the SVC then is

$$B_{SVC} = \frac{1}{X_{SVC}} = j\left(\frac{1}{X_{TSC}} - \frac{1}{X_{TCR}}\right) = j(B_{TSC} - B_L(\sigma)) \tag{9.11}$$

Example 9.1 Consider an SVC with the following parameters:

TCR: $X_L = 0.25$ per unit

TSC: $X_C = 10$ per unit and $n = 5$.

The firing angle α is chosen as 120°. We will find the fundamental frequency equivalent reactance of the SVC when all five possible combinations of the TSC are chosen.

From (9.6), we have

$$\sigma = 360° - 2\alpha = 120° = 2.094 \text{ rad}$$

Then

$$B_L(\sigma) = \frac{\sigma - \sin\sigma}{\pi X_L} = 1.564 \text{ per unit}$$

Therefore,

$$X_{TCR} = \frac{1}{B_L(\sigma)} = j0.6394 \text{ per unit}$$

The equivalent reactances of the SVC as the number of TSC modules vary are listed in Table 9.1.

9.1.4 SVC Characteristics

The characteristics of the different SVC components are shown in Figure 9.9. Figure 9.9 (a) shows the connection of a variable inductor (TCR) with a voltage bus. Note that $|V| = \omega L|I|$, which is an equation of a straight line with a slope of $K_S = \omega L$. Thus, the current can vary between the maximum and the minimum

Table 9.1 Equivalent reactance of the SVC of Example 9.1.

TSC modules	1	2	3	4	5
X_{TSC} (pu)	$-j10$	$-j5$	$-j3.33$	$-j2.5$	$-j2$
X_{SVC} (pu)	$j0.683$	$j0.733$	$j0.791$	$j0.859$	$j0.94$

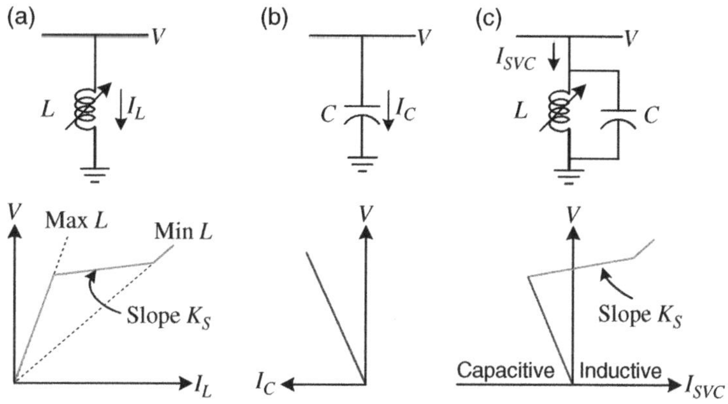

Figure 9.9 SVC characteristics for different connections: with (a) TCR, (b) capacitor and (c) parallel connection of TCR and capacitor.

with a slope K_S. Figure 9.9 (b) shows the capacitor connection, while Figure 9.9 (c) shows the characteristics when both the variable inductor and capacitor are connected. It can be seen that the overall voltage–current characteristic is not a horizontal line.

Consider the power network shown in Figure 9.10 (a), where a voltage source supplies a reactive load through a lossless line. The receiving end voltage is given by

$$V = V_S - jX_S I_S \tag{9.12}$$

The ideal voltage–current relationship is shown in Figure 9.10 (b). This means that the voltage is held constant at V_0 irrespective of the leading or lagging current drawn. However, the practical voltage–current relationship of the network in the presence of an SVC is shown in Figure 9.10 (c).

The voltage V is defined by the slope of the reactance X_{SL} as

$$V = V_0 - jX_{SL} I_S \tag{9.13}$$

Curve-0 of Figure 9.10 (c), which passes through the point $V = V_0$, $I_S = 0$ defines the nominal range without any SVC. Now suppose the supply voltage increases such that V increases to V_1 without any SVC. This is represented by Curve-A. This curve intersects the SVC curve at point A' and the operating point moves to V_A with SVC, where the SVC absorbs inductive current. On the other hand, when the supply voltage drops, the voltage V decreases to V_2 without any SVC (Curve-B). The SVC then supplies capacitive current to hold the voltage to V_B.

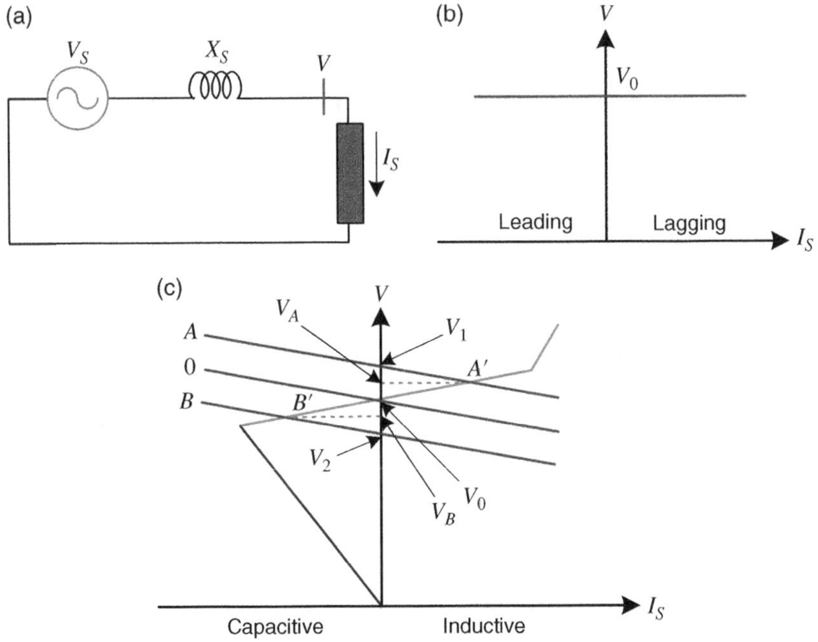

Figure 9.10 (a) A transmission network, (b) ideal characteristic, and (c) with SVC.

9.2 Static Compensator (STATCOM)

The schematic diagram of a STATCOM-compensated power system is shown in Figure 9.11. This is similar to the ideal shunt compensator model presented in Figure 8.11. In this figure, the STATCOM contains a three-phase VSC and a DC capacitor (C_{dc}). The VSC is connected to the HV bus through a transformer. The reactance X_T represents the leakage reactance of the transformer.

Let the midpoint and the VSC output voltages be $V_M = |V_M| \angle \delta_M$ and $V_C = |V_C| \angle \delta_C$ respectively. Then the current I_Q is given by

$$I_Q = \frac{|V_M|\angle\delta_M - |V_C|\angle\delta_C}{jX_C} \tag{9.14}$$

If $\delta_M > \delta_C$, real power will flow from the midpoint to the VSC. On the other hand, if $\delta_M < \delta_C$, the power flow direction will reverse. None of these two situations are desirable, as they will either charge or discharge the capacitor C_{dc}. However, if $\delta_M = \delta_C$, (9.14) can be written as

$$I_Q = \frac{|V_M|\angle\delta_M - |V_C|\angle\delta_M}{jX_C} = \frac{|V_M| - |V_C|}{X_C}\angle(\delta_M - 90°) \tag{9.15}$$

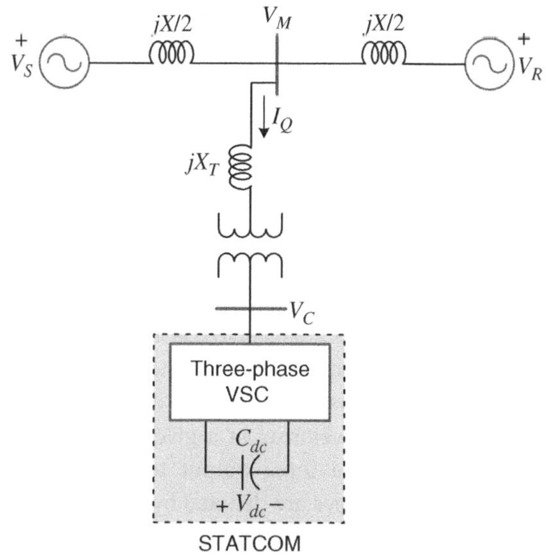

Figure 9.11 Schematic diagram of an SMIB system with an STATCOM connected at the midpoint.

Therefore, a pure reactive current will flow through the transformer. Now if $|V_M| > |V_C|$, reactive current will flow from the AC system to the STATCOM, and this flow will reverse if $|V_M| < |V_C|$. Thus, just by controlling the magnitude of the VSC output voltage, reactive power can be absorbed or generated.

However, if the angles are kept equal, that is, $\delta_M = \delta_C$, the DC capacitor needs to supply the converter and transformer losses. This will cause it to discharge. To avoid this, the angle δ_M should lead δ_C by a small amount. This can be achieved through a proportional plus integral (PI) controller that will regulate the capacitor voltage to a set reference value by drawing the exact amount of power required. Hence, the STATCOM will have two control loops – one for DC capacitor voltage control and the other for VSC output voltage control to vary the reactive power requirement.

The main component of a STATCOM is a high-power voltage source converter that is capable of supporting reactive power in an HV transmission system. Two different classes of VSC structures have been discussed for this application: (i) using gate turn-off thyristors (GTOs) and (ii) using IGBTs using multilevel configurations. These are briefly discussed in Section 9.3.

9.3 High-Power Converters

A multistep inverter is constructed using many six-step inverters and a magnetic circuit to provide phase shifts between the inverters. This principle of construction has been around for quite some time [2]. However, with the advent of the GTOs, the inverter circuits are becoming more complicated. It is therefore important to outline the principle of construction of these converters. It is to be noted that a voltage source converter can both rectify AC voltage to DC or invert DC voltage into AC. Because an STATCOM functions as an inverter, we may use the term "inverter" interchangeably with a converter.

9.3.1 Six-Step Converter

The basic configuration of a six-step inverter is shown in Figure 9.12. It contains six GTOs, each of which is connected in antiparallel with a diode to maintain continuity of current. A $6n$-step inverter is built using n 6-step inverters. For example, two such converters are used to construct a 12-step inverter, and four inverters are required to construct a 24-step inverter. We shall discuss the working principle of a six-step inverter first. The inverter of Figure 9.12 produces a set of three quasi-square voltage waveforms of a given frequency by connecting the DC voltage sequentially to the three output terminals via appropriate inverter switches (GTOs). The switches are denoted by S_1 to S_6 and are fired in the same sequence.

The output voltage waveforms, when each switch is conducting for a period of 180°, are shown in Figure 9.13. The Fourier series of the output voltage waveforms are given by

$$\begin{aligned} f_A(t) &= \sum_{n=1,3,5,\ldots} \frac{4V_{dc}}{n\pi} \sin(n\omega t) \\ f_B(t) &= \sum_{n=1,3,5,\ldots} \frac{4V_{dc}}{n\pi} \sin(n\omega t - 2n\pi/3) \\ f_C(t) &= \sum_{n=1,3,5,\ldots} \frac{4V_{dc}}{n\pi} \sin(n\omega t + 2n\pi/3) \end{aligned} \tag{9.16}$$

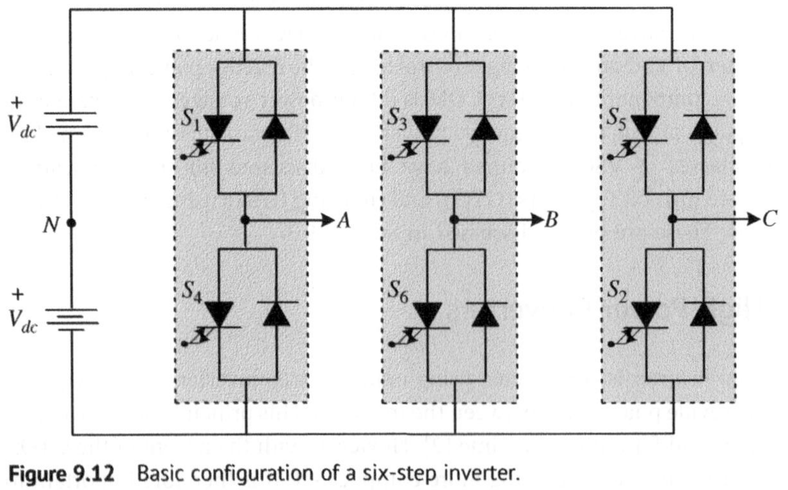

Figure 9.12 Basic configuration of a six-step inverter.

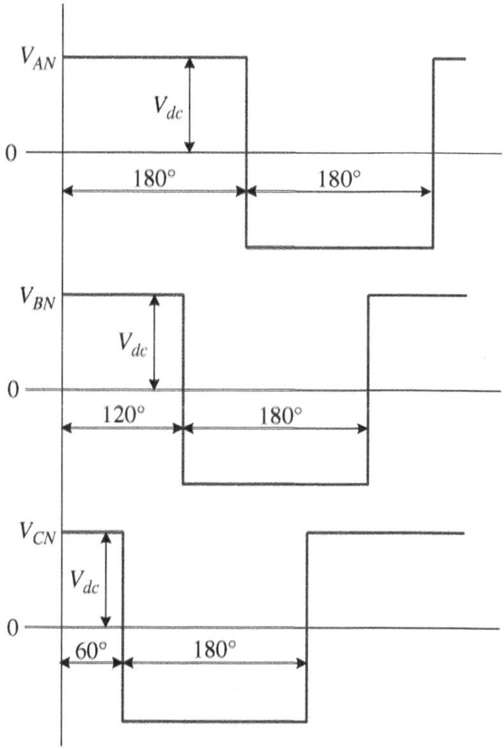

Figure 9.13 Six-step inverter output voltage waveforms.

where ω is the fundamental frequency in rad/s. From the above three equations, it is obvious that there are no even harmonics on the AC side. It can also be seen that for a particular value of n, the three functions are either in phase or appear in the sequence A-B-C or in the sequence A-C-B. Let us denote them as zero sequence, positive sequence, and negative sequence, respectively. It is to be noted that these sequences are not the symmetrical component sequences used for the representation of unbalanced circuits.

The sequences for the various values of n are given in Table 9.2. From this table, it is evident that the triplen harmonics (3, 9, 15, ...) are zero sequence and thus can circulate in the Δ-connected primary windings of the transformer connected to the inverter and therefore do not appear in the transformer secondary. Alternatively, if the transformer primary is connected in an ungrounded star, the zero-sequence components are forced to zero.

The output (AC) side voltage harmonics that appear at the converter output are $6q \pm 1$ where $q = 1, 2, 3, \ldots$. These include both positive and negative sequences. The DC side current harmonics generated are given by $6q$ [3]. Therefore, we can state that the AC side harmonics are complementary to the DC side harmonics.

Table 9.2 Harmonic sequences.

Sequences	Harmonic number *n*				
Zero sequence	3	9	15	21	...
Positive sequence	1	7	13	19	...
Negative sequence	5	11	17	23	...

9.3.2 Twelve-Step Converter

Lower-order voltage harmonics are a menace to a power system. Because the power system is mostly inductive, the currents due to higher-order voltage harmonics get attenuated as the harmonic number increases since the effective impedance increases with it. However, since the increase in the impedance seen by the circuit at lower-order harmonic frequencies is not large, the lower-order harmonic voltages produce significant harmonic current. It is therefore desirable to eliminate the lower-order harmonics. Let us now consider two 6-step inverters that are connected together as shown in Figure 9.14. In this diagram, only the DC side is shown. It can be seen that both these inverters are being supplied by the same DC source, which has a potential of $V = 2V_{dc}$.

Let us consider the sixth harmonic component on the DC side of the first inverter. It can be eliminated if another sixth harmonic component that is shifted by 180° on its frequency is added to it. Their sum will then add up to zero. To effect a phase shift of 180° from the second inverter, firing instants of all the switches of the second inverter must lag the corresponding switches of the first inverter by 30° on the fundamental frequency such that the sixth harmonic component of the second inverter lags that of the first inverter by 180°. Furthermore, such a phase shift

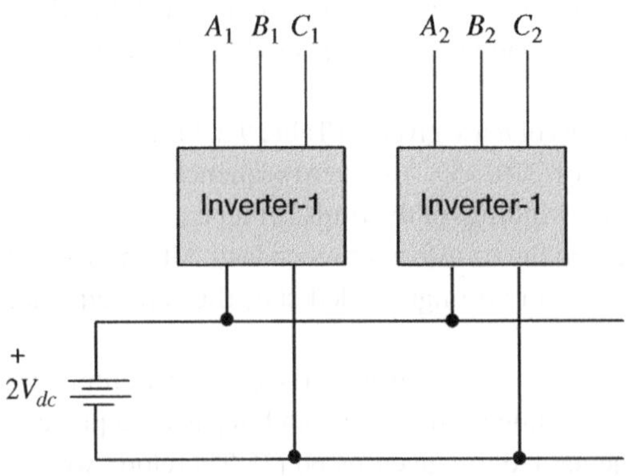

Figure 9.14 DC side connection diagram of two 6-step inverters.

will cancel all the odd multiples of the sixth harmonic components. For example, the 18th harmonic component of the second inverter will lag that of the first inverter by 540°, making them in phase opposition. Thus, only the even multiples of the sixth harmonic will remain on the DC side.

Let us now investigate what happens on the AC side if all the firing angles of the switches of the second inverter are delayed by 30° on the fundamental frequency from the corresponding switches of the first inverter. The respective phase-A output voltage waveforms of the two inverters are shown in Figure 9.15, where the fundamental voltage of one of the inverters (second inverter) lags that of the other (first inverter) by θ. The Fourier expressions of the phase-A voltage waveforms of these two inverters are then given by

$$
\begin{aligned}
f_{A1}(t) &= \frac{4V_{dc}}{\pi}\sin(\omega t) + \sum_{n=5,7,\ldots} \frac{4V_{dc}}{n\pi}\sin(n\omega t) \\
f_{A2}(t) &= \frac{4V_{dc}}{\pi}\sin(\omega t-\theta) + \sum_{n=5,7,\ldots} \frac{4V_{dc}}{n\pi}\sin(n\omega t-n\theta)
\end{aligned}
\tag{9.17}
$$

where θ is equal to 30°.

From (9.17), it is obvious that the fifth and seventh harmonic components of the second inverter are phase shifted by −150° and −210°, respectively, with respect to that of the first inverter. To eliminate these two harmonic components, the fifth harmonic component needs to be further phase shifted by −30° and the seventh

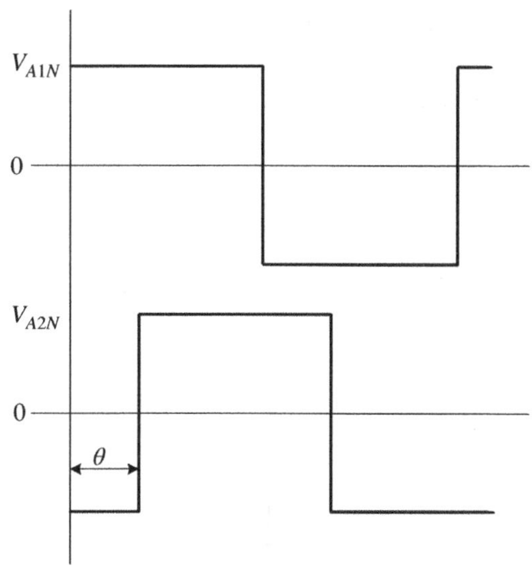

Figure 9.15 Output voltage waveforms of phase-shifted inverters.

harmonic component needs a phase shift of +30°. This means that the negative sequence is phase shifted by −30°, while the positive sequence is by +30°.

We can connect any three single-phase transformers in wye-Δ to provide ±30° phase shift between their primary and secondary windings. In that case, if the positive sequence is phase shifted by +30°, the negative sequence gets phase shifted by −30°, and vice versa. To explain this, consider the wye-Δ-connected transformer that is shown in Figure 9.16. The uppercase letters denote the primary side, while the secondary side is denoted by lowercase letters. Also adding a suffix 1 to the winding names differentiates the terminal names and the windings. The voltage vector diagram for the positive sequence components is shown in Figure 9.17. It is evident from this figure that the induced secondary voltages (e.g., $V_{a1a1'}$) lead the primary line-to-neutral voltage by 30°. This phase shift remains the same for all positive sequence harmonic components (e.g., 7th, 13th, etc.). For the negative sequence harmonics (e.g., 5th, 11th, etc.), the phases B and C exchange positions, and the phase becomes −30°.

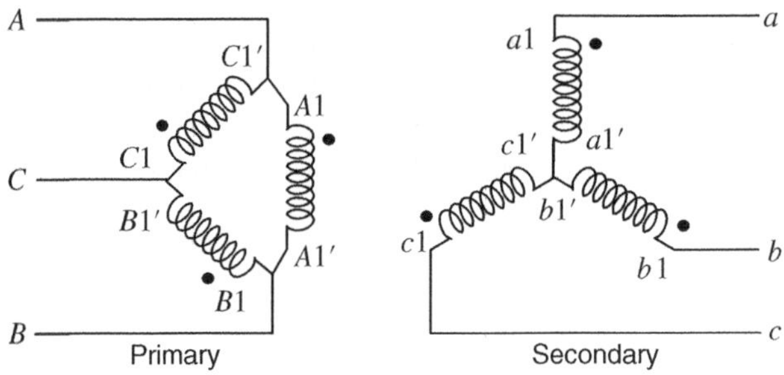

Figure 9.16 Schematic diagram of phase-shifted wye-Δ transformer.

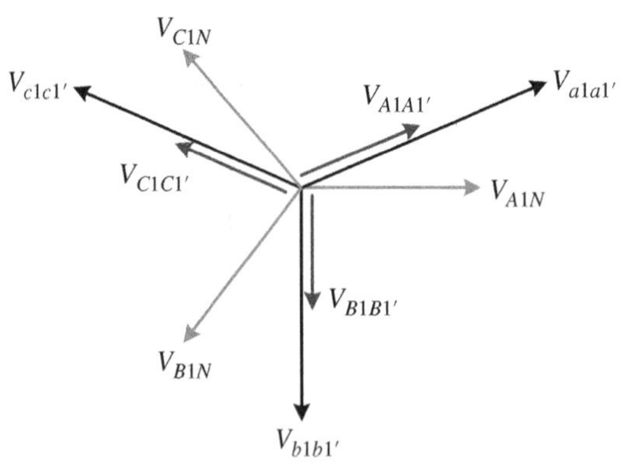

Figure 9.17 Positive-sequenced phasor diagram.

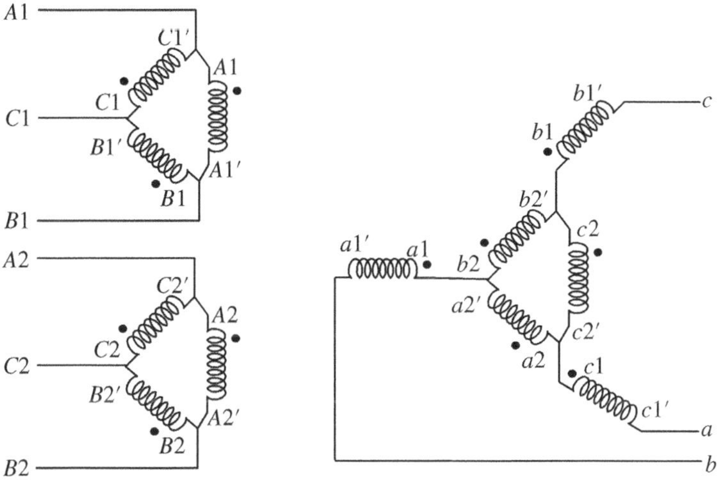

Figure 9.18 Schematic diagram of three single-phase transformers to provide a phase shift of + 30° to the positive sequence.

The transformer connection for the addition of the output waveforms of two inverters is shown in Figure 9.18. The two primary side transformers are denoted by 1 and 2 with uppercase letters *A*, *B*, and *C*. Their corresponding secondary sides are denoted by lowercase letters *a*, *b*, and *c*. The primary sides of Transformers 1 and 2 are connected to the outputs of Inverters 1 and 2, respectively. The inverter output voltages *A*2, *B*2, and *C*2 are arranged to lag output voltages of *A*1, *B*1, and *C*1 by 30° through phase shift in their gating signals. On the secondary side, Transformer 1 has an open star connection, while Transformer 2 is connected in Δ. The turns ratio of Transformer 1 is 1:1, and that of Transformer 2 is $1 : \sqrt{3}$.

The secondary windings $a2a2'$, $b2b2'$, and $c2c2'$ of Transformer 2 are connected in Δ. Thus, there is no phase shift introduced by this transformer. The secondary windings of Transformer 1 are connected in a fashion that the terminals $a1$, $b1$, and $c1$ form an open star. From Figure 9.18, we get

$$V_{ab} = V_{a1a1'} + V_{a2a2'} - V_{c1c1'} \tag{9.18}$$

If we consider only the fundamental components of the inverter–transformer output, then the vector diagram corresponding to (9.18) is shown in Figure 9.19 (a). This shows that V_{ab} is in phase with the primary reference voltage V_{A1N}, which is the fundamental output of Inverter 1. The other voltages can be written by symmetry from Figure 9.18 as

$$\begin{aligned} V_{bc} &= V_{b1b1'} + V_{b2b2'} - V_{a1a1'} \\ V_{ca} &= V_{c1c1'} + V_{c2c2'} - V_{b1b1'} \end{aligned} \tag{9.19}$$

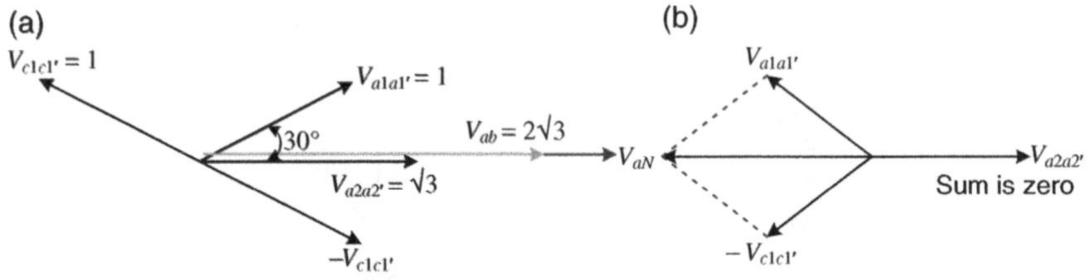

Figure 9.19 Phasor diagram of (a) fundamental and (b) fifth harmonic.

A vector diagram corresponding to the fifth harmonic is shown in Figure 9.19 (b). The phase angles of the three terms in the equation for V_{ab} are such that they cancel out completely. A similar diagram can also be drawn for the seventh harmonic, and it can be shown that this component also becomes equal to zero.

A time domain construction corresponding to the expression V_{ab} is shown in Figure 9.20 (a). The levels produced are 1, $1 + \sqrt{3}$, and $2 + \sqrt{3}$ times the DC bus voltage of the inverter supplying the primary, which is $1/\sqrt{3}$ per unit for this figure. The circuit discussed above is a 12-step inverter. The harmonic spectrum of the output line-to-line voltage is shown in Figure 9.20 (b). The DC side voltage V_{dc} is chosen to be 1.0 per unit to obtain the output voltage. It is evident from Figure 9.20 (b) that the 11th and 13th are the most dominant AC side harmonic components.

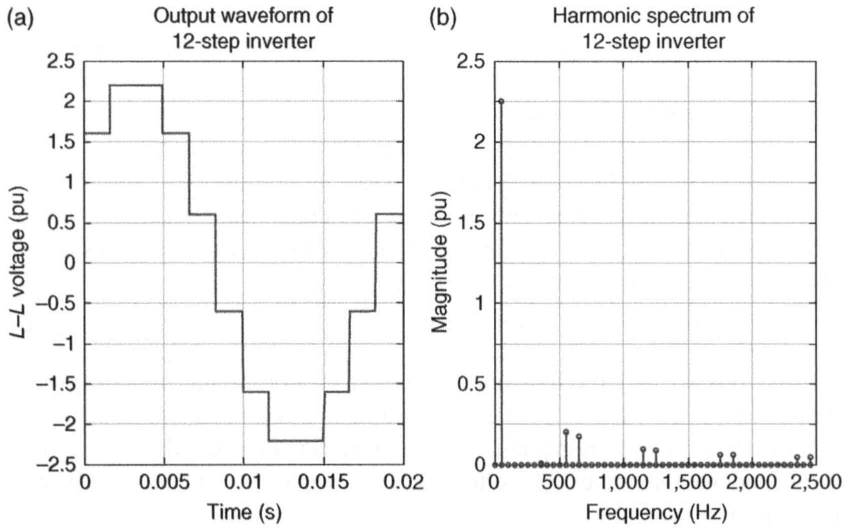

Figure 9.20 (a) Output voltage waveform of a 12-step inverter and (b) its harmonic spectrum.

9.3.3 6q-Step Converter

It is seen that by providing a phase shift of 30° between two inverter–transformer combinations, it is possible to generate a 12-step waveform with a spectrum that contains 11th and higher-order harmonics only. Now, combining two such inverters, we shall be able to get a 24-step inverter. In this case, however, the phase difference between the successive inverter–transformers must be 15°. In a similar way, we can construct a 6q-step inverter by providing a phase shift of 360°/6q between the successive inverter–transformers.

The construction of a multistep inverter requires a number of single-phase transformers that can be connected to provide any feasible degree of phase shift. The development of a 48-step inverter using 18 single-phase three-winding transformers and 6 single-phase two-winding transformers is reported in [4, 5], where a phase shift of 7.5° is provided between the successive inverters. The output voltage waveform of this converter and its harmonic spectrum are shown in Figure 9.21. It can be seen that the highest harmonic components are the 47th and 49th. The modeling of this converter for series compensation is discussed later in the chapter.

9.3.4 Multilevel Converters

Multilevel converters use a series connection of semiconductor IGBT switches to synthesize different voltage levels, which are either a low-frequency staircase or a

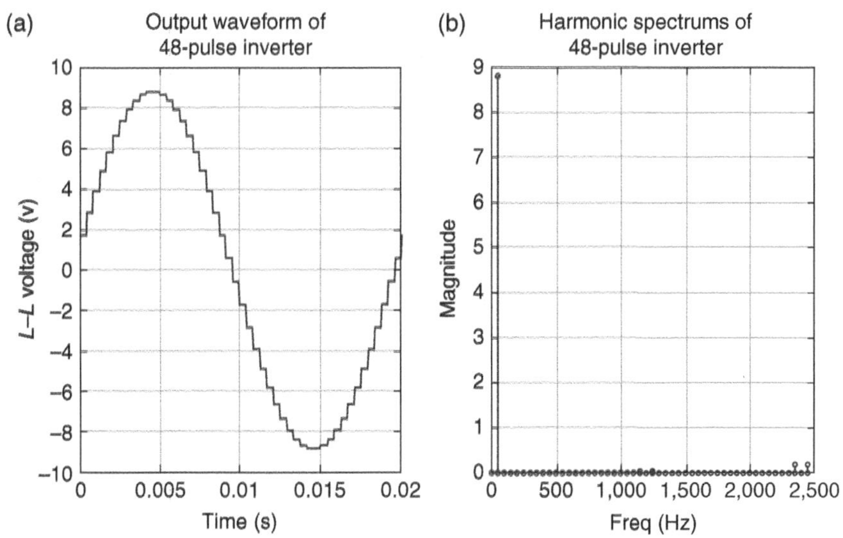

Figure 9.21 (a) Output voltage waveform of a 48-step inverter and (b) its harmonic spectrum.

high-frequency modulated voltage waveform using sinusoidal pulse width modulation (SPWM) or space vector pulse width modulation (SVPWM). By increasing the number of switches, these converters can be configured on different levels, for example, three-level, five-level, and seven-level. As the number of levels increases, the converter output voltage can be closer to sinusoidal voltage waveforms, in the same fashion as the multistep converters. Moreover, multilevel converters can be configured in a way such that the switching losses are minimized. The main drawbacks of these converters are the requirement for a high number of switches and passive components. However, these converters do not need the complex transformer connections that are required for multistep converters.

There are three main types of multilevel converter topologies. These are:

- Diode-clamped
- Flying capacitor
- Cascaded converter

The schematic diagrams of these three converters are shown in Figure 9.22. In a diode-clamped configuration, the DC link is split into several voltage levels using capacitors. The clamping diodes connect the common terminals between two adjacent capacitors and the switches of each leg. In a flying capacitor converter, the capacitors are charged to different voltage levels. Different output voltage levels can be achieved by connecting the capacitors in parallel with the DC link voltage

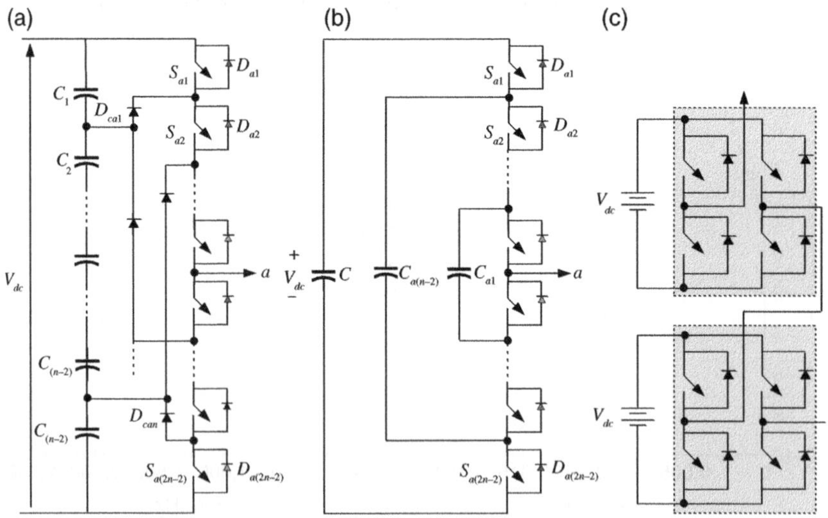

Figure 9.22 One phase of (a) diode-clamped, (b) flying capacitor, and (c) cascaded converter.

based on the switching states of the semiconductor switches. In a cascaded multilevel converter, several full bridge (H-bridge) converters are connected in series to increase the voltage level. The modular multilevel converter (MMC) topology has gained attention for its application in HVDC transmission. In this, several submodules are connected in series. The submodules can be either half bridge or full bridge.

The output voltages of multilevel converters can be synthesized using SPWM or SVPWM. This, together with output LC or LCL filters, makes the output voltage nearly sinusoidal. Thus, the advantage of multilevel converters is that they can be considered as pure voltage sources for system studies. Brief operating principles of multilevel converters and the PWM waveform synthesis are discussed in [6].

9.4 Subsynchronous Oscillations

From the power transfer equation given in (7.6), it is obvious that for a given machine angle δ and terminal voltages, the steady-state power transfer over a transmission line can be enhanced if the line reactance X can be decreased. Unfortunately, however, X is dependent on the transmission line characteristics and cannot be altered. This has prompted some utilities to insert a series capacitor in the transmission line such that the power transfer relation can now be written as

$$P_R = \frac{V_1 V_2}{X - X_C} \sin\delta, \delta = \delta_1 - \delta_2 \tag{9.20}$$

where X_C is the reactance of the series capacitor that gets subtracted from the transmission line reactance. The ratio X_C/X defines the percentage of compensation that is limited to less than 70%.

In 1970 and 1971, two accidents occurred in Southern California Edison Company's generating station at Mojave in Nevada. There were two shafts, one common to the turbine and generator, and the other shared by the generator and exciter. The latter shaft broke. In 1972, the same thing happened at Arizona Public Service's Four Corners station at the junction of Arizona, New Mexico, Colorado, and Utah. It was noticed that all these plants were connected to transmission lines that have been compensated by fixed series capacitors. Further research revealed that the series capacitor altered the natural frequency of the electrical circuit. This altered natural frequency triggered torsional oscillations when it coincided with one of the modes of the mechanical system. Usually, the oscillation frequency is less than the synchronous frequency, and hence this condition is termed as the subsynchronous resonance (SSR).

The accidents of the early 1970s have slowed down the installation of series capacitors in transmission systems. They have also triggered a lot of research in

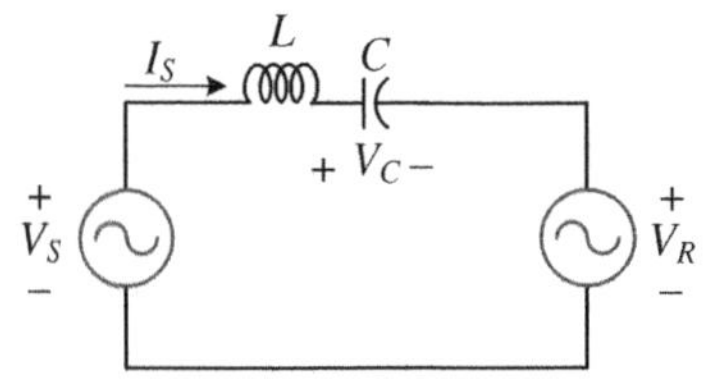

Figure 9.23 A series-compensated SMIB system.

this area. A large body of literature is available for those who want to probe this subject further. Various committees constituted by the Institute of Electrical and Electronic Engineers (IEEE) have proposed terms and definitions of SSR [7]. Two benchmark models for computer simulation of SSR have been prepared by IEEE [8, 9]. Two books that are dedicated to SSR studies only have also been prepared by renowned experts [10, 11]. In this section, the main features of SSR are discussed.

9.4.1 Subsynchronous and Supersynchronous Frequencies

Consider a series-compensated SMIB system, as shown in Figure 9.23. The system voltages are:

$$v_S = V\sin(\omega_s t + \delta) \text{ and } v_R = V\sin(\omega_s t)$$

The line impedance is written in the Laplace domain as

$$Z(s) = \frac{1}{sC} + sL = \frac{s^2LC + 1}{sC} = \frac{s^2 + \omega_n^2}{s/L} \tag{9.21}$$

where $\omega_n = 1/\sqrt{LC}$ is the natural frequency of oscillation of the electrical network.

From Figure 9.23, the line current is given by

$$I_S(s) = \frac{V_S(s) - V_R(s)}{Z(s)} = I_{S1}(s) + I_{S2}(s) \tag{9.22}$$

where $I_{S1}(s)$ is the component of current due to $V_S(s)$ and $I_{S2}(s)$ is due to $V_R(s)$. Now

$$V_S(s) = L[V\sin(\omega_s t + \delta)] = V\frac{s\sin\delta + \omega_s\cos\delta}{s^2 + \omega_s^2} \tag{9.23}$$

Using (9.22) and combining (9.21) and (9.23), we get

$$I_{S1}(s) = \frac{sV}{L}\frac{s\sin\delta + \omega_s\cos\delta}{\left(s^2 + \omega_s^2\right)\left(s^2 + \omega_n^2\right)} \tag*{9.24)}$$

Taking the inverse Laplace transform, the current in the time domain is

$$i_{S1}(t) = \frac{V}{L\left(\omega_s^2 - \omega_n^2\right)}[\cos(\omega_s t + \delta) - \omega_s\cos\delta\cos(\omega_n t) + \omega_n\sin\delta\cos(\omega_n t)] \tag{9.25}$$

It is to be noted that i_{S1} contains two frequencies, ω_s and ω_n. A similar expression can also be obtained for i_{S2}.

The concept of two complementary frequencies, as discussed in [12], will now be used. Assuming (9.25) is for phase-A of the circuit, the current i_{S1} for the other two phases can also be obtained. Then, the three currents will be resolved into a pseudo-stationary 0-d-q axis using Park's transformation, given by

$$\begin{bmatrix} i_0 \\ i_d \\ i_q \end{bmatrix} = \sqrt{\frac{2}{3}} \begin{bmatrix} 1/\sqrt{2} & 1/\sqrt{2} & 1/\sqrt{2} \\ \cos\theta & \cos(\theta - 120°) & \cos(\theta + 120°) \\ \sin\theta & \sin(\theta - 120°) & \sin(\theta + 120°) \end{bmatrix} \begin{bmatrix} i_a \\ i_b \\ i_c \end{bmatrix} \tag{9.26}$$

where $\theta = \omega_s t + \delta$. For a balanced system, i_0 will be zero. However, the d- and q-axis components will not be dependent only on ω_s. For example,

$$\cos(\omega_n t)\cos(\omega_s t + \delta) = \frac{1}{2}[\cos\{(\omega_s + \omega_n)t + \delta\} + \cos\{(\omega_s - \omega_n)t + \delta\}] \tag{9.27}$$

This implies that d- and q-axis currents will have several terms like that given in (9.27). They will have the following two frequencies:

- Supersynchronous frequency of $\omega_{sup} = \omega_s + \omega_n$
- Subsynchronous frequency of $\omega_{sub} = \omega_s - \omega_n$

The subsynchronous frequency is more damaging as it can be close to the modes associated with the mechanical shaft system.

9.4.2 Shaft Torsional Modes

The swing equation of the generator shaft of (7.22) is derived assuming that the rotor shaft is a solid mass. This, however, is not the case since a generator rotor contains mechanical shaft sections that are coupled together. A typical generator shaft system is shown in Figure 9.24. It contains a high-pressure (HP) turbine, an intermediate-pressure (IP) turbine, and a low-pressure (LP) turbine. In practice, there may be more than one LP stage. In Figure 9.24, T_{HP}, T_{IP}, and T_{LP} are the per unit mechanical torques developed in the respective turbines, T_e is the per unit generator electrical torque, and K is the stiffness of the shaft section in per unit torque per electrical radian. The numbers associated with the shaft sections are used for identifying the angles and speeds of these masses. Note that in Figure 9.24, the generator, LP turbine, IP turbine, and HP turbine are numbers 1, 2, 3, and 4, respectively. These numbers are used to represent the inertia

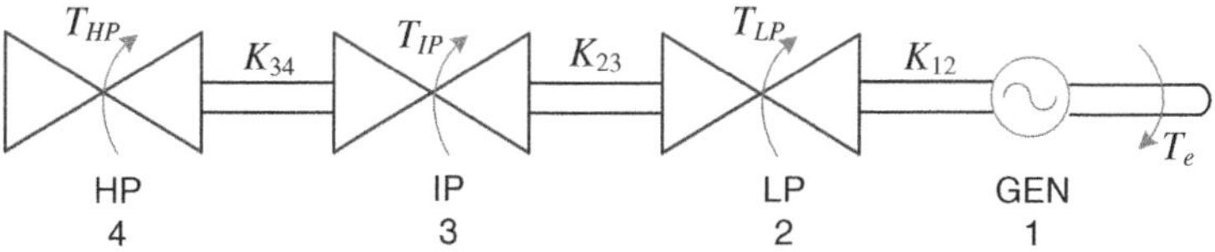

Figure 9.24 Schematic diagram of a typical turbine-generator system.

constant, angular frequency, and the angle in the solutions of state equations, as discussed below.

A second-order differential equation governs the behavior of each shaft section. For example, the equations of motion of the generator rotor are given by [13]

$$\begin{aligned} 2H_1\frac{d(\Delta\omega_1)}{dt} &= K_{12}(\delta_2-\delta_1)-T_e-D_1(\Delta\omega_1) \\ \frac{d(\delta_1)}{dt} &= (\Delta\omega_1)\omega_s \end{aligned} \tag{9.28}$$

In (9.28), the input torque is supplied by the LP turbine, that is, $T_m = K_{12}(\delta_2 - \delta_1)$. For the LP turbine, the input torque is the sum of the torque supplied to the turbine T_{LP} and the torque that is supplied by the IP turbine, that is, $K_{23}(\delta_3 - \delta_2)$. The output torque is supplied to the generator $K_{12}(\delta_2 - \delta_1)$. Then the equation of motion for the LP turbine is

$$\begin{aligned} 2H_2\frac{d(\Delta\omega_2)}{dt} &= T_{LP}+K_{23}(\delta_3-\delta_2)-K_{12}(\delta_2-\delta_1)-D_2(\Delta\omega_2) \\ \frac{d(\delta_2)}{dt} &= (\Delta\omega_2)\omega_s \end{aligned} \tag{9.29}$$

Similar equations can also be written for the other sections. Defining state and input vectors, respectively, as

$$\begin{aligned} \mathbf{x} &= [\Delta\omega_1\ \ \Delta\omega_2\ \ \Delta\omega_3\ \ \Delta\omega_4\ \ \delta_1\ \ \delta_2\ \ \delta_3\ \ \delta_4]^T \\ \mathbf{u} &= [P_e\ \ \ P_{LP}\ \ \ P_{IP}\ \ \ P_{HP}]^T \end{aligned}$$

The equation of motion of the shaft system can be written as

$$\dot{\mathbf{x}} = \mathbf{A}\mathbf{x} + \mathbf{B}\mathbf{u} \tag{9.30}$$

where

$$\mathbf{A}=\begin{bmatrix} -\frac{D_1}{2H_1} & 0 & 0 & 0 & -\frac{K_{12}}{2H_1} & \frac{K_{12}}{2H_1} & 0 & 0 \\ 0 & -\frac{D_2}{2H_2} & 0 & 0 & \frac{K_{12}}{2H_2} & -\frac{K_{12}+K_{23}}{2H_2} & \frac{K_{23}}{2H_2} & 0 \\ 0 & 0 & -\frac{D_3}{2H_3} & 0 & 0 & \frac{K_{23}}{2H_3} & -\frac{K_{23}+K_{34}}{2H_3} & \frac{K_{34}}{2H_3} \\ 0 & 0 & 0 & -\frac{D_4}{2H_4} & 0 & 0 & \frac{K_{34}}{2H_4} & -\frac{K_{34}}{2H_4} \\ \omega_s & 0 & 0 & 0 & 0 & 0 & 0 & 0 \\ 0 & \omega_s & 0 & 0 & 0 & 0 & 0 & 0 \\ 0 & 0 & \omega_s & 0 & 0 & 0 & 0 & 0 \\ 0 & 0 & 0 & \omega_s & 0 & 0 & 0 & 0 \end{bmatrix}$$

$$\mathbf{B} = \begin{bmatrix} -\frac{1}{2H_1} & 0 & 0 & 0 \\ 0 & -\frac{1}{2H_2} & 0 & 0 \\ 0 & 0 & -\frac{1}{2H_3} & 0 \\ 0 & 0 & 0 & -\frac{1}{2H_4} \\ 0 & 0 & 0 & 0 \\ 0 & 0 & 0 & 0 \\ 0 & 0 & 0 & 0 \\ 0 & 0 & 0 & 0 \end{bmatrix}$$

Example 9.2 Consider a mechanical shaft system with system parameters given in Table 9.3. The damping is taken as 0.1 for each mass. There are three complex conjugate eigenvalues of the system. These are the three different modes of rotor mass. These modes are listed in Table 9.4 along with their eigenvalues

Table 9.3 Shaft mechanical system data for Example 9.2.

Mass no.	Rotor mass	Power sharing (pu)	H (MJ/MVA)	Shaft section	Stiffness (pu/rad)
1	GEN	–	0.8	GEN-LP	70.0
2	LP	0.25	0.9	LP-IP	35
3	IP	0.3	0.2	IP-HP	20
4	HP	0.45	0.1		

Table 9.4 Shaft modes and their participation.

Mode no.	Frequency (Hz)	Eigenvalue	Main participation
1	39.93	$-0.16 \pm j250$	Mass 4 (HP) swings against Mass-3 (IP) with the participation of the other two masses being negligible.
2	27.37	$-0.07 \pm j172$	Masses 1 and 4 (GEN & HP) swinging against Mass-2 (LP)
3	18.94	$-0.16 \pm j119$	Masses 3 and 4 (IP & HP) swinging against Mass 1 (GEN)

and participation of rotor mass in these modes. The participation factor is computed through left and right eigenvectors, as has been discussed in [14]. Note that there is a Mode-0, which describes the oscillation of the entire rotor mass against the power system. Usually, it is this mode that is considered in system stability studies through the swing equation. All the different rotor masses participate equally in this mode.

9.4.3 Subsynchronous Frequency Analysis

Consider a shaft oscillation at frequency ω. The angle of the shaft at the generation is given by

$$\theta = \omega_s t + \alpha \cos(\omega t) \tag{9.31}$$

Assuming that the magnitude of the rotor flux is constant, the coupling of the flux to a stator coil is given by

$$\phi = A \cos(\omega_s t + \alpha \cos(\omega t)) \tag{9.32}$$

The above expression however will neglect the induction generator effects discussed in [13]. The flux equation of (9.32) can be closely approximated as [15]

$$\phi = A \cos(\omega_s t) - \frac{A\alpha}{2} [\cos\{(\omega_s + \omega)t\} - \cos\{(\omega_s - \omega)t\}] \tag{9.33}$$

The flux equation of (9.33) includes frequency terms of ω_s, $\omega_s + \omega$, and $\omega_s - \omega$. Substituting $A = n/\omega_s$, where n is the number of turns in the coil, the voltage generated by the phase coil becomes

$$v = n\frac{d\phi}{dt} = \sin(\omega_s t) + \frac{\alpha}{2\omega_s}[(\omega_s + \omega)\sin\{(\omega_s + \omega)t\} - (\omega_s - \omega)\sin\{(\omega_s - \omega)t\}] \tag{9.34}$$

There are several currents that flow, but the dominant term affecting the power modulation is given by the low-frequency term. Note that the phase of the low-frequency term is fixed with respect to the nominal frequency. Assuming that the electrical circuit is resonant, the low-frequency current is dominant for SSR and is given by

$$i_{low} = -\frac{\alpha}{2\omega_s} \frac{(\omega_s - \omega)\sin\{(\omega_s - \omega)t\}}{Z(\omega)} \tag{9.35}$$

where the line impedance at frequencies other than the nominal frequency is expressed as a function of the shaft frequency as

$$Z(\omega) = R + jL(\omega_s - \omega) - \frac{j}{C(\omega_s - \omega)} \tag{9.36}$$

It is to be noted that in the above expression, we have introduced a line drop rather than considering a lossless line as shown in Figure 9.23. Equation (9.36) can be expressed as

$$Z(\omega) = R + j\left\{\frac{(\omega_s - \omega)^2 - \omega_n^2}{(\omega_s - \omega)L}\right\} = R + j\left\{\frac{(\omega_s + \omega_n - \omega)(\omega_s - \omega_n - \omega)}{(\omega_s - \omega)L}\right\} \tag{9.37}$$

where ω_n is the undamped natural frequency of the compensated electrical network. This result indicates that the electrical impedance will become very low for shaft frequencies at $\omega_s - \omega$. For frequencies below this, the impedance is inductive, and above this it becomes capacitive.

We are interested in power flow at the frequency ω. This can arise for a voltage at ω_s and a current at $\omega_s - \omega$ or other combinations. Note that the low-frequency resonant current is dominant here. The power equation is given by

$$P(\omega) = vi_{low} \approx -\sin(\omega_s t)\frac{\alpha}{2\omega_s}\frac{(\omega_s - \omega)\sin\{(\omega_s - \omega)t - \angle Z(\omega)\}}{|Z(\omega)|} \tag{9.38}$$

The above equation can be simplified as

$$P(\omega) \approx -\frac{\alpha(\omega_s - \omega)}{4\omega_s|Z(\omega)|}[\cos\{\omega t - \angle Z(\omega)\} - \cos\{(2\omega_s - \omega)t - \angle Z(\omega)\}] \tag{9.39}$$

Considering only the terms at frequency ω, and noting the shaft angle satisfies $\delta = \alpha\sin(\omega t)$, we finally get

$$P(\omega) = -\frac{\alpha(\omega_s - \omega)\angle Z(\omega)}{4\omega_s|Z(\omega)|}\delta \tag{9.40}$$

Example 9.3 Consider the same system as given in Example 9.2. We examine the case of an electrical resonance frequency of 30 Hz. The coefficient of δ in (9.40) is plotted in magnitude and phase in Figure 9.25. The magnitude peaks exactly at 20 Hz (= 50–30 Hz), and the phase varies from −90° to −270° (i.e., +90°). Damping of the mechanical oscillations in (9.29) is provided when electrical power is aligned with $d\delta/dt$. The damping contribution is thus the imaginary part of the coefficient of δ in (9.40). Figure 9.26 shows the damping contribution. Thus, for an electrical frequency of 30 Hz, there is a strong negative damping coefficient at the shaft oscillation frequency of 18.94 Hz (see Table 9.4).

The above example shows that when the electrical circuit is resonant at a frequency ω_n, for example, 30 Hz, and the system frequency is 50 Hz, then for a shaft frequency around 20 Hz (i.e., $\omega_n = \omega_s - \omega$) both electrical and mechanical systems

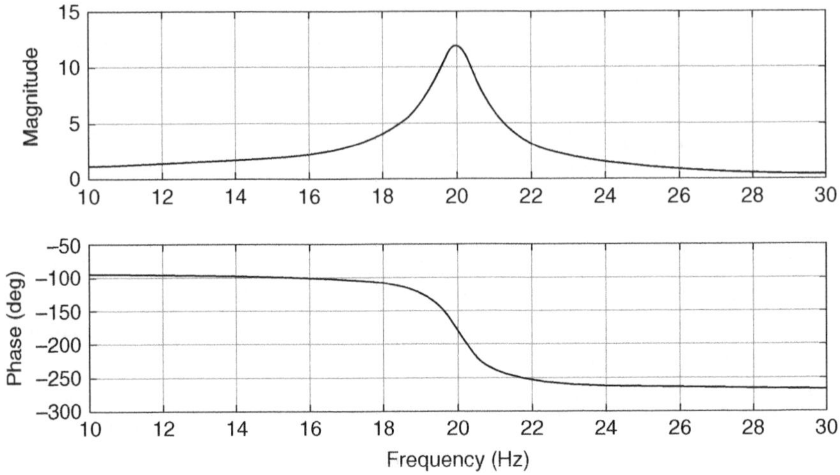

Figure 9.25 Bode plot of electrical power contribution of Example 9.3.

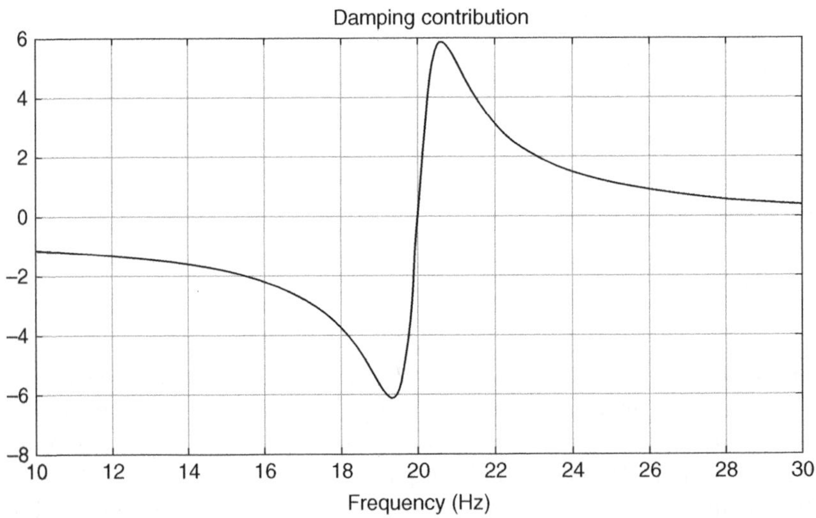

Figure 9.26 Damping contribution from the electrical side of Example 9.3.

will show a strong response. When the resistive portion of the line impedance is low, Z will change phase from $-90°$ to $+90°$ as the frequency changes through resonance. Therefore, for low shaft frequencies, the power will be aligned with $-d\delta/dt$ and will provide negative damping. However, since the impedance is high, there will be little influence. As the frequency increases, the magnitude P will increase and the phase will shift to align with $d\delta/dt$; thus, in the set of frequencies just below the equivalent electrical resonance, the term will be large and provide negative damping to shaft oscillations. If the natural shaft damping is low and the electrical resonance is high $Q = R\sqrt{C/L}$, the presence of the electrical resonance can provide a net negative damping. Shaft oscillations can grow, and this will cause shaft damage.

The key conditions are that the mechanical resonance frequency translates to an electrical frequency closely below electrical resonance. Therefore, mechanical resonance at 19 Hz, which maps to 31 Hz electrical, would be dangerous if the electrical circuit was resonant just below 31 Hz.

Example 9.4 Let us again consider the same system as Example 9.2. Electrical resonance conditions were chosen to match each of the mechanical resonances in turn. The electrical damping coefficients are shown in Table 9.5. The damping ratios are found from the eigenvalues of the different modes of the mechanical system. The electrical damping coefficients are found in graphs of damping contribution as in Figure 9.26.

These damping results are confirmed from time domain simulations. These simulations include the mechanical system of (9.28), the RLC electrical network to obtain i_S, and the linking equations of

$$v_S = -\sqrt{2}\left(1 + \frac{\dot{\delta}_1}{\omega_s}\right)\sin(\omega_s t + \delta_1),\ v_R = -\sqrt{2}\sin(\omega_s t)\text{ and }p_e = v_S i_S$$

Note that in the simulation studies, the instantaneous power p_e is used rather than the average power P_e. The RLC parameters chosen are

$$X_L = \omega L = 0.5, R = 0.025X_L\text{ and }C = 1/\left(\omega_n^2 L\right)$$

with ω_n being chosen according to Table 9.5.

Figure 9.27 shows the time domain simulation results for three different electrical resonant frequencies. With an electrical resonance at 9.5 Hz, the system

Table 9.5 Damping coefficient associated with each mode.

Mechanical frequency (Hz)	Mechanical damping ratio	Electrical resonant frequency (Hz)	Electrical damping coefficient
39.93	0.00064	9.5	-2
27.37	0.0004	22	-4.5
18.94	0.0013	30	-6

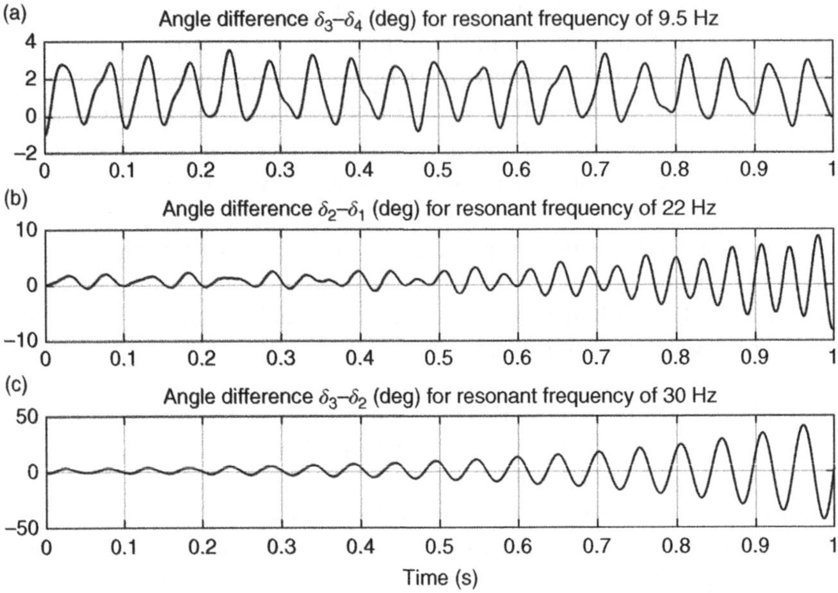

Figure 9.27 System response to resonant frequencies of (a) 9.5 Hz, (b) 22 Hz, and (c) 30 Hz.

exhibits a sustained oscillation indicating that the negative electrical damping is insufficient to fully overcome the shaft mechanical damping. This is shown in Figure 9.27 (a). The system responses for resonant frequencies of 22 Hz and 30 Hz are shown in Figure 9.27 (b) and (c), respectively. The growing oscillations at 19 Hz between angles δ_3 and δ_2 of Figure 9.27 (c) are larger than those at 28 Hz between δ_2 and δ_1 of Figure 9.27 (b). Therefore, the largest potential for shaft damage is for the resonant frequency of 30 Hz.

9.4.4 Countermeasures to SSR

A variety of countermeasures to the SSR problem have been reported in the literature [16]. In an ideal situation, it will be desirable to isolate the electrical system from the mechanical system such that the network resonance frequency does not influence the shaft modes. This, however, is not possible, and alternate means of SSR suppression have been attempted.

Among the power electronic solutions to the problem, the NGH damper needs to be mentioned [17]. In this scheme, a linear resistor is connected in series with a pair of back-to-back thyristors. The resistor–thyristor combination is connected in parallel with a series capacitor, as shown in Figure 9.28. The thyristors are normally blocked such that the series capacitor can enhance the power flow through the electrical network. However, whenever the presence of SSR is detected from the measurement of capacitor voltage, the thyristors are switched on to provide a path for the capacitor to discharge.

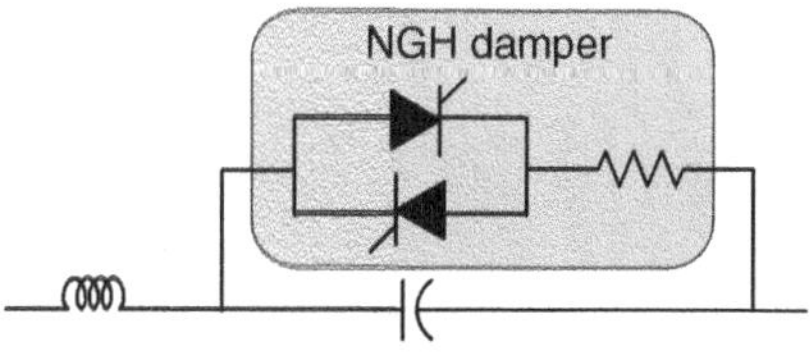

Figure 9.28 Series capacitor-compensated transmission line with NGH damper.

The main objective is to have a series compensation such that the complementary of the network resonance frequency is not near any shaft torsional mode. To achieve this, a variable series compensator that can modulate the fundamental frequency series reactance is desirable. Another approach is to modulate the series voltage to appear resistive at subsynchronous frequency and thus damp SSR. The controlled series compensation is discussed next.

9.5 Thyristor-Controlled Series Compensator (TCSC)

A TCSC is basically a thyristor-controlled reactor that is connected in parallel with a fixed capacitor. The schematic diagram of a TCSC is shown in Figure 9.29, where L is the TCR inductor. The TCSC is supplied by a sinusoidal current source i_L. The current through the TCR and voltage across the fixed capacitor C are denoted by i_P and v_C, respectively.

The steady-state capacitor voltage and TCR inductor current waveforms are shown in Figure 9.30. With respect to this figure, the following time instants are defined:

- Instant t_0: This is the positive going zero crossing of the capacitor voltage. The inductor current is at its negative peak of $-I_{P0}$ at this instant. The firing angle α is measured from this instant.
- Instant t_1: The inductor current becomes zero at this instant.
- Instant t_2: This is the instant of the firing of the thyristor after a delay of α from t_0, such that

$$\alpha = (t_2 - t_1)\omega$$

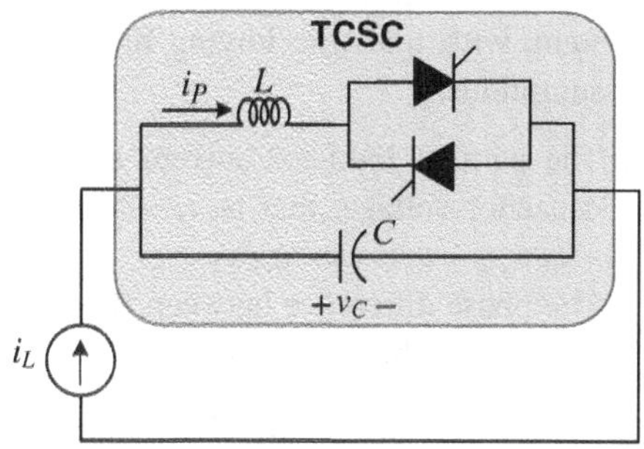

Figure 9.29 Equivalent circuit of a TCSC driven by a current source.

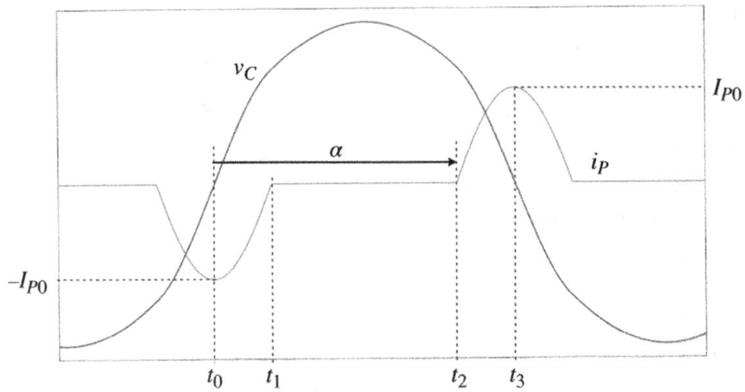

Figure 9.30 Voltage and current waveforms of a TCSC.

- Instant t_3: This is the negative going zero crossing of the capacitor voltage, where the inductor current is at its positive peak of I_{P0}.

It is evident from Figure 9.30 that the capacitor voltage has the following two segments:

- $v_C = (1/C) \int i_L dt$ when none of the thyristors is conducting, for example, between t_1 and t_2. The capacitor voltage is driven by the current source.
- $v_C = (1/C) \int (i_L - i_P) dt$ when one of the thyristors is conducting, for example, between t_0 and t_1 or t_2 and t_3.

To obtain an expression for the capacitor voltage, the above two piecewise linear equations must be combined. Let the injected current through the source be given by

$$i_L = I_m \sin(\omega t + \varphi) \tag{9.41}$$

The piecewise models will be defined in terms of the state equations of the system, with i_L as the forcing function. To do that, let us make the following assumptions:

- The parallel inductor current is symmetric around the zero crossing of the capacitor voltage, that is, $t_1 - t_0 = t_3 - t_2$. Under this assumption, we have $t_1 - t_0 = t_3 - t_2 = (\pi - \alpha)/\omega$.
- The phase difference between the actual and fundamental capacitor voltage is zero. Thus, the angle φ is taken equal to 90°. This implies that $i_L = I_m \cos(\omega t)$.

9.5.1 When One of the Thyristors Is On

The state vector of the system is defined as

$$\mathbf{x} = \begin{bmatrix} v_C & i_P \end{bmatrix}^T$$

The state space equation when the thyristors are on is given by

$$\dot{\mathbf{x}} = \begin{bmatrix} 0 & -1/C \\ 1/L & 0 \end{bmatrix}\mathbf{x} + \begin{bmatrix} 1/C \\ 0 \end{bmatrix} i_L = \mathbf{Ax} + Bi_L \tag{9.42}$$

Let us choose the initial time as t_0 and the final time as t_1. The state vectors for these instants are (see Figure 9.30)

$$\mathbf{x}(t_0) = [0 \quad -I_{P0}]^T \text{ and } \mathbf{x}(t_1) = [v_C(t_1) \quad 0]^T \tag{9.43}$$

Choosing the time instant t_0 as 0, the solution of the state equation of (9.42) is given by

$$\mathbf{x}(t_1) = \mathbf{\Phi}(t_1)x(t_0) + \int_0^{t_1} \Phi(t_1-\tau)\mathbf{B}i_L(\tau)d\tau \tag{9.44}$$

where $\Phi(t)$ is the state transition matrix (STM) defined by

$$\mathbf{\Phi}(t) = e^{At} = L^{-1}(s\mathbf{I}-\mathbf{A}) \tag{9.45}$$

Now

$$s\boldsymbol{I}-\boldsymbol{A} = \begin{bmatrix} s & 1/C \\ -1/L & s \end{bmatrix} \Rightarrow |s\boldsymbol{I}-\boldsymbol{A}| = s^2 + \frac{1}{LC} = s^2 + \omega_n^2$$

where $\omega_n = 1/\sqrt{LC}$ is the natural frequency of the LC pair. Then,

$$\mathbf{\Phi}(t_1) = L^{-1}\left\{\frac{1}{s^2+\omega_n^2}\begin{bmatrix} s & -1/C \\ 1/L & s \end{bmatrix}\right\} = \begin{bmatrix} \cos(\omega_n t_1) & -\dfrac{\sin(\omega_n t_1)}{C\omega_n} \\ \dfrac{\sin(\omega_n t_1)}{L\omega_n} & \cos(\omega_n t_1) \end{bmatrix} \tag{9.46}$$

Therefore,

$$\begin{aligned} &\int_0^{t_1} \mathbf{\Phi}(t-\tau)\mathbf{B}i_L(\tau)d\tau \\ &= \int_0^{t_1} \begin{bmatrix} \dfrac{\cos\{\omega_n(t_1-\tau)\}}{C} \\ \dfrac{\sin\{\omega_n(t_1-\tau)\}}{LC\omega_n} \end{bmatrix} i_L(\tau)d\tau = I_m\int_0^{t_1} \begin{bmatrix} \dfrac{\cos\{\omega_n(t_1-\tau)\}}{C} \\ \dfrac{\sin\{\omega_n(t_1-\tau)\}}{LC\omega_n} \end{bmatrix} \cos(\omega\tau)d\tau \end{aligned} \tag{9.47}$$

9.5.2 When Both Thyristors Are Off

During this time, the current through the inductor is off, and we have a first-order differential equation for the capacitor voltage as

$$\dot{v}_C = \frac{1}{C} I_m \sin(\omega t + 90^\circ) = \frac{I_m}{C} \cos \omega t \tag{9.48}$$

For an arbitrary initial time t_i, the solution to the equation for a final time t_f is written as

$$\begin{aligned} v_C(t_f) &= v_C(t_i) + \frac{I_m}{C} \int_{t_i}^{t_f} \cos \omega t \, dt \\ &= v_C(t_i) + \frac{I_m}{\omega C} \{ \cos \omega t_f - \cos \omega t_i \} \end{aligned} \tag{9.49}$$

9.5.3 Estimating the Fundamental Impedance of a TCSC

Using the derivations presented in Sections 9.5.1 and 9.5.2, the fundamental characteristics of the TCSC will now be derived for a given value of α. The step-by-step approach for the determination of the fundamental impedance is discussed below.

Step-1: (Estimation of I_{P0}): Using (9.41), (9.43), (9.44), (9.46), and (9.47), the following solution of the state space equation during the switch-on state is obtained:

$$\begin{bmatrix} v_C(t_1) \\ 0 \end{bmatrix} = \begin{bmatrix} \dfrac{\sin(\omega_n t)}{C\omega_n} \\ -\cos(\omega_n t) \end{bmatrix} I_{P0} + I_m \int_0^{t_1} \begin{bmatrix} \dfrac{\cos\{\omega_n(t_1-\tau)\}}{C} \\ \omega_n \sin\{\omega_n(t_1-\tau)\} \end{bmatrix} \cos(\omega\tau) d\tau \tag{9.50}$$

Now, the term under the integral of the second row is expanded as

$$\begin{aligned} \sin\{\omega_n(t_1-\tau)\}\cos(\omega\tau) &= \frac{1}{2}[\sin\{\omega_n(t_1-\tau)-\omega\tau\} + \sin\{\omega_n(t_1-\tau)+\omega\tau\}] \\ &= \frac{1}{2}[\sin\{\omega_n t_1 - (\omega_n+\omega)\tau\} + \sin\{\omega_n t_1 - (\omega_n-\omega)\tau\}] \end{aligned}$$

Thus,

$$\begin{aligned} &\frac{1}{2}\int_0^{t_1} [\sin\{\omega_n t_1 - (\omega_n+\omega)\tau\} + \sin\{\omega_n t_1 - (\omega_n-\omega)\tau\}] d\tau \\ &= \frac{1}{2}\left[\frac{\cos\{\omega_n t_1 - (\omega_n+\omega)\tau\}}{\omega_n+\omega} - \frac{\cos\{\omega_n t_1 - (\omega_n-\omega)\tau\}}{\omega_n-\omega}\right]_0^{t_1} \\ &= \frac{1}{2}\left[\frac{\cos(\omega t_1) - \cos(\omega_n t_1)}{\omega_n+\omega} - \frac{\cos(\omega t_1) - \cos(\omega_n t_1)}{\omega_n-\omega}\right] \\ &= \frac{\omega_n \cos(\omega_n t_1) - \omega_n \cos(\omega t_1)}{\omega_n^2 - \omega^2} \end{aligned}$$

Therefore, the second row of (9.50) is expanded as

$$0 = -I_{P0}\cos(\omega_n t_1) + I_m\omega_n^2\left[\frac{\cos(\omega_n t_1) - \cos(\omega t_1)}{\omega_n^2 - \omega^2}\right] \tag{9.51}$$

Rearranging the above equation I_{P0} is obtained as

$$I_{P0} = \frac{I_m\omega_n^2}{\omega_n^2 - \omega^2}\left[\frac{\cos(\omega_n t_1) - \cos(\omega t_1)}{\cos(\omega_n t_1)}\right] \tag{9.52}$$

Now, because $t_1 = (\pi - \alpha)/\omega$, (9.52) is written as

$$I_{P0} = \frac{I_m\omega_n^2}{\omega_n^2 - \omega^2}\left[\frac{\cos(\beta) - \cos(\pi - \alpha)}{\cos(\beta)}\right], \quad \beta = \frac{\omega_n}{\omega}(\pi - \alpha) \tag{9.53}$$

Step-2: (Obtaining the capacitor voltage between t_0 and t_3): Once I_{P0} is known, the capacitor voltage can be completely characterized.

- Between t_0 and t_1: Use (9.43) and (9.44) to determine $v_C(t_1)$.
- Between t_1 and t_2: Using $v_C(t_1)$ calculated above, use (9.49) to determine $v_C(t_2)$.
- Between t_2 and t: Choosing $\mathbf{x} = \begin{bmatrix} v_C(t_2) & 0 \end{bmatrix}^T$, use (9.44) to determine $v_C(t_3)$.

The different segments of capacitor voltages are computed using a suitable differential equation solver, for example, "ode45" of MATLAB. In the steady state, the capacitor voltage has a mirror image symmetry.

Step-3: (Fundamental frequency model of TCSC): Once the capacitor voltage is completely characterized for a particular value of α, its fundamental component magnitude and phase are computed using Fourier analysis. Let the obtained fundamental component be given by

$$v_{Cf} = V_{Cm}\sin(\omega t + \theta)$$

Then the fundamental frequency impedance of the TCSC for a particular value of α is given by

$$Z_{TCSC}(\alpha) = \frac{V_{Cm}}{I_m}\angle(\theta - 90^\circ) = R_{TCSC}(\alpha) + jX_{TCSC}(\alpha) \tag{9.54}$$

Example 9.5 Consider a TCSC with the following parameters:

$$C = 79.5\,\mu\text{F and } L = 50\,\text{mH}$$

The magnitude of the forcing current (I_m) is chosen as 1 A. Figure 9.31 depicts the variation in the fundamental reactance as α changes from 90° to 180°. In the region $90^\circ < \alpha < 120^\circ$, the reactance has positive values, indicating that in this region, the TCSC inserts a series inductance. That is why this region is termed the inductive region. On the other hand, the reactance is negative in the region $126^\circ < \alpha < 180^\circ$, indicating that this is the capacitive region. The region $120^\circ < \alpha < 126^\circ$ is the forbidden region. In this region, the resonance between the

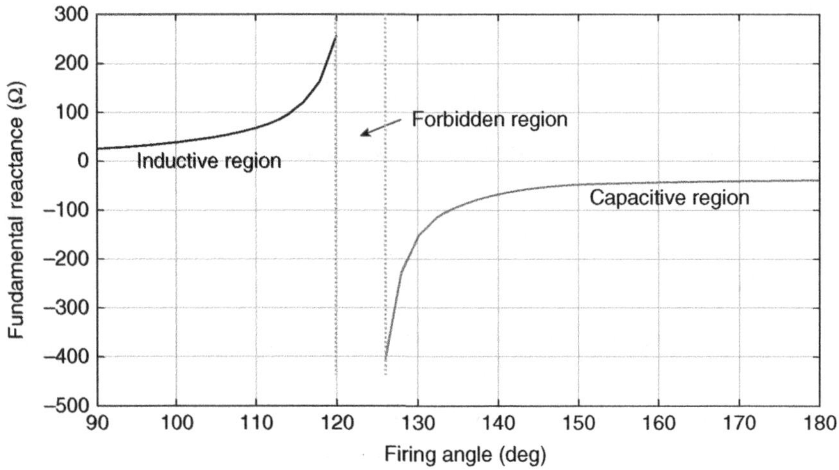

Figure 9.31 Variations in the fundamental reactance with α.

capacitor and the variable inductor occurs. The TCSC is prevented from operating in this region.

Table 9.6 lists the values of I_{P0}, V_{Cm}, and $Z_{TCSC} = R_{TCSC} + jZ_{TCSC}$ as α changes from 125° to 175°, that is, in the capacitive mode. All the quantities become large as the system moves toward resonance. It is also interesting to note that the TCSC impedance is not a pure reactance, a resistance gets added due to the switching action. Figure 9.32 (a) and (c) depicts the waveforms obtained using the procedure

Table 9.6 Solutions of the system parameters for different values of α.

α (deg)	I_{P0} (A)	V_{Cm} (V)	X_{TCSC} (Ω)	R_{TCSC} (Ω)
125	23.05	698.46	–698.26	16.68
130	4.34	155.20	–155.17	3.29
135	2.09	90.88	–90.86	1.67
140	1.22	66.83	–66.82	1.04
145	0.76	54.89	–54.88	0.71
150	0.48	48.22	–48.22	0.51
155	0.30	44.34	–44.34	0.38
160	0.18	42.09	–42.09	0.31
165	0.09	40.86	–40.86	0.27
170	0.04	40.27	–40.27	0.26
175	0.01	40.06	–40.07	0.27

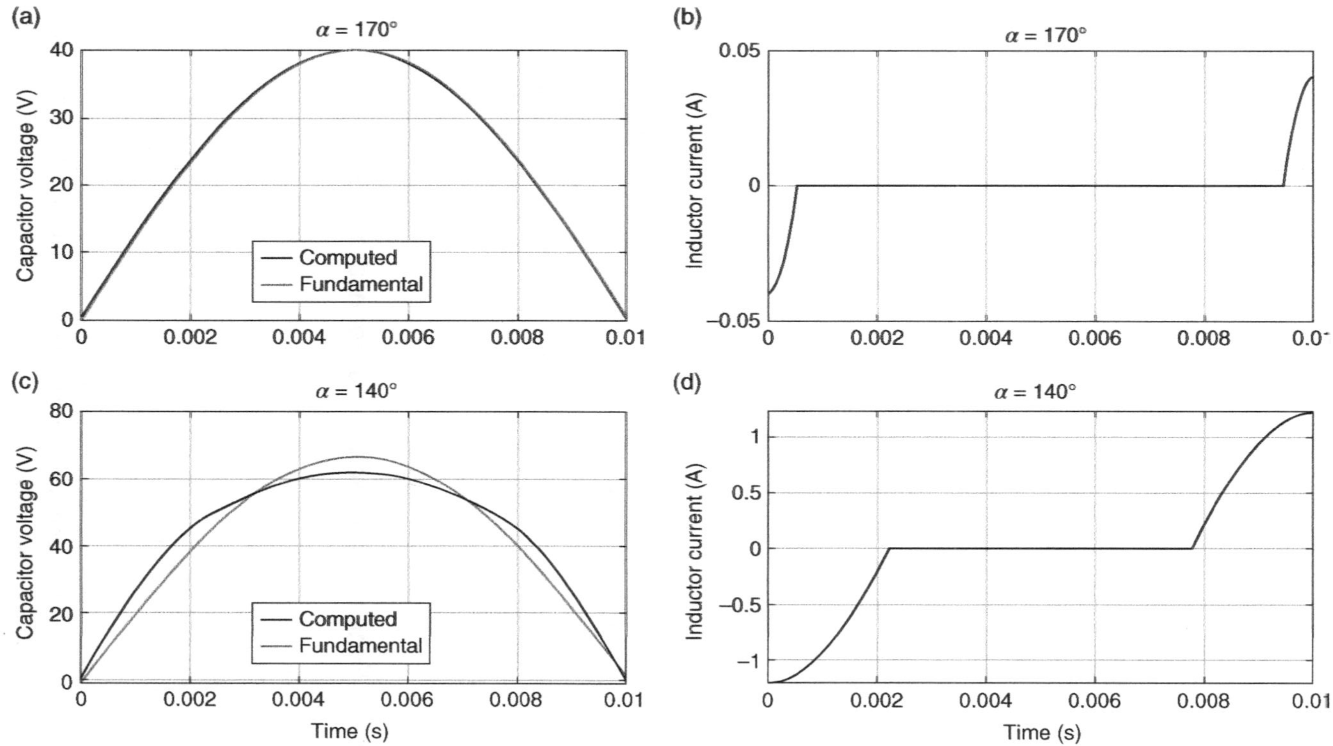

Figure 9.32 Computed voltage and current waveforms for two different values of (a) $\alpha = 170°$, (b) $\alpha = 170°$, (c) $\alpha = 140°$ and (d) $\alpha = 140°$.

given in Section 9.5.3 and the extracted fundamental capacitor voltage for two different values of α. It is interesting to note that as α gets closer to 180°, the estimated voltage matches the fundamental voltage. This is obvious since the peak of the inductor current reduces as α is close to 180° (Figure 9.32 b and d).

In the derivation above, it has been assumed that the inductor coil L is ideal, that is, it has no associated resistance (infinite quality factor). In practice, however, the quality factor is finite, and the estimation of the fundamental characteristics is more involved. This is discussed in [18], where both simulation and experimental verifications are presented. Modeling the actual behavior of a TCSC is complicated and is very challenging. In [19], the TCSC reactance, for a capacitive reactance of X_C and inductive reactance of X_P, is given as

$$Z_{TCSC} = X_C - \Psi_1(X_C - \Psi_2) + \Psi_4\Psi_5 \tag{9.55}$$

where

$$\beta = \pi - \alpha, \Psi_1 = \frac{2\beta + \sin(2\beta)}{\pi}, \Psi_2 = \frac{X_C X_P}{X_C - X_P}, \Psi_3 = \sqrt{\frac{X_C}{X_P}},$$

$$\Psi_4 = \Psi_3 \tan(\beta\Psi_3) - \tan(\beta) \text{ and } \Psi_5 = \frac{4\Psi_2^2 \cos^2(\beta)}{\pi X_P}$$

Because the model given in (9.55) is nonlinear, it is estimated and linearized for control design as reported in [20]. An accurate discrete-time model of TCSC is derived in [21], which combines the piecewise linear models, as discussed in this section. The linear models are then combined, and the overall linear model of the system is derived by taking a partial derivative of the system with respect to the state vector and the firing angle α.

9.6 Static Synchronous Series Compensator (SSSC)

An ideal series compensator is shown in Figure 8.12. An SSSC is a series compensating device, which constitutes a voltage source converter and a connecting transformer, as shown in Figure 9.33 (a). The leakage inductance of the connecting transformer is denoted by L_T. The DC side of the VSC is supplied by the DC storage capacitor C_{dc}. The converter losses are represented by the DC resistance R_{dc}. The SSSC injects the voltage V_Q in series with the line. The equivalent circuit of the SSSC is shown in Figure 9.33 (b), where the resistance R_T represents the transformer losses, and i_L is the series current flowing in the circuit. The SSSC model is derived in [4, 5] and is discussed below.

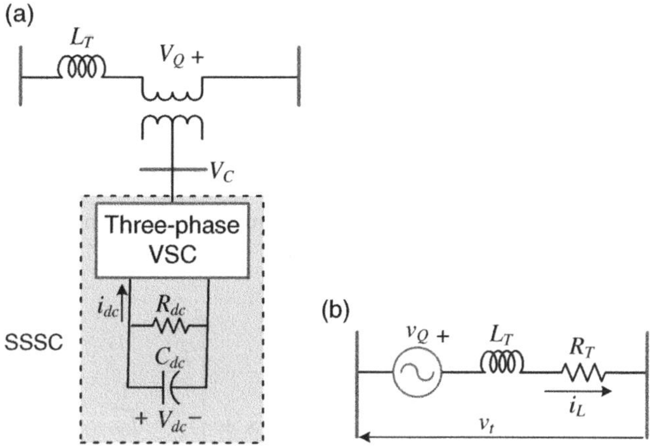

Figure 9.33 (a) SSSC connection in series with a transmission line and (b) its equivalent circuit.

The DC side current is referred to the high-tension side of the transformer as

$$-i_{dc}V_{dc} = i_{La}v_{Qa} + i_{Lb}v_{Qb} + i_{Lc}v_{Qc} \tag{9.56}$$

For a multistep inverter, the fundamental component of the instantaneous voltage of phase-A is given by [4, 5]

$$v_{Qa} = K_{dc}V_{dc}\sin\left(\omega t + \varphi_f\right) \tag{9.57}$$

where K_{dc} is a constant and φ_f is the phase displacement of the fundamental voltage from a reference bus voltage. A similar expression can also be written for the other two phases by phase displacing them by 120°. Combining (9.56) and (9.57), the following expression for the DC current is obtained.

$$i_{dc} = -K_{dc}(\mu_a i_{La} + \mu_b i_{Lb} + \mu_c i_{Lc}) \tag{9.58}$$

where

$$\mu_a = \sin\left(\omega t + \varphi_f\right), \mu_a = \sin\left(\omega t + \varphi_f - 120^\circ\right) \text{ and } \mu_c = \sin\left(\omega t + \varphi_f + 120^\circ\right) \tag{9.59}$$

From the DC side of Figure 9.33 (a), the DC voltage is given by the differential equation

$$\frac{dV_{dc}}{dt} = -\frac{1}{C_{dc}}i_{dc} - \frac{1}{R_{dc}C_{dc}}V_{dc} \tag{9.60}$$

Using (9.59), the injected voltages can be written as

$$v_{Qa} = K_{dc}V_{dc}\mu_a, v_{Qc} = K_{dc}V_{dc}\mu_b \text{ and } v_{Qc} = K_{dc}V_{dc}\mu_c \tag{9.61}$$

Let us now define the following state and input vectors:

$$\mathbf{x} = \begin{bmatrix} i_{La} & i_{Lb} & i_{Lc} & V_{dc} \end{bmatrix}^T \text{ and } \mathbf{u} = \begin{bmatrix} v_{ta} & v_{tb} & v_{tc} \end{bmatrix}^T$$

Then, the following state space equation is formed from Figure 9.33 and equations (9.58), (9.60), and (9.61)

$$\dot{\mathbf{x}} = \begin{bmatrix} -\frac{R_T}{L_T} & 0 & 0 & \frac{K_{dc}\mu_a}{L_T} \\ 0 & -\frac{R_T}{L_T} & 0 & \frac{K_{dc}\mu_b}{L_T} \\ 0 & 0 & -\frac{R_T}{L_T} & \frac{K_{dc}\mu_c}{L_T} \\ -\frac{K_{dc}\mu_a}{C_{dc}} & -\frac{K_{dc}\mu_a}{C_{dc}} & -\frac{K_{dc}\mu_a}{C_{dc}} & -\frac{1}{R_{dc}C_{dc}} \end{bmatrix} \mathbf{x} + \begin{bmatrix} \frac{1}{L_T} & 0 & 0 \\ 0 & \frac{1}{L_T} & 0 \\ 0 & 0 & \frac{1}{L_T} \\ 0 & 0 & 0 \end{bmatrix} \mathbf{u} \tag{9.62}$$

The d-q transformation of the above equation is then computed as discussed in [4, 5]. This is given by

$$\frac{d}{dt}\begin{bmatrix} i_{Ld} \\ i_{Lq} \\ V_{dc} \end{bmatrix} = \begin{bmatrix} -\frac{\omega R_T}{X_T} & -\omega & -\rho_1 \sin\varphi_f \\ \omega & -\frac{\omega R_T}{X_T} & -\rho_1 \cos\varphi_f \\ \rho_2 \sin\varphi_f & \rho_2 \cos\varphi_f & -\frac{\omega X_{dc}}{R_{dc}} \end{bmatrix} \begin{bmatrix} i_{Ld} \\ i_{Lq} \\ V_{dc} \end{bmatrix} + \begin{bmatrix} \frac{\omega}{X_T} & 0 \\ 0 & \frac{\omega}{X_T} \\ 0 & 0 \end{bmatrix} \begin{bmatrix} v_{td} \\ v_{tq} \end{bmatrix} \tag{9.63}$$

where

$$X_T = \omega L_T,\ X_{dc} = \frac{1}{\omega C_{dc}},\ \rho_1 = \frac{\sqrt{3}\omega}{X_T} \text{ and } \rho_2 = \sqrt{3}\omega X_{dc}$$

The DC voltage dynamics is presented by the last row of (9.63) as

$$\frac{dV_{dc}}{dt} = \rho_2 \sin\left(\varphi_f\right) i_d + \rho_2 \cos\left(\varphi_f\right) i_q - \frac{\omega X_{dc}}{R_{dc}} V_{dc} \tag{9.64}$$

The main aim of the converter control is to regulate the DC voltage around a prespecified value by controlling the angle φ_f. This means that the phase shift φ_f is so adjusted that the required amount of power flows toward the DC bus to replenish the converter losses to hold the DC capacitor voltage constant. Now, note that (9.63) is nonlinear. For control design (9.63), it is linearized as

$$\frac{d\Delta V_{dc}}{dt} = \gamma_1 \Delta i_d + \gamma_2 \Delta i_q - \gamma_3 \Delta V_{dc} + \gamma_4 \Delta \varphi_f \tag{9.65}$$

where Δ and the subscript 0 denote the perturbed variables and the nominal steady state variables, respectively, and

$$\gamma_1 = \rho_2 \sin\left(\varphi_{f0}\right),\ \gamma_2 = \rho_2 \cos\left(\varphi_{f0}\right),$$
$$\gamma_3 = \frac{\omega X_{dc}}{R_{dc}},\ \gamma_4 = \rho_2\left\{\cos\left(\varphi_{f0}\right) i_{d0} - \sin\left(\varphi_{f0}\right) i_{q0}\right\}$$

A proportional PI controller is then designed for the DC voltage control as

$$\Delta\varphi_f = K_P\left(\Delta V_{dc}^* - \Delta V_{dc}\right) + K_I \int\left(\Delta V_{dc}^* - \Delta V_{dc}\right)dt \tag{9.66}$$

where ΔV_{dc}^* is the perturbation from the prespecified DC voltage. In (9.65), the terms Δi_d and Δi_q can be taken as external disturbances. Therefore, neglecting these two terms (9.65) can be written in the Laplace domain as

$$\Delta V_{dc}(s) = \frac{\gamma_4}{(s + \gamma_3)} \Delta\varphi_f(s) \tag{9.67}$$

Again, the Laplace transform of (9.66) is

$$\Delta\varphi_f(s) = \frac{K_P s + K_I}{s}\left[\Delta V_{dc}^*(s) - \Delta V_{dc}(s)\right] \tag{9.68}$$

The closed-loop transfer function of the system is then derived by combining (9.67) and (9.68) as

$$\frac{\Delta V_{dc}(s)}{\Delta V_{dc}^*(s)} = \frac{\gamma_4(K_P s + K_I)}{s^2 + (\gamma_3 + \gamma_4 K_P)s + \gamma_4 K_I} \tag{9.69}$$

The characteristic equation of the system is then

$$s^2 + (\gamma_3 + \gamma_4 K_P)s + \gamma_4 K_I = 0 \tag{9.70}$$

The closed-loop system will be stable provided that $\gamma_4 > 0$ when

$$\gamma_3 + \gamma_4 K_P > 0$$

Reference [4] shows the control application for the damping of SSR on the IEEE first benchmark model. An addition of a series capacitor along with the SSSC further improves the SSR damping as reported in [22].

9.7 Other FACTS Devices

There are several other FACTS devices. In this section, these are discussed briefly.

9.7.1 Unified Power Flow Controller (UPFC)

A UPFC is a versatile device that contains a shunt compensator and a series compensator that are joined together through a DC bus. Figure 9.34 shows the schematic diagram of a UPFC, where the shunt compensating VSC and the series compensating VSC have a common storage capacitor on their DC sides. Both the VSCs are connected to the transmission line through two separate transformers. In this configuration, the series VSC performs the main task of injecting a voltage in series with the line. The shunt VSC supplies or absorbs the real power that is demanded by the series VSC to inject voltage of a specified magnitude and phase angle. Simultaneously, the shunt VSC can also supply or absorb reactive power as may be necessary to provide shunt reactive compensation. Therefore, the UPFC can be considered a flexible series compensator that is not restricted by real power absorption or supply to hold the DC capacitor voltage constant. However, a UPFC can perform different tasks, as discussed below.

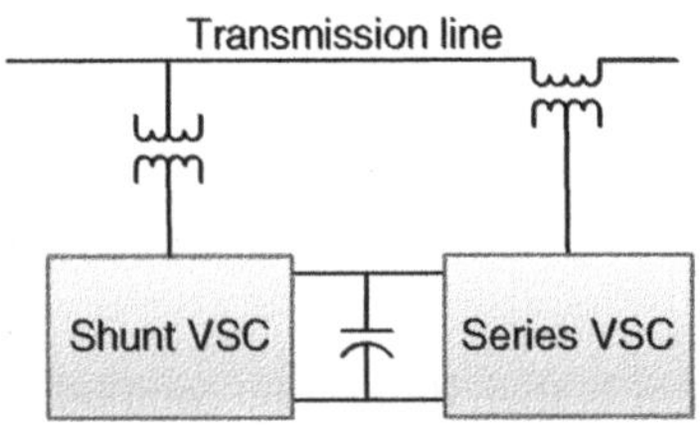

Figure 9.34 Schematic diagram of a UPSC.

The equivalent circuit of a UPFC-connected transmission line is shown in Figure 9.35 (a). Here, V_U is the injected voltage, while V_P is the voltage at the PCC bus. The phasor diagram of the system is shown in Figure 9.35 (b), where the angle between V_S and V_U is denoted by ρ and that between V_P and V_R is denoted by φ.

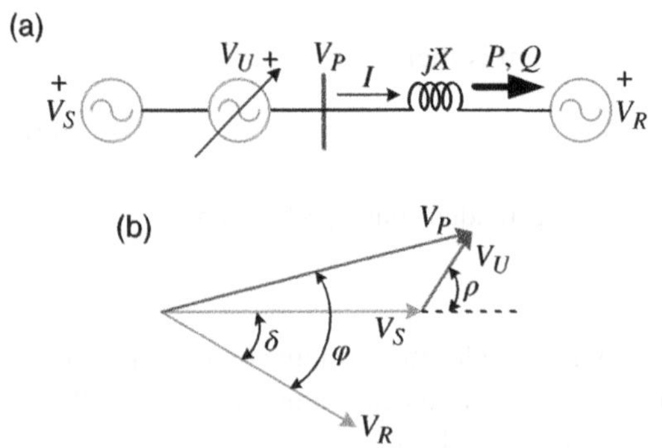

Figure 9.35 (a) Equivalent circuit of a UPSC-connected SMIB system and (b) its phasor diagram.

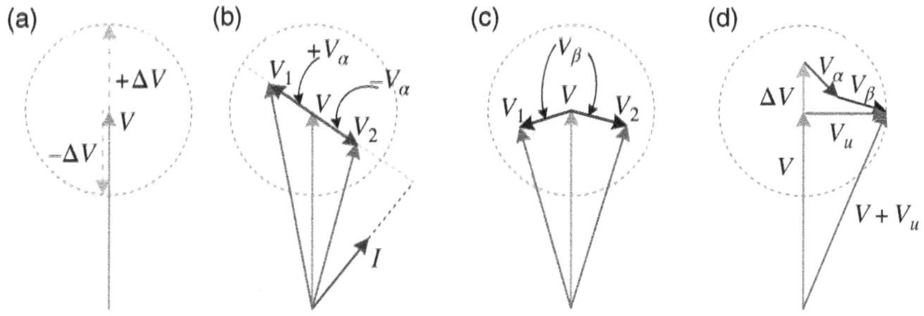

Figure 9.36 Different operating modes of UPFC: (a) voltage regulation, (b) series compensation, (c) phase shifting, and (d) simultaneous control.

A UPFC can perform four different functions [23], which are shown in Figure 9.36. With respect to Figure 9.35 (a), we denote $V_S = V\angle 0°$ and $V_U = |V_U| \angle \rho$. Then, these four different functions are discussed below.

1) *Voltage Regulation*: This can either increase or decrease the PCC voltage V_P magnitude vis-à-vis that of V_S by choosing $\rho = 0°$. The voltage regulation function is shown in Figure 9.36 (a), where $V_U = \Delta V\angle 0°$.
2) *Series Compensation*: This is shown in Figure 9.36 (b). Let the line current be given by $I = |I| \angle -\phi$. The UPQC injects a series voltage in phase quadrature with the current, that is, $\rho = \phi + 90°$ and we denote $|V_U| = V_\alpha$. Then, the capacitive and inductive modes of operation are given, respectively, as $V_1 = V + V_\alpha$ and $V_2 = V - V_\alpha$.
3) *Phase Shifting*: By changing the phase angle ρ of the injected voltage, the angle between the sending-end and the receiving-end voltages can be changed. At the same time, the magnitude of V_U can be changed to increase or decrease the desired power flow, as shown in Figure 9.36 (c). Denoting $|V_U| = V_\beta$, we get $V_1 = V + V_\beta$ and $V_2 = V - V_\beta$.
4) *Simultaneous Control*: This is combined control where the UPFC can simultaneously perform voltage regulation, series compensation, and phase shifting, as shown in Figure 9.36 (d). No other device can perform these multiple functions individually [23].

With respect to Figure 9.35 (a), let us define the following:

$$V_S = V\angle\delta, V_R = V\angle 0° \text{ and } V_U = |V_U|\angle(\delta + \rho)$$

Then

$$V_P = V_S + V_U = V(\cos\delta + j\sin\delta) + |V_U|\{\cos(\delta + \rho) + j\sin(\delta + \rho)\} \tag{9.71}$$

The line current is

$$I = \frac{V_P - V}{jX} = \frac{|V_U|\{\cos(\delta+\rho) + j\sin(\delta+\rho) + V(\cos\delta + j\sin\delta)\} - V}{jX}$$

The complex power at the receiving end then is

$$P - jQ = V^* I = V\left[\frac{V(\cos\delta + j\sin\delta) + |V_U|\{\cos(\delta+\rho) + j\sin(\delta+\rho)\} - V}{jX}\right] \tag{9.72}$$

Separating the real and imaginary components, the real and reactive power are given by

$$P = \frac{1}{X}\left[V^2 \sin\delta + |V_U| V \sin(\delta+\rho)\right] \tag{9.73}$$

$$Q = \frac{1}{X}\left[V^2 \cos\delta + |V_U| V \cos(\delta+\rho) - V^2\right] \tag{9.74}$$

From (7.6) and (7.7) of Chapter 7, the receiving real and reactive power, which are dependent only on the load angle, are written as

$$P_R(\delta) = \frac{V^2 \sin\delta}{X} \text{ and } Q_R(\delta) = \frac{V^2\cos\delta - V^2}{X}$$

Then, (9.73) and (9.74) can be rewritten as follows:

$$P = P_R(\delta) + \frac{|V_U| V \sin(\delta+\rho)}{X} \tag{9.75}$$

$$Q = Q_R(\delta) + \frac{|V_U| V \cos(\delta+\rho)}{X} \tag{9.76}$$

Because the phase shift ρ can be varied between 0 and 2π, the controllable ranges for the real and reactive power are [23]

$$P_R(\delta) - \frac{|V_{Umax}| V}{X} \leq P_R(\delta) \leq P_R(\delta) + \frac{|V_{Umax}| V}{X} \tag{9.77}$$

$$Q_R(\delta) - \frac{|V_{Umax}| V}{X} \leq Q_R(\delta) \leq Q_R(\delta) + \frac{|V_{Umax}| V}{X} \tag{9.78}$$

where $|V_{Umax}|$ is the maximum voltage range of the UPQC. These equations indicate that a UPQC can provide a controllable bidirectional power flow that is independent of the load angle δ. The different operating modes of UPQC and their advantages are discussed in [23].

9.7.2 Thyristor-Controlled Braking Resistor (TCBR)

Each phase of a TCBR comprises a high-power resistor that is connected in series with an antiparallel thyristor pair, as shown in Figure 9.37 (a). Here the TCBR with resistance R is connected to the generator (G) terminals. Because power transmission lines are mostly inductive, a line fault can drastically reduce the real power transfer, causing the generator to accelerate. If the TCBR is switched on at an appropriate time, real power will dissipate through the resistor, thereby reducing the machine acceleration and enhancing the transient stability.

An alternative simpler TCBR topology was proposed in [24] using a diode rectifier bridge having block/deblock (B/DB) capability, as shown in Figure 9.37 (b). Nominally, the diode bridge remains blocked during normal power transfer. However, once the fault is detected, the diodes are deblocked, and the power is dissipated in the resistor to arrest machine acceleration.

A natural choice for braking resistor (BR) control is the accelerating power. However, as discussed in Chapter 7, the mechanical power input almost remains constant during large transients. Machine acceleration primarily depends on the output electrical power. Therefore, the output power signal can be used as the BR trigger signal. A simplistic approach for the breaker trigger signal is to compare the output power P_e with a critical threshold P_{cr} and switch on the BR as follows:

$$\text{“If } P_e < P_{cr}, \text{turn the BR on, else switch off BR.”} \tag{9.79}$$

The threshold value is so chosen that spurious triggering due to load change is avoided.

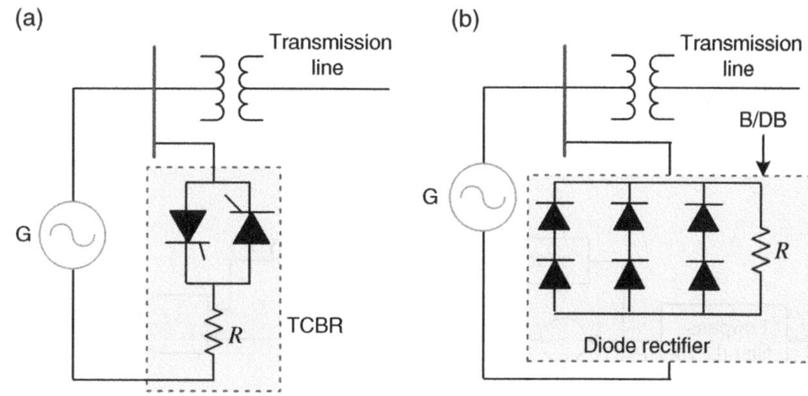

Figure 9.37 Schematic diagram of braking resistor connection: (a) using thyristors and (b) using diode bridge.

Following any large transient, the output power will go through high-frequency oscillations. These may cause a chattering effect if the logic of (9.79) is used, that is, with a dip in power, the BR will close, and when the power swings to the opposite direction, the BR will open. The high-frequency oscillations are eliminated using a lowpass filter to obtain filtered output P_{filt} and use it in a logic similar to that given in (9.79). This, however, will not solve the chattering problem since the low-frequency power oscillations will persist. During any transient, the rate of change of P_{filt} will be high immediately following the fault. Once the oscillations subside, the rate of change will reduce. Therefore, the triggering logic must incorporate both these signals, as given below.

$$\text{"If } P_e < P_{cr}, \text{ and } dP_{filt}/dt > dP_{cr}/dt, \text{ turn the BR on, else switch off BR."} \tag{9.80}$$

The value of dP_{cr}/dt is chosen such that it is less than dP_{filt}/dt during small oscillations. The BR triggering logic is shown in Figure 9.38.

9.7.3 Thyristor-Controlled Voltage Regulator (TCVR)

The maximum power that can be transmitted over a line is given by V_1V_2/X. Therefore, this can be enhanced if the voltage at the sending end can be boosted, which is the purpose of a TCVR. The schematic diagram of a TCVR is shown in Figure 9.39 (a). This contains three-phase or three single-phase tap-changing transformers. The primary side of the transformer is connected in a shunt with the transmission line. The secondary side is connected in series with the line to inject voltage in series. By controlling the transformer taps, the output voltage can be varied, as per

$$\begin{aligned} V_{a1} &= V_a + \Delta V_a \\ V_{b1} &= V_b + \Delta V_b \\ V_{c1} &= V_c + \Delta V_c \end{aligned} \tag{9.81}$$

The resulting phasor diagram is shown in Figure 9.39 (b).

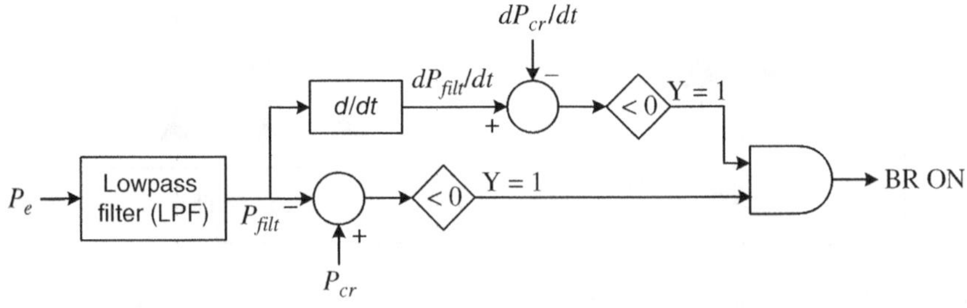

Figure 9.38 BR triggering logic.

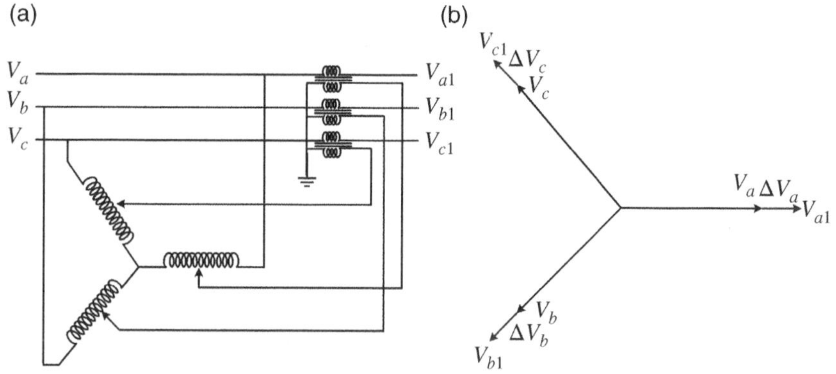

Figure 9.39 (a) Schematic diagram of a TCVR and (b) its phasor diagram.

Ideally, it would be desirable to have a continuously variable secondary tap setting. This, however, is not feasible. Discrete tap settings are possible using thyristor switches, as shown in Figure 9.40. In this figure, either of the four switches can be connected at a time. The injected voltage will decrease as the switch number goes up, that is, SW_1 will have maximum injected voltage while SW_4 will inject the minimum voltage.

Figure 9.40 Thyristor-based transformer tap changer.

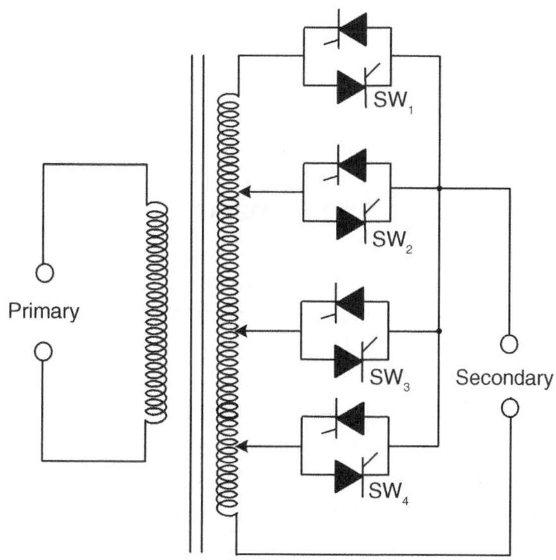

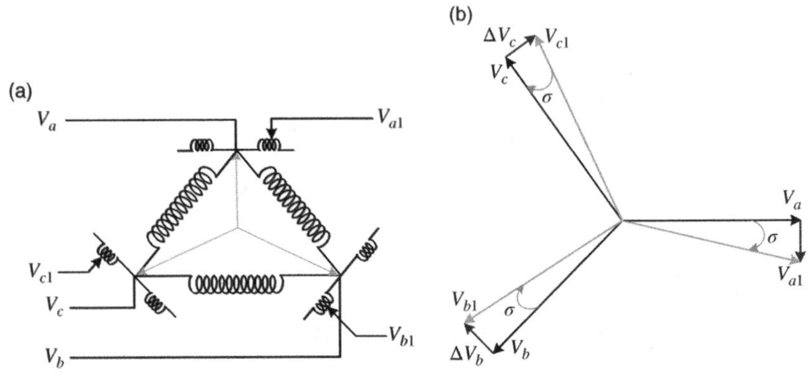

Figure 9.41 (a) Schematic diagram of TCPAR connection and (b) its phasor diagram.

9.7.4 Thyristor-Controlled Phase Angle Regulator (TCPAR)

We know that the power flow depends on the relative angle difference between the sending- and receiving-end voltages. Therefore, the maximum power transfer can be enhanced if the angle can be advanced. The maximum power flow can also be reduced by reducing the angle difference. TCPAR also uses thyristor-based tap-changing transformers. The schematic diagram of the TCPAR connection is shown in Figure 9.41 (a). Here, the TCPAR injects voltages in quadrature with the sending-end voltages, as shown in Figure 9.41 (b). The resulting power transfer equation is given by

$$P_e = P_{\max} \sin(\delta \pm \sigma) \tag{9.82}$$

The power transfer using TCPAR is shown in Figure 9.42.

9.8 Concluding Remarks

Different FACTS devices have been discussed in this chapter. Some of these devices are used for enhancing power flow over a line directly, such as TCVR and TCPAR, or indirectly, such as TCBR. However, other devices that provide shunt and series compensation, as well as joint series and shunt compensation, can not only enhance power transfer but also provide power oscillation damping. Some of these aspects have also been discussed in this chapter.

Subsynchronous oscillations in series capacitor-compensated systems can occur. However, series compensation using TCSC or SSSC does not suffer from this problem. They can switch over from capacitive to inductive mode at the onset of SSR

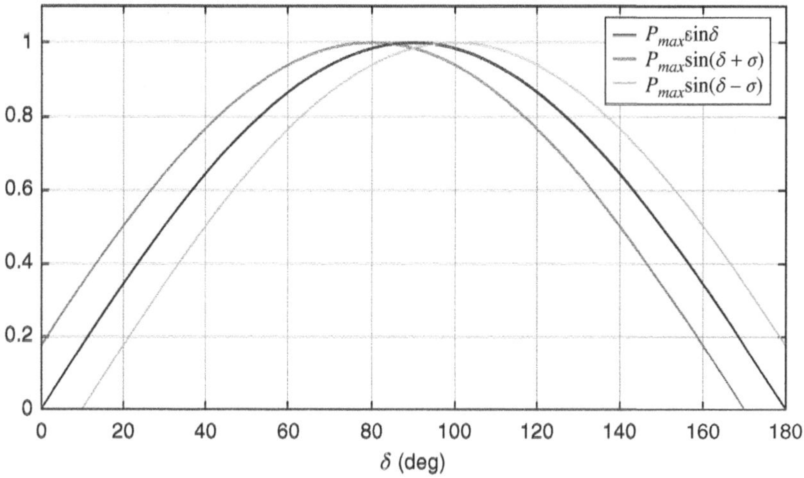

Figure 9.42 Power-angle characteristics using TCPAR.

oscillations. Moreover, these devices can also change the percentage compensation in order to avoid or suppress SSR oscillations. A STATCOM can also be used in a series capacitor-compensated system in order to dampen SSR oscillations. In [25], it has been shown that by placing a STATCOM at the electrical center of a series capacitor-compensated IEEE first benchmark model, SSR oscillations can be damped through the STATCOM control.

References

1 T. J. E. Miller, Ed., *Reactive Control of Electric Systems*, John Wiley, New York, 1982.

2 B. D. Bedford and R. G. Hoft, *Principles of Inverter Circuits*, John Wiley, New York, 1964.

3 L. Gyugyi, "Advanced static VAR compensator using gate turn-off thyristors for utility applications," Paper No. 23-203, *CIGRE*, 1990 Session.

4 L. Sunil Kumar and A. Ghosh, "Modeling and control design of a static synchronous series compensator,"*IEEE Transactions on Power Delivery*, Vol. 14, No. 4, pp. 1448–1453, 1999.

5 L. Sunil Kumar and A. Ghosh, "Static synchronous series compensator – Design, control, and applications," *Electric Power Systems Research*, Vol. 49, pp. 139–148, 1999.

6 A. Ghosh and F. Zare, *Control of Power Electronic Converters with Microgrid Applications*, IEEE Press-Wiley, Hoboken, New Jersey, 2023.

7 IEEE SSR Working Group, "Terms, definitions, and symbols for subsynchronous oscillations," *IEEE Transactions on Power Apparatus & Systems*, Vol. PAS-104, No. 6, 1985.

8 IEEE Committee Report, "First benchmark model for computer simulation of subsynchronous resonance," *IEEE Transactions on Power Apparatus & Systems*, Vol. PAS-96, pp. 1565–1570, 1977.

9 IEEE Committee Report, "Second benchmark model for computer simulation of subsynchronous resonance," *IEEE Transactions on Power Apparatus & Systems*, Vol. PAS-104, pp. 1057–1066, 1985.

10 P. M. Anderson, B. L. Agrawal and J. E. Van Ness, *Subsynchronous Resonance in Power Systems*, IEEE Press, New York, 1990.

11 K. R. Padiyar, *Analysis of Subsynchronous Resonance in Power Systems*, Kluwer Academic Publishers, Boston, 1999.

12 P. M. Anderson, "Power system oscillations, summary of utility experience," *IEEE Power Engineering Society Publication on Inter-Area Oscillations in Power Systems*, No. 95 TP 101, pp. 43–54, October 1994.

13 P. Kundur, *Power System Stability and Control*, McGraw-Hill, New York, 1994.

14 B. Pal and B. Chaudhuri, *Robust Control in Power Systems*, Springer Science +Business Media, New York, 2005.

15 M. S. Roden, *Analog and Digital Communication Systems*, Prentice Hall, Englewood Cliffs, 1979.

16 IEEE Committee Report, "Countermeasures to subsynchronous resonance," *IEEE Transactions on Power Apparatus & Systems*, Vol. PAS-99, pp. 1810–1817, 1980.

17 N. G. Hingorani, "A new scheme for subsynchronous resonance damping of torsional oscillations and transient torque – Part 1," *IEEE Transactions on Power Apparatus & Systems*, Vol. PAS-100, pp. 1852–1855, 1981.

18 A. Ghosh, A. Joshi and M. K. Mishra, "State space simulation and accurate determination of fundamental impedance characteristics of a TCSC," *Proceedings IEEE-PES Winter Meeting-2001*, Columbus, Ohio, 2001.

19 N. Christl, R. Hedin, P.E. Krause and S.M. McKenna, "Advanced series compensation (ASC) with thyristor controlled impedance", *CIGRE Regional Meeting*, Paris, Session 1992.

20 P. C. Srivastava, A. Ghosh and S. V. Jayaram Kumar, "Model-based control design of a TCSC compensated power system," *Electric Power & Energy Systems*, Vol. 21, No. 4, pp. 299–307, 1999.

21 A. Ghosh and G. Ledwich, "Modelling and control of thyristor-controlled series compensators," *Proceedings of IEE – Generation, Transmission & Distribution*, Vol. 142, No. 3, pp. 297–304, 1995.

22 G. N. Pillai, A. Ghosh and A. Joshi, "Torsional interaction studies on a power system compensated by SSSC and fixed capacitor," *IEEE Transactions on Power Delivery*, Vol. 18, No. 3, pp. 988–993, 2003.

23 N. G. Hingorani and L. Gyugyi, *Understanding FACTS: Concepts and Technology of Flexible AC Transmission Systems*, Wiley-IEEE Press, New Jersey, 2000.
24 B. Das, A. Ghosh and Pr. Sachchidanand, "A novel control strategy for braking resistor," *Electric Power & Energy Systems*, Vol. 20, No. 6, pp. 391–403, 1998.
25 K. R. Padiyar and N. Prabhu, "Design and performance evaluation of subsynchronous damping controller with STATCOM," *IEEE Transactions on Power Delivery*, Vol. 21, No. 3, pp. 1398–1405, 2006.

Problems

P9.1 The fixed resistance of a TCR is 0.3 per unit. Determine the firing angle α in degree such that the fundamental frequency of the TCR reactance is 40 per unit.

P9.2 Consider a TCR that is supplied by the voltage $v = 100\sin(100\pi t)$. The TCR reactance is 10 mH. Then determine the magnitude of the odd harmonic currents up to 11th order with respect to the fundamental current when the firing angle $\alpha = 100°$.

P9.3 Consider the system shown in Figure P9.3, where an SVC is placed at the middle of a transmission line with $V_S = 1\angle 20°$per unit. The SVC has the following parameters:

TCR: $X_L = 0.66$ per unit
TSC: $X_C = 1$ per unit with the identical number of five modules, that is, $n = 5$.

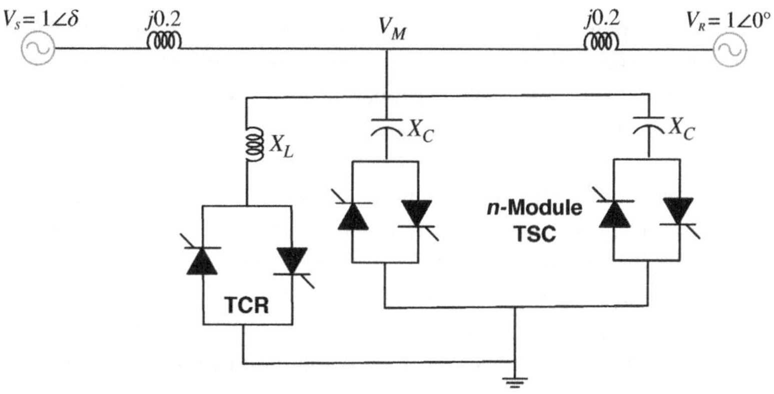

Figure P9.3 SVC circuit with five TSC modules of Problem 9.3.

The firing angle α of the TCR is chosen as 80°. Determine the number of TSC modules that must be chosen such that the midpoint voltage is $V_M = 1\angle 10°$ per unit.

P9.4 Consider again the circuit of Figure P9.3. The system parameters are as follows:

Source voltage (V_S): $1\angle 30°$ per unit
TSC reactance (X_C): 0.5 per unit with four number of units.
TCR reactance (X_C): 0.1188 per unit
TCR firing angle (α): 94.06°
Midpoint voltage (V_M): $1\angle \varphi$ per unit

If all four TCR units are used, determine the angle (φ) of the midpoint voltage.

P9.5 Consider the series-compensated transmission system shown in Figure P9.5. Determine the frequencies of network currents that will appear at the generator terminals when the fundamental frequency is 50 Hz.

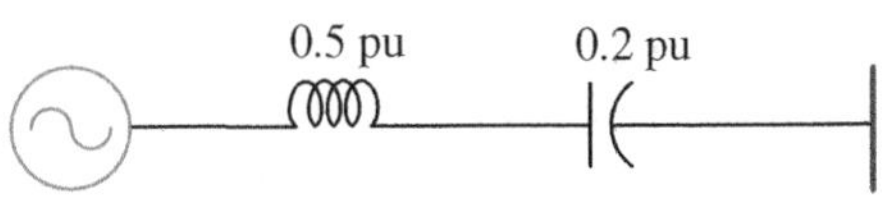

Figure P9.5 Series-compensated transmission line of Problem 9.5.

P9.6 Consider a 4-mass turbine rotor model with the following parameters:

$$K_{12} = 38.75, K_{23} = 32.5 \text{ and } K_{34} = 28.475$$

The power supplied by the generator is 0.95 per unit. The LP and IP turbines share 30% power each, while the HP turbine shares 40% power. If the machine angle is 25°, determine the angles of the LP, IP, and HP turbines when the system is in a steady state.

P9.7 Consider the UPFC-compensated power system shown in Figure 9.7, where the shunt compensator only supplies power P_{pq} to the series compensator. The system has the following parameters in per unit: $V_S = 1\angle 22°$, $V_R = 1\angle 0°$, and $X = 0.5$. The nominal power transfer without the UPFC is denoted as P_0. Determine the voltage V_{PQ} when it is working as a voltage regulator to raise the power transfer to $1.2P_0$. What will be the value of P_{pq}?

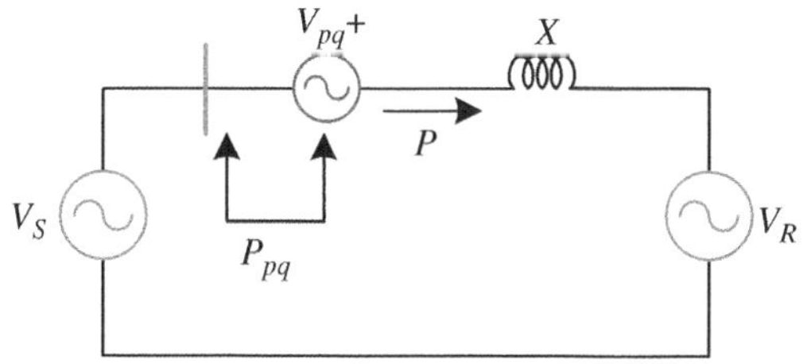

Figure P9.7 UPQC-compensated transmission line of Problem 9.7.

P9.8 Consider an SMIB system that has $V_S = 1\angle 30°$ per unit, $V_R = 1\angle 0°$ per unit, and $X = 0.5$ per unit, while it transfers a maximum power of P_{max}. A TCPAR is connected at the sending end which adds voltage $V_{TCPAR} = \Delta V \angle \sigma$ per unit with the sending end voltage (see Figure 9.42). Then find the magnitude and angle of V_{TCPAR} when the power transfer equation is increased by 10% as per (9.82).

P9.9 Consider again the TCPAR-connected system of Problem 9.8. The voltage injected by the TCPAR is in quadrature with the source voltage, that is, $V_{TCPAR} = \Delta V \angle (\delta + 90°)$. Then, find the power transfer relationship and plot the power angle curves for $\Delta V = \pm 1, \pm 0.5, \pm 0.25$.

10

High-Voltage DC (HVDC) Transmission Systems

By the dawn of the twentieth century, “the war of the currents” was over, and AC had won. This does not necessarily mean that DC was totally forgotten. DC has bounced back in the form of high-voltage transmission and low-voltage microgrid. High-voltage DC (HVDC) transmission made a modest beginning in 1954 when a 100 kV, 20 MW DC link was established between the Swedish mainland and the island of Gotland on the Baltic Sea using 98 km long submarine DC cables. For converter stations, bulky mercury arc valves were used. Around 1957, thyristors started appearing in the market. The successful use of thyristors for the control of industrial drives encouraged its adoption in HVDC converters. The Gotland project (called Gotland 1) was upgraded to include thyristor valves to enhance the transmission capacity to 30 MW. This was the first time the thyristors were used for HVDC transmission.

Traditionally, HVDC systems were used for point-to-point power transmission. They were usually used for transmitting power between two AC systems through rectification and inversion. Usually, 12-step converters (see Section 9.3.2) are used for such point-to-point transmission. These converters use thyristors without gate turn-off capacity. They are called line commutated converters (LCCs) as they depend on the AC supply voltage crossing zero. They do not need an external circuit to turn the thyristors off. However, with the advancement of insulated gate bipolar transistor (IGBT) technology, high-power, voltage source converters (VSCs) can now be constructed. Typically, multilevel converters (see Section 9.3.4) and modular multilevel converters (MMCs) are used for this purpose. The VSC-based HVDC systems, which are referred to as VSC-HVDC, are slowly replacing LCC-HVDC systems in many places.

A multiterminal HVDC (MTDC) system can have more than two terminals. Such a system can be placed in large geographical areas and is a particularly

Modern Power System, First Edition. Arindam Ghosh.

Published 2026 by John Wiley & Sons, Inc.
Companion website: www.wiley.com/go/ghoshpowersystem

attractive solution for offshore wind farms. An MTDC system uses VSC technology that can change the direction of power flow easily. With the increased penetration of intermittent wind power technology, a rapid development in the MTDC system is expected to take place in the near future. In this chapter, all three technologies (i.e., LCC-HVDC, VSC-HVDC, and MTDC) will be discussed.

10.1 Attributes of DC Systems

An HVDC system will require converter stations – for a point-to-point system, a converter at each end. This is not a requirement for an AC transmission system. On the other hand, DC lines require fewer conductors than AC lines, and DC line supporting towers are smaller than those of the AC systems. Even if the initial cost of HVDC systems is higher than an HVAC system, the overall cost of DC systems becomes less than AC systems as the length of the lines increases. A typical cost comparison between the two systems is shown in Figure 10.1. Usually, the break-even point is between 600 and 800 km.

10.1.1 Advantages and Disadvantages of HVDC Systems

There are several advantages of DC systems over AC systems. Some of these are listed below.

- In an AC circuit, current density is at maximum near the surface of the conductor. This is called the *skin effect*. This effect reduces the cross section of

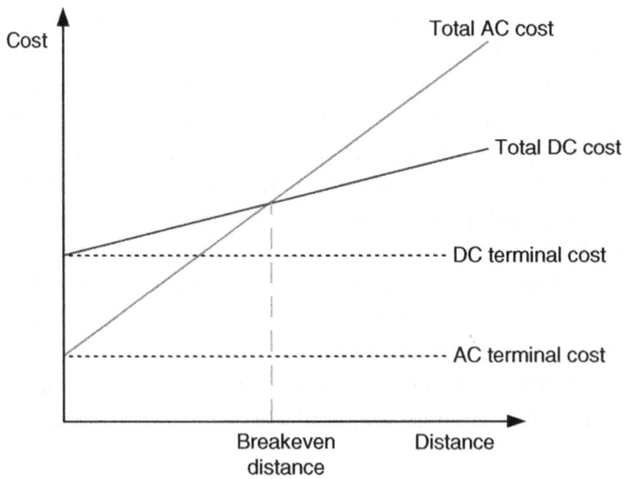

Figure 10.1 Cost of HVDC system versus cost of HVAC system.

the conductor, thereby increasing the resistance and power loss. This is absent in DC circuits.

- HVDC systems require less copper than AC systems.
- DC systems do not have reactive power problems.
- AC systems have more capacitive loss, especially when the conductors are close to the ground, for example, for submarine power transfer. The breakeven distance for subsea cables is about 50 km, after which the DC is more efficient. NorNed link currently is the longest underwater HVDC system, spanning 580 km, with a capacity of 700 MW. It connects the Netherlands with Denmark.
- An HVDC system can asynchronously connect two AC systems, that is, a 50 Hz system can be connected to a 60 Hz system through an HVDC line or back-to-back configuration. Itaipu Dam is located on the border of Brazil and Paraguay. It has a generation capacity of 14 GW, which supplies power to both Brazil (at 60 Hz) and Paraguay (at 50 Hz). Since the load demand in Paraguay is much less than the power transmitted from the Dam, the power is exported to the Brazilian side through two 800 km, 600 kV HVDC lines.
- DC systems have an inherent short time overload capacity that can be used for power system oscillations. Two systems connected by AC lines can become unstable. Linking the AC systems through a DC link will overcome this problem.

In an AC circuit, the circuit breakers (CBs) can open at the zero crossings of the currents that occur naturally twice in a cycle. However, this is not feasible for DC systems since the current does not have zero crossings. The lack of DC circuit breakers (DCCBs) is regarded as a limitation to HVDC systems. In general, for LCC systems, a rectifier and an inverter connect the DC line at the two ends. The rectifier converter is forced into full inversion mode when a fault is detected on the DC line. Since the line is forced to have two inverters at the two ends, the current will become zero. There are several other DC system protection systems in addition to the AC system protection that are summarized in [1]. A review of DCCBs is given in [2].

10.1.2 Types of HVDC Systems

Figure 10.2 shows a monopolar HVDC system. These are the simplest and least expensive HVDC systems, which are used for moderate power transfer [3]. This system uses only two converters – one at each end. It can have either a ground (or sea) return path or a metallic return path. In congested areas or areas with high earth resistivity, the metallic return path is used. The ground connections are used for reference potential [3]. When a fault occurs, the entire transmission system must be shut down. This topology is not used in practice; it is used in the development stages to test converter capacity.

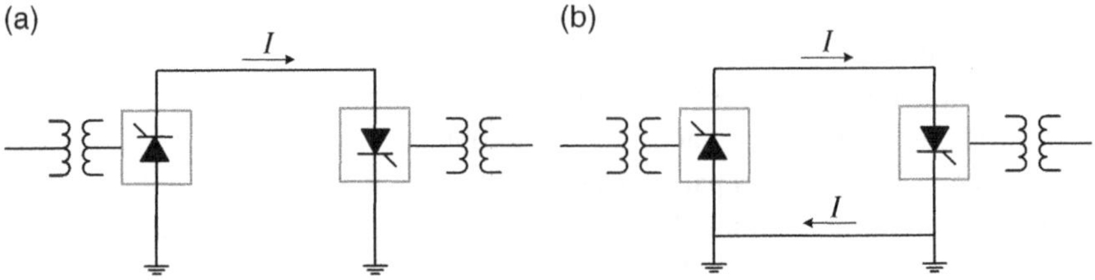

Figure 10.2 Monopolar HVDC system: (a) ground return and (b) metallic return.

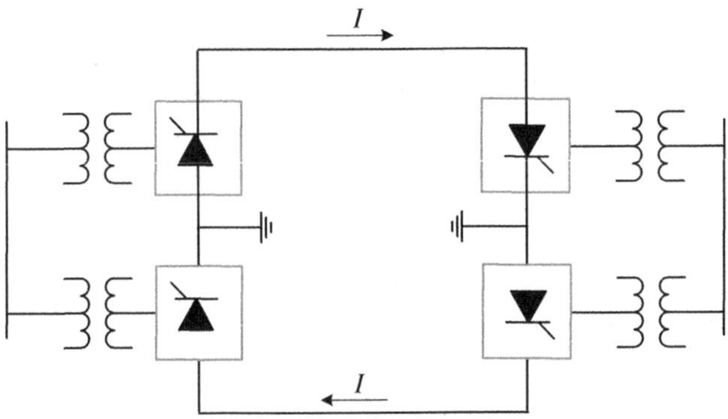

Figure 10.3 Bipolar HVDC system.

Bipolar systems use 12-pulse converters as shown in Figure 10.3. This has two conductors – one positive and one negative. Each of the terminals has a set of two identical converters. The neutral point at the junctions between the converters is solidly grounded. In the 12-pulse mode, both the poles operate with equal current, and hence the ground current is expected to be zero. When a fault occurs in one of the converters, these links can operate as a monopolar link using the ground as the return path. The bipolar HVDC systems are the most popular for point-to-point systems.

A homopolar HVDC system has two conductors of the same polarity (usually negative). It always operates with earth or metallic return. Since the poles operate in parallel, it has lower insulation cost. However, this is not a very popular concept.

Replacing the transmission line is by a short cable, the back-to-back connection is formed, as shown in Figure 10.4 (a). These can be used to connect two AC networks that operate at different frequencies. It can also be used for isolating two AC systems such that oscillations in one system do not propagate onto the other. The schematic diagram of an MTDC system is shown in Figure 10.4 (b). As the name signifies, it will have three or more converter stations connected

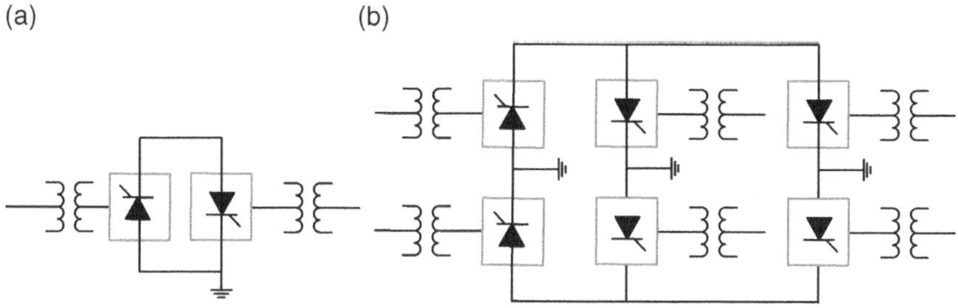

Figure 10.4 (a) Back-to-back and (b) multiterminal HVDC system.

through transmission lines. These converters can be connected in either series or in parallel. Some of the converters will act like rectifiers, while others will act like inverters. However, there can be several configurations of MTDC systems, as discussed in Section 10.4.

10.2 LCC-HVDC Systems

Consider the six-step converter discussed in Section 9.3.1 and depicted in Figure 9.12. This is redrawn in Figure 10.5. The converter is supplied by a three-phase voltage source and is connected to a load on the DC side. The thyristors, which are denoted by T, are numbered in the sequence in which they are turned on. The thyristors of each leg are complementary, that is, when one is

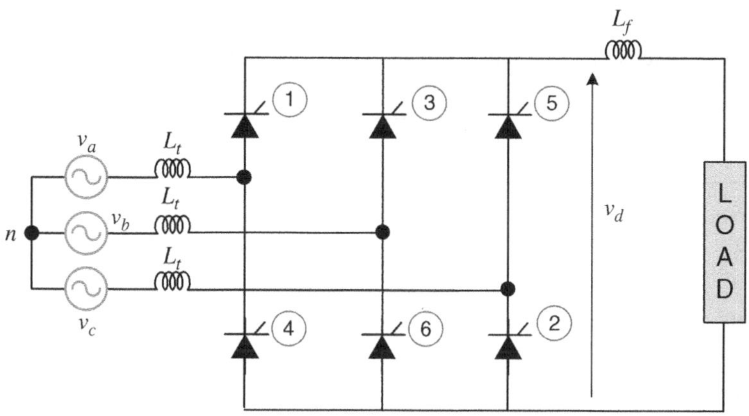

Figure 10.5 A six-step converter supplying a DC load.

on the other is off (e.g., when T_1 is on, T_4 is off, and vice versa). Let us assume that the voltages are given by

$$\begin{aligned} v_a &= \sqrt{2}V_M \sin(\omega t)\ \text{V} \\ v_b &= \sqrt{2}V_M \sin(\omega t - 120^\circ)\ \text{V} \\ v_c &= \sqrt{2}V_M \sin(\omega t + 120^\circ)\ \text{V} \end{aligned} \tag{10.1}$$

where ω is chosen as 100π rad/s (50 Hz) and V_M is the magnitude of the line-to-neutral (RMS) voltage that is chosen as 1,000 V. The line-to-line voltages are then given by

$$\begin{aligned} v_{ab} &= \sqrt{2}V_{LL} \sin(\omega t + 30^\circ)\ \text{V} \\ v_{bc} &= \sqrt{2}V_{LL} \sin(\omega t - 90^\circ)\ \text{V} \\ v_c &= \sqrt{2}V_{LL} \sin(\omega t + 150^\circ)\ \text{V} \end{aligned} \tag{10.2}$$

where V_{LL} is the RMS line-to-line voltage that is given by $\sqrt{3}V_M$.

10.2.1 System Characteristics with Zero Ignition Angle

In this section, it has been assumed that the thyristor bridge works like a diode bridge. This implies that a pair of thyristors is switched on as soon as the voltage across them becomes positive. We further assume that the line inductance L_f is zero. Each thyristor conducts for a period of 60° in a 360° cycle. The thyristor switching sequence and the output DC voltage are then listed in Table 10.1. The line-to-line voltage waveforms and the DC voltage are shown in Figure 10.6.

Table 10.1 Thyristor switching sequence of a six-step converter without ignition delay.

Period	Thyristors conducting	DC voltage (v_d)
0° to 30°	and T_6	v_{cb}
30° to 90°	T_1 and T_6	v_{ab}
90° to 150°	T_1 and T_2	v_{ac}
150° to 210°	T_3 and T_2	v_{bc}
210° to 270°	T_3 and T_4	v_{ba}
270° to 330°	T_5 and T_4	v_{ca}
330° to 360°	T_5 and T_6	v_{cb}

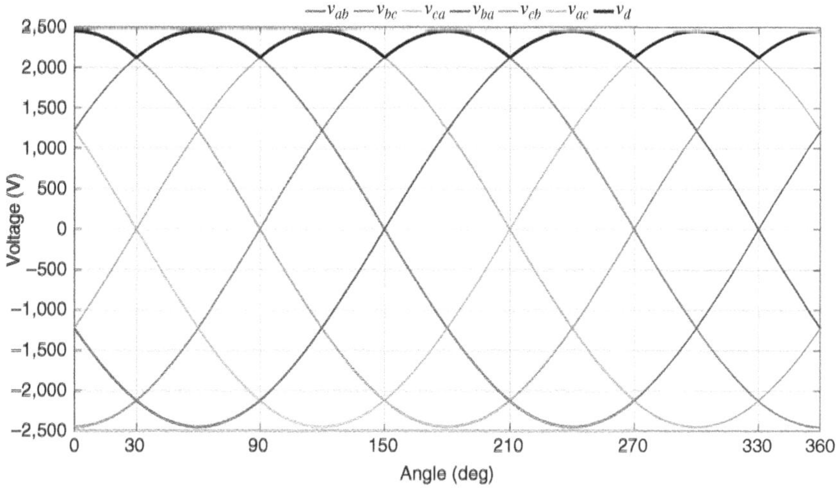

Figure 10.6 Line-to-line input voltages and the output DC voltages of a six-step converter without ignition delay.

Note from Figure 10.6 that the DC voltage varies periodically every 60°. The average DC voltage is then computed by averaging this voltage over any 60° period, that is, one-third of half a cycle. Consider the period between 30° and 90°. The average voltage when the system is not loaded is given by

$$V_{d0} = \frac{3}{\pi}\int_{30^\circ}^{90^\circ} v_d d\theta = \frac{3}{\pi}\int_{30^\circ}^{90^\circ} v_{ab} d\theta \tag{10.3}$$

Using (10.2), (10.3) is written as

$$\begin{aligned} V_{d0} &= \frac{3\sqrt{2}V_{LL}}{\pi}\int_{30^\circ}^{90^\circ} \sin(\theta + 30^\circ)d\theta \\ &= \frac{3\sqrt{2}}{\pi}V_{LL}[-\cos(120^\circ) + \cos 60] = \frac{3\sqrt{2}}{\pi}V_{LL} = 1.35V_{LL} \end{aligned} \tag{10.4}$$

For the DC voltage waveform shown in Figure 10.6, V_{d0} is 2,338.3 V.

10.2.2 System Characteristics with Nonzero Ignition Angle

For the uncontrolled operation of Section 10.2.3, the DC output voltage will remain fixed and will only depend on the line-to-line voltage, as given in (10.4). This is the maximum voltage that the bridge can produce. To have control over the DC voltage output, the firing is shifted by an angle α, which is called the *delay angle* or the *firing angle*. The maximum value of the firing angle is limited to 180°.

Table 10.2 Thyristor switching sequence of a six-step converter with ignition delay.

Period	Thyristor fired	Thyristor conducting	DC voltage (v_d)
$30° + \alpha$ to $90° + \alpha$	T_1	T_6	v_{ab}
$90° + \alpha$ to $150° + \alpha$	T_2	T_1	v_{ac}
$150° + \alpha$ to $210° + \alpha$	T_3	T_2	v_{bc}
$210° + \alpha$ to $270° + \alpha$	T_4	T_3	v_{ba}
$270° + \alpha$ to $330° + \alpha$	T_5	T_4	v_{ca}
$330° + \alpha$ to $30° + \alpha$	T_6	T_5	v_{cb}

For $0° \leq \alpha < 90°$, the converter produces positive DC voltage, while it produces negative DC voltage during $90° \leq \alpha < 180°$. The firing sequence is given in Table 10.2. Notice that the top thyristors are fired in sequences 1, 3, and 5 with a delay of 120°. Similarly, the bottom thyristors are also fired in sequences 2, 4, and 6 with a delay of 120°. The voltage waveforms given by (10.2) and for $\alpha = 30°$ are shown in Figure 10.7.

The average DC voltage is then given by

$$V_d = \frac{3}{\pi}\int_{30° + \alpha}^{90° + \alpha} v_{ab} d\theta = 1.35 V_{LL} \int_{30° + \alpha}^{90° + \alpha} \sin(\theta + 30°) d\theta \tag{10.5}$$

$$= V_{d0}[-\cos(120° + \alpha) + \cos(60° + \alpha)] = V_{d0} \cos\alpha$$

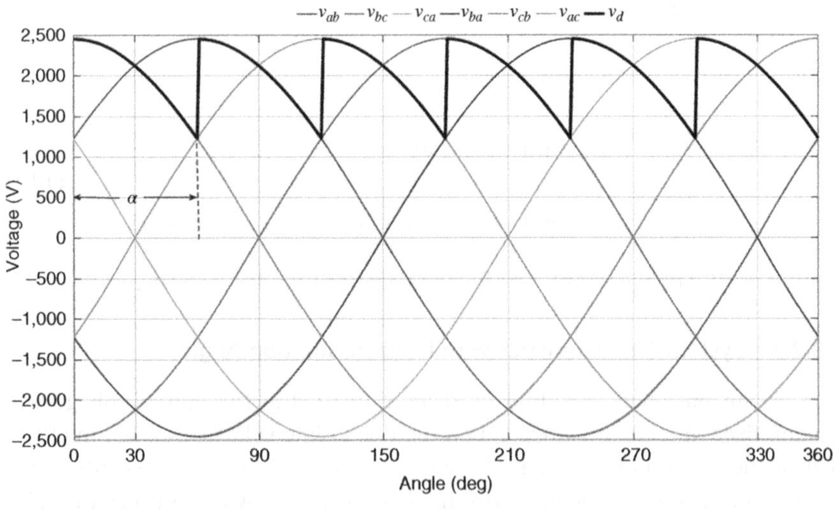

Figure 10.7 Line-to-line input voltages and the output DC voltages of a six-step converter without ignition delay.

The average voltage of Figure 10.7 is 2,025 for $\alpha = 30°$. The fundamental frequency line current is given by the relation [4]

$$I_{Lf} = \frac{\sqrt{6}}{\pi} I_d = 0.78 I_d \tag{10.6}$$

10.2.3 Overlap Angle

So far, we have assumed that the inductance L_t in Figure 10.5 is zero. Also, in the steady state $L_f = 0$. This means that the line current commutates as soon as the voltage across the thyristor reverses polarity. However, an HVDC converter is connected through a converter transformer with its leakage inductance being present in the path of the current flow. Since the current through an inductor cannot change instantaneously, the current cannot commutate as soon as the voltage across a thyristor reverses polarity.

Given any delay angle α, the current commutates with a finite angle μ, which is called the *overlap angle*. Consider the case where T_5 and T_6 are conducting, and T_1 is switched on after a delay angle α. It is expected that T_5 will commutate following the overlap period. This is shown in Figure 10.8. The resulting waveforms are shown in Figure 10.9, where δ is called the *extinction angle* and is equal to

$$\delta = \alpha + \mu \tag{10.7}$$

During the overlap period, there will be a voltage drop, given by

$$v_L = L_t \frac{di_a}{dt} \tag{10.8}$$

Therefore, the reduction in the average DC voltage is

$$V_R = \int_{\alpha}^{\alpha + \mu} v_L d\theta \tag{10.9}$$

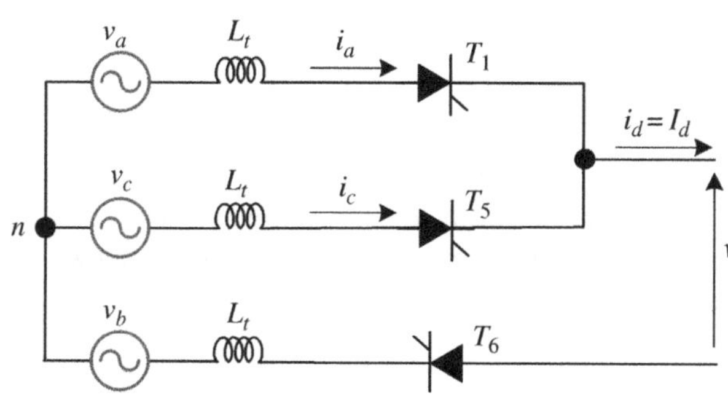

Figure 10.8 Equivalent circuit when T_5 is commutating.

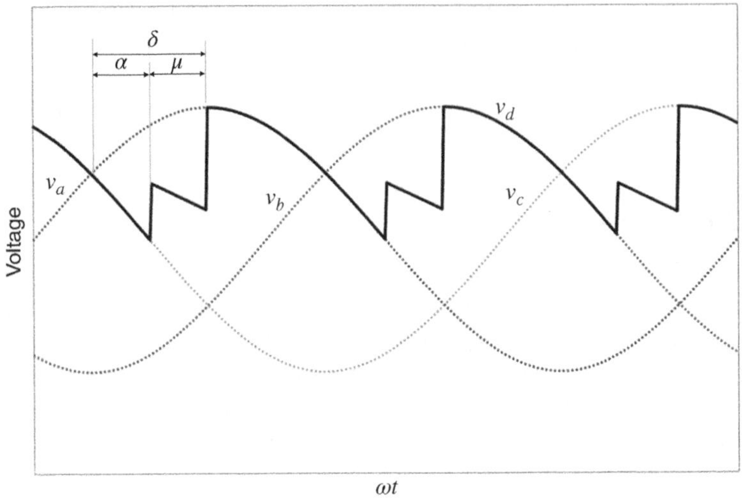

Figure 10.9 Voltage waveforms showing the commutation process during rectification.

The reduction in the average output DC voltage due to the overlap angle is given as [4]

$$V_d = V_{d0} \cos\alpha - \Delta V_d \tag{10.10}$$

where

$$\Delta V_d = \frac{V_{d0}}{2}(\cos\alpha - \cos\delta) = R_t I_d \tag{10.11}$$

In (10.11), the resistance R_t is defined in terms of the commutation reactance X_t as

$$R_t = \frac{3}{\pi}X_t = \frac{3}{\pi}\omega L_t$$

10.2.4 Inverter Operation

From (10.5), we note that $V_d = V_{d0} \cos\alpha$ when there is no overlap. Therefore, V_d becomes negative for $\alpha > 90°$. However, the angle reduces from 90° with overlap [4]. The timing information for rectifier and inverter operations is shown in Figure 10.10. From this figure, we define the following for the inverter:

Ignition angle: $\beta = \pi - \alpha$
Extinction angle: $\gamma = \pi - \delta$
Overlap angle: $\mu = \beta - \gamma$

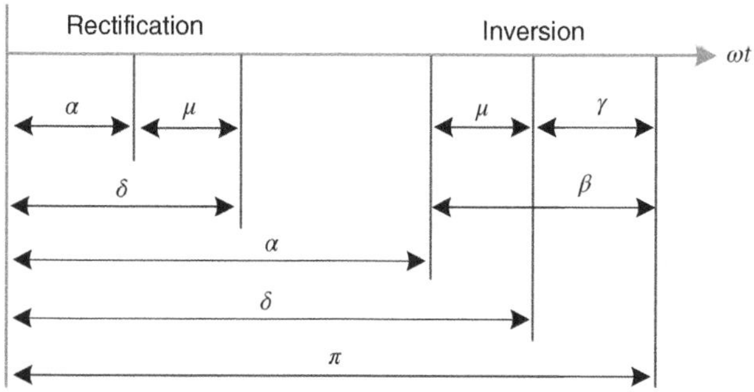

Figure 10.10 Rectification and inversion timing information.

The DC side voltage is then defined as

$$V_d = V_{d0}\cos\beta + R_t I_d \tag{10.12}$$

10.2.5 Active Power

Substituting (10.11) into (10.10), we have

$$V_d = V_{d0}\cos\alpha - \frac{V_{d0}}{2}(\cos\alpha - \cos\delta) = \frac{V_{d0}}{2}(\cos\alpha + \cos\delta) \tag{10.13}$$

Assuming that the converter is lossless, the power on both sides is equal. The DC side power is

$$P_{dc} = V_d I_d = \frac{V_{d0}}{2}(\cos\alpha + \cos\delta)I_d \tag{10.14}$$

Using (10.4), (10.14) is rewritten as follows:

$$P_{dc} = V_d I_d = \frac{3\sqrt{2}}{\pi}V_{LL}\frac{(\cos\alpha + \cos\delta)}{2}I_d \tag{10.15}$$

Let the angle between the fundamental frequency line-to-neutral voltage and line current be ϕ. Then, the power on the AC side is

$$P_{ac} = \sqrt{3}V_{LL}I_{Lf}\cos\phi \tag{10.16}$$

Using (10.6), (10.16) is rewritten as follows:

$$P_{ac} = \frac{3\sqrt{2}}{\pi}V_{LL}I_d\cos\phi \tag{10.17}$$

Equating (10.15) with (10.17), we have

$$\frac{(\cos\alpha + \cos\delta)}{2} = \cos\phi \tag{10.18}$$

Note that the power factor relationship is an approximate one since it is assumed that the overlap angle is zero. However, the error is minimal, and the DC voltage then can be written from (10.13) as

$$V_d = V_{d0}\cos\phi \tag{10.19}$$

Note that, for the inverter side, the power factor angle is given by

$$\frac{(\cos\beta + \cos\gamma)}{2} = \cos\phi \tag{10.20}$$

Example 10.1 A 400 kV, 50 Hz, transmission line is connected to an HVDC system. The delay angle $\alpha = 30°$ and the commutation angle $\mu = 20°$. We shall determine the DC output voltage. From (10.4), the no-load DC voltage is given by

$$V_{d0} = 1.35V_{LL} = 540\text{ kV}$$

From (10.7), the extinction angle is calculated as

$$\delta = \alpha + \mu = 30 + 20 = 50°$$

The reduction in voltage is then computed from (10.11) as

$$\Delta V_d = \frac{V_{d0}}{2}(\cos\alpha - \cos\delta)$$
$$= 270(\cos 30° - \cos 50°) = 270 \times 0.2232 = 60.27\text{ kV}$$

The DC output voltage is then

$$V_d = V_{d0}\cos\alpha - \Delta V_d = 407.38\text{ kV}$$

Example 10.2 In Example 10.1, the current delivered by the rectifier is 2,000 A. We shall calculate the commutation reactance (X_t), power factor, and the real and reactive power at the AC side.

From (10.11), we have

$$R_t = \frac{\Delta V_d}{I_d} = 30.14\,\Omega$$

Then,

$$X_t = \frac{\pi R_t}{3} = 31.56\,\Omega$$

The power flowing in the system is

$$P_{ac} = P_{dc} = V_d I_d = 814.76\,\text{MW}$$

Now, the power factor is given from (10.19) by

$$\cos\phi = \frac{V_d}{V_{d0}} = \frac{407.38}{540} = 0.754$$

Therefore, the phase angle is given by

$$\phi = \cos^{-1}(0.754) = 41.03^\circ$$

Then the reactive power at the AC side is

$$Q_{ac} = P_{ac}\tan\phi = 708.92\,\text{MVAr}$$

10.2.6 Twelve-Pulse Converter

The 12-pulse or 12-step converter has been discussed in Section 9.3.2. The calculations for different quantities given in Sections 10.2.3–10.2.5 are almost the same for this converter, except that the transformer turns ratio and the number of bridges must be considered. For example, the open load DC voltage is given by [4]

$$V_{d0} = 1.35 \times 2 \times T \times V_{LL} \tag{10.21}$$

where T is the transformer turns ratio.

One of the most common abnormal operations is commutation failure, which typically occurs during inverter operation. This can occur when the DC current is very high or when the AC voltage is low. This phenomenon occurs when one (or more than one) thyristor valve fails to complete commutation before the commutating voltage reverses. Then this valve continues to conduct without transferring its current to the valve that comes next in the firing sequence. This will cause unnecessary stress on the converter values and will cause interruptions in the power transmission. There are several other issues with the LCC-HVDC system such as power flow control and converter control. These are discussed in [4, 5].

10.3 VSC-HVDC Systems

In this system, the rectifiers and inverters are replaced by VSCs. Some of the multilevel converter structures are discussed in Section 9.3.4. These, along with MMCs, can achieve voltage levels of ±650 kV while transmitting 3,000 MW of power over a 2,000 km distance [6]. In 1997, ABB (now Hitachi Energy) demonstrated the world's first VSC-HVDC project in a place called Hällsjön in Sweden.

ABB called it *HVDC Light* before the term VSC-HVDC evolved. Such systems are relatively easy and quicker to install. They are used for connecting offshore wind farms to onshore power grids through subsea cables or for underground power transmission. Because these systems use VSCs, they can provide rapid control of both active and reactive power. Furthermore, changing the direction of the power flow can easily be reversed by changing the direction of the current flow.

10.3.1 Control of a Voltage Source Converter (VSC)

Consider the three-phase VSC shown in Figure 10.11. There are several modulation techniques, including sinusoidal pulse width modulation (SPWM). In SPWM, the switching pulses are generated by comparing a sinusoidal modulating waveform with a triangular waveform, called the carrier waveform. The switches S_1 to S_6 are turned on or off at the intersecting points of these two waveforms [7]. The RMS value of the fundamental line-to-neutral voltage is given by

$$V_{LN} = m_a \frac{V_{dc}}{2\sqrt{2}}, \quad m_a < 1 \tag{10.22}$$

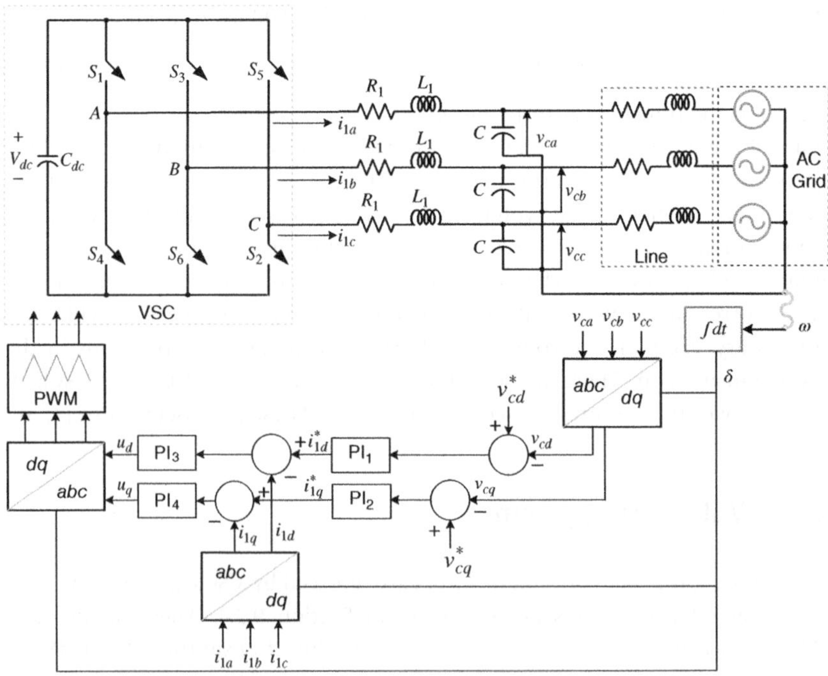

Figure 10.11 Control scheme of a VSC connected to an AC grid.

where m_a is called the modulation index or ratio and is given by the ratio of the peak modulating voltage to the peak of the carrier waveform. It is desirable to keep the modulation index between 0 and 1. In general, SPWM works at a fixed switching frequency, and therefore, is a preferred option for high-power converters. Space vector modulation is another modulation and averaging technique that takes into consideration the eight switching states of the three-phase VSC. This, however, is a complex procedure, and we shall restrict our attention to SPWM. It is to be noted that most HVDC converters use either multilevel or MMCs. SPWM for multilevel converters is presented in [7]. However, we shall stick to the simplest converter structure shown in Figure 10.11.

The SPWM is essentially an open-loop controller. The fundamental voltage, in the under-modulated range, follows the modulating waveform. However, the converter produces a square wave output voltage, which obviously will introduce harmonics in the AC system. To eliminate the harmonics, a passive filter is connected to the output of the converter. The filter can be L, LC, or LCL type, where the leakage inductance of a transformer can be one of the inductors. In Figure 10.11, an LC-type filter is used with the assumption that the leakage inductance of the transformer is subsumed in the line inductance. The purpose of the converter is to produce a balanced set of voltages at grid frequency according to some specified values.

The control structure of Figure 10.11 includes four finely tuned proportion-plus-integral (PI) controllers. First, the reference d- and q-axis voltages (v_{cd}^*, v_{cq}^*) are obtained. The d- and q-axis voltages and currents are obtained through dq transformation using a phase-locked loop (PLL) to extract the angle δ. The PI controllers are then used to obtain the control signals u_d and u_q. These are then used to generate the firing signals in the ABC frame. The PI controllers are tuned such that the modulation index is kept in the under-modulated range. There are, however, superior control algorithms that can be found in [7].

A typical response of the converter is shown in Figure 10.12. The DC side voltage is chosen as 800 kV, and the grid voltage is 400 kV (L-L). The converter output voltages are phase advanced from the grid voltages by 20°. The line-to-neutral converter output voltages and the power supplied to the grid are shown in Figure 10.12, where a cold start has been assumed.

10.3.2 VSC-HVDC Configuration

A typical VSC-HVDC system is shown in Figure 10.13, which connects two terminals, denoted by the subscripts 1 and 2. In this, V_G is the grid side voltage and V_C is the converter side voltage (e.g., voltage across the filter capacitor v_c shown in Figure 10.11). It has been assumed that the grid side and the converter side voltages are connected through the impedance X, which can be the leakage reactance

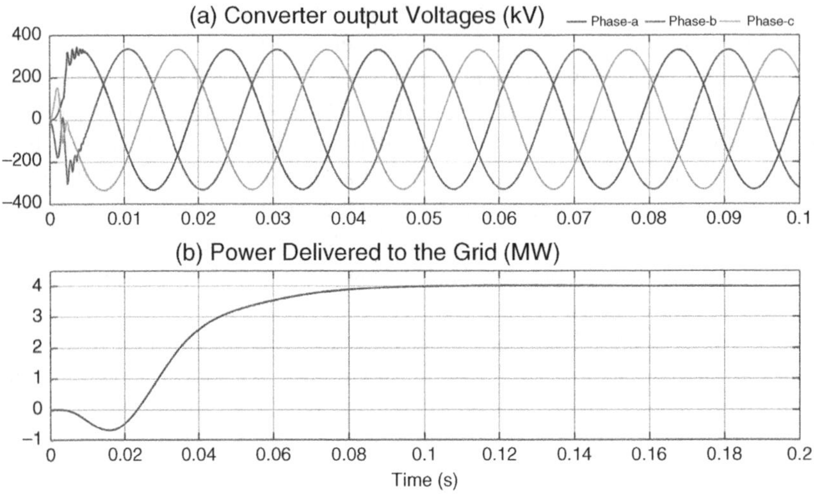

Figure 10.12 VSC control system response: (a) output voltage and (b) power delivered.

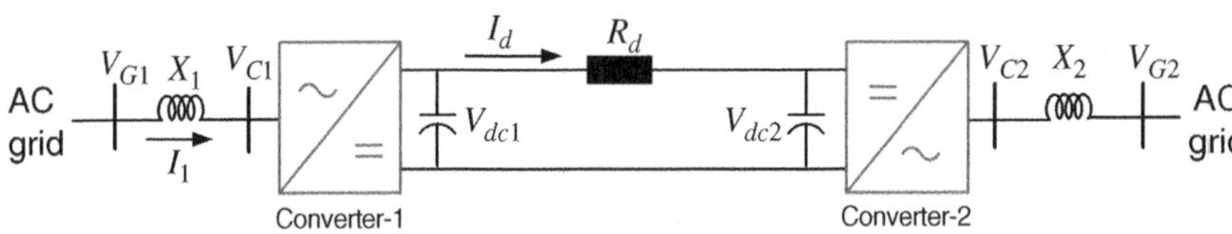

Figure 10.13 Two-terminal VSC-HVDC system.

of the connecting transformer. The DC voltages are denoted by V_{dc1} and V_{dc2}, while the DC line resistance is denoted by R_d. It is obvious that when $V_{dc1} > V_{dc2}$, the current I_d will flow from Converter-1 (or Treminal-1) to Converter-2, while the current will flow in the reverse direction when $V_{dc1} < V_{dc2}$.

It is to be noted that if the DC transmission line is missing and there is a single capacitor on the DC side such that $V_{dc1} = V_{dc2} = V_{dc}$, then it will be called a back-to-back connection. The power flow can then be controlled in either direction as explained in [8]. For HVDC transmission, assuming the power flows from Converter-1 to Converter-2, the DC line current I_d is given by

$$I_d = \frac{V_{dc1} - V_{dc2}}{R_d} \tag{10.23}$$

Then, the power supplied and received by the converters is given by

$$P_{dc1} = V_{dc1} I_d \text{ and } P_{dc2} = P_{dc1} - R_d I_d^2 \tag{10.24}$$

It is, therefore, obvious that the DC power flow can be controlled by controlling the voltages V_{dc1} and V_{dc2}.

Because the operating principles of both the VSCs are the same, we shall neglect subscripts 1 and 2 henceforth. With respect to Figure 10.13, let us define the following:

$$V_G = |V_G|\angle\delta_G, V_C = |V_C|\angle\delta_C \text{ and } \delta_G - \delta_C = \delta$$

Then from (7.3) and (7.4), we have

$$P = \frac{|V_G||V_C|}{X}\sin\delta \tag{10.25}$$

$$Q = \frac{|V_G|^2 - |V_G||V_C|\cos\delta}{X} \tag{10.26}$$

Let us assume that δ is small, such that $\cos\delta = 1$. Then, (10.26) is rewritten as

$$Q = \frac{|V_G|^2 - |V_G||V_C|}{X} = \frac{|V_G|}{X}\{|V_G| - |V_C|\} \tag{10.27}$$

Therefore, if $|V_G| > |V_C|$, the reactive power flows into the converter from the AC side. On the other hand, if $|V_G| < |V_C|$, the reactive will flow out of the converter to the AC side. Moreover, when $|V_G| = |V_C|$, no reactive power exchange takes place between the converter and the AC system. Therefore, the reactive power can be controlled by adjusting the magnitude of the converter voltage V_C vis-à-vis that of the AC side voltage V_G. Also note that the reactive power exchange does not depend on the direction of the DC power flow. Furthermore, as can be seen from (10.22), the DC output voltage can be controlled through the modulation index, and the DC output voltage can be maintained to a desired specified value irrespective of the change in $|V_C|$. The DC power flow will also be independent of the reactive power control. The power and reactive power capability diagram is given in [9]. It depends on three factors – maximum current, maximum DC voltage, and maximum power limit.

10.3.3 Direct Control of VSC-HVDC Systems

This is the simplest form of control, where the real power and AC voltage are controlled in the sending (rectifier) end, while the DC and AC voltages are controlled in the receiving (inverter) end [10]. The schematic diagram of direct control is shown in Figure 10.14. For a VSC, the DC and AC side voltages are related by that given in (10.22) using a simple SPWM scheme. If the DC side voltage is not sufficient, then the AC side voltage will see a tracking failure, resulting in a distorted voltage waveform of a reduced magnitude. Now consider (10.25). Given the magnitude of V_{G1}, the power flow can be controlled by the voltage magnitude $|V_{C1}|$ and

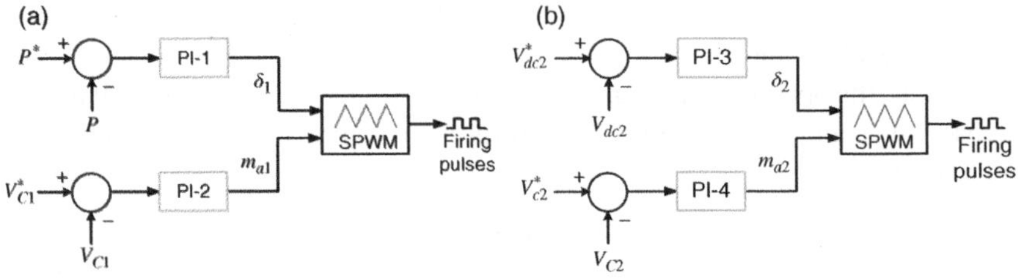

Figure 10.14 VSC-HVDC direct control: (a) rectifier side and (b) inverter side.

the angle δ_1. Therefore, the sending end in this scheme is controlled by two PI controllers – one to determine the angle δ_1, and the other to determine the modulation index m_{a1}. These are then used in an SPWM scheme to generate the firing sequences. This is shown in Figure 10.14 (a), where the power flow is controlled to obtain δ_1, while the modulation index is obtained from the control of the voltage magnitude.

At the receiving end, the power flow will depend on $|V_{C2}|$ and the angle δ_2. Furthermore, the DC power flow will depend on the potential difference between V_{dc1} and V_{dc2}. Therefore, adjusting V_{dc2} vis-à-vis V_{dc1}, the required flow will flow out the receiving end terminal. Therefore, the DC voltage control will generate the angle δ_2, and the AC voltage control will generate the modulation index m_{a2}. This is shown in Figure 10.14 (b).

Even though this is a simple scheme, there are strong interactions between the real and reactive power, as evident from (10.25) and (10.26). Therefore, it will not be possible to control these two quantities independently. Moreover, it will require a communication channel to control the power flow, which depends on the difference between the voltages V_{dc1} and V_{dc2}.

10.3.4 Vector Control of VSC-HVDC Systems

In the vector control scheme, the active and reactive power are decoupled, and the angle δ is not dependent on the modulation index [9]. The vector control technique has several control layers, such as:

- Inner Current Controller
- AC Voltage Controller
- DC Voltage Controller
- Real and Reactive Power Controller

The operations of these controllers are synchronized through a PLL.

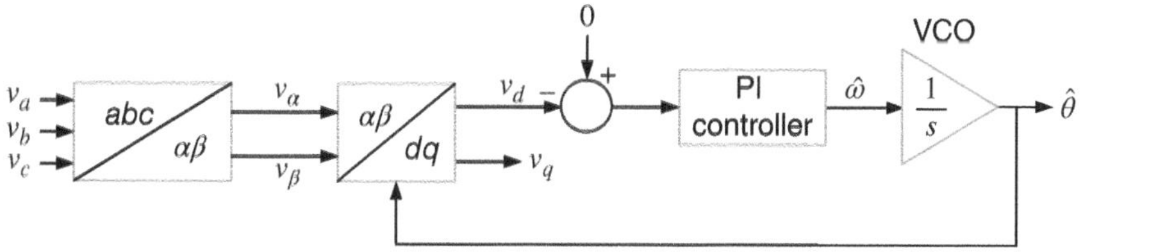

Figure 10.15 A typical phase-locked loop.

Phase-Locked Loop (PLL): The basic aim of a PLL is to make a signal track another signal. The block diagram of a PLL is shown in Figure 10.15. Let a set of voltages be given by

$$\begin{bmatrix} v_a \\ v_b \\ v_c \end{bmatrix} = V_m \begin{bmatrix} \sin(\theta) \\ \sin(\theta - 120°) \\ \sin(\theta + 120°) \end{bmatrix}$$

Then, they are converted into αβ frame using the following transform:

$$\begin{bmatrix} v_\alpha \\ v_\beta \end{bmatrix} = \frac{2}{3} \begin{bmatrix} 1 & -\frac{1}{2} & -\frac{1}{2} \\ 0 & \frac{\sqrt{3}}{2} & -\frac{\sqrt{3}}{2} \end{bmatrix} \begin{bmatrix} v_a \\ v_b \\ v_c \end{bmatrix} \tag{10.28}$$

These are then converted in the dq domain using the following transform:

$$\begin{bmatrix} v_d \\ v_q \end{bmatrix} = \begin{bmatrix} \cos\hat{\theta} & \sin\hat{\theta} \\ -\sin\hat{\theta} & \cos\hat{\theta} \end{bmatrix} \begin{bmatrix} v_\alpha \\ v_\beta \end{bmatrix} \tag{10.29}$$

where $\hat{\theta}$ is the estimated value of θ. To achieve this, the d-axis voltage is forced to 0 through a PI controller to obtain the estimate of the speed. This is then passed through a voltage-controlled oscillator (VCO) to obtain $\hat{\theta}$ [7]. The VSC controllers are designed based on d- and q-axis components derived based on the PLL output.

Mathematical Model: Consider the VSC-HVDC system of Figure 10.13. With respect to the figure, we can write

$$v_{Gk} - v_{Ck} = Ri_k + L\frac{di_k}{dt}, k = a, b, c \tag{10.30}$$

where the lowercase letters indicate the instantaneous quantities, $L = X/\omega$, and R is the resistance associated with the inductor L. The dq transformation of (10.30) results in the following equations:

$$\begin{aligned} \frac{di_d}{dt} &= \omega i_q + \frac{1}{L}(v_{Gd} - v_{Cd}) - \frac{R}{L}i_d \\ \frac{di_q}{dt} &= -\omega i_d + \frac{1}{L}(v_{Gq} - v_{Cq}) - \frac{R}{L}i_q \end{aligned} \tag{10.31}$$

The VSC input AC voltage and the DC voltage are related through the modulation index, similar to that given in (10.22) as [11]

$$V_C = m_a \frac{V_{dc}}{2} \sin(\omega t + \delta) \tag{10.32}$$

where δ is the angle between V_G and V_C. Eq. (10.32) is resolved in the d- and q-axis domain as [7]

$$v_{Cd} = m_a \frac{V_{dc}}{2} \cos\delta \text{ and } v_{Cq} = m_a \frac{V_{dc}}{2} \sin\delta \tag{10.33}$$

The duty ratio and the phase angle are given from (10.33) as given below.

$$m_a = \frac{2\left(\sqrt{v_{Cd}^2 + v_{Cq}^2}\right)}{V_{dc}} \tag{10.34}$$

$$\delta = \tan^{-1}\left(\frac{v_{Cq}}{v_{Cd}}\right) \tag{10.35}$$

Decoupling of Real and Reactive Power: By placing the d-axis in phase with the a-axis using the PLL, we can have

$$v_{Gd} = |V_G| \text{ and } v_{Gq} = 0 \tag{10.36}$$

The real and reactive power is then given as follows:

$$\begin{aligned} P &= \frac{3}{2} v_{Gd} i_d = \frac{3}{2} |V_G| i_d \\ Q &= -\frac{3}{2} v_{Gd} i_q = -\frac{3}{2} |V_G| i_q \end{aligned} \tag{10.37}$$

Given the reference active power P^* and reactive power Q^*, the reference currents are then given by

$$\begin{aligned} i_d^* &= \frac{2P^*}{3|V_G|} \\ i_q^* &= -\frac{2Q^*}{3|V_G|} \end{aligned} \tag{10.38}$$

It can be seen from (10.38) that the d-axis current reference can be calculated from the active power reference, while the q-axis current reference can be calculated from the reactive power reference. This forms the outer current controller, as shown in Figure 10.16 (a), where the errors between the reference and measured quantities are passed through PI controllers to generate d- and q-axis current references.

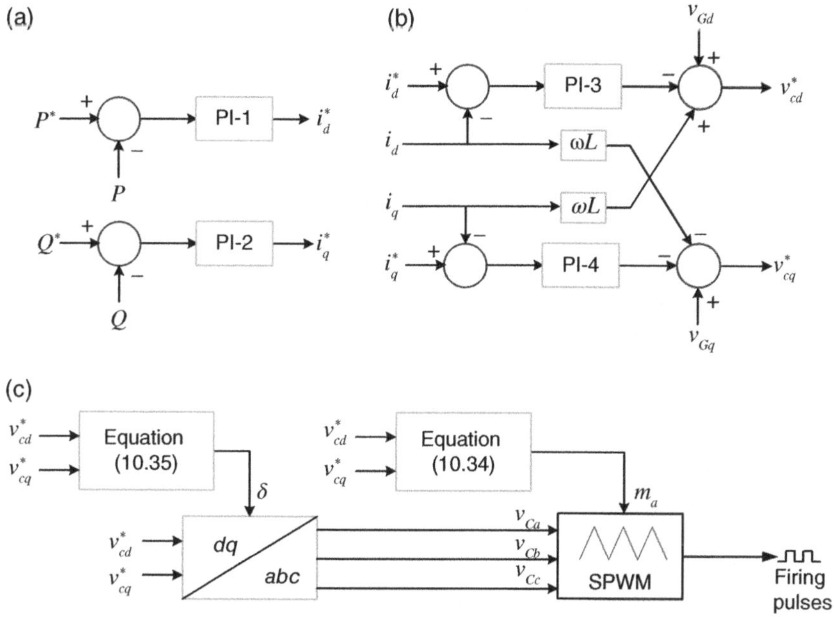

Figure 10.16 Block diagrams of VSC-HVDC control system: (a) outer current controller, (b) inner current controller, and (c) firing pulse generation.

The inner current controllers are also PI controllers, as shown in Figure 10.16 (b). Eq. (10.31) is rewritten to compute the reference for the converter output voltage as [12]

$$\begin{aligned} v_{cd}^* &= -\left(L\frac{di_d}{dt} + Ri_d\right) + \omega L i_q + v_{Gd} \\ v_{cq}^* &= -\left(L\frac{di_q}{dt} + Ri_q\right) - \omega L i_d + v_{Gq} \end{aligned} \tag{10.39}$$

Two PI controllers (PI-3 and PI-4) are used to obtain the d- and q-axis currents and their derivatives as given below.

$$\begin{aligned} L\frac{di_d}{dt} + Ri_d &= K_{P3}\left(i_d^* - i_d\right) + K_{I3}\int\left(i_d^* - i_d\right)dt \\ L\frac{di_q}{dt} + Ri_q &= K_{P4}\left(i_q^* - i_q\right) + K_{I4}\int\left(i_q^* - i_q\right)dt \end{aligned} \tag{10.40}$$

Once the d- and q-axis voltage references are obtained, (10.34) and (10.35) are used to compute the modulation index m_a and phase angle δ. These are then used to generate switching pulses, as shown in Figure 10.16 (c). In this scheme, VSC-1 controls the active and reactive power, while VSC-2 controls the DC voltage and reactive power [10, 11]. It is, however, to be noted that there are several other possible control strategies for inverter control. Refer to [7] for various converter control techniques.

10.4 Multiterminal HVDC Systems

MTDC can be of both LCC- or VSC-type systems, even though the latter is gaining prominence these days. A collaborative project between Hydro Québec and the National Grid in the United States set up the first MTDC in the world. The converters for this project were LCC types. This started in 1986 as a two-terminal bipolar HVDC link between Des Catons in Québec and Comerford in New Hampshire (see Figure 10.17). Each of these converter stations is rated 690 MW at ±450 kV, carrying power over a distance of 275 km. This is called Phase I of the project.

Phase II of the project was commissioned in 1990 when the HVDC line was extended 1,100 km north from Des Cantons to a 2,250 MW converter station at Radisson in the James Bay area of Québec, which is very rich in hydroelectric power. The line from Comerford was extended 214 km south to a 1,800 MW converter station at Sandy Point (near Boston). In 1992, a 2,138 MW converter station at Nicolet (near Montreal) was added, making it a five-terminal HVDC system. From north to south, the converter stations were at Radisson, Nicolet, Des Catons, Comerford, and Sandy Point. However, the stations at Des Cantons and Comerford were disconnected subsequently.

The Northeast Agra project in India is the first ultrahigh voltage DC (UHVDC) installation in the world with ratings of ±800 kV and 6,000 MW [13, 14]. The schematic diagram of the project is shown in Figure 10.18, where the converter and rectifier stations use LCC. The main aim of the project is to transmit mostly hydroelectric power generated in the northeastern part of India to Agra, which is close to the major load center in New Delhi and surrounding areas. The first pooling point is at Biswanath Chariali in the state of Assam, and the second pooling point is at Alipurduar in West Bengal, which is about 432 km from Biswanath Chariali. Each of the rectifiers that are placed at both these sites can add 3,000 MW to the HVDC line. The power is then transmitted over a distance of 1,296 km to Agra, where there are two inverter stations, as shown in Figure 10.17.

Multiterminal VSC-HVDC systems were installed in Nan'ao island (in 2013) and Zhoushan archipelago in China. This archipelago contains nearly 2,000 islands. A five-terminal HVDC system was established here, where the converters were placed in the islands of Yangshan, Sijiao, Qushan, Daishan, and Dinghai with the ratings of 100 MW, 100 MW, 100 MW, 300 MW, and 400 MW, respectively.

Figure 10.17 Schematic diagram of Hydro Quebec-New England multiterminal project.

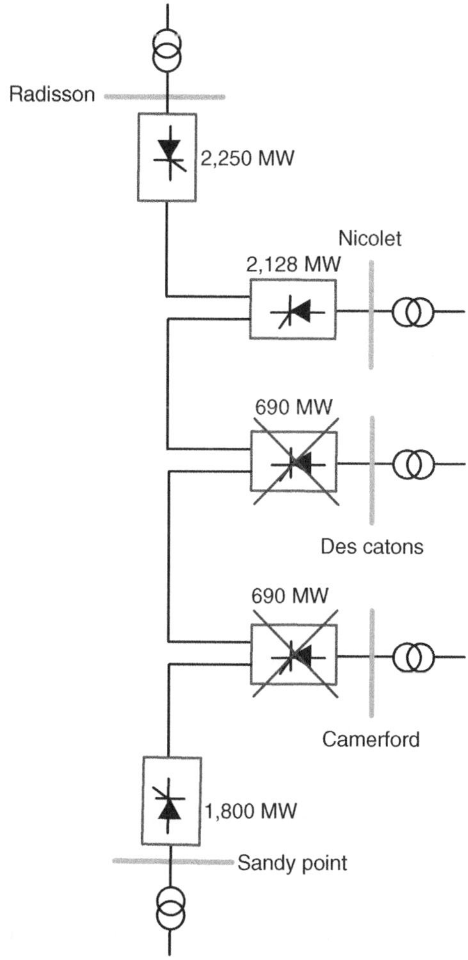

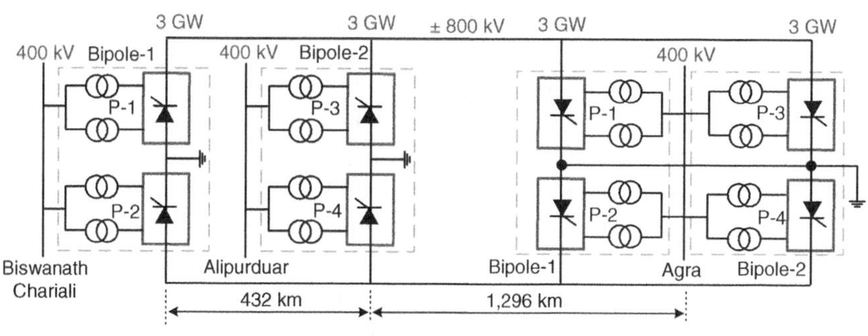

Figure 10.18 Schematic diagram of the northeast Agra UHVDC scheme.

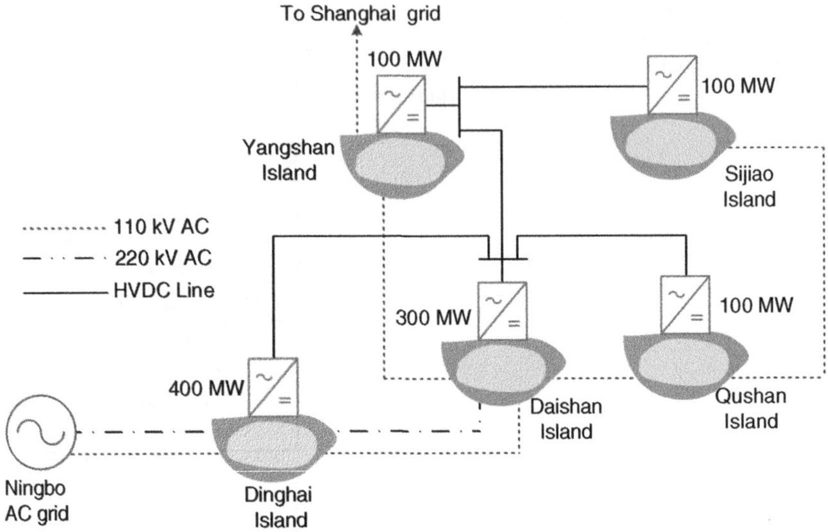

Figure 10.19 Zhoushan five-terminal HVDC scheme.

The schematic diagram of the Zhoushan project is shown in Figure 10.19. The MTDC system is connected to both Ningbo and Shanghai power grids at either 110 kV or 220 kV. The voltage rating of the HVDC line is ±200 kV [15].

10.4.1 Multiterminal System Configurations

The MTDC system can be monopolar or bipolar. Again, the monopolar systems can be symmetric or asymmetric. In general, an HVDC line will have two converters at two ends, or several converters in the case of an MTDC. These are called monopolar grids. A symmetric monopolar grid is particularly suitable for VSC-HVDC transmission [9], in which the DC sides of the converters are connected through high-voltage conductors. In this configuration, either the DC side is not grounded or the DC capacitors have a grounded middle, as shown in Figure 10.20 (a). An asymmetric monopolar grid can either have a metallic return or a ground return, as shown in Figure 10.20 (b and c), respectively. In metallic return, a dedicated low-voltage conductor is used for the ground path, which is grounded at one terminal only. On the other hand, the return path is the ground path through the earth as each converter is grounded. The monopolar grounded return is most cost-effective for systems that use cable conductors, as the ground return results in cost savings. However, environmental concerns may make the large current flow of the earth unacceptable in many parts of the world.

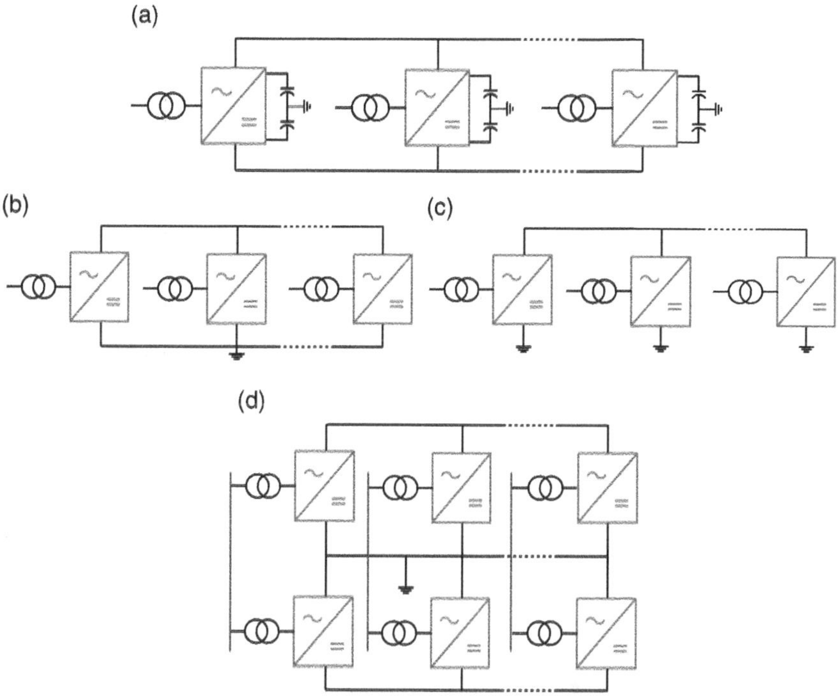

Figure 10.20 Multiterminal configurations: (a) symmetric monopolar, (b) asymmetric monopolar with metallic return, (c) asymmetric monopolar with ground return, and (d) bipolar with metallic return.

Monopolar grids can suffer from the lack of redundancies. For example, if there is a fault in a converter terminal, the power flow through the converter will be zero. Bipolar configurations provide redundancy by using two converters at each terminal, as shown in Figure 10.20(d). In this, each converter is connected to a common earth wire. The other end of the converter can be connected to the positive pole or the negative pole. If any of the converters goes out of service, the power transmission can continue between the positive pole and ground or through the negative pole and ground. However, the power flow can be reduced in such a situation.

10.4.2 MTDC Control

There are two methods that have been proposed for MTDC control – the voltage margin (VM) method and the voltage droop (VD) method. These are discussed briefly below.

Voltage Margin (VM) Method: Let us first consider a two-terminal DC link, as shown in Figure 10.21. Here, it has been assumed that Terminal-A controls the DC voltage through a PI controller. The function of the controller is to hold the DC voltage around its specified value V_{dc}^* by adjusting P_A within its lower and upper limits [16]. The V_{dc} - P characteristic at Terminal-A is shown in Figure 10.22 (a). If

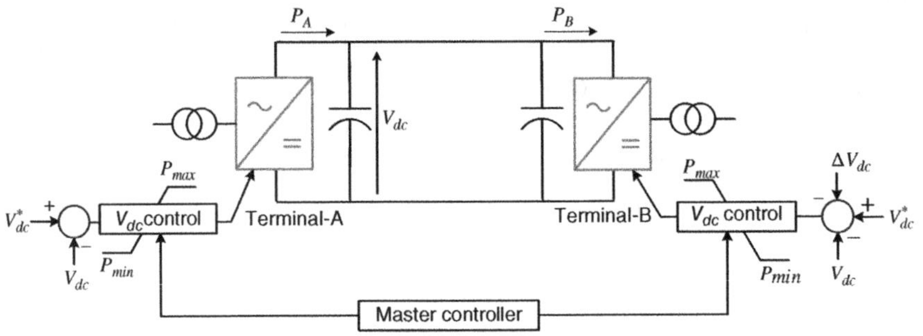

Figure 10.21 Schematic diagram of voltage margin control of two-terminal HVDC system.

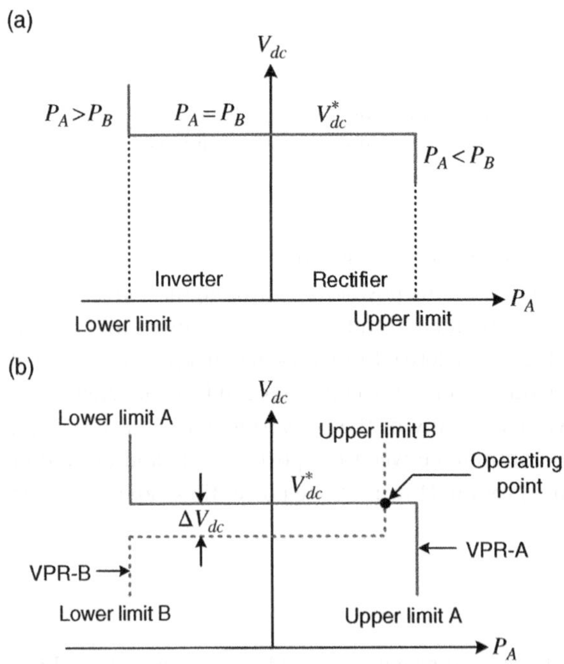

Figure 10.22 (a) V_{dc}-P characteristic at Terminal-A and (b) operating point in voltage margin method.

V_{dc} is lower than V_{dc}^*, the PI controller increases P_A until it becomes equal to P_B. On the other hand, if V_{dc} is higher than V_{dc}^*, the controller decreases P_A to make it equal to P_B. When $P_A = P_B$, V_{dc} is held constant at V_{dc}^*. Similarly, when P_B exceeds the upper limit, V_{dc} will decrease, and when P_B falls below the lower limit, V_{dc} will increase.

The problem with the scheme shown in Figure 10.22 (a) is that the DC voltage controllers can interfere with each other. To avoid this, a voltage margin ΔV_{dc} is introduced, which is the difference between the DC reference voltage at the two terminals. At Terminal-B, ΔV_{dc} is subtracted from V_{dc}^*, as shown in Figure 10.21. The voltage margin characteristic is shown in Figure 10.22 (b), where VPR-A and VPR-B are V_{dc}-P characteristics at Terminal-A and Terminal-B, respectively. The operating point is located at the intersection of these two characteristics. In this method, Terminal-A controls the DC voltage, and Terminal-B controls the power flow.

For a multiterminal system, a two-stage DC voltage control has been proposed in [16]. The same principle as discussed above, that is, one terminal controls the DC voltage, while the other terminals control the power flow, is applied here as well. The schematic diagram of the control scheme is shown in Figure 10.23 (a), which contains three controllers. In the voltage-determining terminal, the upper and lower limits are given by P_{max} and P_{min}, respectively. For the power flow terminals, the lower power limit is adjusted by V_{dc} Control-H, which has a lower limit P_{min}. The upper power limit for this block is set by the P control block.

Figure 10.23 (b) shows typical V_{dc}-P characteristics of a three-terminal system, in which the DC system voltage is determined by Terminal-C [16]. Terminal-B operated in rectifier mode with rated power. Terminal-A operates as an inverter. Terminal-C acts like a slack bus in AC power systems, where it supplies the rest of the power requirement and system losses.

Voltage Droop (VD) Method: In the VM method, only one terminal is used for DC voltage regulation at a time. This can put a lot of stress on the terminal. Moreover, when the transition from one voltage-regulating terminal to another occurs, the voltage change is abrupt, resulting in added stress in the system [17]. In the VD method, more than one terminal participates in DC voltage control simultaneously. The DC voltage can be determined by either DC voltage versus power droop or DC voltage versus current droop [7]. Assuming that the DC voltages of all the buses are equal, the V-P droop characteristics for a two-terminal system are shown in Figure 10.24 (a). This is, however, not a valid assumption in reality. There may be voltage drops due to the line flows that the droop controller must consider.

An interesting example of a four-terminal system is given in [17], in which two offshore wind farms are connected to an offshore platform load for oil or gas and

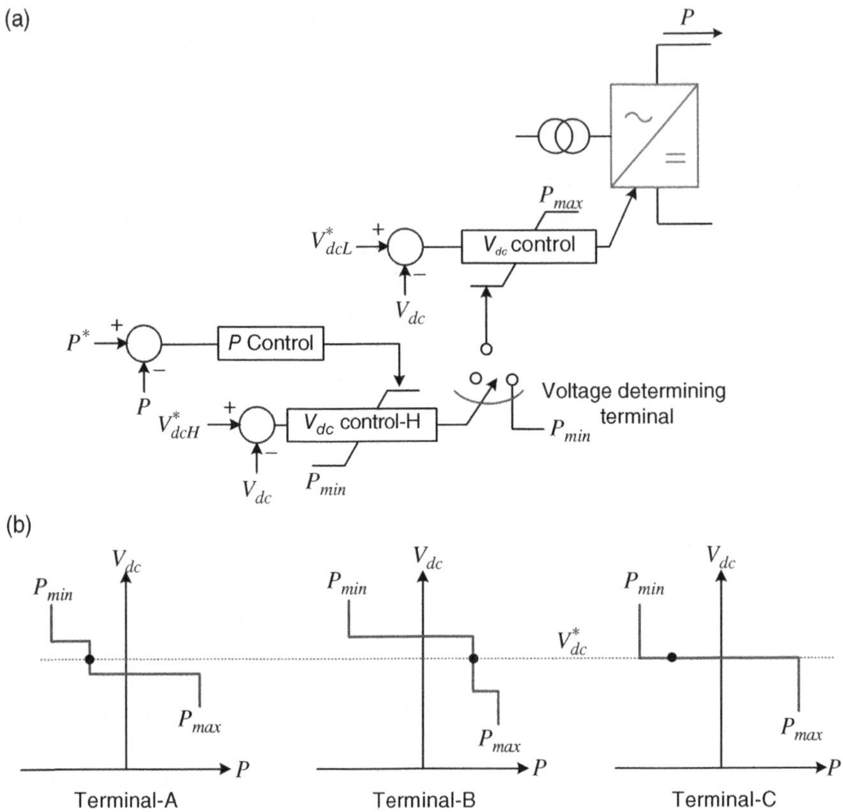

Figure 10.23 (a) Two-stage voltage control system and (b) V_{dc}-P characteristics of a three-terminal system.

an onshore power grid. The schematic diagram of the droop-sharing scheme is shown in Figure 10.24 (b). In this, the following are defined:

- Windfarm-1: Maximum power = P^*_{W1}, maximum voltage = V^*_{W1}.
- Windfarm-2: Maximum power = P^*_{W2}, maximum voltage = V^*_{W2}.
- Grid: Maximum power supply capability = P_G^{max}, maximum power absorbing capability = P_G^{min}.

Suppose the load demand from the oil and gas platform is P_L. The wind farms supply P_{W1} and P_{W2} amount of power. The rest of the power generated by the windfarms P_G is fed to the grid. The droop control of a DC microgrid has been discussed in [7]. The same principle can also be applied here.

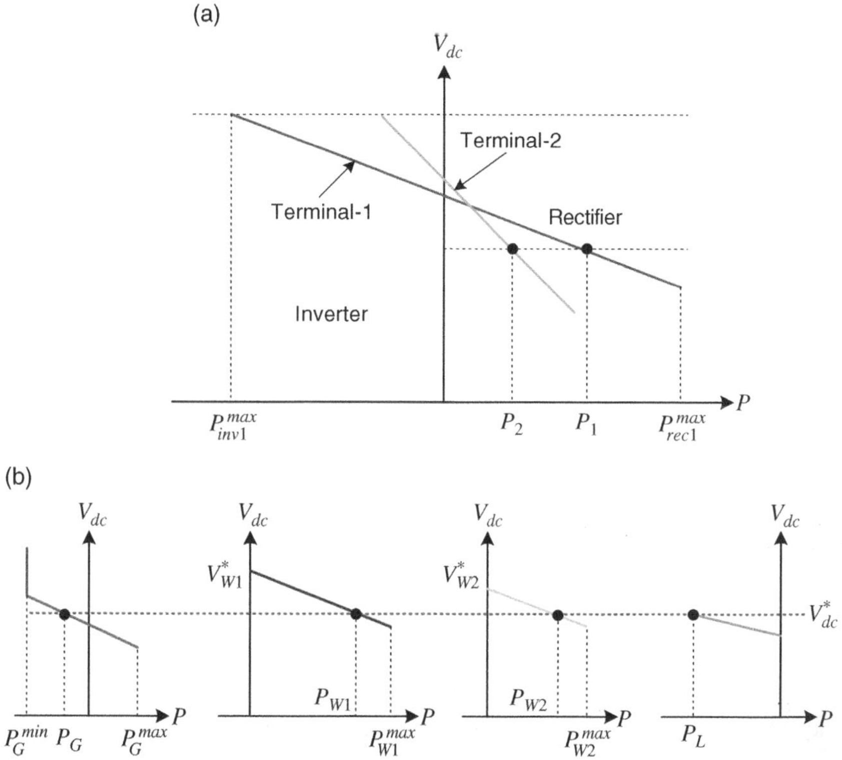

Figure 10.24 (a) V-P droop lines for a two-terminal system and (b) droop control of a four-terminal system.

10.5 DC Protection Systems

Alternating current CBs take advantage of the periodic zero-crossing of the current waveforms for breaking any faulted circuit. If the breaker opens at the current zero-crossing, the resulting arc may not be significant during the interruption and can be extinguished effectively. However, the current is continuous in a DC circuit. Therefore, breaking the circuit during faults will result in persistent arcs that may be difficult to extinguish. Careful consideration must be given to the design of DCCBs.

The basic scheme of the DCCB is shown in Figure 10.25. It contains three paths – nominal path, commutation path, and energy absorption path [2]. The nominal path has the switch S_1. The current through this path is I_n. The commutating path has a switch S_2 that is connected in series with an inductor L_c and capacitor C_c. The

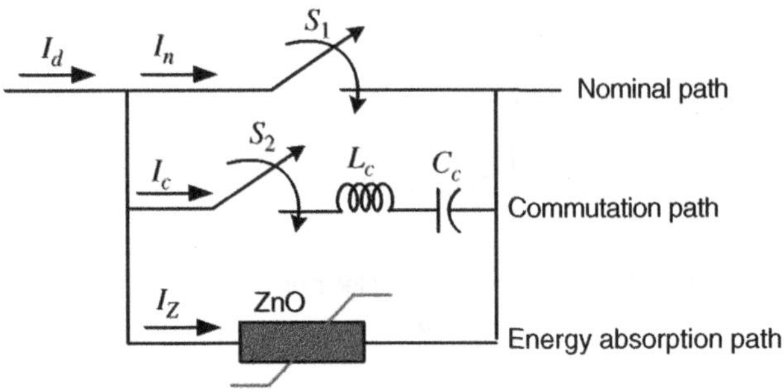

Figure 10.25 Basic configuration of a DCCB.

current through this path is I_c. The energy absorption path contains a zinc oxide varistor, which is used to protect the circuit against voltage surges. During normal operation, the varistor has a very high resistance, and negligible current I_Z flows through it. When a voltage surge occurs, the resistance of the varistor decreases rapidly, allowing it to conduct large currents. The excess energy is then shunted away from the protected circuit, clamping the voltage to a safer level.

The operation of this circuit has been illustrated in [2]. The commutation path has a series-connected inductor and capacitor. Therefore, when the switch S_2 is closed, the current I_c oscillates at the resonant frequency, given by $\sqrt{1/L_cC_c}$ rad/s. When the amplitude of this oscillating current is higher than the DC current I_c, a zero-crossing will occur in the nominal path. The current can be interrupted by the opening of the switch S_1. However, the current I_c will continue to flow, thereby charging the capacitor. When the capacitor voltage crosses a certain threshold, the energy absorption will become active, where the varistor will clamp the voltage and its resistance will decrease, allowing the current to flow in this path. The current flow will then discharge the capacitor. When this current becomes zero, the current I_d will become zero.

10.6 Concluding Remarks

High-voltage DC transmission is becoming an essential component of power grids all over the world, especially due to the interconnections of offshore wind farms in massive supergrids. In this chapter, two different types of HVDC transmission technologies are discussed. For standard point-to-point transmission, the LCC type is still very suitable. Moreover, multiterminal systems using this technology are also operating in several places. An excellent reference for this technology is the book by Padiyar [5].

VSC-based HVDC has been gaining popularity currently due to the integration of offshore wind farms. Usually, they use MMCs, which have high-voltage and high-current capabilities. The control of such converters is more involved. In this chapter, only the control philosophy has been discussed, and actual converter control will depend on the type of converter and the output filters chosen.

MTDC systems are evolving these days. The modeling, analysis, and control of such systems is reported in [18]. One of the critical aspects of a multiterminal DC system is the load flow of AC–DC combined systems. This topic is covered in detail in [9, 19]. Finally, the HVDC CBs and protection systems are very crucial for the successful development of multiterminal systems. An excellent review of this aspect can be found in [20].

References

1 Romero Engineering Company, *Introduction to HVDC Protection*. Available. https://www.romeroengineering.co/blog/introduction-to-hvdc-protection.

2 C. M. Franck, "HVDC circuit breakers: a review identifying future research needs," *IEEE Transactions on Power Delivery*, Vol. 26, No. 2, pp. 998–1007, 2011.

3 M. P. Bahrman and B. K. Johnson, "The ABCs of HVDC transmission technologies," *.369*, pp. 32–44 March/April 2007.

4 P. S. Kundur, *Power System Stability and Control*, McGraw-Hill, 1994.

5 K. R. Padiyar, *HVDC Power Transmission Systems*, New Age International, New Delhi, 2016.

6 M. Callavik, P. Lundberg, et al., "Evolution of HVDC Light," *ABB Review*, pp. 60–67 January 2018.

7 A. Ghosh and F. Zare, *Control of Power Electronic Converters with Microgrid Applications*, IEEE Press-Wiley, Hoboken, New Jersey, 2023.

8 R. Majumder, A. Ghosh, G. Ledwich and F. Zare, "Power management and power flow control with back-to-back converters in a utility connected microgrid," *IEEE Transactions on Power Systems*, Vol. 25, No. 2, pp. 821–834, 2010.

9 M. Eremia, C. C. Liu and A. A. Edris, *Advanced Solutions in Power Systems: HVDC, FACTS and Artificial Intelligence*, IEEE Press-Wiley, Hoboken, New Jersey, 2016.

10 V. K. Sood and H. Patel, "Comparison between direct and vector control strategy for VSC-HVDC system in EMTP-RV,"*Joint International Conference on Power Electronics,* Drives and Energy Systems & Power India, New Delhi, 2010.

11 H. Patel and V. K. Sood, "Modeling of voltage source converter based HVDC system in EMTP-RV," *IEEE Electrical Power and Energy Conference*, Halifax, Canada, 2010.

12 E. Kontos, *Control and Protection of VSC-based Multi-terminal DC Networks*, M.Sc. Thesis, TU Delft, 2013.

13 Hitachi Energy, *North East Agra: The World's first Multi-Terminal HVDC Link*, Uno Lamm HVDC Center, Sweden, 2022.

14 S. Barman, A. R. Paul, and H. G. Bahirat, "*Control strategy for ± 800 kV 6000 MWNER-Agra multi-terminal HVDC project*," National Power Systems Conference (NPSC), Tiruchirappalli, 2018.

15 M. Barnes, D. Van Hertem, S. P. Teeuwsen, and M. Callavik, "HVDC systems in smart grids," Proceedings of IEEE, Vol. 105, No. 11, pp. 2082–2098, 2017.

16 T. Nakajima and S. Irokawa, "*A control system for HVDC transmission by voltage sourced converters*," IEEE Power Engineering Society Summer Meeting, Edmonton, 1999.

17 T. Haileselassie, K. Uhlen and T. Undeland, "Control of multiterminal HVDC transmission for offshore wind energy," *Nordic Power Conference*, Bornholm, Denmark, September 2011.

18 N. Ray Chaudhuri, B. Chaudhari, R. Majumder, A. Yazdani, *Multi-terminal Direct-Current Grids: Modeling, Analysis and Control*, IEEE Press-Wiley, Hoboken, New Jersey, 2014.

19 S. Khan and S. Bhowmick, *Power Flow Modeling of HVDC Transmission Systems*, CRC Press, Boca Raton, Florida, 2023.

20 M. Barnes, D. S. Vilchis-Rodriguez, X. Pei, et al., "HVDC circuit breakers: a review," *IEEE Access*, Vol. 8, pp. 211829–211848, 2020.

Problems

P10.1 A 220 kV, 50 Hz transmission line is connected to a line-commutated converter.

(a) Neglecting the overlap, find the firing angle α when the output DC voltage is 260 kV.

(b) If the converter transfers 260 MW of power, find the fundamental frequency line current.

P10.2 The converter of Problem 10.1 is connected through a rectifier transformer. Find the firing angle α when the output DC voltage is 250 kV, and the converter supplies 1 kA of current. The commutating reactance (X_t) is given as 6 Ω.

P10.3 A 400 kV, 50 Hz transmission line is connected to an HVDC system. The rectifier side delay angle $\alpha = 20°$ and the commutation angle $\mu = 10°$. The line is supplying 500 A of current to the rectifier. Neglecting the converter losses, determine the power supplied to the rectifier and the output voltage of the rectifier.

P10.4 Consider an LCC inverter with the following parameters:

DC voltage (V_{do}) = 500 kV
Extinction angle (γ) = 20°
Overlap angle (μ_i) = 12°

Calculate the voltage drop ΔV and the DC voltage on the inverter side.

P10.5 Consider the two-terminal VSC-HVDC system shown in Figure 10.13. With respect to this figure, the following parameters are chosen:

$$V_{dc1} = 800 \text{ kV}$$

$$V_{dc1} = 780 \text{ kV}$$

$$V_{C1} = V_{C2} = 400 \text{ kV (L} - \text{L)}$$

Line resistance (R_d) = 10 Ω

Find

(a) The power supplied from the sending end and the power received from the receiving end.
(b) Assuming SPWM is applied in the open loop to both the converters, find the modulation indices m_{a1} and m_{a2}.

P10.6 In Problem 10.5, let us assume that the line-to-neutral grid voltages are given by

$$V_{G1} = V_{G1} = \frac{220}{\sqrt{3}} \angle 0 \text{o}$$

The line-to-neutral converter side voltages are given by

$$V_{C1} = \frac{400}{\sqrt{3}} \angle \delta_1, \quad V_{C2} = \frac{400}{\sqrt{3}} \angle \delta_2$$

The grid side and the converter side are connected through two 220/400 kV transformers with a 10% leakage inductance for a 2,000 MVA base (these are the inductances X_1 and X_2). Then, find δ_1 and δ_2.

P10.7 In Problem 10.6, find the per unit the reactive power flow from the converters to the grid for a base of 400 kV and 2,000 MVA.

11

Renewable Energy

According to the Intergovernmental Panel on Climate Change (IPCC) report of 2022 [1], "human-induced climate change, including more frequent and intense extreme events, has caused widespread adverse impacts and related losses and damages to nature and people, beyond natural climate variability." It is stated that the rise in temperature has caused disruptions to human, animal, and plant lives and the environmental damage is irreversible unless immediate actions are taken.

Earth's temperature has risen by about 0.08 °C per decade since 1880, that is, about 1 °C since then. However, the rate of warming since 1981 is about 0.18 °C per decade. The IPCC report states that by 2040, a temperature rise of 1.5 °C may be disastrous. About 3.8 billion people are vulnerable to climate change and so is the ecosystem. Therefore, the consensus is emerging among almost all the nations in the world to limit global warming to below 1.5 °C. The IPCC report is very extensive – it covers almost all climate change-related facts.

The gases that trap heat in the atmosphere are called greenhouse gases. The main component (about 80%) of greenhouse gases is carbon dioxide (CO_2). The main cause of CO_2 emission is the burning of fossil fuels, for example, coal, petroleum products, and natural gas. Both power generation and transportation sectors, together with the industrial sectors, are responsible for emitting nearly 75% CO_2 into the atmosphere. At the same time, large-scale deforestation hinders the biological carbon cycle, in which CO_2 is absorbed by plants. Methane (CH_4) and nitrous oxide (N_2O) produce about 18% of greenhouse gases. During the production and transport of coal, natural gas, and oil, methane is emitted. In addition, livestock, and the decay of organic waste in landfills also contribute to methane emission. Nitrous oxide is also produced through the combustion of fossil fuels and other industrial activities.

Several climate change mitigation pathways have been mentioned in [1]. Because the power sector is one of the largest emitters, the use of renewable energy has been the main focus of development in the last two decades. This has picked up

Modern Power System, First Edition. Arindam Ghosh.

Published 2026 by John Wiley & Sons, Inc.
Companion website: www.wiley.com/go/ghoshpowersystem

momentum and it is expected that several renewable technologies will mature in the next decade or two such that they can be used efficiently. The goal is to have a 100% green power grid by 2050, where renewable generation and storge will not only provide the baseload but also the intermediate and peak loads.

In this chapter, the different renewable energy generation technologies will be discussed first. Specifically, hydropower, solar, and wind, along with hydrogen and other new technologies will be covered briefly. This will be followed by a discussion of the power electronic interface that can be used in some of the renewable energy generators and the issues that are faced in the integration of renewable energy in power transmission and distribution systems.

11.1 Waterpower

Water is an inorganic compound that is made up of hydrogen and oxygen and can be found in solid, liquid, and gaseous states. It is the most essential component of life forms on the planet earth, be it humans, animals, and plants. Water covers about 71% of the earth's surface – mostly in the oceans and seas. Water moves continuously through a *water cycle*, in which the liquid form of water evaporates into water vapor. It then condenses to form clouds and then precipitates back to earth in the form of rain and snow. Due to its abundance and cycle, water is an obvious choice for renewable energy. There are various forms of waterpower such as hydropower, tidal power, and wave power. In this section, these are discussed briefly.

11.1.1 Hydropower

This is the most common form of hydropower or hydroelectricity, which accounts for about 17% of electricity generation in the world. In fact, it is the third largest form of power generation after coal and natural gas [2]. The largest hydroelectric power generating facility in the world is the *Three Gorges Dam* on the Yangtze River in Yichang, Hubei, China. It has a generating capacity of 22.5 GW. *Itaipu Dam* on the Parana River on the border between Brazil and Paraguay produces 14 GW of power. These two dams produce over 100 TWh of electricity every year.

Conventional hydropower is the production of electrical energy through the gravitational force of falling water. When water falls from a height, its potential energy is converted into kinetic energy (KE) by placing turbines in the path of water flow. The different components of hydropower are as follows:

- An upstream reservoir for collecting water.
- A dam controls the flow of water and prevents the water stored in the reservoir from flowing downstream in an uncontrolled manner.

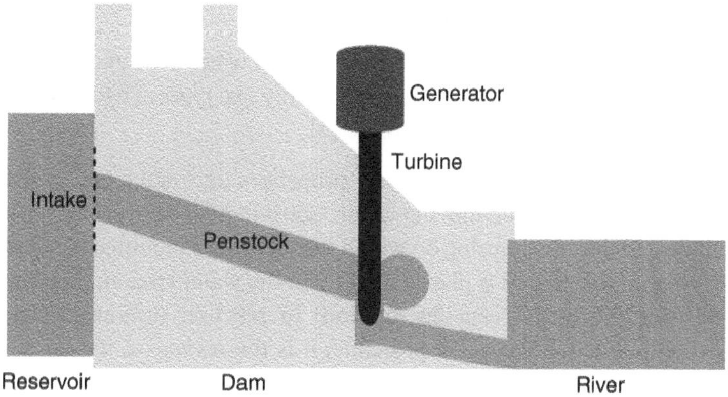

Figure 11.1 A typical hydroelectricity generation facility.

- Penstocks for controlling the flow of water. These are sluices or sliding gates that are placed in the path of the water flow through an enclosed path.
- Water turbines and generators for producing electricity.

A typical hydropower-generating facility is shown in Figure 11.1.

The amount of power that can be produced from hydropower is given by the following equation.

$$P = \eta\rho Qgh \text{ Watts} \tag{11.1}$$

where η is the efficiency of the turbine, ρ is the density of water in kg/m^3, Q is the flow in m^3/s, g is the acceleration due to gravity in m/s^2, and h is the height of falling water, that is, between the inlet and outlet, which is also often termed as water head.

Example 11.1 Consider a hydropower-generating plant with 5 turbine-generators, each having the following parameters:

$$\eta = 90\%,\ \rho = 998\ \text{kg/m}^3,\ Q = 100\ \text{m}^3\text{/s},\ g = 9.8\ \text{m/s}^2,\ \text{and}\ h = 150\ \text{m}$$

Then the output power from each generator is

$$P = \frac{0.9 \times 998 \times 100 \times 9.8 \times 150}{10^6} = 132\ \text{MW}$$

Therefore, the total output power of the plant is $132 \times 5 = 660$ MW.

11.1.2 Types of Hydropower Turbines

Hydropower turbines are chosen based on factors such as water head and volume of water flowing. Based on these factors, there are mainly two types of hydropower turbines – reaction turbine and impulse turbine [3].

Reaction Turbine: Usually, these turbines use propellors and make use of the combined forces of pressure and moving water. Consider it to be a turbine with propeller blades. In this, the housing on which the blades are mounted is called the runner. Now, if the runner is placed directly in the water stream, then the water flows over the blades causing rotation. In reaction turbines, water does not strike individual blades. In principle, it is the reverse of the screws (propellers) in a ship, where they are mechanically rotated for forward movement. In propeller turbines, the propellers are placed in pipes to maintain constant pressure. There are different types of reaction turbines, for example, bulb turbines, Straflo turbines, and Kaplan turbines. The different reaction turbines are discussed in [3]. The reaction turbines are generally used at sites with lower water heads, but higher flow of water, like the run of the river hydropower. A typical Kaplan turbine is shown in Figure 11.2. The usual efficiency of Kaplan turbines is up to 90% because it can regulate both water flow and the angle of the blades.

Impulse Turbine: An impulse turbine generally uses the velocity of the water to move the runner at atmospheric pressure. The water stream hits each bucket on the runner. There is no suction on the downside of the turbine. The water flows out the bottom of the turbine housing after hitting the runner. There are two types of impulsive turbines – the Pelton wheel and the crossflow turbine. In a Pelton wheel (see Figure 11.3), one or more water jets discharge water on the buckets. It is generally suitable for high-head, low-flow applications. Crossflow turbines allow water to flow through the blades twice (see [3] for details). They can be used for large water flow, but at lower heights than the Pelton wheel.

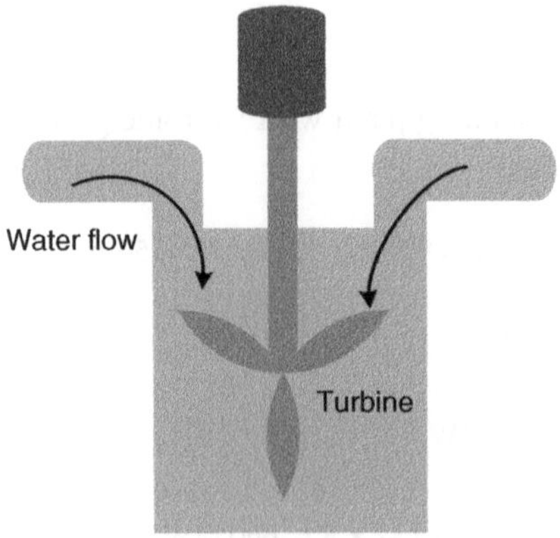

Figure 11.2 A typical Kaplan turbine.

11.1.3 Pumped Hydro Storage (PHS)

Pumped hydro is basically a hydropower storage system. In this, there are two reservoirs – one at a higher elevation and the other at a lower

Figure 11.3 Impulse turbines – Pelton wheel.

elevation. It has two stages – discharge and recharge. During the peak hours, when the demand for electricity is high, the water released from the higher reservoir is discharged to the lower reservoir. The water passes through a turbine to generate electricity in the same manner as hydroelectricity generation. On the other hand, when the demand for electricity is low and/or generation from renewable sources is high, the water is pumped back to the higher reservoir from the lower reservoir in the recharge phase. Usually, the same equipment is used for generation and pumping. During recharge, the synchronous generator works as a motor and the turbine works as a pump. However, this will require a change in the rotational direction of the rotation between the discharging and charging modes. In cases where the water head is high, there may be a requirement for a separate turbine and pump.

A ternary pumped hydro storage (TPHS) unit is an advanced pumped storage system that consists of a separate turbine and a pump on a single shaft [4]. An electric machine (synchronous) is placed on the shaft that can operate as either a generator or a motor. The advantage of a TPHS is that it can simultaneously operate in both discharge and recharge modes. A typical TPHS is shown in Figure 11.4. It consists of a turbine (T), a pump (P), and a clutch. The function of the clutch is to disconnect the pump during turbine operation.

Both the turbine and pump are equipped with *guide vanes*, which are fixed grooves on the turbines to direct water flow. By closing the guide vanes, the surface area around the device is reduced, thereby increasing efficiency. When the TPHS is operating in the turbine (discharge) mode, the clutch is disengaged, and the guide vanes of the pump are closed. During the pumping mode, the clutch is engaged, and the turbine guide vanes are closed. In the ternary mode, the guide vanes for both the turbine and the pump are open, and the clutch is engaged. In this mode, both the turbine and pump function simultaneously, allowing the system to regulate the power level while performing a net pumping action. This is usually called a *hydraulic short circuit*.

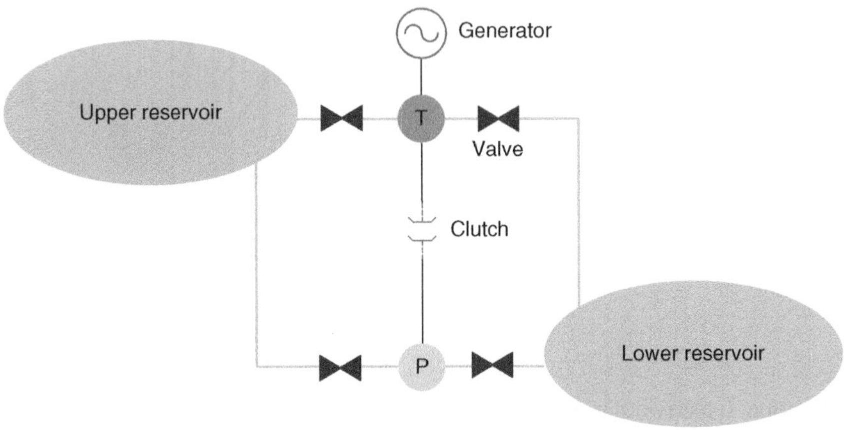

Figure 11.4 Schematic diagram of a typical ternary pumped hydro storage system.

An example of a ternary operation is given in [4]. This is explained below.

- The turbine and the pump are mounted on the same shaft, and therefore, while the pump is operational, the generator will also be operational in ternary mode.
- Now, consider the case when the flow through the pump causes an electrical energy of 150 MW to be delivered to the pump.
- The guide vanes of the turbine are so adjusted that the water flowing through the turbine results in a generation of 100 MW by the generator.
- Therefore, the power consumption from the grid is only 50 MW.
- However, the flow to the upper reservoir is controlled.

11.1.4 Tidal Energy

The two celestial bodies in the neighborhood of our planet (sun and moon) have maximum effect on our ocean and seawater through their gravitational attractions. Because the water moves continuously, the earth cannot hold it firmly against the gravitational pull of these two celestial objects. That is why tides regularly occur twice in a period of about 25 hours. During the high tide, the sea level rises, while it falls during the low tides. These movements of water alter the potential energy, which can be harnessed to produce electricity.

Two consecutive high tides occur 12 hours and 25 minutes apart. Again, the switchover from high tide to low tide takes about 6 hours and 12.5 minutes. Therefore, in a period of 24 hours and 50 minutes, there are two high and two low tides. The height of the tides depends on the position of the moon vis-à-vis that of the sun. The motion of the moon when it orbits around the earth is shown in

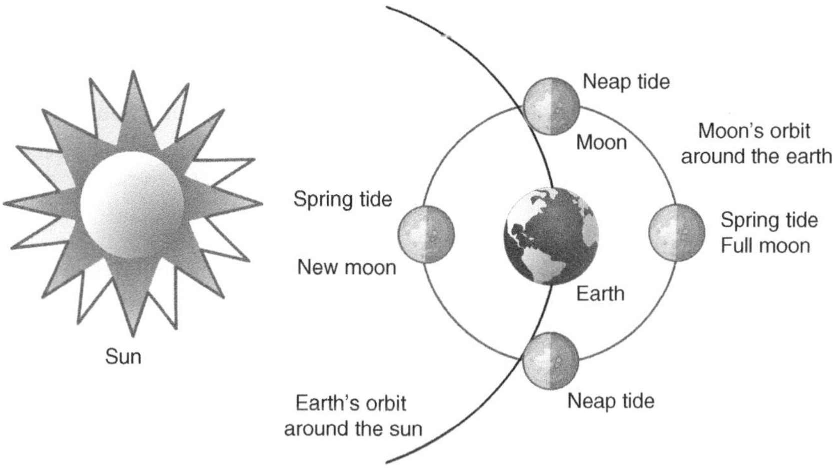

Figure 11.5 Motion of the moon during its orbit around the earth (not to the scale).

Figure 11.5. Moon completes one rotation around the earth in about 28 days (lunar month). When the moon is between the sun and the earth, it is called the new (dark) moon, while when the earth is between the sun and the moon, it is called the full moon. The *spring tides* occur during these two days, when the sun, earth and, moon are almost in a line, which occurs twice every lunar month. During these days, the gravitational pull of the sun is added to the gravitational pull of the moon. Therefore, the high tides are higher during the spring tides. Also, the low tides are lower during the spring tides. *Neap tides* occur seven days after the spring tides when the sun and the moon are almost at right angles to each other. During this time, the gravitational attractions of the sun and the moon are partially canceled. The high tides are lower, and low tides are higher during these days. The height of the tides varies between the spring and neap tides depending on the relative positions of the sun and the moon with respect to the earth. It is important to note that the tides are much more predictable than wind or solar energy. However, harnessing this energy is a challenge.

Tidal power is generated by converting the potential energy of rising and falling tides to kinetic energy, which is then converted into electricity through turbine-generators. To capture tidal energy, a dam is created to enclose a part of the sea. This dam is called a *barrage*, which also houses turbines. During high tides, water enters the enclosed reservoir and passes through turbines to produce electric energy. The process continues till the water inside the reservoir is almost as high as the seawater outside the dam. At low tides, the process is reversed. The water inside the reservoir exits into the sea through the turbines to produce electricity.

However, there is a time gap of roughly two hours between the two processes. The water level must be at a minimum level; otherwise, the energy dissipated through the turbines cannot be compensated. The largest tidal power plant is the Sihwa Lake Tidal Power Station, which is located in Gyeonggi Province, South Korea. It was installed in 2011 with a maximum capacity of 254 MW.

Several aspects of tidal power need mentioning. Some of these are mentioned below.

- The power generation of flowing water is variable depending on the state of the water level in the reservoir. Furthermore, there is a dead time between the high and low tides. Therefore, tidal power output is not constant and can only be used to supplement power generation.
- Tides are predictable. Depending on the positions of the moon and the sun, the amount of water flow can be estimated accurately. Therefore, the amount of power that can be generated can also be estimated accurately.
- The amount of power generated will depend on the height of the tidal waves. From economic considerations, the tides need an altitude of 3 m from the base level. Therefore, the regions that have huge tidal waves are suitable for tidal power.

Tidal energy harvesting is still not a mature technology. Based on economic considerations, the amount of power produced does not justify power generation through this technology yet. The turbines and the penstocks are to be hardened for use in highly corrosive seawater. Moreover, there are concerns about disruptions of marine life and land ownership. However, this is a predictable, unlimited source of energy that will be too beneficial to be ignored. Therefore, there is a concerted effort to improve this technology so that it can be made economically viable and environmentally justifiable.

11.1.5 Wave Energy

Surfing is a water sport that has a massive following. A surfer uses a surfboard to move toward the shore on moving waves of water. The waves are caused by wind blowing on the ocean's surface. Waves can travel in different directions and at different speeds depending on the wind direction and speed. The potential energy of the waves can be captured to generate electricity.

There are several advantages of wave energy that have been listed in [5]. Some of these are:

- Waves have higher energy density (2–3 kW/m^2) than wind or solar.
- Wave can travel long distances without much energy dissipation.
- Wave is more predictable.

- Because about 40% of the world's population live within 100 km of the coast, wave energy need not be transmitted over long distances.

However, there are some disadvantages that have been listed in [5], for example:

- The frequency of oscillations is very low (of the order of 0.1 Hz). Therefore, an energy conversion stage will be required.
- Both the amplitude and period of the waves vary. Therefore, the output power level will vary.
- The directions of the wave vary. Therefore, the power-capturing device will have to align itself to get the maximum energy output.
- Finally, the obvious problem seawater corrosion problem will exist, which will require careful design of the components.

There are several methods for capturing wave energy for electricity generation [6]. Some of these are given below.

- *Point Absorber*: It consists of a floating structure that uses the vertical motion of waves to generate electricity. These devices typically consist of a floating buoy attached to the seabed. As the buoy moves up and down with the waves, it drives a mechanical system or hydraulic pump that converts the motion into energy. The electricity is generated when the potential energy of the waves is converted to kinetic energy, which is then used in generators. The construction is not simple as there are several factors that need to be considered, as discussed in [5, 6].
- *Oscillating Water Column (OWC)*: This is only suitable for coastal regions, specifically those with rocky cliffs that have natural caves where the waves are relatively higher. These have the advantages of easier installation and maintenance. Moreover, the dispatching of the captured power need not be evacuated through subsea cables. A typical OWC is shown in Figure 11.6. As shown in the figure, this will require a partially submerged rocky cliff or concrete structure. Incident waves will

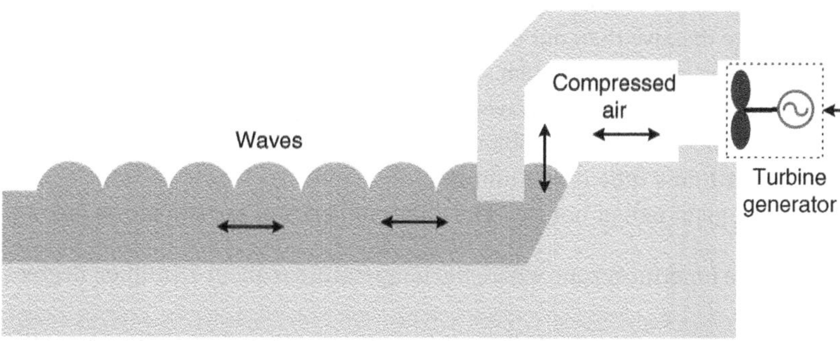

Figure 11.6 Oscillating water column wave energy conversion systems.

compress air inside the structure. The air will flow out of the structure through a wind turbine. When the water recedes, the air will come inside through the turbine. To capture energy both during air compression and decompression, a Wells turbine is used. This is a low-pressure turbine that rotates in the same direction independent of the direction of air flow. Because a Wells turbine rotates in the same direction for both these actions, the generator connected to the turbine need not change direction. The OWC is basically a wind-based power generation that uses moving waves.

There are several other possible wave energy conversion systems, such as attenuators, oscillating wave surge converters, and submerged pressure differential. The different types of generators used, including linear generators and subsea electricity transmission are discussed in [5].

11.2 Solar Power

A galaxy is a gigantic collection of stars and their planets, dust, gas, and backholes that are held together by gravity. It is estimated that there are about 2 trillion galaxies in the universe, with each of them having about 100 billion stars. However, it is to be remembered that these are estimates and our knowledge about the universe is still limited. James Webb Space Telescope (JWST), which is placed at Lagrange point L_2 at about 1.5 million kilometers from the earth, is going to enhance our knowledge about the history of the universe and search for other solar systems that are capable of sustaining life.

Our star, the sun, is one of the stars in the Milky Way galaxy. The sun is unique in the sense that it is the only star in our solar system. Many other solar systems have two or more stars. The sun orbits the core of the Milky Way galaxy once in 250 million years. It is a medium-sized star with a diameter of 1,392,700 km. However, it contains about 99.86% of the total mass of our entire solar system. In fact, it is 330,000 more massive than our earth. The gravity of the sun holds the solar system together. The length of our solar system is about 1.58 light years. A light year is the measure of the length of light traveling in a year (1 light year = 9.461×10^{12} km). Because our planet is about 150 million kilometers away from the sun, the sunlight reaches the earth in a little over 8 minutes.

The sun is a giant ball of hot gas. The different layers of the sun are as follows:

- *Core*: It is the innermost zone where the temperature is about 15 million degrees Celsius.
- *Radiative*: This zone surrounds the core and covers about 70% of the way to the surface of the sun.

- *Convective*: This zone is above the radiative zone and transports energy through the convection of hot plasma.
- *Photosphere*: This is the visible surface of the sun from which most of the sun's light and heat are emitted. The temperature here is 5,500 °C.

The other two layers are the chromosphere and corona. Of these, corona extends to millions of kilometers and is extremely hot (1–2 million °C).

In the core of the sun, thermonuclear reactions take place in which hydrogen atoms, due to tremendous heat and pressure, fuse together to form helium atoms. The reaction is called proton-proton (p-p) chain reaction. Two protons collide to form deuterium releasing a positron and a neutrino. The deuterium nucleus fuses with another proton forming helium-3 nucleus and releasing gamma rays. When two helium-3 nuclei collide, they form a helium-4 nucleus. Through this p-p chain reaction, four hydrogen nuclei are turned into one helium nucleus. This process releases a tremendous amount of energy that is transferred through the solar system.

In this process, energy is released to keep the sun hot. The energy then radiates from the core of the sun and moves across the solar system. It is interesting to note that 99% of the energy is produced in the sun's core. The rest of the sun is heated by energy transferred outward from the core. It is estimated that our sun will run out of its fuel in another 5 billion years. It will then expand into a red giant incinerating everything in its vicinity including our planet. Following this, it will shade its outer layers and will turn into a white dwarf with a very dense core. It is reassuring to think that our sun will not turn into a black hole as it is not large enough.

11.2.1 Solar Tracking

Earth's orbit around the sun is shown in Figure 11.7. The orbit is slightly elliptical where the distance between the sun and the earth varies between 147 million kilometers and 152 million kilometers. The earth spins around an axis that is tilted about 23.45° along its polar axis. Earth makes one revolution around the sun in 365.25 days and rotates 360.99° in a period of 24 hours. To account for the slight imbalance in revolution and rotation, one day is added to the month of February every 4 years (this is not quite accurate though as there are other rules regarding the leap years). The tilt angle is very important for the seasons that we experience both in the northern and southern hemispheres. The labeling in Figure 11.7 is given from the northern hemisphere perspective, where summer and winter solstices occur on June 21 and December 21, respectively (again this is not quite accurate). Equinoxes are the days when the sun crosses earth's equator. These roughly occur on April 21 and September 21.

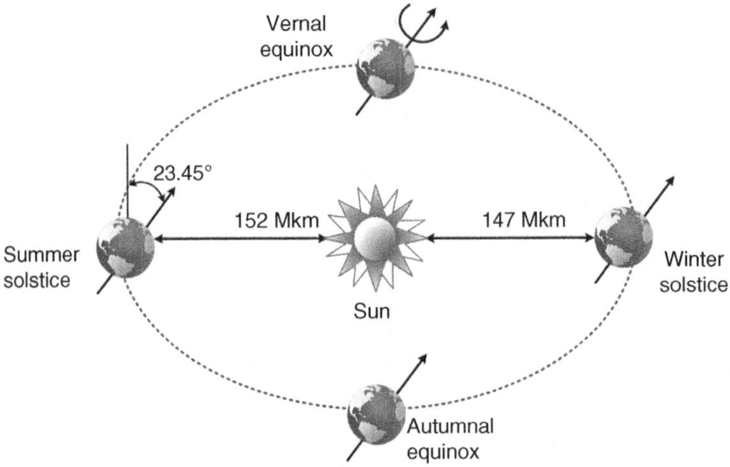

Figure 11.7 Earth's orbit around the sun.

The earth's orbit plays an important role in the calculation of solar irradiance at a particular time on a clear day. It is well known that the sun rises in the east and sets in the west. Furthermore, it reaches its highest point sometime in the middle of the day. But the position of the sun vis-à-vis that of a particular location on the earth is not constant due to the tilt angle of the earth's spin. Therefore, careful calculations are to be made to predict where in the sky the sun will be on a particular day of the year at a particular time. This is important for the orientation of the solar photovoltaic (PV) panel positioning. A particularly good source for solar resources and their calculations is the Book by Masters ([7], Chapter 7).

We may have seen rooftop-mounted PV panels that are placed on slanted roofs facing either east or west direction. It is even common to see panels separated and placed both east facing and west facing. Ideally, the PV panels are placed so that they remain perpendicular to the sun's rays. This also ensures that less sunlight is reflected, thereby capturing the maximum possible energy from the sunlight. Obviously, this is not possible with roof-mounted PV panels. However, for large solar installations, tracking of the sun's rays is performed by either single-axis or dual-axis trackers.

The schematic diagram of a single-axis tracker is shown in Figure 11.8 (a). It moves on from east to west on the north-south axis. A dual-axis tracker can move from north to south and from east to west, as shown in Figure 11.8 (b). These trackers can follow the sun vertically and horizontally, and therefore, can get the maximum possible irradiance. Obviously, both tracking systems will need electric motors to change their angles. Furthermore, they will need sensors and intelligent

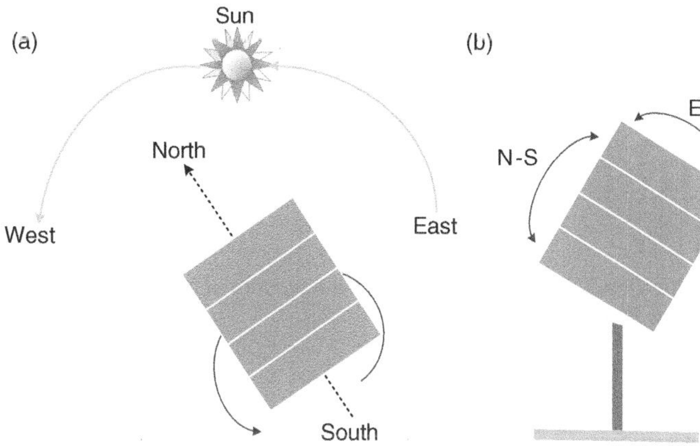

Figure 11.8 (a) Single-axis tracker and (b) dual-axis tracker.

algorithms to track the sun irrespective of the seasons and changes in the height of the sun with respect to the earth. The algorithms can be based on solar resource calculations, as given in [7]. Moreover, machine learning-based methods can also be used to improve the accuracy of the tracking systems.

11.2.2 Solar Photovoltaic (PV) Systems

Solar PVs, in their basic form, are called cells. A PV cell is a p-n junction semiconductor device that captures energy from photons and converts them to electricity. The most common PV cells are constructed from silicon. There are two types of crystalline PV cells available – monocrystalline and polycrystalline. As the name suggests, a monocrystalline has a uniform lattice that allows easy electron flow, whereas a polycrystalline cell has multiple crystalline structures that can impede the flow of electrons. The monocrystalline cells are more efficient than the polycrystalline cells, but they are more expensive than the polycrystalline cells. For the PV characteristics refer to [7].

The equivalent circuit of a diode is shown in Figure 11.9 (a), V_d is the voltage impressed across the diode and I_d is the current flowing through the diode, which is given by the following equation:

$$I_d = I_0\left(e^{qV_d/kT} - 1\right)\ \text{A} \tag{11.2}$$

where I_0 is the reverse saturation current in A, which is of the order of 10^{-12} A/cm^2; q is the electron charge, which is equal to 1.602×10^{-19} Coulomb; k is Boltzmann's constant, which is equal to 1.380649×10^{-23} Joules per Kelvin; and T is the temperature in Kelvin (0 °C = 273.15 K).

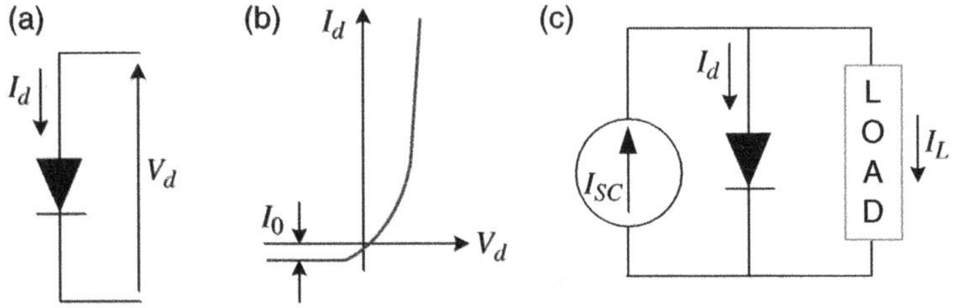

Figure 11.9 (a) A diode circuit, (b) voltage–current characteristics of a diode, and (c) equivalent circuit of a PV cell.

The voltage–current characteristics of a diode are shown in Figure 11.9 (b).

The equivalent circuit of a PV cell is shown in Figure 11.9 (c) [7]. In this, I_{SC} is the short circuit current of the PV cell and I_L is the output load current of the PV. From KCL, we have

$$I_L = I_{SC} - I_0$$

Substituting (11.2) in the above equation, we have

$$I_L = I_{SC} - I_0\left(e^{qV_d/kT} - 1\right) \tag{11.3}$$

To determine the open circuit voltage of the PV cell, we assume that the output current I_L is zero. Then, the following equation is written from (11.3) taking $V_{OC} = V_d$.

$$e^{qV_{OC}/kT} = \frac{I_{SC}}{I_0} + 1$$

$$\Rightarrow \frac{qV_{OC}}{kT} = \ln\left(\frac{I_{SC}}{I_0} + 1\right)$$

Therefore, the open circuit voltage is

$$V_{OC} = \frac{kT}{q}\ln\left(\frac{I_{SC}}{I_0} + 1\right) \tag{11.4}$$

In (11.3) and (11.4), the short circuit current I_{SC} depends on the solar insolation and therefore can vary throughout the day.

Example 11.2 Consider a PV system with the following parameters:

Cell size: 50 cm^2
Reverse saturation current (I_0): 1×10^{-12} A/cm

Short circuit current (I_{SC}): 50 mA/cm
Temperature: 20 °C

We define the following:

$$\beta = \frac{q}{kT} = \frac{1.602 \times 10^{-19}}{1.380649 \times 10^{-23} \times 293.15} = 39.5812 \text{ per V}$$

Then

$$V_{OC} = \frac{1}{\beta} \ln\left(\frac{50 \times 10^{-3}}{1 \times 10^{-12}} + 1\right) = 0.6224 \text{ V}$$

Now, for 50 cm^2 PV cell, we have the following:

$$I_0 = 1 \times 10^{-12} \times 50 = 5 \times 10^{-11} \text{ A}$$

$$I_{SC} = 5 \times 10^{-3} \times 50 = 2.5 \text{ A}$$

The PV output current I_L as the output voltage V_d varies from 0 to V_{OC} is shown in Figure 11.10. The output power is given by

$$P_L = V_d I_L = V_d I_{SC} - V_d I_0 \left(e^{qV_d/kT} - 1\right) \text{ W} \tag{11.5}$$

The relationship between the output power and V_d is also shown in Figure 11.10.

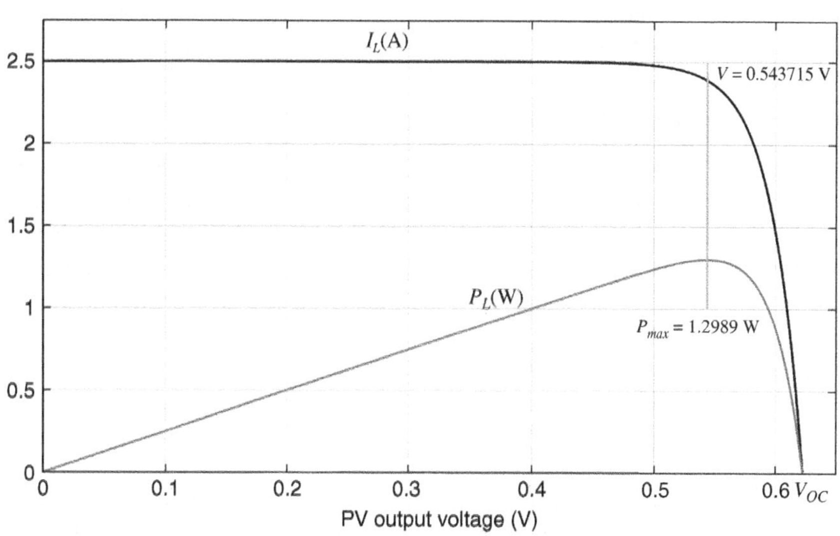

Figure 11.10 PV voltage–current and voltage–power characteristics.

Taking derivatives of both sides of (11.5) with respect to V_d, and equating it to zero, the maximum power is obtained as

$$\begin{aligned} \frac{\partial P_L}{\partial V_d} &= I_{SC} + I_0 - I_0\left(e^{\beta V_d} + \beta V_d e^{\beta V_d}\right) = 0 \\ \Rightarrow e^{\beta V_d} + \beta V_d e^{\beta V_d} &= \frac{I_{SC}}{I_0} + 1 \end{aligned} \tag{11.6}$$

Solving the above nonlinear equation, we get the following:

The maximum power (P_{max}) = 1.2989 W.
The maximum power occurs when $V_d = 0.543715$ V.

These are indicated by the straight line in Figure 11.10.

Usually, an individual PV cell output voltage is about 0.5 V. To boost this voltage, several cells are connected in series. For example, to have an output voltage of 12 V, 36 cells must be connected in series in a module. Again, assume that four such modules are connected in series. The output voltage is then 48 V. Moreover, to increase current, several modules must be connected in parallel as well. Assume that four solar modules are connected in series to produce and output voltage of 48 V. The output current of the combined module for a particular insolation is say 4 A. Therefore, the output power is 192 W. Now assume that five such combined modules are connected in parallel as well to form a solar array. Then, the output current of the array is 20 A, and the output power is 20×48 ($= 192 \times 5$) = 960 W. Therefore, a solar array will have modules connected in series to increase voltage and in parallel to increase current such that the overall output power is increased. Figure 11.11 shows a cell, a module, and an array (see [7] for more details).

11.2.3 Maximum Power Point Tracking (MPPT)

Consider the system of Example 11.2. The voltage–current curves for different values of I_{SC} (insolation) are shown in Figure 11.12. The maximum power points (MPPs) for different curves are indicated in the figure. It is obvious that they do not follow a linear relationship. However, it is desirable to obtain the maximum possible output power for any insolation. This is achieved through a maximum power point tracking (MPPT) system. It is to be noted that MPPT is not dependent on the single or dual axis tracking system, which is used for aligning the solar arrays in such a way that they are perpendicular to the sun's rays at any given time. These are achieved by servo motors, whereas the MPPT is achieved through power converter control based on some logic.

There are several MPPT algorithms that have been proposed (see [8–10]). The most popular method is the perturb and observe (P&O) method. This method is simple to implement. However, it is not very accurate. Equation (11.6) indicates that the derivative of power with respect to the voltage is zero at maximum power.

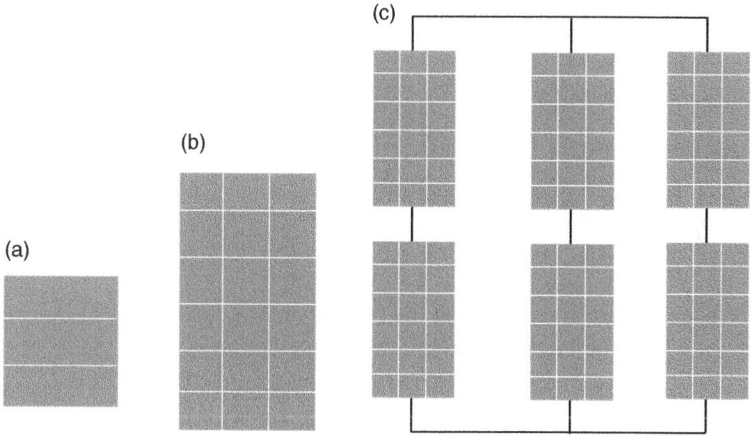

Figure 11.11 Solar PV (a) cell, (b) module, and (c) array.

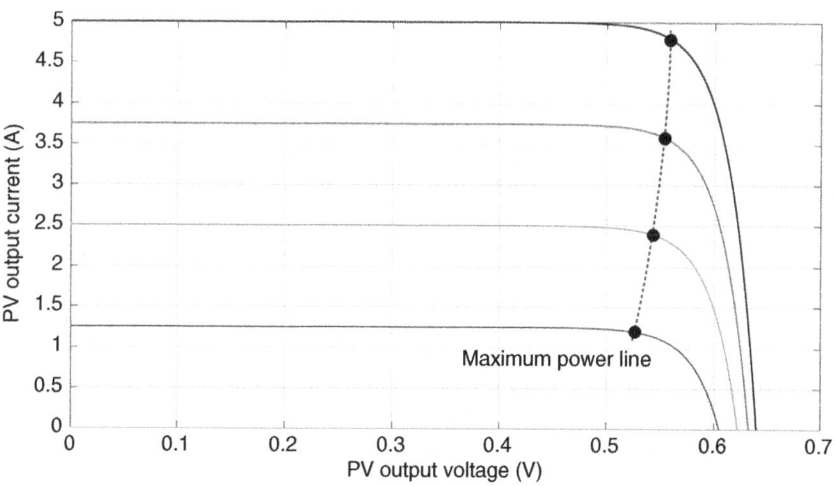

Figure 11.12 Maximum power line for different insolation.

In the P&O method, as the name suggests, the operating voltage is perturbed. If the output power increases, then the voltage is further perturbed in the same direction. On the other hand, if the output power decreases, the direction of voltage perturbation is reversed. This process continues till the MPP is reached. One of the drawbacks of this method is the chattering around the maximum power due to the perturbations.

Let us denote the following:

Output power at MPP: P_{LMP}
Voltage at MPP: V_{dMP}

Then, at MPP, we have

$$\left.\frac{\partial P_L}{\partial V_d}\right|_{\substack{P_L = P_{LMP} \\ V_d = V_{dMP}}} = 0 \tag{11.7}$$

Let us consider an arbitrary point on the I–V curve. The mathematical expression based on which the P&O decides the next perturbation is

$$\partial P_{PO} = \partial P_L / \partial V_d \tag{11.8}$$

The P&O method then decides the direction of the next perturbation based on the sign of ∂P_{PO}. The discrete form of (11.8) is written as [11]

$$\nabla P_{PO}(k) = \frac{P_L(k) - P_L(k-1)}{V_d(k) - V_d(k-1)} = \frac{\Delta P_L(k)}{\Delta V_d(k)} \tag{11.9}$$

where k is the sampling instant and ∇P_{PO} is the discrete-time equivalent of ∂P_{PO}.

Table 11.1 shows the relationship between the voltage perturbation and the change in power [8]. As shown in the table, if the signs of the voltage perturbation and the direction of the change in power are the same, the next perturbation will be positive. However, if their signs are opposite, the next perturbation needs to be negative. The accuracy of the algorithm depends on the frequency of perturbation. If a slower frequency is used, then the accuracy can be affected if there is a rapid irradiance change that occurs between two successive perturbations [11].

Consider for example a PV system with an initial voltage of 30 V and power of 200 W. We increase the voltage by 1 V such that the voltage is 31 V, and the power is measured as 210 W. Since $\nabla P_{PO} > 0$ and $\Delta P_L > 0$, we increase the voltage to 32 V. The output power is now 215 W. Since both power and voltage differences have increased, we increase the voltage again to 33 V. The new power is 212 W. This

Table 11.1 P&O correction directions.

Present perturbation	Change in power	Next perturbation
$\nabla P_{PO}(k) > 0$	$\Delta P_L(k) > 0$	$V_d(k+1) = V_d(k) + \Delta V_d(k)$
$\nabla P_{PO}(k) > 0$	$\Delta P_L(k) < 0$	$V_d(k+1) = V_d(k) - \Delta V_d(k)$
$\nabla P_{PO}(k) < 0$	$\Delta P_L(k) > 0$	$V_d(k+1) = V_d(k) - \Delta V_d(k)$
$\nabla P_{PO}(k) < 0$	$\Delta P_L(k) < 0$	$V_d(k+1) = V_d(k) + \Delta V_d(k)$

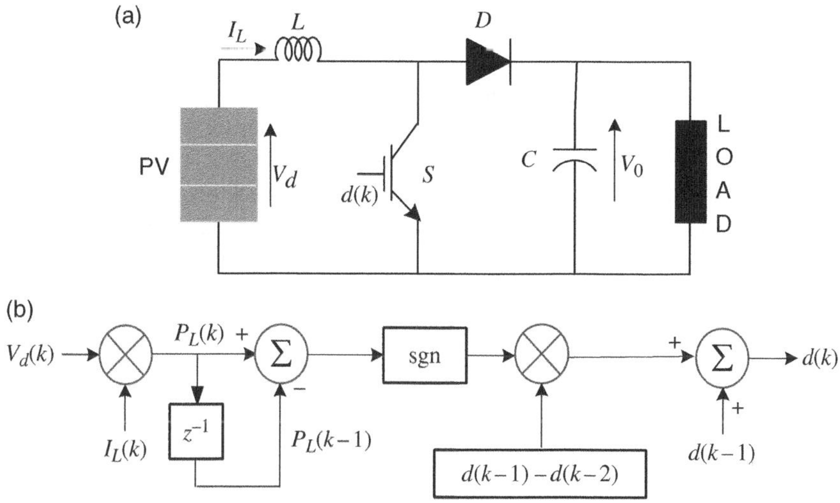

Figure 11.13 (a) DC–DC boost converter and (b) its MPPT implementation.

means that the power has decreased, and we have passed the MPP. We then decrease the voltage to 32 V where the power output is 215 W, which is the MPP.

Usually, the MPPT is implemented through a DC–DC converter. To increase the PV output voltage to a reasonable level, a boost converter is used, the schematic diagram of which is shown in Figure 11.13 (a). For the PV output voltage of V_d, the boost converter output voltage of V_0 is given by the following equation.

$$\frac{V_0}{V_d} = \frac{1}{1-d} \tag{11.10}$$

where $0 < d < 1$ is the duty ratio of the converter [12]. In the P&O method, the duty ratio at k^{th} instant is decided based on the sign of the difference between $P_L(k)$ and $P_L(k-1)$, given by [13]

$$d(k) = d(k-1) + [d(k-1) - d(k-2)] \times \text{sgn}\{P_L(k) - P_L(k-1)\} \tag{11.11}$$

where sgn is the signum function, defined as

$$\text{sgn}(x) = \begin{cases} +1 & \text{if } x > 0 \\ 0 & \text{if } x = 0 \\ -1 & \text{if } x < 0 \end{cases} \tag{11.12}$$

The signum function block is used in Figure 11.13 (b).

11.2.4 Concentrated Solar Power (CSP)

Similar to solar PVs, CSPs also use solar irradiance to generate electricity. However, unlike solar PVs that convert solar energy to electricity directly, CSP converts solar energy to heat, which is then used to generate electricity using a turbine-generator system. There are different CSP structures [14]. The most predominant among them is the central receiver system, as shown in Figure 11.14. This contains arrays of sun-tracking solar reflectors. These reflectors, which are usually called heliostats, are flat and slightly concave mirrors. The heliostats reflect the sunlight to concentrate on the solar tower. The heat in the solar tower is absorbed by a heat-transferring fluid (HTF). The heat in the fluid is then used to generate steam, which is then used to generate electricity using a steam turbine. The advantage of CSP is that its power is dispatchable, that is, it can be controlled and used as and when required. For it to be successful, the choice of HTF is very crucial. Molten salt is one of the most commonly used HTFs. It is a combination of 60% sodium nitrate ($NANO_3$) and 40% potassium nitrate (KNO_3). This is a stable mixture and can be used in the temperature range of 260–621 °C. However, it crystallizes at 238 °C and solidifies at 221 °C [14]. There are several other HTFs that are been investigated these days, such as oils, alcohol, liquid metals, and metal hydride.

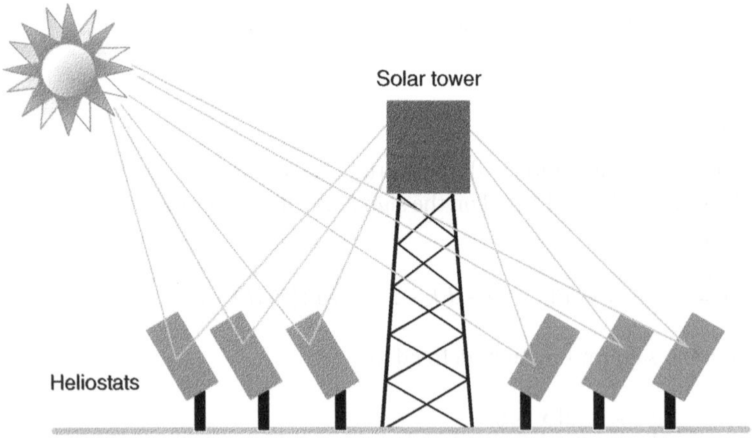

Figure 11.14 The schematic diagram of a central receiving CSP system.

11.3 Wind Power

About 71% of our planet is covered in water, be it in oceans, rivers, or lakes. The rest of the planet is covered in landmass. It is the interaction between the water and land that creates winds. This has the following two phases:

- *Daytime Heating of Land*: Land heats up faster than water during the day because land has a lower heat capacity than water. This is why the air above land warms up more quickly, causing it to rise. As the warm air rises over land, cooler air from the sea (which heats more slowly) moves in to replace it, creating wind. This is known as a *sea breeze* during the day.
- *Nighttime Wind Reversal*: At night, the reverse happens. The land cools down faster than water, and the warmer air over the sea rises while cooler air from the land moves toward the sea. This forms a *land breeze*. It is to be noted that water retains heat longer because of its higher heat capacity. Thus, the air over water remains warmer for longer periods during the night.

Because the circulation of wind happens due to the differential in heating caused by the sun, wind power is an indirect form of solar power.

Trade winds flow from east to west consistently in regions around the equator. Usually, they are found between 30° north and 30° south. In the northern hemisphere, Trade winds blow from the northeast, while it blows from the southeast in the southern hemisphere. Near the equator, the warm air rises due to intense solar heating. As this warm air rises, it cools and moves away from the equator at high altitudes. This air descends at around 30° latitudes in both hemispheres, creating high-pressure zones. The descending air then flows back toward the equator at the surface, creating Trade winds. Due to their consistent nature, Trade winds are good sources of wind power. They are less variable and, therefore, have predictable power output. However, they are restricted between 30° north and 30° south. Usually, the Caribbean, parts of Africa, northern Australia, and Pacific islands experience strong trade winds. Unfortunately, these regions are prone to tropical storms and cyclones (or hurricanes), which can damage wind turbines. Moreover, most of these locations are usually remotely located without proper infrastructure to capture and evacuate the generated wind power.

The first use of wind power was reported in 1891 when the Danish physicist and engineer Poul la Cour erected a windmill in a place called Askov in Denmark. The windmill could produce constant power to drive a generator. He used the power to produce hydrogen through the electrolysis of water. This process is called the production of green hydrogen (see the next section). The hydrogen was stored in a 12 m^3

tank. A high school in Askov was illuminated by a mixture of hydrogen and oxygen between 1895 and 1902 without a single day of failure. Due to his pioneering work in wind power, Wind Europe established the Poul la Cour Prize in 1992 for outstanding contributions to the wind energy industry.

11.3.1 Wind Turbine Types

There are two types of wind turbines – vertical axis and horizontal axis. In a vertical axis wind turbine (VAWT), the rotor is arranged vertically. It has the advantage in sites in which the wind direction changes frequently since such a turbine does not have to face the wind to be effective. A VAWT can be placed on the ground or on top of buildings and will not need high towers. Therefore, generators and gearboxes can be placed on the ground so that they can be easily accessible for maintenance and repairs. Moreover, VAWT is aesthetically pleasing (check any images of VAWT). However, these turbines have some critical disadvantages; the most prominent of which are the poor efficiency. Thus, most commercial wind turbines are of horizontal axis type.

The blades in a horizontal-axis wind turbine (HAWT) rotate along the horizontal axis and are placed on a tower. Figure 11.15 shows the schematic diagram of HAWT wind converter systems. A HAWT is mounted on a tower. Small turbines can be placed almost anywhere like water pumping stations. They can generate up to 100 kW of power. Medium-sized wind turbines can sit on towers that are 80 m in height. Their blades can be as long as 40 m while generating up to 1.8 MW of power. Larger wind turbines can generate anywhere from 4.8 MW to 9.5 MW of power with blade lengths of 160 m that are mounted on 240 m towers. Each HAWT

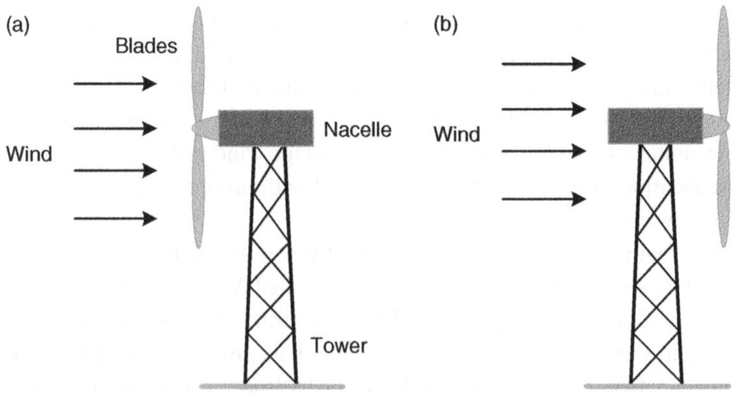

Figure 11.15 The schematic diagram HAWT placement: (a) upwind and (b) downwind.

contains blades (usually either two or three) that are connected to nacelles. The nacelle of the turbine houses the gearbox, the generator, the step-up transformer, and the power converter (if required). It can be very heavy.

As shown in Figure 11.15, a HAWT can face the wind (upwind) or face away from the wind. In a downwind configuration, the wind itself can control the yaw (i.e., left-right motion) such that it can orient itself without external control of the flow of wind direction [7]. However, the tower is placed between the wind direction and the blades, and therefore, can have "shadowing" effect when a blade swings behind the tower. This will flex the blades and lead to blade failures. The upwind turbines face the wind and can operate smoothly while generating more power. However, these turbines will need yaw control. Most modern wind turbines are the upwind type.

Wind speed on the sea is higher than that on the land. Moreover, they flow smoothly as there is no interference of nature-made or human-made obstacles in their path. Therefore, offshore wind farms are gaining importance these days. In a wind farm, a large number of wind turbines are placed strategically, while their combined power is dispatched in the same fashion as that of a power plant. The main difference between an offshore wind turbine and an onshore wind turbine is how they are structurally supported. The tower supporting onshore wind turbines is placed solidly on the ground. However, offshore wind turbines can be placed on fixed structures or floating structures. Generally, the maximum seabed depth where a fixed structure can be placed is 60 m. The limiting factor for such installations is the size of the turbine blades. They are limited to within 40 m while generating only up to 1.8 MW of power.

There are three types of floating structures, where wind turbines sit on a platform. Mooring lines that are deeply embedded on the sea floor through anchors prevent the platforms from drifting away. There are three floating structures (see Figure 11.16). These are:

- *Spar Buoy*: It is a hollow cylindrical structure that holds the turbine tower. It floats vertically in deep water, but heavy weights are placed at the bottom to lower the center of gravity (see Figure 11.16 (a)).
- *Submersible Platform*: It has large floating hulls with a tower at the center. Multiple wind turbines can be placed on such hulls.
- *Tension Leg Platform*: As the name suggests, this platform is held in place with high-tension moor lines (see Figure 11.16 (b)). These are usually smaller platforms and are susceptible to damage in case of large waves like tsunamis.

The advantage of floating platforms is that they can support large wind turbines with a power rating of 10 MW, which is not possible with fixed structure platforms.

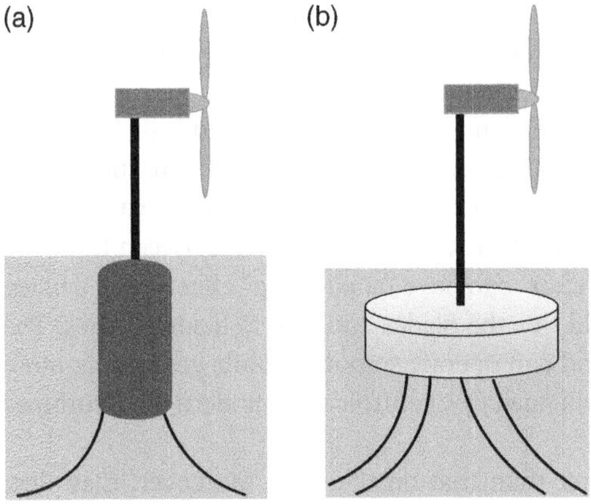

Figure 11.16 Offshore wind platforms: (a) spar buoy and (b) tension leg.

11.3.2 Wind Power Calculations

The KE of a mass m moving at a velocity v is given by

$$KE = \frac{1}{2}mv^2 \tag{11.13}$$

If the mass can produce power, it will then be given by

$$\text{Power} = \frac{KE}{\text{Time}} = \frac{1}{2}\left(\frac{m}{\text{Time}}\right)v^2 \tag{11.14}$$

Now suppose, a mass m of cross-sectional through a cross-sectional area A, with a velocity of v, then the flow rate is given by

$$\text{Flow rate through area } A = \frac{m}{\text{Time}} = \rho A v \tag{11.15}$$

where ρ is the air density at kg/m^3. Combining (11.14) and (11.15), we have power contained in the wind as

$$P_W = \frac{1}{2}\rho A v^3 \tag{11.16}$$

Now, the unit on the right-hand side of (11.16) is

$$\frac{\text{kg}}{\text{m}^3} \times \text{m}^2 \times \frac{\text{m}^3}{\text{s}^3} = \frac{\text{kg m}^2}{\text{s}^3} = \text{W}$$

Therefore, the wind power is

$$P_W = \frac{1}{2}\rho A v^3 \text{ W} \tag{11.17}$$

From (11.17), the following can be observed:

- The power output is proportional to the velocity of the wind cubed, that is, for the same area and air density, a 10 m/s wind speed will produce 1,000 times more power than a wind speed of 1 m/s and 37 times more power than a wind speed of 3 m/s.
- Wind power is proportional to the swept area of the turbine rotor, that is, the longer the length of the turbine, the more the power is generated. The area of a circle is given by πr^2, where r is the radius of the circle. Therefore, doubling the blade size will increase power output by four times.
- Wind power is proportional to air density, which is related to temperature. The warm air is less dense than cooler air because the gas molecules in warm air have a greater velocity and are farther apart than in cooler air. Therefore, wind power production can be variable between day and night and between onshore and offshore wind turbines.

When wind flows through turbine blades, only a fraction of the power is extracted by the turbine. If the turbine extracted all the KE contained in the wind, the air would stop completely behind the turbine. This will prevent any further wind flow through the turbine. Therefore, the downward velocity of the wind cannot be zero, but it will be less than the upward velocity. The efficiency of a wind rotor depends on two factors – blade pitch angle and tip speed ratio (TSR). The pitch angle is the angle between the rotor blades and the direction of the oncoming wind. It is usually regulated externally by twisting the blades to align them toward the wind. The TSR is the ratio between the speed of the tip of the wind turbine blade to that of the wind, that is,

$$\text{TSR} = \frac{\text{Rotor tip speed}}{\text{Wind speed}}$$

Denoting the TSR as λ, it is given by

$$\lambda = \frac{v_{tip}}{v} \tag{11.18}$$

where $v_{tip} = \omega r$, ω being the rotational speed in rad/s, and r is the radius of the rotor blades in m and v is the wind speed in m/s. The rotor efficiency C_P depends on both TSR (λ) and the rotor pitch angle (β) and is denoted by $C_P(\beta, \lambda)$. The power output of the wind turbine is then given by

$$P_{out} = C_P(\beta, \lambda) \times P_W \tag{11.19}$$

The maximum efficiency of the rotor blade of the wind turbine is derived to be 59.259% as derived by the German physicist Albert Betz. This is called Betz's limit.

11.3.3 Pitch Angle Control

The wind turbine operation is divided into four regions, as shown in Figure 11.17 for a typical 5 MW wind power system [15]. If the wind speed is less than the "cut-in" speed, the generator is switched off or remains off in Region-1. Region-2 is above the cut-in speed, where the generator starts producing power. However, the speed is not sufficient to generate the maximum amount of power. The generator attains the rated speed and produces the rated power in Region-3. The turbine shuts down above the "cut-out" speed caused by high winds to prevent damage to the turbine blades (Region-4). Beyond the cut-out speed, the turbine blades are *stall regulated*, where the turbine blades are placed in such a fashion that they generate less aerodynamic force at high wind speed, thereby stalling such that the turbine power output becomes zero.

In the operating regions, the speed and power of the turbine are controlled by the generator load torque and the pitch angles of the turbine blades. A scheme for pitch angle control is shown in Figure 11.18. The error between the measured

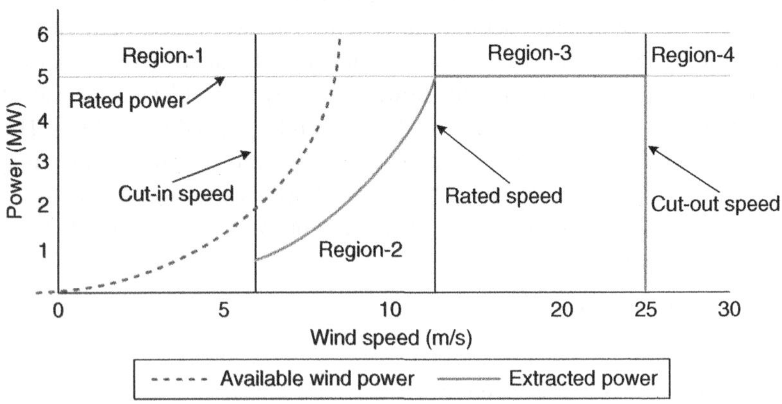

Figure 11.17 Wind turbine operating regions.

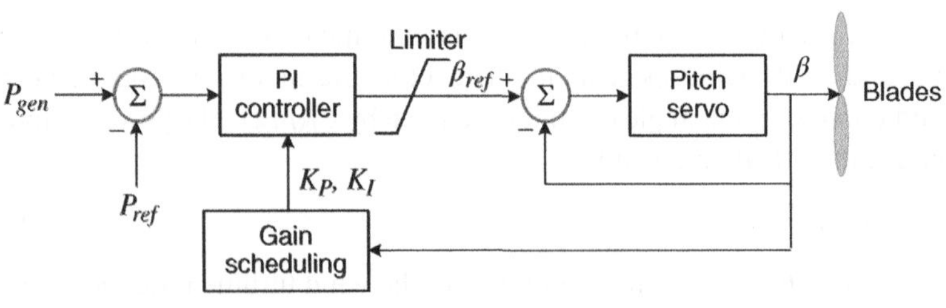

Figure 11.18 Gain scheduling pitch angle controller.

power generated P_{gen} and the rated power P_{ref} is passed through a proportional-plus-integral (PI) controller to produce the reference pitch angle β_{ref}. A limiter is placed at the output of the PI controller to limit the variations in the reference angle. The reference angle is compared with the actual pitch angle β and the error is corrected through the pitch angle servo motor. There is a drawback in this scheme because the wind speed cannot be measured precisely. The control parameters cannot be chosen as a function of the wind speed. Therefore, a gain scheduling controller is implemented, which produces the proportional gain K_P and the integral gain K_I based on the knowledge of the pitch angle [16]. The gain schedular compensates for the nonlinear aerodynamic characteristics of the wind turbine. A set of PI gains is computed a priori linearizing the system around several operating values of the pitch angle β. These gains are then chosen depending on the values of the pitch angle (see [16] for details).

11.3.4 Types of Wind Power Collectors

IEEE PES (Power and Energy Society) Wind Plant Collector System Design Working Group classifies wind turbines into five different types [17]. These are:

- *Type-1*: Fixed speed.
- *Type-2*: Limited variable speed.
- *Type-3*: Variable speed with partial power electronic conversion.
- *Type-4*: Variable speed with full-power electronic conversion.
- *Type-5*: Variable speed drive train connected to torque–speed converter coupled with a synchronous generator.

Of these, the first three types of wind turbines use induction generators. Type-4 turbines usually use permanent magnet synchronous generators (PMSGs).

Induction motors are cheaper and are rugged in nature. The rotor of the motor usually rotates at a speed slightly less than the synchronous speed. The difference in speed is called the slip s, defined by

$$s = \frac{\omega_s - \omega_r}{\omega_s} = \frac{n_s - n_r}{n_s} \tag{11.20}$$

where ω_s and ω_r, respectively, are the synchronous speed and the rotor speed radians per second and n_s and n_r are the respective speed in rpm.

The IEEE-recommended equivalent circuit of an induction machine is shown in Figure 11.19 (a) [18]. Here, subscript 1 defines the stator side quantities, while subscript 2 defines rotor side quantities, when referred to the stator side, and X_m is the magnetizing reactance. The Thevenin equivalent circuit is shown in Figure 11.19 (b). Because R_1 is smaller than $(X_1 + X_m)$, the Thevenin voltage and reactances are given by [18]

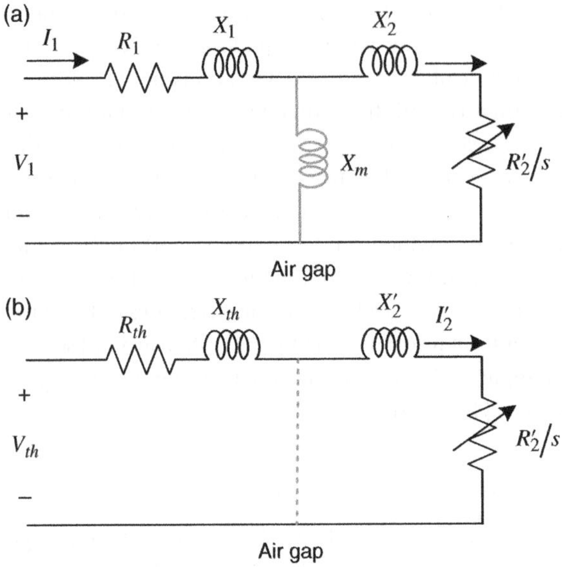

Figure 11.19 (a) IEEE-recommended equivalent circuit of induction machine and (b) its Thevenin equivalent.

$$V_{th} \cong \frac{X_m}{X_1 + X_m} V_1 \tag{11.21}$$

$$R_{th} \cong \left(\frac{X_m}{X_1 + X_m}\right)^2 R_1 \tag{11.22}$$

$$X_{th} \cong X_1 \tag{11.23}$$

From the Thevenin equivalent circuit of Figure 11.19 (b), the torque is given by [19]

$$T = \frac{3}{\omega_s} \times \frac{V_{th}^2 (R_2'/s)}{\left(R_{th} + R_2'/s\right)^2 + \left(X_{th} + X_2'\right)^2} \tag{11.24}$$

The typical torque–speed characteristics of an induction machine are shown in Figure 11.20, where the per unit speed is defined by the ratio $n_r : n_s$. The machine operates in the motoring mode when $0 < s < 1$. However, when the slip (s) is negative, the machine operates as a generator. When an induction machine is connected to a wind turbine, the machine will start its operation as an induction motor till it reaches synchronous speed. When the wind turbine forces the machine to operate at a speed higher than the synchronous speed, it starts working as an induction generator. As has been explained in [19], in the induction

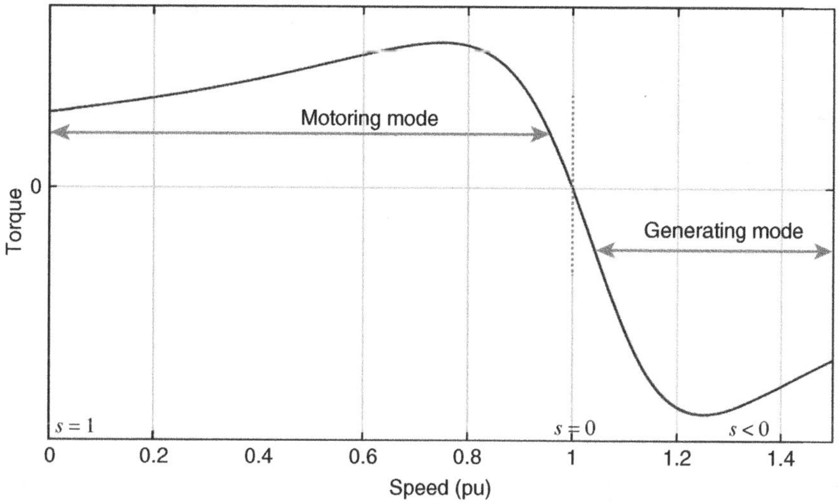

Figure 11.20 Typical torque–speed characteristics of an induction machine.

generator mode, the machine draws a current that lags the voltage by more than 90°. This means that the machine will require reactive power to produce real power. An induction generator can draw reactive current when it is connected to a power grid. Additionally, capacitor banks can also be connected in parallel.

Type-1: This is shown in Figure 11.21. The wind turbine is connected to a squirrel cage induction generator (SCIG) through a gearbox. The machine is equipped with a soft starter such that it does not draw a large current during starting. Furthermore, a set of capacitors is connected in shunt at the machine terminals. A combination of them can be selected in the same fashion as a TSC depending on the reactive power requirements.

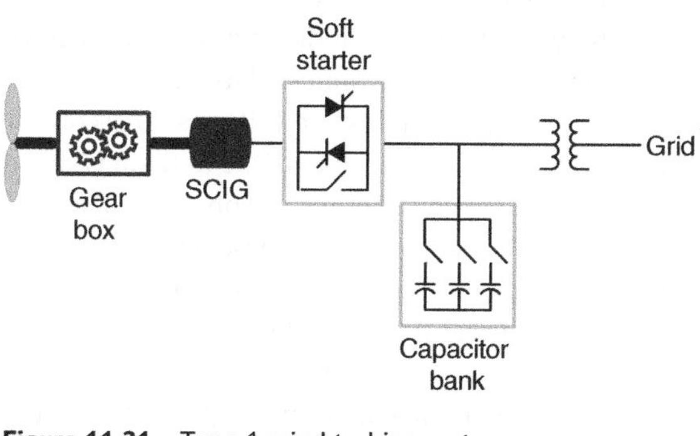

Figure 11.21 Type-1 wind turbine system.

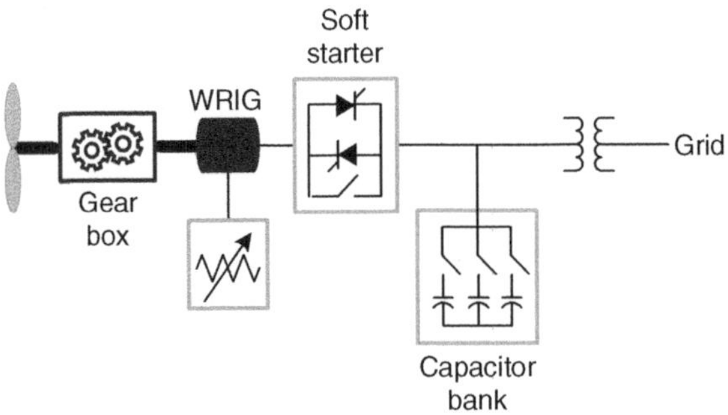

Figure 11.22 Type-2 wind turbine system.

Type-2: This is shown in Figure 11.22. This is similar to Type-1 wind turbine system, except that a wound rotor induction generator (WRIG) is used here. A variable resistor is inserted on the rotor circuit that can control the rotor current rapidly to keep the power output constant during wind fluctuations [17]. The speed of the machine can be controlled to some extent such that the TSR can be changed to capture maximum power.

Type-1 and Type-2 wind turbine systems have no speed control or limited speed control capabilities. However, modern advanced wind turbine systems can adjust the speed. By controlling the generator speed, pitch control time constants become longer, thereby pitch complexity is reduced [20]. It is to be noted that the pitch angle is fixed at lower wind speeds, while the pitch control mechanism limits the maximum output power at higher wind speeds. The other advantages of adjustable speed wind turbines are that they reduce mechanical stresses on the turbine and improve system efficiency [20]. Type-3 and Type-4 wind turbine systems are adjustable speed types.

Type-3: This type of wind turbine uses a doubly fed induction generator (DFIG). The schematic diagram of a DFIG-based wind collector system is shown in Figure 11.23. This contains a WRIG with its stator windings directly connected to the grid through a transformer. The rotor circuit is connected to the grid through a back-to-back power converter system, which contains two voltage source converters (VSCs) that are joined together through a DC capacitor. One of them is called the rotor side converter, while the other is called the grid side converter. The converters control the magnitude of the rotor voltage and its phase angle to control both active and reactive powers. The rotor side converter compensates for the difference between the mechanical and electrical frequency by injecting a variable frequency rotor current. A small amount of power is injected into the rotor circuit – the closer the speed of the generator to the synchronous speed, the smaller

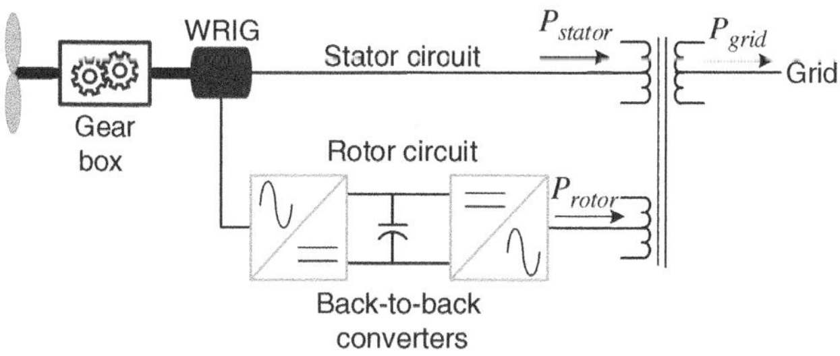

Figure 11.23 Type-3 wind turbine system (DFIG).

the power to be injected. Thus, the back-to-back converter size is about 30% of the generator size [17]. This is an advantage of the DFIG system. The stator side and the grid side converters are connected through a three-winding transformer, where one primary winding is connected to the stator, while the other is connected to the rotor. The turns ratio of the rotor side winding is chosen such that the converters can operate at lower DC voltage.

The power flowing through the stator, rotor, and grid is indicated in Figure 11.23. Neglecting all losses, we can write the following expression from this diagram.

$$P_{grid} = P_{stator} + P_{rotor} \tag{11.25}$$

The power through the rotor circuit, which is also known as *slip power*, is expressed as the slip multiplied by the stator power, that is, [16]

$$P_{rotor} = -sP_{stator} \tag{11.26}$$

This means that the power to the grid can be written by substituting (11.26) in (11.25) as

$$P_{grid} = (1-s)P_{stator} \tag{11.27}$$

When the slip is positive $s > 0$, the machine is operating at sub-synchronous frequency, and the rotor power is negative, that is, $P_{rotor} < 0$. On the other hand, P_{rotor} is positive when the machine is operating in the super-synchronous mode ($s < 0$). However, the grid power is positive in either case.

The main advantage of Type-3 system is that it can control the real and reactive power separately while maintaining synchronism with the power grid. One such control approach is outlined in [16].

Type-4: This type employs a full-rated back-to-back converter, as shown in Figure 11.24. Usually, PMSGs are connected to the wind turbines. It is known

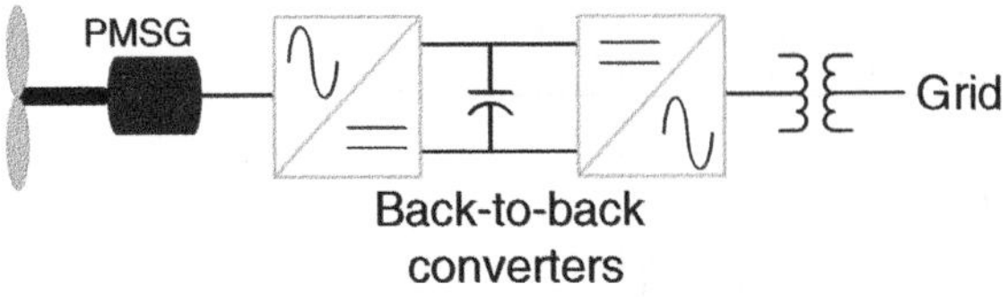

Figure 11.24 Type-4 wind turbine system.

that the frequency of the output voltage depends on the speed of rotation $120f/p$ rpm, where f is the frequency in Hz and p is the number of poles. For example, when a two-pole machine rotates at 3,000 rpm, the output frequency will be 50 Hz. If the turbine is allowed to operate at its optimum aerodynamic speed, then the frequency will be variable depending on the speed. However, the wind side converter will convert this variable frequency AC voltage into DC, which will then be converted into fundamental frequency (50 or 60 Hz) AC voltage by the grid side converter. This will eliminate the use of the bulky gearbox, and hence this type of collection is most suitable for offshore wind power systems. The main drawbacks of this type are the cost and losses involved with the full-rated converters. However, these are less bulky and have full control of the real and reactive power.

Type-5 employs a torque–speed converter with a synchronous generator. This type is not very popular and is not discussed here. See [17] for its functionality.

11.4 Hydrogen

We start this section with a quote from Jules Vern, the French visionary and author of *Twenty Thousand Leagues Under the Seas* (1870), *Around the World in Eighty Days* (1872), and other masterpieces. In *Mysterious Island* (1884), Vern wrote the following:

> *"I believe that water will one day be employed as a fuel, that hydrogen and oxygen which constitute it, used singly or together, will furnish an inexhaustible source of heat and light."*

That vision will come to fruition in the near future when the real breakthrough in green hydrogen production is achieved.

Even though hydrogen can be used as fuel and has been known for a long time, there was hesitancy in using it due to the Hindenburg air disaster of 1937 (see Figure 11.25). Hindenburg was a big airship built by Zeppelin Company for

Figure 11.25 The Hindenburg disaster.

transatlantic air travel. The airship was 245 m long and had four 1,100 horsepower diesel engines – hydrogen was not used as fuel. It has a maximum speed of 135 km per hour. It was used for luxury travel at that time. Ideally, it should have been filled with helium. Even though hydrogen is lighter than helium, helium is a preferred medium for air balloons as it is lighter than air and is not flammable. However, due to the export restriction imposed by the USA on Germany during that period, it was filled with hydrogen. Hindenburg took off from Frankfurt, Germany on May 3, 1937, with 36 passengers and 61 crewmen. It approached its landing site of Lakehurst, New Jersey on May 6, 1937. However, during the landing process, hydrogen leaked and caught fire in which 35 people lost their lives but the rest survived. The cause of ignition is still unknown, but there are several hypotheses. More information about the disaster can be found on various websites, such as [21].

Despite the flammable nature of hydrogen, it has now been considered as one of the plausible sources of renewable energy – first for automotive purposes and then for power generation. There are several attributes of hydrogen as a fuel. Some of these are:

- When burned, hydrogen produces a small amount of nitrogen oxides (NO_x) at high combustion temperatures that are high enough for nitrogen and oxygen in the air to combine.
- When used in a fuel cell, the only byproduct is water.

- Because it has very low density, it can readily escape from confined environments and is less likely to concentrate like petrol fumes, notwithstanding the Hindenburg experience.
- Hydrogen, however, is not a natural energy source. It is high-quality energy carrier that is not naturally available.

Currently, Hydrogen is mainly used for industrial purposes, such as refining petroleum, metal treatment, fertilizer production, and food processing. However, there is immense potential for the use of hydrogen for power generation and the automotive industry. These will be discussed briefly in this section.

11.4.1 Hydrogen Production

As mentioned before, hydrogen is not the primary source of energy – it does not exist in its own form. Therefore, hydrogen must be sourced from other sources. The hydrogen production type is usually color coded. Some of these are:

- *Shades of Brown or Black*: In these, hydrogen is produced from coal (brown or black) that converts carbon-rich materials into hydrogen and carbon dioxide: the latter is released into the atmosphere.
- *Grey*: Hydrogen is extracted from natural gas through steam reforming. However, the resulting CO_2 gas is released into the atmosphere.
- *Blue*: This process is the same as grey hydrogen, except that the greenhouse gases are captured using carbon capture and storage technologies.
- *Green*: In this, renewable energy sources such as solar or wind are used to electrolyze water to produce hydrogen and oxygen. Needless to say, this is the most environmentally friendly desired option.

Natural gas is essentially a fossil fuel, the largest component of which is methane CH_4. A process called steam methane reforming (SMR), which is also known as the Bosch process is used to convert methane into hydrogen and carbon dioxide. This process currently accounts for 48% of all hydrogen production. The chemical reaction for this process is as follows:

1) At high temperatures (700–1,100 °C), water vapor (steam) is mixed with natural gas. Then the following reactions take place:
2) $CH_4 + H_2O \rightarrow CO + 3H_2$
3) $CO + H_2O \rightarrow CO_2 + H_2$
4) Carbon dioxide is released into the atmosphere.

The process is shown in Figure 11.26.

The blue hydrogen is produced using the methane pyrolysis process, in which methane is passed through a molten metal catalyst at a high temperature

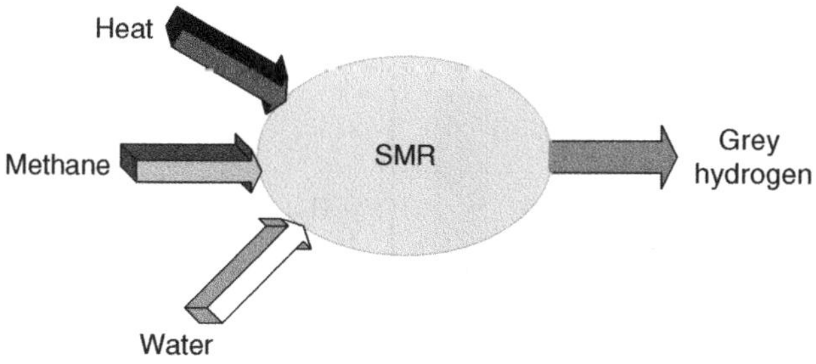

Figure 11.26 Grey hydrogen production through the SMR process.

(1,650 °C) in the absence of oxygen. The resulting product is solid carbon and hydrogen, as per

$$CH_4 \rightarrow C + 2H_2$$

The solid carbon can be used for various processes or simply for landfill. The methane pyrolysis process is shown in Figure 11.27.

Green hydrogen is the most promising technology from a climate change point of view. In this, renewable energy sources such as wind and solar are used to electrolyze water to produce hydrogen. There are different types of electrolysis technologies that are currently available [22]. Of these, proton exchange membrane (PEM) electrolysis is the most promising one. It consists of a solid polymer electrolyte that can operate at high current density. It has an efficiency of over 80%. The schematic diagram of a PEM electrolysis process is shown in Figure 11.28. The governing chemical reactions are as follows:

Anode: $H_2O \rightarrow 2H^+ + \frac{1}{2} O_2 + 2e^-$
Cathode: $2H^+ + 2e^- \rightarrow H_2$
Total reaction: $H_2O \rightarrow H_2 + \frac{1}{2} O_2$

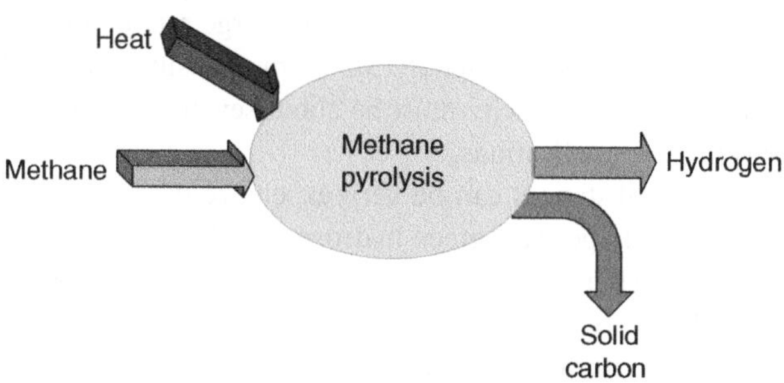

Figure 11.27 Blue hydrogen production through methane pyrolysis process.

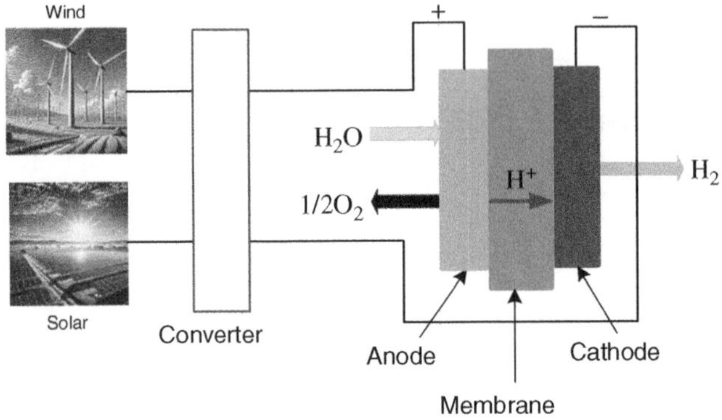

Figure 11.28 Green hydrogen production through PEM electrolysis.

11.4.2 Hydrogen Storage and Transmission

Large-scale storage of hydrogen is mostly geographically dependent. A salt cavern is a large underground cavity that has been created by deposits of dried-up seas. Salt caverns are usually found deep below the earth's surface. Since the 1970s, they have been used as long-time storage of various chemicals, such as natural gas, propane, butane, and ethane [23]. These caverns exist in nature and have an efficiency retention of about 98%. Hydrogen can be stored in such caverns under high pressure enabling a high discharge rate. Therefore, they are suitable for power generation. Moreover, these caverns offer a low risk of contamination of hydrogen as opposed to natural gas or oil reservoirs. However, the walls of the caverns must be 75% of the cavern diameter.

Depleted oil or gas reservoirs can also be used for hydrogen storage. Even though these are larger in size than salt caverns, they can contaminate hydrogen with oil or gas residues, which will have to be removed before hydrogen can be used in fuel cells.

For automotive or small applications, hydrogen in compressed or liquefied forms can be stored in storage tanks. These have a high discharge rate of around 99% efficiency. However, hydrogen has only about 15% energy density compared to petrol. Therefore, the tank storage capacity must be about seven times that of gasoline fuel pumps for vehicular applications.

Hydrogen transport over long distances can be very expensive due to its low energy density. At the same time, large-scale green hydrogen production will usually take place in remote areas where large solar farms can be placed or where there are large wind farms. There are a few such areas that have been identified in Australia. In some countries, natural gas pipelines are already existing, which can be used, or new infrastructure can be developed. As has been pointed out in [23], pipelines are a suitable option for a distance of less than 1,500 km. For transporting hydrogen over this distance, it must be (a) liquified, (b) converted into ammonia, or (c) converted into liquid organic hydrogen carriers (LOHC).

The main problem with the liquification of hydrogen is that it has to be cooled at −253° C, which will require a large amount of energy. Converted hydrogen into ammonia and reconverting it back to pure hydrogen will probably use 15–30% of the energy contained in hydrogen. Nevertheless, ammonia can be liquified at −33 °C and can contain 1.7 times more hydrogen than liquified hydrogen. Moreover, ammonia can also be directly used for fertilizers and other industrial applications. One of the promising methods of hydrogen storage and transportation is to use LOHC. This process involves a *hydrogeneration* process through which hydrogen molecules are bound together with LOHC, thereby storing hydrogen in liquid form. The hydrogen then can be extracted from this liquid through the *dehydrogenation* process where the hydrogen is released, and the liquid becomes dehydrogenated. This is a mode of hydrogen transport that has already been established in Japan.

11.4.3 Utilization of Hydrogen

Hydrogen can be used as storage or bulk power generation and, at the same time, can also be used in automotive industries. In the Netherlands, a 440 MW combined cycle gas turbine (CCGT) plant has been converted into hydrogen from natural gas [23]. Fuel cells offer flexible power generation and storage options for smaller generation capacity (up to 2 MW), currently.

A fuel cell can generate electricity from hydrogen through a chemical reaction. Hydrogen is fed to the anode of a fuel cell, where a chemical reaction strips the electrons from the hydrogen atoms. The ionized hydrogen atoms now carry positive electrical charges. The current can then flow if the output of a fuel cell is connected to a load in a direction that is opposite to the negatively charged electrons. Oxygen from the air is fed to the fuel cell at the cathode. There, it combines with the electrons returning from the electrical circuit and hydrogen ions that have traveled through the electrolyte from the anode. The electrolyte in a fuel cell only allows appropriate ions to pass between the anode and the cathode. It prevents electrons from traveling through them so as not to disrupt chemical reactions. A typical fuel cell is shown in Figure 11.29. There are several types of fuel cells [7]. Of these, the PEM fuel

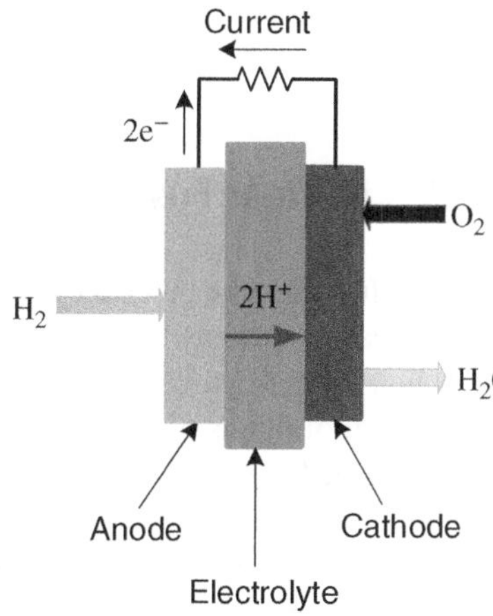

Figure 11.29 A typical fuel cell.

cell has an efficiency of 40–50% and an operating temperature of 80 °C. Solid oxide fuel cell has a higher efficiency (60%) but must operate at about 1,000 °C. The advantage of this type of fuel cell is that the waste heat can also be used for power generation.

One of the main attractions of fuel cells is their usage in the automotive industry. However, except for Toyota, no other major manufacturer has started building fuel cell cars yet. Toyota Mirai has been sold since 2014. One of the main advantages of fuel cell cars is that their tanks can be filled quickly like any petrol (gasoline) car. However, that will require a large hydrogen refueling infrastructure, which is not present right now. Moreover, fuel cell cars operate very efficiently and effectively at steady speeds on flat roads. They cannot provide enough acceleration to merge on fast-moving highways. Toyota has added high-voltage, low-capacity batteries (like their hybrid electric vehicles) to supplement power during acceleration for a relatively short period of time. Hopefully, it is expected that these shortcomings will be overcome as fuel cell cars become popular in the future when other major manufacturers enter the market.

Wärtsilä is a Finnish multinational corporation known for its innovative approaches to marine and energy systems. Wärtsilä 31 is a medium-sized, four-stroke engine that holds the Guinness World Record for the most efficient diesel engine with 165 g fuel consumption per kW-h of energy production. In 2024, Wärtsilä announced the development of Wärtsilä 31SG-H2 and Wärtsilä 31H2 engines [24]. Wärtsilä 31SG-H2 is a hydrogen-ready engine that can run on natural gas with a 25% hydrogen blend. With a larger volume of hydrogen intake, it can run in 100% carbon-free mode. Wärtsilä 31H2 runs purely on hydrogen. These engines are rated between 4.5 MW and 10 MW. However, with bigger engine capacity, such engines have the potential to solve the baseload problem.

11.5 Nuclear Fusion

Nuclear fusion is the process by which two light atomic nuclei combine to form a single heavier nucleus, releasing massive amounts of energy. Our sun and other stars fuse lighter hydrogen atoms to form heavier helium atoms. When hydrogen is heated to millions of degrees Celsius, it forms into plasma in which the negatively charged electrons are separated from the positively charged atomic nuclei. Usually, positively charged nuclei repel each other. However, under the massive gravitational force of the sun, and due to the high temperatures, the ions move faster and eventually, come closer to each other. The nuclei then fuse releasing huge amounts of energy [25].

Nuclear fusion can be called the *Holy Grail* of energy production. Once a technical breakthrough is achieved, the electric power problem of the entire world will be solved. There are several technical barriers that must be overcome before nuclear fusion can be energy positive, that is, produce more energy than given as an input. The massive gravitational force of the sun makes fusion possible, which is not possible on the earth. However, nuclear fusion is an attractive proposition – it produces four times more energy per kilogram than nuclear fission and nearly four million times more energy than fossil fuels such as coal or gas [25]. Moreover, there is no possibility of catastrophic accidents to occur. If the reactor is damaged, heat generation will rapidly cease, and the production of plasma will stop. There is also no possibility of a divergent chain reaction like a fission reactor. It will also not have the trouble of sorting or storing nuclear waste as well.

The most pursued fusion process currently is called deuterium–tritium (D–T) fusion This is the fusion of the two isotopes of hydrogen – deuterium and tritium [25]. The atoms of each of these contain extra neutrons. Deuterium can be extracted from seawater, through an inexpensive process. Tritium is not naturally available on the Earth but can be found in the upper atmosphere. It can be produced artificially when lithium reacts with fusion-generated neutrons. Usually, it is a byproduct of the operation of nuclear reactors where ceramic pebbles containing lithium are present. The fusion process is shown in Figure 11.30.

The fusion reaction temperature required is more than 100 million degrees Celsius, which is hotter than the sun. This temperature will make deuterium and tritium fuse if the pressure and magnetic forces are maintained [25]. Currently, the largest experimental facility is constructed by International Thermonuclear Experimental Reactor (ITER) in Southern France. This is a multinational project with the goal of producing 10 times more thermal output power than the input. This project uses the so-called tokamak reactor. In this, the plasma is confined in a donut shape (torus) using magnetic fields. Magnetic fields are used because the separated ions and electrons follow the magnetic field lines. Also, the

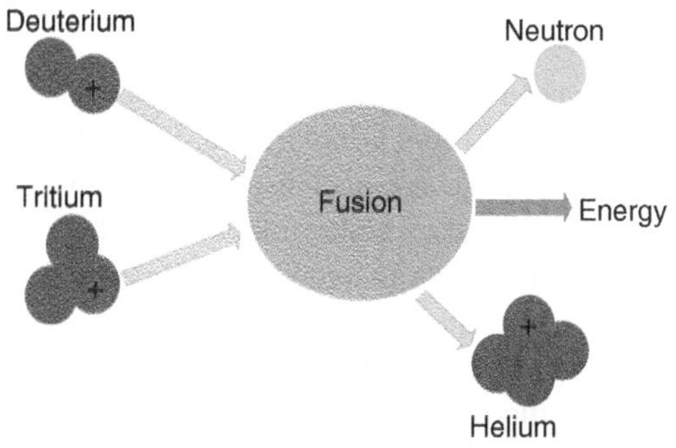

Figure 11.30 Typical nuclear fusion process.

strong magnetic attraction prevents the particles from coming in contact with the reactor walls, thereby dissipating heat. A stellarator reactor also works on the principle of plasma confinement requires less injected power than a tokamak. The fusion research has gained much traction currently. There are several experimental facilities that have been constructed around the world. Several other technologies are being considered currently such as lasers and X-rays. For a review of these technologies and projects refer to [26].

11.6 Renewable Energy in Power Transmission Systems

Renewable energy generated from hydropower plants and geothermal power plants have been integrated into power transmission systems for a long time. They act more like thermal or nuclear power plants due to the large inertia of the associated generators. Their energy can be termed as dispatchable (generated on demand), even though there may be seasonal variations due to rainfall in hydro plants. Moreover, pumped hydro storage (PHS) systems have been operational for a long time and their operational characteristics are well-known. Among the newer technologies, wave energy, tidal energy, or hydrogen are still in the experimental stages and impacts on transmission systems cannot be evaluated. Furthermore, nuclear fusion, which can be the ultimate goal, is still decades away from coming to fruition.

Wind turbines and farms and PV-based solar farms are the most mature renewable energy technologies that are prevalent currently. However, the sun does not always shine, and the wind does not always blow. Furthermore, most of these systems are connected to power grids through power converters. Usually, power converters cannot supply the required inertia, as opposed to synchronous generators. As the penetration level of renewable energy sources increases in a power grid, the overall inertia of the grid reduces. Low inertia can lead to grid instability resulting from grid faults and other disturbances. A system with low inertia responds more rapidly to changes in generation or load, which can lead to several challenges:

- *Frequency Instability*: Inertia helps to regulate the frequency of the power system. When there is low inertia, the system is less able to maintain a stable frequency, and even small disturbances can cause significant frequency deviations. This can lead to under-frequency or over-frequency events, which can trigger protective relays to issue trip commands to circuit breakers and can potentially result in widespread outages.
- *Voltage Fluctuations*: Rapid changes in generation or load can lead to voltage fluctuations, affecting the quality and reliability of power delivery to consumers.

- *Grid Resilience*: Systems with higher inertia can better withstand and recover from disturbances such as faults or generator trips. Low inertia systems may struggle to absorb and mitigate the impact of such events, increasing the risk of cascading failures and widespread blackouts.

To address the challenges associated with low inertia, various measures need to be implemented such as:

- deployment of energy storage units such as batteries or flywheels,
- deployment of synchronous condensers,
- implementation of advanced monitoring and control technologies, and
- the addition of flexible generation resources that can ramp up or down to compensate for the variability of renewable generation.

It is imperative that the low inertia problem can be addressed effectively such that grid stability is maintained as the renewable energy penetration increases. A synchronous condenser is a synchronous motor that runs without a load. When it is overexcited, it supports the voltage level by injecting reactive power. It can absorb reactive power to reduce the voltage when it operates in an underexcited mode. Moreover, a synchronous condenser, with its rotating mass, can also contribute to system inertia during transients.

There are several measures that have been considered lately to have the stable operation of a grid with less inertia [27], for which there will be short-term solutions and long-term solutions. The short-term solutions include volt/var support through droop control, active power control through frequency regulation, ancillary services, and fault ride through. The long-term solutions will depend on the converter control and configurations. In general, there are two types of converters that are used for utility applications – grid following and grid forming [28]. A grid-following (or grid-tied) converter injects current into the power grid, thereby controlling the real and reactive power supply. They are synchronized through a phase-locked loop (PLL). A grid-forming converter (GFC), on the other hand, acts like a controllable voltage source. This is discussed below.

11.6.1 Grid Forming Converter (GFC)

A GFC has the capability of controlling the voltage and frequency of the grid, more or less in the same manner as a synchronous generator. Some of the key features of GFCs are:

- *Voltage and Frequency Regulation*: GFCs can regulate both the voltage magnitude and frequency, even in the absence of external grid support.

- *Inertia Emulation*: A GFC can emulate the inertia of a traditional synchronous generator. This is called a virtual synchronous generator (VSG), which is discussed in the next section.
- *Islanding Capability*: GFCs can supply power to local loads even when disconnected from the main grid. This feature is essential for microgrid applications and enhancing grid resilience.
- *Fast Response Time*: GFCs have fast response time, which allows them to quickly adapt to changes in grid conditions and support dynamic grid operations.

GFCs are often used in renewable energy integration. The schematic diagram of a GFC is shown in Figure 11.31. In this figure, the asterisk denotes the reference quantities, which are power, reactive power, voltage, and frequency. The real and reactive powers are calculated from the measurements of voltage and current. They are then passed through power–frequency (P–f) and reactive power–voltage magnitude (Q–V) droop to calculate the desired voltage magnitude and frequency. These are then used in PWM to control the converter switches. The P–f droop control has been discussed in Section 4.4. A similar approach can also be applied for Q–V droop. A review of GFCs can be found from [29].

11.6.2 Virtual Synchronous Generator (VSG)

A VSG is a concept that emulates the behavior of a traditional synchronous generator through a GFC. Through this, a GFC can control both the voltage

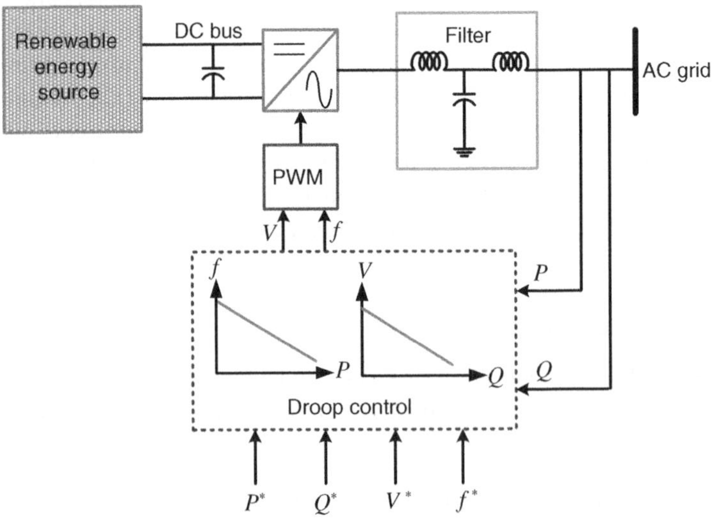

Figure 11.31 Renewable energy connection through grid forming converter.

magnitude and frequency of the grid, providing stability like traditional generators. The VSG emulates the inertia of a synchronous generator, whereby it operates in the same manner as a synchronous generator. However, a VSG typically has fast response times, allowing it to quickly adjust its output in response to changes in grid conditions. This rapid response helps maintain grid stability and ensures reliable power delivery.

The schematic diagram of a VSG is shown in Figure 11.32. Apart from active power–frequency and reactive power–voltage droop equations, it has a governor model and a swing equation [30]. The governor model generates the mechanical power that will be used in the swing equation. The governor equations are given by

$$\begin{aligned} P_m &= \frac{G}{1+sT}P^* - D\Delta\omega \\ \Delta\omega &= \omega - \omega^* \end{aligned} \tag{11.28}$$

where D is the droop gain, and P^* is the reference power. The term $\Delta\omega = \omega - \omega^*$ is obtained from the droop equation, where ω^* is the reference frequency in rad/s and ω is the output of the droop equation.

The swing equation, given in (7.37), is modified to include a damping term K_D. This is expressed as

$$\frac{d\delta}{dt} = \omega^* \times \Delta\omega_{sw} \tag{11.29}$$

$$\frac{d\Delta\omega_{sw}}{dt} = \frac{1}{2H}(P_m - P_e) - \frac{K_D}{2H}\Delta\omega_{sw} \tag{11.30}$$

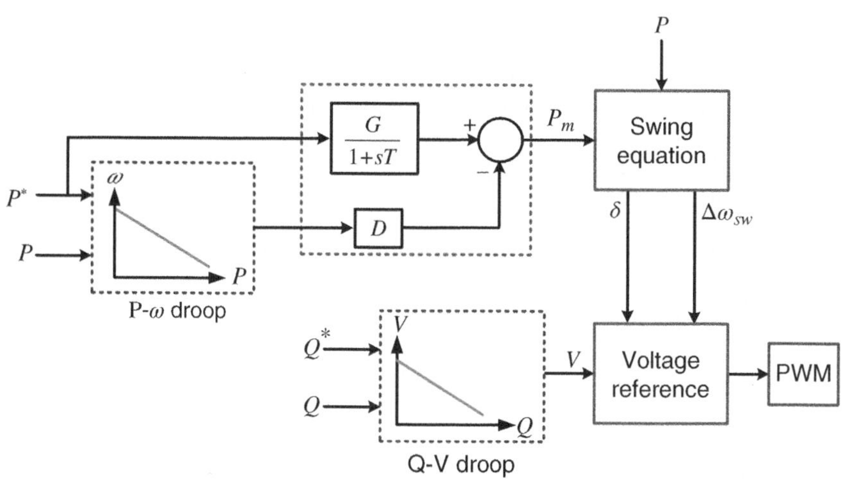

Figure 11.32 Schematic diagram of a virtual synchronous generator.

$$\omega_{CONV} = \omega + \omega_{sw} \tag{11.31}$$

where ω_{CONV} is the desired frequency of the converter output voltages. The voltage reference is obtained from the Q–V droop equation and the instantaneous reference voltages of the converter are then computed as

$$\begin{aligned} v_a &= V\sin(\omega_{CONV}t + \delta) \\ v_b &= V\sin(\omega_{CONV}t + \delta - 120°) \\ v_c &= V\sin(\omega_{CONV}t + \delta + 120°) \end{aligned} \tag{11.32}$$

These are then tracked through a feedback control mechanism using pulse width modulation (PWM).

A typical P-f droop equation is given by

$$\omega = \omega^* + n(P^* - P) \tag{11.33}$$

Usually, the power measurement will contain ripples, which are removed by passing the signal through a lowpass filter (LPH) with a transfer function of $1/1 + s\tau$. The droop equation of (11.33) can then be modified in the Laplace domain as follows:

$$\omega = \omega^* + n\left(P^* - \frac{1}{1 + s\tau}P\right) \tag{11.34}$$

The rearrangement of (11.34) results in the following equation:

$$\begin{aligned} P &= \frac{1 + s\tau}{n}(\omega^* - \omega) + (1 + s\tau)P^* \\ &= \frac{1 + s\tau}{n}\omega^* - \frac{1 + s\tau}{n}\omega + P^* + s\tau P^* \end{aligned} \tag{11.35}$$

Note that $s = d/dt$, and because ω^* and P^* are constant values, their derivatives will be zero, that is, $s\omega^* = 0$ and $sP^* = 0$. Equation (11.35) is then written as

$$P = \frac{1}{n}\omega^* - \frac{1}{n}\left(\omega - \tau\frac{d\omega}{dt}\right) + P^*$$

Rearranging the above equation, the following equation is obtained.

$$\frac{\tau}{n}\frac{d\omega}{dt} = P^* - P - \frac{1}{n}(\omega - \omega^*) \tag{11.36}$$

Since $d\omega/dt = d^2\delta/dt^2$, (11.36) is modified as

$$\frac{\tau}{n}\frac{d^2\delta}{dt^2} = P^* - P - \frac{1}{n}\Delta\omega \tag{11.37}$$

Equation (11.37) is equivalent to the swing equation [31].

11.6.3 Fault Ride Through (FRT)

Fault ride through (FRT) is the capability of electrical equipment, particularly power converters, and grid-connected devices, to remain operational and stable during transient faults or disturbances in the power grid. These disturbances could be caused by short circuits, voltage sags, or other irregularities in the grid. When a fault occurs in the power grid, there is a sudden change in voltage and/or current levels. Previous inverter-interfaced renewable generators could disconnect from the power system under such conditions. However, with the increased penetration of wind and solar farms, this is not possible as the disconnections during a fault may cause the grid to be unstable, resulting in widespread blackouts. Therefore, the modern grid codes that the renewable generators stay connected to ride through grid disturbances.

Equipment with FRT capability can detect these changes and respond appropriately to ride through the fault without shutting down or causing further disruptions to the system. This capability is crucial for maintaining the stability and reliability of the electrical grid, as it helps prevent cascading failures and minimizes downtime. For example, according to National Grid in the United Kingdom, grid code CC.6.3.15 defines the FRT as follows [32]:

- The generating plants will remain stably connected to the power system for a fault (balanced or unbalanced) up to 140 ms.
- The generating plants will remain stably connected to the power system for balanced voltage dips of more than 140 ms.

FRT is often a requirement for grid-connected renewable energy systems, such as wind turbines and solar inverters, to ensure that they can continue to operate seamlessly even in the presence of grid disturbances. A widespread gird disturbance and blackout occurred in South Australia (SA) on September 28, 2016. Prior to the event, electricity demand in SA was 1,826 MW, of which 883 MW was generated by wind, 330 MW was generated through gas turbines, and 613 MW was imported from Victoria through two interconnectors [33]. The extreme weather conditions resulted in five faults in 87 seconds, after 4 PM. These faults resulted in six voltage disturbances and 456 MW of wind power being lost. Also, there was a transient reduction of 42 MW of wind power. These events required that 900 MW be supplied from Victoria through the Heywood interconnector. This is beyond the capacity of the interconnector, and it tripped. There was no immediate load shedding, and, consequently, due to the mismatch between the generation and load, the frequency collapsed resulting in a system-wide blackout. As per the Australian Energy Market Operator (AEMO), the software settings prevented frequent ride through [33]. A group of wind turbines that could not handle repeated dropouts were disconnected.

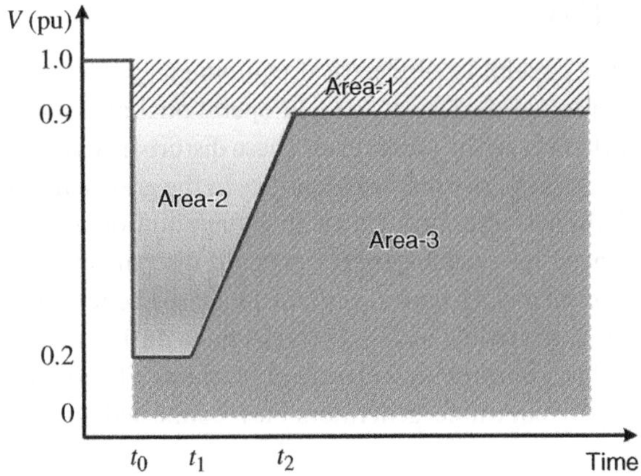

Figure 11.33 A typical low-voltage ride through curve.

A typical low-voltage ride through curve for wind generators is shown in Figure 11.33. In this, the fault occurs at t_0, where the power drops 0.2 per unit. The fault is cleared at t_1 and the power ramps up to 0.9 per unit till time t_2. There are three areas [34]:

- *Area-1*: The wind turbine must remain connected to the power system maintaining normal operation.
- *Area-2*: The wind turbine must remain connected to the grid. During this time, the wind turbine must supply the required reactive power to provide maximum voltage support.
- *Area-3*: The wind turbine may disconnect from the grid.

Type 1 and Type 2 wind turbines cannot supply sufficient reactive power that will be required for the turbines to provide maximum voltage support. To avoid this problem, STATCOM can be employed. Different jurisdictions have different grid standards for FRT. In [35], some of these low voltage ride-through (LVRT) curves are given, along with discussions on how to develop FRT criteria.

11.7 Renewable Energy in Power Distribution Systems

The are several rooftop PVs that are getting connected to power distribution systems. Usually, these are connected to one of the three phases of a three-phase system. They cause several problems in the power distribution systems such as

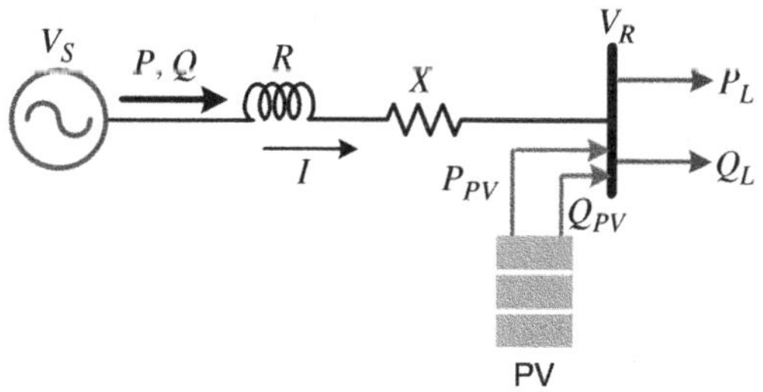

Figure 11.34 A single-phase distribution system with a PV connection.

voltage rise, increased line losses, voltage unbalance, and unbalanced reverse power flow. Some of these aspects are discussed in this section.

11.7.1 Voltage Rise and Line Loss

Consider the single-phase distribution system that has a rooftop PV connection at the load bus, as shown in Figure 11.34. The feeder impedance is given by $R + jX$. The load is assumed to be constant PQ type drawing real power P_L and reactive power Q_L. The PV is injecting real power P_{PV} and reactive power Q_{PV}. The total real and reactive power entering or leaving the load bus is then

$$P = P_L - P_{PV} \text{ and } Q = Q_L - Q_{PV}$$

Let the sending and the receiving end voltages be denoted as

$$V_S = |V_1|\angle 0^\circ \text{ and } V_R = |V_2|\angle -\delta$$

Then the current flowing through the feeder is

$$I = \frac{|V_1|\angle 0^\circ - |V_2|\angle -\delta}{R + jX} \tag{11.38}$$

The complex power at the load bus is then given by

$$\begin{aligned} P + jQ &= |V_2|\angle -\delta \times I^* = |V_2|\angle -\delta \left[\frac{|V_2|\angle 0^\circ - |V_2|\angle \delta}{R - jX}\right] \\ &= \frac{|V_1||V_2|\angle -\delta - |V_2|^2}{R - jX} \end{aligned} \tag{11.39}$$

Separating the real and imaginary components, we have the following two equations:

$$PR + QX = |V_1||V_2|\cos\delta - |V_2|^2 \tag{11.40}$$

$$QR - PX = -|V_1||V_2|\sin\delta \tag{11.41}$$

Knowing P, Q, R, X, and V_1, the quantities V_2 and δ can be determined by Newton–Raphson method, where the Jacobean matrix is given below.

$$J = \begin{bmatrix} |V_1|\cos\delta & -|V_1||V_2|\sin\delta \\ |V_1|\sin\delta & |V_1||V_2|\cos\delta \end{bmatrix}$$

Example 11.3 For the system of Figure 11.34, the following parameters are chosen:

$$V_S = \frac{415}{\sqrt{3}} = 239.6\text{ V}, R = 0.2\ \Omega, X = 0.1\ \Omega$$

The constant PQ load is assumed as 800 W, with a lagging power factor of 0.95, that is,

$$P_L = 800\text{ W}, Q_L = \tan\left(\cos^{-1}(0.95)\right) = 262.95\text{ Var}$$

A few specific cases are considered. They are discussed below.

Case-1 (No PV Injection): In this case, $P = P_L$ and $Q = Q_L$. The results are as follows:

$$V_L = 238.82\angle -0.027^\circ\text{ V}$$
$$I = 3.526\angle -18.22^\circ\text{ A}$$

The line loss ($|I|^2R$) is equal to 2.487 W.

Case-2 (Low PV Injection at Unity Power Factor): Let us assume that the PV is injecting 800 W of power at unity power factor, such that $P = 0$ and $Q = Q_L$. This implies that the entire load demand is supplied by the PV, while the source only supplies the reactive power requirement of the load. The results are as follows:

$$V_L = 239.49\angle -0.053^\circ\text{ V}$$
$$I = 1.098\angle -89.95^\circ\text{ A}$$

The line loss is equal to 0.24 W. Since the feeder current is almost reactive, the line loss is very small.

Case-3 (High PV Injection at Unity Power Factor): Let us assume that the PV is injecting 4 kW of power at unity power factor, such that $P = -3200$ W and $Q = Q_L$. This implies that 3,200 W power is being feedback to the source. The results are as follows:

$$V_L = 242.13\angle 0.36^\circ\text{ V}$$
$$I = 13.26\angle -174.93^\circ\text{ A}$$

The line loss is equal to 35.17 W. Since the current has increased, the line loss has also increased significantly. Also, the load voltage leads the source voltage as the power is feedback toward the source.

Case-4 (High PV Injection at Non-unity Power Factor): Let us assume that the PV is injecting 4 kW of power at unity power factor and supplying the total reactive power of the load, such that $P = -3200$ W and $Q = 0$. The results are as follows:

$$V_L = 242.24\angle -0.32^\circ \text{ V}$$
$$I = 13.21\angle -179.68^\circ \text{ A}$$

The line loss is equal to 34.9 W. There is no significant improvement in the system performance vis-à-vis that of Case-3.

With the P_{PV} remaining constant at 4 kW, the reactive power Q_{PV} injected by the PV varied from −4 kVar to +4 kVar. The line loss and the load bus voltage are shown in Figure 11.35. It can be seen that the line loss is higher as the reactive power absorption increases, even though it reduces the load bus voltage. The line loss is minimal when the reactive power is about 10% of the real power. Increased line loss not only causes wastage of power generation but also can cause heating in the distribution feeder.

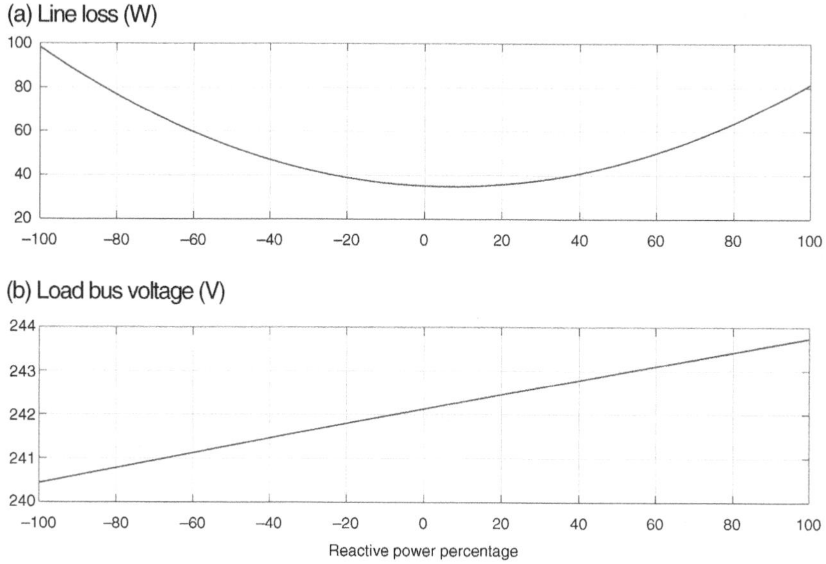

Figure 11.35 Line loss and load bus voltage with the variation in reactive power.

Example 11.4 Consider the distribution system shown in Figure 11.36. The system parameters are as follows:

$$V_S = 230\,\text{V}, Z_f = 0.02 + j0.01\,\Omega, Z_S = 5Z_f$$

Note that the distribution feeders are mostly resistive and hence the resistance has been chosen to be twice the reactance. The loads are assumed to be PQ type, as indicated in Figure 11.36, where the subscripts indicate the bus number. There are two load conditions – light and heavy load. These are listed in Table 11.2. The loads are assumed to have a power factor of 0.90. In this example, we shall consider that the PVs inject power at the unity power factor, where the PV ratings are listed in Table 11.2.

The PVs are connected one at a time. First, the PV at bus-5 is connected, then the PV at bus-4 is connected, followed by the PV at bus-3 and bus-2. We shall consider the heavy load case first, where the PVs are connected one at a time. The results are listed in Table 11.3. Similarly, the results for light load are listed in Table 11.4.

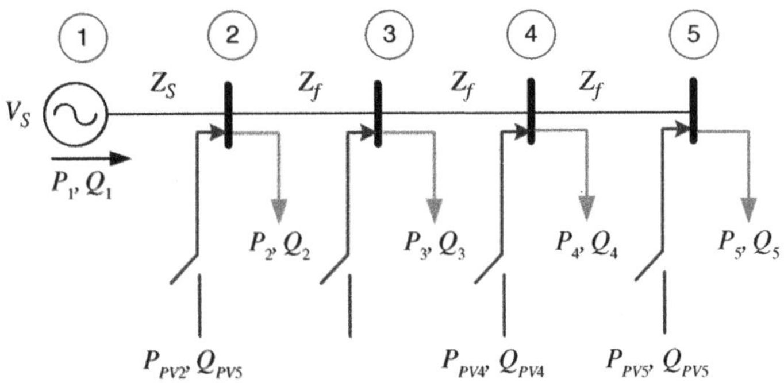

Figure 11.36 A five-bus radial system with PV injection.

Table 11.2 Loads and PV injection for the system of Figure 11.33.

Conditions	Bus-2 (kW)	Bus-3 (kW)	Bus-4 (kW)	Bus-5 (kW)
Light load	1	1.2	1.5	2
Heavy load	4	4.2	5	5.5
PV injection	2	3	3	5

Table 11.3 Impact of PV connection during heavy load.

PV connection	$\|V_2\|$ (V)	$\|V_3\|$ (V)	$\|V_4\|$ (V)	$\|V_5\|$ (V)	Source power P_1 (kW)	Line loss (W)
No PV	219.39	217.61	216.41	215.77	19.80	1104
PV-5	221.73	220.52	219.79	219.62	14.34	641.14
PV-5,4	223.14	222.21	221.76	221.59	11.15	452.96
PV-5,4,3	224.52	223.86	223.42	233.25	8.02	323.97
PV-5,4,3,2	225.42	224.76	224.32	224.16	5.97	268.29

Table 11.4 Impact of PV connection during light load.

PV connection	$\|V_2\|$ (V)	$\|V_3\|$ (V)	$\|V_4\|$ (V)	$\|V_5\|$ (V)	Source power P_1 (kW)	Line loss (W)
No PV	226.87	226.35	225.97	225.75	5.79	96.99
PV-5	229.09	229.01	229.07	229.29	0.72	23.26
PV-5,4	230.38	230.57	230.88	231.10	−2.26	42.68
PV-5,4,3	231.66	232.11	232.42	232.64	−5.21	94.96
PV-5,4,3,2	232.51	232.96	233.27	233.48	−7.16	140.67

The results are displayed graphically in Figures 11.37–11.39, where the case numbers are denoted as follows:

- *Case-1*: No PV is connected.
- *Case-2*: PV connected to bus-5.
- *Case-3*: PVs connected to buses 5 and 4.
- *Case-4*: PVs connected to buses 5, 4, and 3.
- *Case-5*: PVs connected to buses 5, 4, 3, and 2.

The magnitudes of voltages at bus-4 are shown in Figure 11.37. The trends during both the heavy and light load cases are similar – as the PV injection increases, the voltage magnitude increases. However, the magnitude remains lower than the nominal voltage of 230 V during the heavy load. On the other hand, the magnitude rises above the nominal level as the PV injection increases.

The power flowing out from bus-1 (i.e., from the source) is shown in Figure 11.38. When there is no PV, bus-1 supplies the entire power. When one PV has been placed in bus-5, the power dispatched from bus-1 reduces by

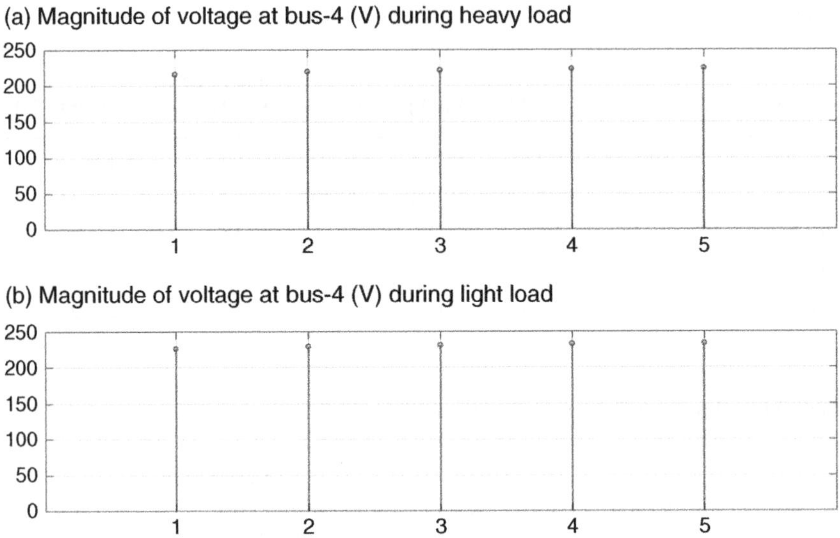

Figure 11.37 Voltage magnitudes of bus-4 during (a) heavy load and (b) light load.

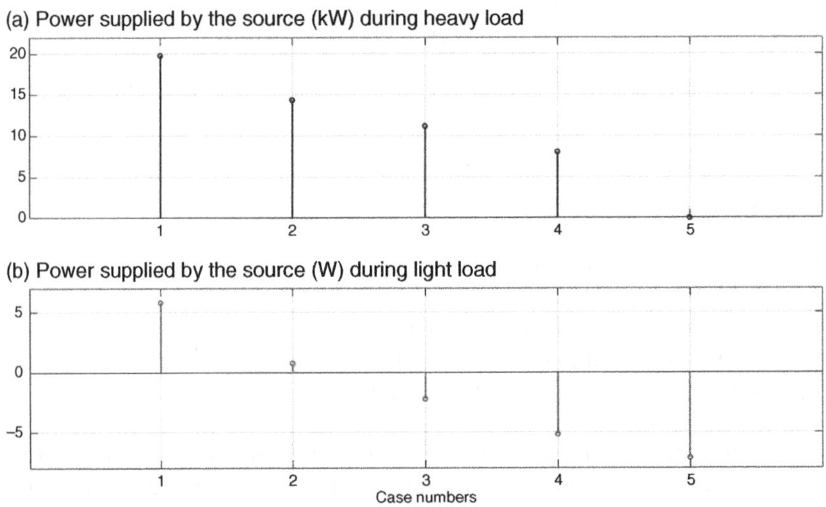

Figure 11.38 Power supplied by the source during (a) heavy load and (b) light load.

about 5 kW, which is equal to the amount of PV injection. Subsequently, as more PVs are added, the amount of power flowing from bus-1 decreases. During the light load condition, the power flow reverses when more than one PV is connected to the system.

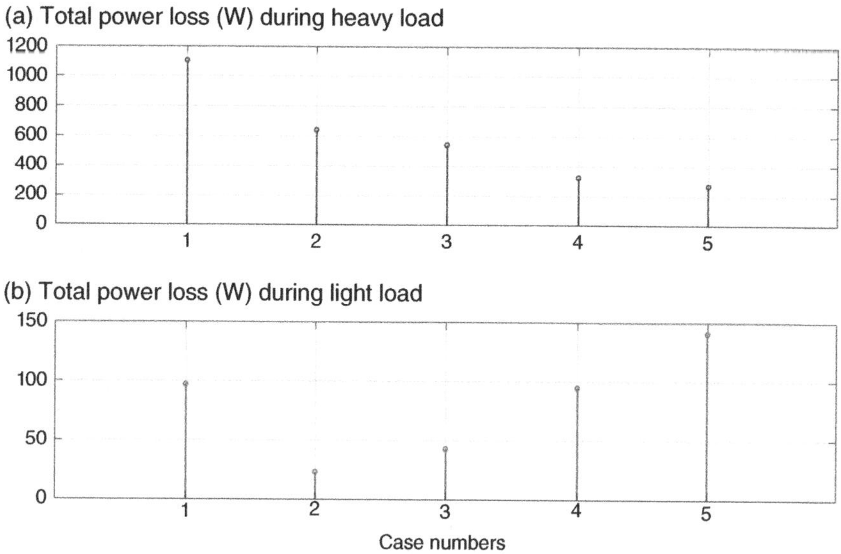

Figure 11.39 Line loss during (a) heavy load and (b) light load.

The line loss is shown in Figure 11.39. Generally, it decreases as more PVs are connected during heavy load conditions since the voltage difference is reduced. It increases during light load conditions. However, the amount of line loss is significantly higher during the heavy load condition than during the light load condition.

The above example shows that PV power injection at the unity power factor causes the bus voltages to rise, especially during light load conditions. The voltage rise is inversely proportional to the bus length from the substation, that is, the further away the bus is from the substation, the more the voltage rise is. Also, the voltage rise increases as the number of PV connected to the feeder increases. Remember that this is a small feeder and therefore the voltage rise may not be that significant. However, when there are 40 or so nodes connected to the feeder, the voltage rise may cause the PV inverters to trip. The example below shows the effect of non-unity power factor injection of PVs on the voltage rise and line loss.

Example 11.5 The voltage rise is a problem during the light load conditions. Therefore, we shall only consider this condition for this example. Let us assume that a converter injects I amount of current into a system at a unity power factor with a voltage magnitude of V. Then the power injected is $P = VI$. Let us assume that the voltage magnitude remains constant. We now assume that the converter absorbs (not injects) Q amount of power. Then the current will be

$$I = \frac{\sqrt{P^2 + Q^2}}{V}$$

This implies that the converter current must be rated 11.8% higher if 50% reactive power (with respect to 100% real power) needs to be absorbed. Also, if the converter absorbs 100% reactive power, the converter rating must increase by 41.4%. Both these levels of reactive power absorptions are used in this study. The results are tabulated in Tables 11.5 and 11.6. It can be seen that with reactive power absorption, the voltage rise problem is arrested – the more the absorption, the less the voltage rise is. However, this occurs at the expense of the line loss, that is, the line loss increases with reactive power absorption.

11.7.2 Reverse Power Flow and Voltage Unbalance

As can be seen from Figure 11.38 (b) when there are several PVs in a distribution feeder and the load is not sufficient to absorb the power generated, reverse power flow can occur. Now consider the case of a three-phase distribution system. Since the installation of rooftop PVs is strictly dependent on the customer's willingness

Table 11.5 Results of different conditions for 50% reactive absorption.

PV connection	$\lvert V_2 \rvert$ (V)	$\lvert V_3 \rvert$ (V)	$\lvert V_4 \rvert$ (V)	$\lvert V_5 \rvert$ (V)	Line loss (W)
PV-5	228.52	228.33	228.28	228.39	78.74
PV-5,4	229.48	229.49	229.63	229.74	144.07
PV-5,4,3	230.42	230.63	230.77	230.84	246.43
PV-5,4,3,2	231.05	231.26	231.40	231.51	324.75

Table 11.6 Results of different conditions for 100% reactive absorption.

PV connection	$\lvert V_2 \rvert$ (V)	$\lvert V_3 \rvert$ (V)	$\lvert V_4 \rvert$ (V)	$\lvert V_5 \rvert$ (V)	Line loss (W)
PV-5	227.94	227.64	227.47	227.47	173.89
PV-5,4	228.54	228.37	228.33	228.33	338.79
PV-5,4,3	229.12	229.08	229.05	229.05	558.01
PV-5,4,3,2	229.51	229.47	229.44	229.44	715.48

and financial situation, there might be unequal numbers of PVs in different phases. Moreover, the rooftop PVs are usually connected to the distribution grids through single-phase inverters. The problems with this arrangement are:

- One of the phases sees reverse power flow, that is, power flowing from downstream toward upstream substation. The other two phases may draw power from the upstream network.
- Two of the phases are feeding back power toward the upstream network, while the third phase is drawing power from the upstream network.
- All three phases are feeding back power to the upstream network, but not in equal quantities.

These unequal forward or reverse power flows through the phases will cause a severe unbalance in the upstream network, which will, in turn, be connected to a transmission system. It is desirable that a transmission system works in a balanced fashion. However, if several distribution substations feedback power in an unbalanced fashion, then the transmission system voltages will be affected. One way to alleviate the problem is to use a distributed static compensator (DSTATCOM). The following example illustrates the idea.

Example 11.6 Consider the 4-bus distribution system shown in Figure 11.40.

In this, the feeder is the backbone 11 kV system from which different low voltage feeders emanate. The low-voltage loads and feeders are aggregated. It is assumed that all the distributed energy resources (DERs) are connected only to one of the phases (phase-a). This is intentionally chosen to consider the worst-case scenario. A DSTATCOM is connected at the point of common coupling (PCC). It will be

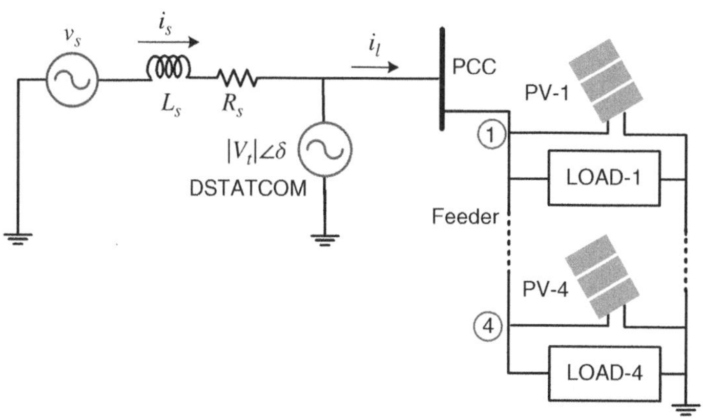

Figure 11.40 A 4-bus distribution system containing loads, PVs, and a DSTATCOM.

shown that the DSTATCOM can circulate power through its DC bus, thereby balancing the power to/from the upstream network.

A DSTATCOM is a VSC that is connected in a shunt with a distribution system through passive filters. It can perform various functions, such as harmonic filtering, power factor correction, load balancing, and voltage regulation. Usually, the DC side of the DSTATCOM is supplied by a DC storage capacitor. However, in this example, we assume that the DC side is connected to a battery energy storage system (BESS). The advantages of using a BESS are that the BESS can (a) store energy when the renewable generation is high and the load is light, and (b) supply energy when the load is high, and the renewable generation is low or is not available. It has been assumed that the DSTATCOM operates in the voltage control mode where it regulates the PCC voltage to a prespecified fixed value in a balanced fashion. It has been assumed that the source voltage is balanced. Then, once the PCC voltage gets balanced, the source current will also get balanced, irrespective of the direction of power flow.

When the DSTATCOM operates in the voltage control mode, a prespecified voltage magnitude of $|V_t|$ is chosen. The VSC control determines the voltage angle δ that is required for the amount of power flow in and out of the BESS. The VSC switching control generates a balanced voltage with magnitude $|V_t|$ and phase angle δ. DSTATCOM structure and control are discussed in detail in [36, 37].

The system parameters are given in Table 11.7. With the PVs injecting power in the steady state, the DSTACOM is connected at 0.1 seconds. It is desired that the

Table 11.7 System parameters of the 4-bus system of Figure 11.36.

System quantities		Values
Bus-1	Load impedance	Phase-a: $400 + j131.48\ \Omega$ Phase-b: $640 + j210.36\ \Omega$ Phase-c: $2400 + j788.85\ \Omega$
	PV	300 kW
Bus-2	Load impedance	Phase-a: $300 + j98.605\ \Omega$ Phase-b: $560 + j184.06\ \Omega$ Phase-c: $2200 + j723.11\ \Omega$
	PV	250 kW
Bus-3	Load impedance	Phase-a: $200 + j65.737\ \Omega$ Phase-b: $480 + j157.77\ \Omega$ Phase-c: $2000 + j657.37\ \Omega$
	PV	400 kW
Bus-4	Load impedance	Phase-a: $100 + j38.868\ \Omega$ Phase-b: $400 + j131.47\ \Omega$ Phase-c: $1800 + j591.63\ \Omega$
	PV	350 kW

DSTATCOM supplies 100 kW power once it is connected and the DSTATCOM desired voltage is 11 kV (i.e., 6.35 kV line neutral). The results are shown in Figs. 11.41 and 11.42. The power flowing from/to the upstream network for the three phases is shown in Figure 11.41 (a). Before the DSTATCOM connection, phase-a feeds back about 530 kW of power to the network, while phases b and c draw 252.5 and 67.4 kW from the network, respectively. Once the system settles down after the DSTATCOM connection, all three phases send about 25.5 kW of power to the network in a nearly balanced fashion. The power supplied by the DSTATCOM is shown in Figure 11.41 (b). It is 100 kW, as desired.

Figure 11.42 depicts the various attributes of the PCC voltage, where rms magnitudes of the three-phase voltages are shown in Figure 11.39 (a). They are 6.34 kV. 5.92 kV and 6.25 kV, respectively, for the three phases before the DSTATCOM connection. After the DSTATCOM is connected, they become nearly equal to 6.35 kV, as desired. The phase angles of the PCC voltages are shown in Figure 11.39 (b). They are displaced by 120° from each other once the system reaches a steady state. The voltage unbalance of a circuit is defined in terms of the ratio of the negative sequence to the positive sequence magnitude, given by [36]

$$\text{VUF} = \left|\frac{V_{a2}}{V_{a1}}\right| \times 100\%$$

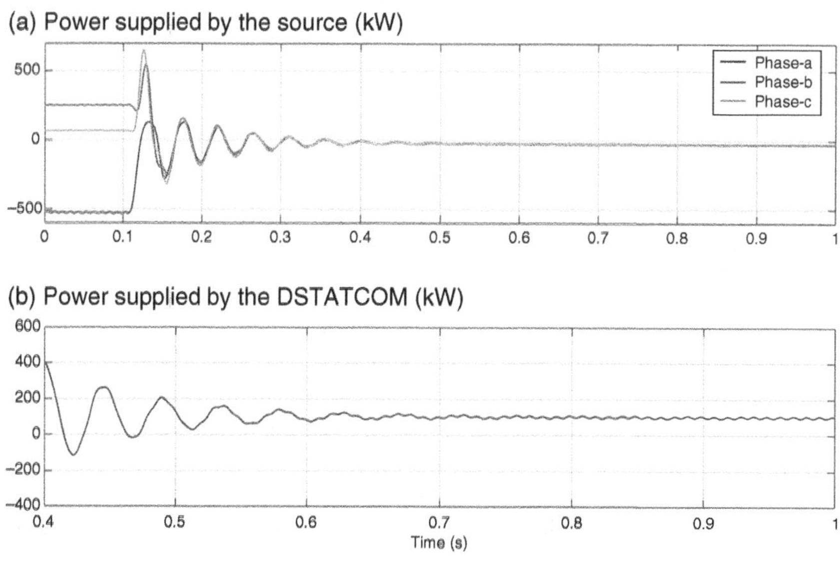

Figure 11.41 (a) Power flowing from the source and (b) power flowing out of the DSTATCOM.

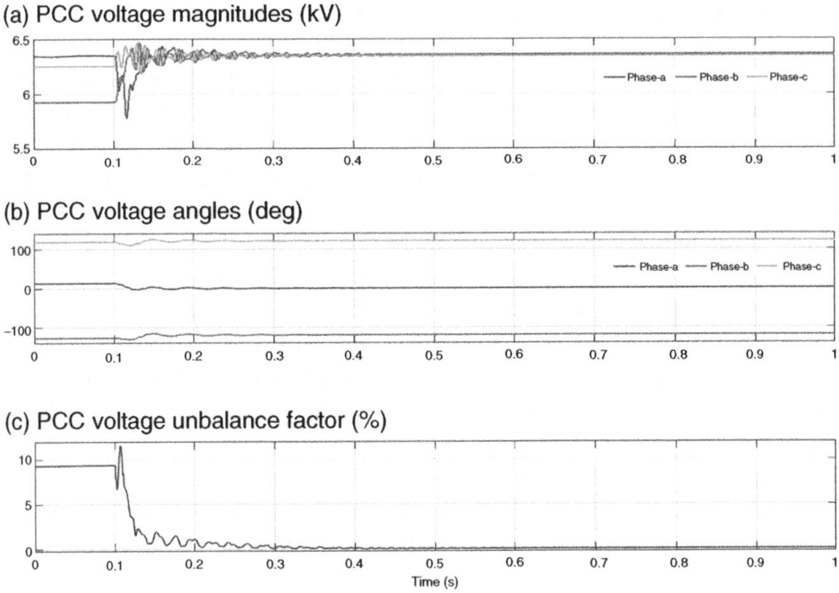

Figure 11.42 Attributes of the PCC voltage (a) rms magnitudes, (b) phase angles, and (c) voltage unbalance factor.

The phasor positive voltage and the negative-sequence components are computed from their instantaneous measurement [36]. The voltage unbalance factor (VUF) that is computed thereafter is shown in Figure 11.39. The VU before the DSTATCOM connection is 9.35%. However, it becomes 0.07% once the system reaches the steady state.

11.8 Concluding Remarks

Three different renewable electric power generation technologies have matured so far. These are hydropower, wind power, and solar PV-based power generation. Other than these, there are several geothermal power plants around the world that trap energy from the earth's mantle. However, these are location dependent, and their penetration level is low. Nevertheless, there are many technologies and possibilities to diversify electric power supply systems. This chapter provided a glimpse of different technologies that are being considered these days. It is to be remembered that some of these technologies are in the experimental stages. A significant amount of investment is required before they can become commercially viable.

It is interesting to note that when these technologies mature or with the present state of the art, some of them can provide baseload, and some of them will be non-dispatchable. For example, hydro (even though it can be seasonal) and geothermal can supply the baseload. On the other hand, solar PV, wind, tidal, and wave energy are intermittent. A fully developed CSP can supply power day or night using thermal storage if it is situated in a proper location. Hydrogen can be used for both power generation and automotive applications, even though it is not a primary source of energy. Finally, and obviously, with mature nuclear fusion technology, most of the other technologies will become obsolete.

There are several problems that are been faced currently as the interconnection of renewable generators increases in both transmission and distribution systems. Some of these aspects are discussed in this chapter. One of the problems in power transmission systems is the lack of inertia from renewable generators, which has been discussed in Section 11.6. However, there are other problems that occur due to the intermittency of wind and solar generation. To address this problem, energy storage systems need to be deployed in strategic locations. The most common storage of pumped hydro has a slow response time and may not address the problem of the rapid ramping rate of renewable generators. There are other storage systems that are been developed these days like compressed air storage systems that are not fast enough to address this issue either. Therefore, very large-scale battery storage systems have been deployed or planned these days. Most of them use lithium-ion technology and are situated in the United States, where California leads the deployment. Moss Landing in California battery storage system has a capacity of 750 MW, with 3,000 MWh energy storage capacity. Australia is also heavily involved in large-scale battery deployment. Victorian Big Battery near Geelong has a capacity of 300 MW, 450 MWh.

Some of the problems facing distribution systems are discussed in Section 11.7. The integration of renewable energy is very prevalent in low-voltage power networks. Moreover, microgrids are also getting set up that join the distribution networks. Therefore, a more holistic approach is needed to address how to plan and utilize the small-scale and distributed renewable generators. Also, almost all these distributed generators are connected to the power grid through power electronic converters. In general, these converters generate harmonics in the frequency range of 150 Hz–2 kHz. However, wide band gap devices that are getting cheaper can have switching frequencies in the range of 100 kHz. Therefore, the frequency range of interest has been changed to 2–150 kHz [37]. A new filtering methodology and protection philosophy must be created for addressing the harmonic problems.

References

1 IPCC Sixth Assessment Report, 2022. Available https://www.ipcc.ch/report/ar6/wg2/

2 International Energy Agency, *Hydropower Power Special Market Report*. Available https://www.iea.org/reports/hydropower-special-market-report/executive-summary

3 US Department of Energy, *Types of Hydropower Turbines*. Available https://www.energy.gov/eere/water/types-hydropower-turbines

4 Argonne National Laboratory, *Modeling Ternary Pumped Hydro Storage Units*, 2013. Available https://ceeesa.es.anl.gov/projects/psh/ANL_DIS-13_07_Modeling_Ternary_Units.pdf

5 Lopez, I., Andreu, J., Ceballos, S. et al. (2013). Review of wave energy technologies and the necessary power-equipment. *Renewable and Sustainable Energy Reviews* 27: 413–434.

6 Falcao, A.F. (2010). Wave energy utilization: a review of technologies. *Renewable and Sustainable Energy Reviews* 14: 899–918.

7 Masters, G.M. (2004). *Renewable and Efficient Electric Power Systems*. New Jersey: John Wiley.

8 Verma, D., Nema, S., Shandilya, A.M., and Dash, S.K. (2016). Maximum power point tracking (MPPT) techniques: recapitulation in solar photovoltaic systems. *Renewable and Sustainable Energy Reviews* 54: 1018–1034.

9 de Brito, M.A.G., Galotto, L. Jr., Sampaio, L.P. et al. (2013). Evaluation of the main MPPT techniques for photovoltaic applications. *IEEE Transactions on Industrial Electronics* 60 (3): 1156–1167.

10 Bendib, B., Belmili, H., and Krim, F. (2015). A survey of the most used MPPT methods: conventional and advanced algorithms applied for photovoltaic systems. *Renewable and Sustainable Energy Reviews* 45: 637–648.

11 Sera, D., Mathe, L., Kerekes, T. et al. (2013). On the perturb-and-observe and incremental conductance MPPT methods for PV systems. *IEEE Journal of Photovoltaics* 3 (3): 1070–1078.

12 Mohan, N., Undeland, T.M., and Robbins, W.P. (2002). *Power Electronics: Converters, Applications and Design*, 3rde. New York: John Wiley.

13 Femia, N., Petrone, G., Spagnuolo, G., and Vitelli, M. (2005). Optimization of perturb and observe maximum power point tracking method. *IEEE Transactions on Power Electronics* 20 (4): 963–973.

14 Zhang, H.L., Baeyens, J., Degrve, J., and Caceres, G. (2013). Concentrated solar power plants: review and design methodology. *Renewable and Sustainable Energy Reviews* 22: 466–481.

15 J. Aho, A. Buckspan and J. Laks, et al., “A tutorial of wind turbine control for supporting grid frequency through active power control,” *American Control Conference*, Montréal, Canada, June 2012.

16 Hansen, A.D., Sorensen, P., Iov, F., and Blaabjerg, F. (2004). Control of variable speed wind turbines with doubly-fed induction generators. *Wind Engineering* 28 (4): 411–434.

17 IEEE PES Wind Plant Collector System Design Working Group, "Characteristics of wind turbine generators for wind power plants," *IEEE PES General Meeting*, Calgary, Canada, July 2009.

18 Sen, P.C. (1997). *Principles of Electric Machines and Power Electronics*, 2nde. New York: John Wiley.

19 Kothari, D.P. and Nagrath, I.J. (2010). *Electric Machines*, 4the. New Delhi: Tata McGraw-Hill.

20 Muller, S., Deicke, M., and De Doncker, R.W. (2002). Doubly fed induction generator systems for wind turbines. *IEEE Industry Applications Magazine* 8 (3): 26–33.

21 Airships.net, "The Hindenburg Disaster." Available https://www.airships.net/hindenburg/disaster/

22 Shiva Kumar, S. and Himabindu, V. (2019). Hydrogen production by PEM water electrolysis – a review. *Materials Science for Energy Technologies* 2: 442–454.

23 International Energy Agency Report prepared for G20, Japan, *Future of Hydrogen: Seizing Today's Opportunity*, 2019.

24 Wärtsilä, *Hydrogen: A Key Component of Net Zero Energy Systems*, Available https://www.wartsila.com/energy/sustainable-fuels/hydrogen-power-plant

25 International Atomic Energy Agency (IAEA), *What is Nuclear Fusion?* Available https://www.iaea.org/newscenter/news/what-is-nuclear-fusion

26 World Nuclear Association, *Nuclear Fusion Power*, Available https://world-nuclear.org/information-library/current-and-future-generation/nuclear-fusion-power.aspx

27 Ackermann, T., Prevost, T., Vittal, V. et al. (2017). Paving the wat: a future without inertia is closer than you think. *IEEE Power and Energy Magazine* 15: 61–69.

28 Lasseter, R.H., Chen, Z., and Pattabiraman, D. (2020). Grid forming inverters: a critical asset for power grids. *IEEE Journal of Emerging and Selected Topics in Power Electronics* 8 (2): 925–935.

29 Rathnayake, D.B., Akrami, M., Phurailatpam, C. et al. (2021). Grid forming inverter modeling, control, and applications. *IEEE Access* 9: 114781–114808.

30 M. Goyal, A. Ghosh and F. Zare, "Power sharing control with frequency droop in a hybrid microgrid," *IEEE Power Engineering Society General Meeting*, Vancouver, 2013.

31 D'Arco, S. and Suul, J.A. (2014). Equivalence of virtual synchronous machines and frequency-droops for converter-based microgrids. *IEEE Transactions on Smart Grid* 5 (1): 394–395.

32 A. Johnson, *RfG Implementation Fault Ride through*, National Grid – Technical Policy. Available https://www.nationalgrid.com/sites/default/files/documents/8589934611-GC0048-T%20FRT%20slides%20February%2016.pdf

33 Australian Energy Market Operator (AEMO), *Black System South Australia 28 September 2016*, Melbourne, March 2017.
34 Energinet DK, *Technical Regulation 3.2.5 for Wind Power Plants above 11 kW*. Available https://en.energinet.dk/media/3h4fzmqb/technical-regulation-325-for-wind-power-plants-above-11-kw-revision-4.pdf
35 Nair, N.C. and Quereshi, W.A. (2014). Fault ride through criteria development," Chapter 3. In: *Renewable Energy Integration: Challenges and Solutions* (ed. J. Hossain and A. Mahmud), 41–67. Singapore: Springer Science+Business Media.
36 Ghosh, A. and Ledwich, G. (2002). *Power Quality Enhancement Using Custom Power Devices*. Norwell, MA: Kluwer Academic Publishers.
37 Ghosh, A. and Zare, F. (2023). *Control of Power Electronic Converters with Microgrid Applications*. Hoboken, New Jersey: IEEE Press-Wiley.

Problems

P11.1 The density of water changes with thermal expansion or contraction. For a given temperature, the density of water is given by the following empirical formula.

$$\rho(T) = \rho_0 \times [1 - \beta(T - T_0)]$$

where $\rho(T)$ is the water temperature at T°C; ρ_0 is the reference density of water, which can be approximated as 1,000 kg/m^3 at 4 °C; T_0 is the reference temperature of 4 °C; and β is the thermal expansion of water, which is approximately equal to 2.14×10^{-4}/°C.

Then find the output power of a 95% efficient water turbine when at the water falls from a height of 90 m at a temperature of 50 °C, where the penstock flow is 80 m^3/s.

P11.2 Suppose a hydropower plant has 10 generators with 95% efficiency, each with a water height of 90 m. These are operating at temperature of 50 °C. Then regulate the penstock flow of water such that these generators together produce a power output of 600 MW.

P11.3 Consider the equivalent circuit of a solar cell shown in Figure 11.9 (c). Determine the percentage efficiency of the cell if it produces 0.40 W at $V_d = 0.5$ V. Given $I_0 = 10^{-9}$ A and $I_{SC} = 1$ A at 25 °C.

P11.4 Suppose a PV panel is supplying a 20 Ω load. The PV output voltage V_d is 36 V and its current I_L is 5.5 A at the maximum power point. Find the duty ratio of the boost converter to deliver this power to the load.

P11.5 Air density (ρ) at a height (h) can be calculated using the following equation.

$$\rho = \rho_0 e^{-h/H} \text{ kg/m}^3$$

where ρ_0 is the air density at the sea level, which is 1.225 kg/m^3; H is the scale height equal to 8,400 m.

Then calculate the wind power output when a wind turbine with a diameter of 20 m is placed (a) at sea level, (b) at a height of 500 m and (c) at a height of 1 km at a wind speed of 12 m/s.

P11.6 A 50 m wind turbine has a TSR of 5 at wind speed of 15 m/s. Find (a) rotor tip speed in m/s and (b) rotor speed in rpm.

P11.7 If the wind turbine of Problem 11.6 is placed at sea level, what is the maximum power can it produce at a wind speed of 15 m/s?

P11.8 A wind farm consists of 30 wind turbines, each with a diameter of 50 m. The wind farm, which is placed at the ground level, is connected to a power grid through a transmission line to an infinite bus. The power unit impedance of the transmission line is 0.2 on a base MVA of 20 and a base voltage of 66 kV. The wind farm voltage is boosted to 60 kV and its per unit voltage is denoted by $V_W \angle 0°$. The infinite bus voltage is given by $V_\infty = 1 \angle -\delta$. If the efficiency of all the wind turbines is 37.942%, then find the angle δ when each turbine operates at a wind speed of 15 m/s.

12

Fundamentals of Smart Grid

As mentioned in Chapter 1, the aim of the smart grid is to apply innovative information and communication technologies to improve the reliability of power systems. The conceptual diagram of a smart grid is shown in Figure 12.1. The different components of a smart grid are listed below:

- A smart grid will have generation that can be traditional as well as renewable.
- Since most renewable generations are dependent on weather, a smart grid will need weather forecasting algorithms.
- The transmission systems can be AC or DC. AC systems will be equipped with energy management systems, including phasor measurement units (PMUs) and FACTS controllers.
- The DC system can be two-terminal or multiterminal.
- Most cyberattacks can occur during power transmission. To prevent these attacks, cybersecurity is an important aspect of the smart grid as well.
- Power distribution systems can have traditional distribution systems and microgrids (MGs).
- Demand response (DR), substation automation, smart load shedding, and self-healing are techniques implemented at this level.
- Digital communication systems are the backbone of the smart grid operation, and different communication technologies can be integrated with the smart grid.

There are four building blocks of a smart grid [1]. These are defined as follows:

- *Sensor System*: CT, PT, phasor measurement unit (PMU), smart meter, etc.
- *ICT Infrastructure*: Advanced metering infrastructure (AMI), home area network (HAN), smart grid communications, cybersecurity, etc.

Modern Power System, First Edition. Arindam Ghosh.

Published 2026 by John Wiley & Sons, Inc.
Companion website: www.wiley.com/go/ghoshpowersystem

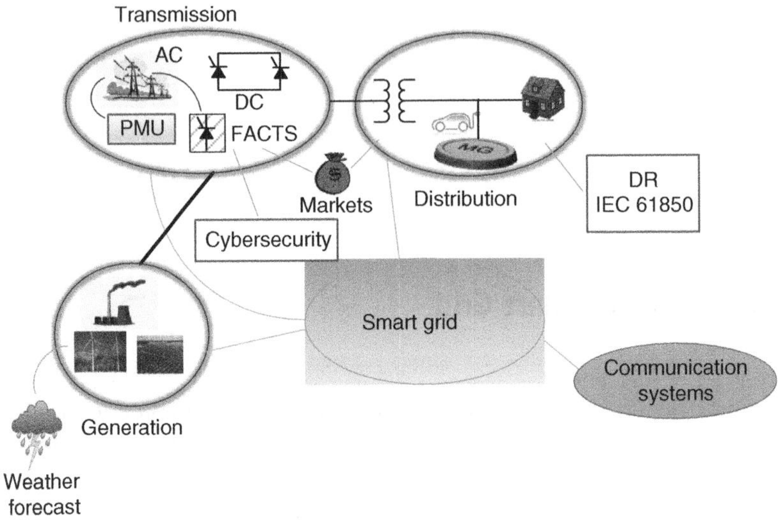

Figure 12.1 Conceptual diagram of smart grid.

- *Control Algorithms*: Wide-area monitoring and control, distribution system load management and reconfiguration, MG, DR, optimal power flow, fault detection and recovery, interarea oscillation damping, self-healing process, etc.
- *Actuators*: HVDC, FACTS, distributed generation, battery energy storage system (BESS), breakers and reclosers, etc.

Of these, basic sensor systems have already been discussed in Chapter 6. Furthermore, the actuators have been discussed in Chapters 8–10. More advanced sensors will be discussed in this chapter. We shall also discuss some of the control algorithms and (information and communication technology (ICT) infrastructure in this chapter.

Smart grid is an evolving area where uniform standards are required to ensure interoperability, reliability, and security of operations. Some of the key aspects of smart grid standards are:

- *Interoperability*: This is a very critical issue that ensures that different devices from different vendors can communicate and work together seamlessly.
- *Security*: To protect the grid and data from cyberattacks.
- *Reliability and Scalability*: This is to maintain the stability of supply, while allowing the integration of new technologies and grid expansion.

There are several organizations that are working on different smart grid standards, such as the International Electrotechnical Commission (IEC), the Institute of Electrical and Electronics Engineers (IEEE), and the National Institute of Standards and Technology (NIST). Some of the smart grid standards are listed below:

- IEEE 1547 for interconnecting distributed resources with electric power systems.
- NIST Framework and Roadmap for Smart Grid Interoperability Standards.
- Distributed Network Protocol (DNP3), which is a set of communication protocols for process automation.
- Open Automated Demand Response (OpenADR) for automating DR.
- IEC 61968/61970 for integration of energy management system applications.

12.1 Sensor Systems

Basic sensors such as CT and PT are discussed in Chapter 6. In this section, we shall discuss advanced sensor systems that are used both for power transmission system monitoring and control and for distribution system load management.

12.1.1 Computation of Phasors from Instantaneous Measurements

If the measurements of the phasors are available, then the power and reactive power flowing through the line can be determined. Real power determines angle stability, while reactive power determines voltage stability. Moreover, such measurements are required for both power flow studies and state estimation.

Let a balanced set of three-phase voltages be given by

$$\begin{aligned} v_a(t) &= 1{,}000 \times \sqrt{2}\,\sin(\omega t - 30^\circ)\ \mathrm{V} \\ v_b(t) &= 1{,}000 \times \sqrt{2}\,\sin(\omega t - 150^\circ)\ \mathrm{V} \\ v_c(t) &= 1{,}000 \times \sqrt{2}\,\sin(\omega t + 90^\circ)\ \mathrm{V} \end{aligned} \tag{12.1}$$

where $\omega = 100\pi$ rad/s. Obviously, the negative- and zero-sequence voltages will be zero, and the positive-sequence voltage will be $V_a = 1,000\angle 0^\circ$ V. The main aim is to find the sequence components from the measurements of the instantaneous voltages (or currents).

Instantaneous voltages are sampled. A sampling frequency of 600 Hz, which is 12 times the fundamental frequency of 50 Hz, is sufficient to capture the signal properties [2, 3]. This implies that the waveforms need to be sampled 12 times per cycle. Let us consider phase-a voltage. Let the instantaneous sampled measurements be denoted by $v_a(k)$, $k = 0, 1, \ldots$. Let us define:

$$x_a(k) = v_a(k)e^{j90^\circ} \tag{12.2}$$

Table 12.1 Moving average filter characteristics.

	Samples					
Instant	**1**	**2**	**3**	**4**	**5**	**6**
k	$x(k-5)$	$x(k-4)$	$x(k-3)$	$x(k-2)$	$x(k-1)$	$x(k)$
$k+1$	$x(k-4)$	$x(k-3)$	$x(k-2)$	$x(k-1)$	$x(k)$	$x(k+1)$
$k+2$	$x(k-3)$	$x(k-2)$	$x(k-1)$	$x(k)$	$x(k+1)$	$x(k+2)$

Note that, for cosine waveforms, the multiplication by e^{j90° will not be necessary. Then, the Fourier coefficient for the fundamental can be obtained using half-cycle averaging to produce the phasor voltage of phase-a, as follows [2, 3]:

$$V_a(k) = \frac{1}{3}\sum_{k=0}^{5} x_a(k)e^{-jk\pi/6} \tag{12.3}$$

Similarly, the phasor voltages of the other two phases are obtained as follows:

$$\begin{aligned} x_b(k) &= v_b(k)e^{j90^\circ}, \quad V_b(k) = \frac{1}{3}\sum_{k=0}^{5} x_b(k)e^{-jk\pi/6} \\ x_c(k) &= v_c(k)e^{j90^\circ}, \quad V_c(k) = \frac{1}{3}\sum_{k=0}^{5} x_c(k)e^{-jk\pi/6} \end{aligned} \tag{12.4}$$

The phasor components can be computed recursively from (12.2) to (12.4) [2, 3]. However, noting from (12.3) and (12.4) that only six samples are required for fundamental computation, a moving average filter (MAF) can be used for continuously extracting the phasor components. The MAF is explained in Table 12.1. In this, at any given time, only six samples are retained for the averaging process, with the most recent sample numbered as the sixth sample. Once a new measurement is available, the last measurement is discarded. The measurements of samples 1–5 are shifted left. The current measurement is then placed at the sixth sample. In this way, only the six most recent samples are kept in the buffer.

Example 12.1 Consider the waveforms given in (12.1)

At the beginning, the MAF buffer is filled for the first half cycle (0.01 seconds). Then, (12.2) to (12.4) are used to estimate the phasor components. The results are shown in Figure 12.2. It can be seen that the estimates are accurate.

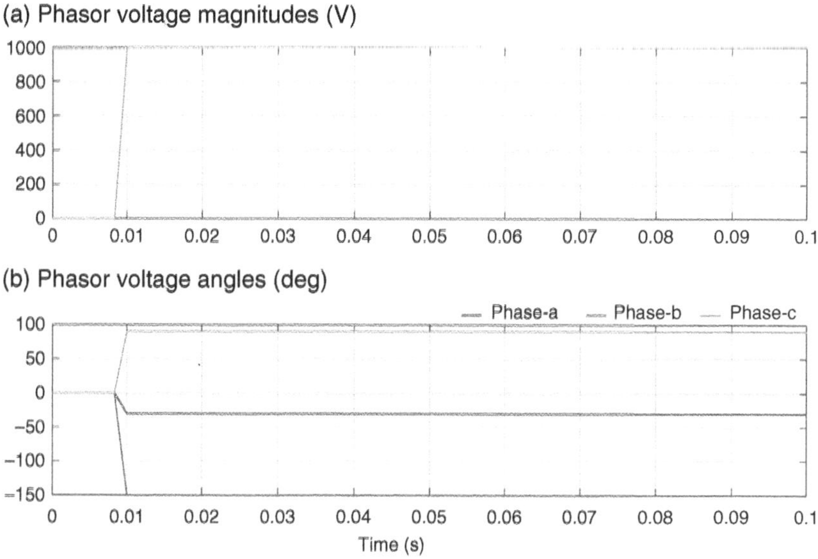

Figure 12.2 Estimation of phasor components from measured data for a balanced system. (a) Voltage magnitudes and (b) angles.

Example 12.2 Consider now a set of unbalanced three-phase voltages, given by

$$\begin{aligned} v_a &= 1{,}000 \times \sqrt{2}\,\sin(\omega t - 30^\circ)\ \mathrm{V} \\ v_b &= 900 \times \sqrt{2}\,\sin(\omega t - 140^\circ)\ \mathrm{V} \\ v_c &= 1{,}000 \times \sqrt{2}\,\sin(\omega t + 80^\circ)\ \mathrm{V} \end{aligned} \tag{12.5}$$

From (5.29) of Chapter 5, the phasor sequence components are computed as follows:

$$\begin{bmatrix} V_{a0} \\ V_{a1} \\ V_{a2} \end{bmatrix} = \frac{1}{3}\begin{bmatrix} 1 & 1 & 1 \\ 1 & a & a^2 \\ 1 & a^2 & a \end{bmatrix}\begin{bmatrix} 1{,}000\angle-30^\circ \\ 900\angle-140^\circ \\ 1{,}200\angle80^\circ \end{bmatrix} = \begin{bmatrix} 132.86\angle15^\circ \\ 1022.8\angle-31^\circ \\ 139.53\angle-177^\circ \end{bmatrix} \mathrm{V}$$

Once the phasor components of the voltages are obtained, they are then used in (5.27) to obtain the phasor sequence components at every sampling instant as given below:

$$\begin{bmatrix} V_{a0}(k) \\ V_{a1}(k) \\ V_{a2}(k) \end{bmatrix} = \frac{1}{3}\begin{bmatrix} 1 & 1 & 1 \\ 1 & a & a^2 \\ 1 & a^2 & a \end{bmatrix}\begin{bmatrix} V_a(k) \\ V_b(k) \\ V_c(k) \end{bmatrix} \tag{12.6}$$

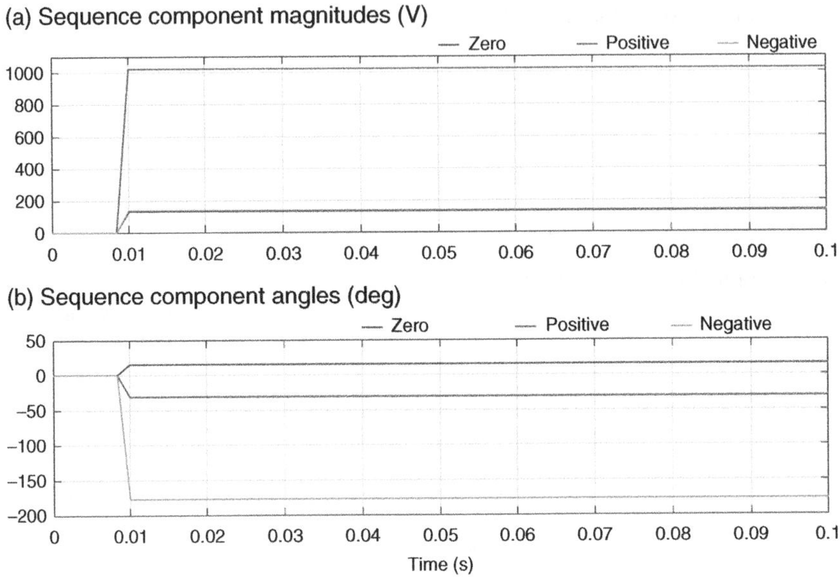

Figure 12.3 Sequence component (a) voltages and (b) their angles.

The traces of the sequence components as the time evolves are shown in Figure 12.3.

In the phasor estimation process, the frequency is assumed to be constant at the fundamental frequency of 50 Hz. However, the prevailing frequency may not be 50 Hz. The positive sequence vector rotates in the complex plane at a frequency that is the difference between the fundamental frequency and the prevailing frequency [4]. Let the positive sequence component be given by

$$V_{a1}(k) = |V_{a1}|e^{j\varphi} \tag{12.7}$$

Then, the prevailing power system frequency (ω_p) can be computed from [4]

$$\omega_p = \omega + \frac{d\varphi}{dt} \tag{12.8}$$

Figure 12.4 shows the estimated frequency using (12.8) when the prevailing frequency is 51 Hz. In the frequency estimation process, the data buffer is filled with samples for the first half cycle and, therefore, the estimate is zero. However, once the buffer is filled, the estimation process converges quickly.

The computation of instantaneous symmetrical components from the instantaneous measurements and extracting the phasor sequence components from them using MAF are developed in [5]. It has been shown that half-cycle averaging can eliminate odd harmonics. However, for the elimination of both even and odd harmonics, the averaging period of the full cycle needs to be used.

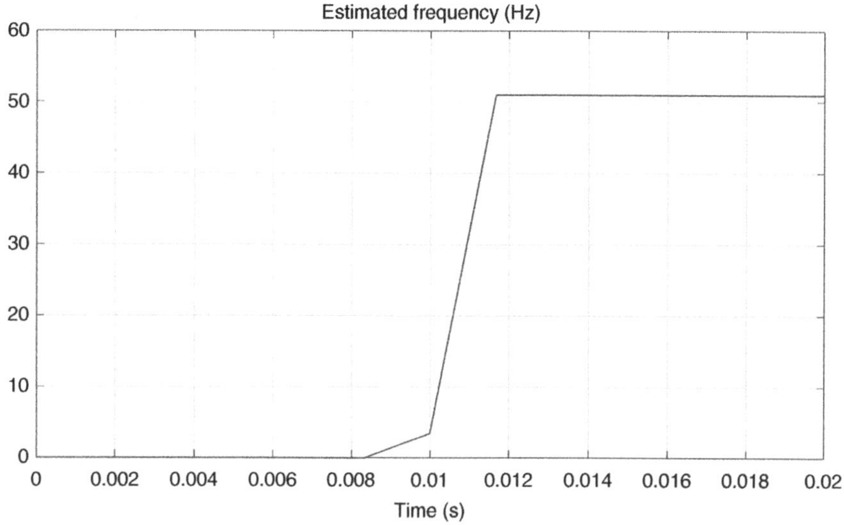

Figure 12.4 Estimated frequency from positive-sequence component.

12.1.2 Phasor Measurement Unit (PMU)

The schematic diagram of a typical PMU is shown in Figure 12.5. It contains an anti-aliasing filter, analog-to-digital converter (ADC), and a microprocessor that computes the phasors. It is expected that several PMUs will be located in a power network at different locations. All these need to be synchronized together. For example, in a power system containing two PMUs, PMU-1 starts its AD conversion at time t_1, while PMU-2 starts its AD conversion at time t_2. The dataset then will not be helpful as the power system voltage or current can change within milliseconds. Therefore, to have meaningful data, both PMUs must start their AD conversion at time t_1. Moreover, several PMUs can be located in a large geographical area. Therefore, all the PMUs must be synchronized with a global clock. The computed phasor data are time stamped, that is, the instant of sampling is included

Figure 12.5 Schematic diagram of a PMU.

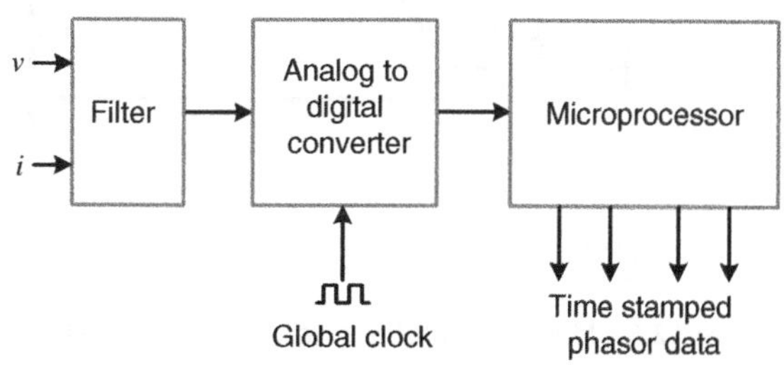

with voltage and current phasor measurements. The time tag can include year, day of the month, hour, minute, and second. The time can be local time or universal time coordinate (UTC).

To coordinate the PMUs through a global clock, a constellation of satellites is used, for example, the (Navigation System with Timing and Ranging (NAVSTAR) satellite system. This is commonly known as the global positioning system (GPS) and is operated by the United States Space Force. This constellation has more than 24 satellites that perform two orbits per day from an altitude of approximately 20,000 km above the earth. The constellation ensures that at least four satellites are visible from any point on the earth's surface at any point in time. It has worldwide coverage with a high degree of accuracy. There are other such satellite constellations around the world. Figure 12.6 shows how the PMUs are coordinated.

PMU is a critical component in modernizing the electrical grid and ensuring its reliable and efficient operation. It has several critical applications. Some of these overlapping applications are listed below:

- *Wide-Area Monitoring and Control*: PMUs provide high-resolution, time-synchronized measurements of electrical quantities, allowing operators to monitor the grid in real time. Moreover, PMUs can improve the accuracy of state estimation by providing more accurate data.

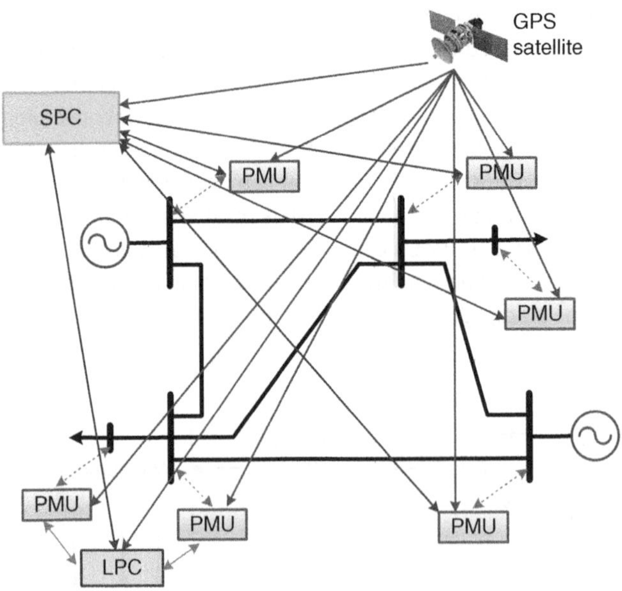

Figure 12.6 Multilayered GPS-synchronized PMU network.

- *Enhancing System Protection*: PMUs supply accurate data that can be used for fault detection and localization. By providing a wide-area view of the power system, PMUs help in quickly identifying the location of a fault, even in complex grid topologies. Furthermore, PMU data can be used for dynamically adjusting the relay settings based on real-time grid conditions, paving the way for adaptive protection schemes.
- *Grid Stability and Reliability*: PMUs can alleviate dynamic stability problems of power systems by monitoring power system oscillations and identifying potential instability issues. Furthermore, PMUs can also improve voltage stability by monitoring bus voltages to prevent voltage collapse.
- *Operational Efficiency*: PMUs can improve operational efficiency by managing congestion and implementing an effective load-shedding strategy to maintain grid stability. Furthermore, PMUs can help in synchronizing renewable energy sources with the power grid by supplying them with frequency and phase information.

Some of these aspects are discussed in [6], which discusses various advantages of PMU applications. In [6], a multilayered PMU network is discussed. The structure of the network is shown in Figure 12.6. The bottom layer consists of the PMUs, which are synchronized through the global clock signal that is sent via GPS satellite. The next layer consists of several local protection centers (LPCs). Each LPC coordinates with a number of PMUs. The top layer is for the data concentrator and coordinator called the system protection center (SPC). This layer is responsible for overall coordination and is interfaced with the LPCs and some of the PMUs directly (see Figure 12.6). The SPC can also act as a data concentrator and can be used in data centers for supervision, state estimation, and scenario analysis. Different aspects of PMUs are covered in [7]. This book covers wide-ranging topics related to power system load estimation and wide-area control.

12.1.3 Smart Meter

Before the advent of microprocessor technology, power consumed by individual customers was measured by *electromechanical energy meters*. At the heart of such a type of meter is a thin induction disk that rotates at a speed that is proportional to energy consumption. Usually, there are two coils: a current coil that is connected in series with the load and a voltage coil that is connected in parallel with the load. The interacting magnetic field of these two coils induces an eddy current on the disk, producing torque. Then, the disk rotates at a speed that is proportional to the product of the current and voltage, that is, power. Such energy meters were robust and inexpensive. Traditional electromechanical energy meters had several dials (usually between four and six). Some of the dials rotated in a clockwise direction, and some in a counterclockwise direction. The reading of these meters requires expert knowledge, and the readings are prone to errors. Obviously, these meters are inadequate for the modern power system under the smart grid paradigm.

Digital energy meters were the next stage of energy meter development. The heart of a digital meter is a microprocessor that computes the energy used and displays it digitally. Voltage and current sensors are embedded in these meters. The measurements from these sensors are converted into digital data through ADCs. A digital meter data can be remotely monitored. Obviously, a communication medium will be required for remote meter reading. There are several advantages of digital meters. Some of these are listed below:

- Digital meters provide high-precision energy consumption data for precise billing.
- They offer real-time data on energy usage, allowing consumers and utility companies to monitor consumption patterns and make informed decisions.
- They can monitor power quality issues such as voltage sag/swell, harmonics, transients, and interruptions.
- They are tampering resistant, which can help in avoiding electricity-related fraud.

Smart meters are digital meters that are equipped with two-way communication systems, that is, they can send information from customers to utility and from utility to customers. Smart meters include all the functionalities of digital meters while including advanced features such as two-way communication, real-time data transmission, remote control, and support for dynamic pricing and DR. They are integral to modern smart grids and energy management systems.

The smart meters are used in AMI, which is not a single entity. It is a collection of smart meters that can transmit energy usage at regular intervals to utilities through either wired or wireless communication mediums. The data are then stored and analyzed for billing, load forecasting, and DR. Therefore, an AMI system must include smart meters, HANs, communication technologies, and data centers with software infrastructure. AMI is the main link between the grid and the consumer loads, distributed generation, and storage. Basically, AMI connects consumers with their distributed energy resources (DERs) and loads to distribution system management and its ancillary functions [8]. The design of an AMI system has been discussed in [9].

12.2 Demand Response

With the continued prosperity around the world, businesses and industries consume more electricity, and households can afford more electrical appliances and devices. Furthermore, with the growth in population, the demand for electricity keeps growing. Electricity demand growth has slowed in many developed countries due to the use of energy-efficient devices and the shift toward less energy-intensive industries. The load demand growth in these countries is limited to below 2% per year. On the other hand, the growth in energy demand in developing countries has been high, between 5% and 10% per year. This is due to rapid urbanization, industrialization, and a rise in the standard of living. Moreover, some of these countries manufacture goods that have been exported to various parts of the world. Therefore, the choice is between setting up more power generation and transmission infrastructure or setting up mechanisms for demand curtailment or shifting to take advantage of renewable energy generation peaks.

DR is a paradigm that uses a set of strategies and programs designed to adjust the power demand rather than adjusting the supply. The customers participate in a DR program with utility companies, where their electricity use can have a set plan which is somewhat similar to a plan for cellular service. Such plans can be either price-based or incentive-based.

The price-based DR program can be of the following types:

- *Time of Use Pricing*: Different rates are set for different times of the day to encourage consumers to shift their usage to off-peak times. For example, power distributor Synergy in Western Australia has four different tariff structures for energy usage. These, ranging from maximum to minimum prices, are Peak (3–9 PM), Off Peak (6–9 AM and 9–11 PM), Overnight (11 PM to 6 AM), and Super Off Peak (9 AM to 3 PM). The Super Off Peak tariff takes advantage of rooftop solar PV generation during this time and is about 16% of Peak tariff. The peak tariff is imposed when the energy consumption is high, and the grid is under maximum stress.
- *Real-Time Pricing*: This tariff can vary continuously based on real-time cost of generation. This requires dynamic adjustment of load control.

The incentive-based programs are implemented with contracts between the utilities and their customers. The different incentive-based programs are as follows:

- *Direct Load Control*: In this scheme, the utilities can control some specific appliances, such as air conditioners. Even though such schemes have been trialed in different places, they are not preferred by customers.
- *Interruptible Programs*: In this, the large consumers agree to reduce their demand upon request in exchange for lower rates.
- *Demand Bidding*: Large consumers or load aggregators can offer bids to reduce their demands at specific times. They are compensated if their bids are accepted.

Basically, DR is about controlling the consumer loads. The load management categories are shown in Figure 12.7.

Not all household loads can, however, participate in a DR program. For example, cooking appliances such as electric stoves, ovens, microwave ovens, or air

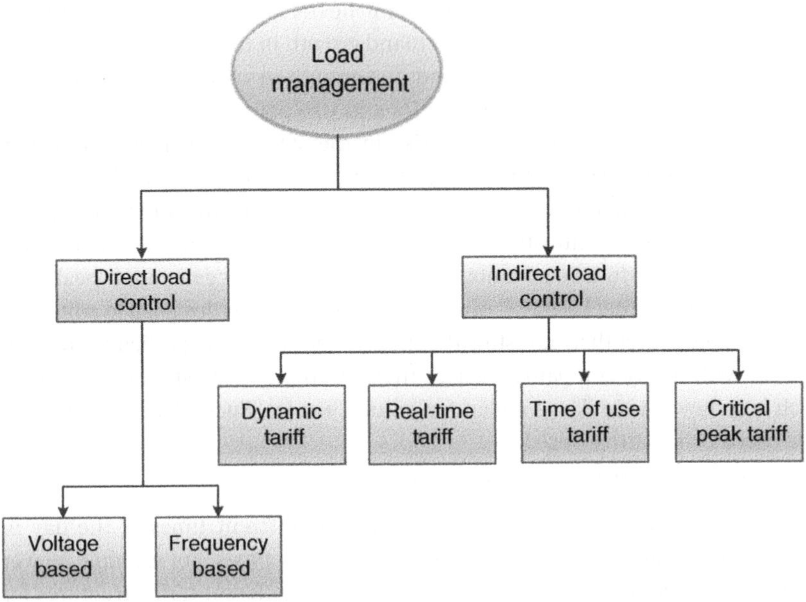

Figure 12.7 Load management categories.

fryers cannot be a part of a DR regime because they have the maximum priority for domestic customers. The loads that can participate in DR program have the following characteristics [10]:

- Controlling those loads can result in minor inconvenience to the customers, for example, air conditioners and nonessential lighting.
- Controlling those loads will go mostly unnoticed by the customers, for example, electric vehicle (EV) charging and swimming pool pumps.

Some of the good candidates for DR program are as follows [10]:

- Dishwasher.
- Washer and dryer.
- Electric hot water system.
- Some aspects of refrigerator operation such as defrost cycle.
- Battery charging for consumer electronics and EVs.

For example, dishwasher, washer, dryer, and defrost cycle of a refrigerator can be switched on when the price of electricity is lower, that is, during the off-peak hours when demand for electricity is less. Moreover, they can also be cycled and do not have to run continuously, provided that they finish the task before a specified time. Battery chargers can be scheduled such that the charge is finished by the required time. An electric hot water system can store warm water for a long period of time. Therefore, water can be heated when the demand for electricity is low.

An interesting perspective on load demand management has been given in [10]. In general, in a traditional power system operation, the "generation follows the load," that is, generation scheduled to cater for the peaks, which are usually known a priori. Under the DR paradigm, utilities are able to curtail some of the loads to maintain the generation-consumption balance without adding excess generation capacity. However, with the increased penetration of intermittent renewable energy sources, the best strategy is to use the "load following generation" approach. This is termed *demand dispatch* in [10], in which loads can be switched on or off depending on generation availability.

However, demand dispatch does not target each individual load, but a sum total of aggregated loads. The aggregator can be several service areas in a large power distribution system. The increase/decrease in generation can be detected based on the system frequency. Suppose a distribution system has a total of N aggregated loads. At a given time, the loads are defined as follows:

$$P_{Li} = P_{LEi} + P_{LCi}, i = 1, \dots, N \tag{12.9}$$

where P_{LE} is the essential load and P_{LC} is the controllable load that can be used for demand dispatch. The total load demand at that time is then defined by $P_L = \sum_{i=1}^{N} P_{Li}$.

Assume that the system frequency is f_s Hz when the load supplied is P_L. Due to a sudden drop in generation, the frequency will drop. Therefore, to recover the frequency, some amount of load needs to be curtailed. Let the total amount of load reduction to recover the frequency be ΔP_L. Since the essential loads cannot be switched off, the controllable loads must be reduced, that is, $\Delta P_L = \Delta P_{LC}$. The quantity ΔP_{LC} can be estimated from the frequency error using the following proportional-plus-integral (PI) controller:

$$\Delta P_{LC} = K_P(f_s - f_m) + K_I \int (f_s - f_m)dt \tag{12.10}$$

where f_m is the measured frequency. Then the controllable loads can be reduced proportionally as per the following equation:

$$\Delta P_{LCi} = \Delta P_{LC} \times \frac{P_{LCi}}{\sum_{k=1}^{N} P_{LCk}}, \quad i = 1, \dots, N \tag{12.11}$$

The new set points for the controllable loads are then given by

$$P_{LCi}^{new} = P_{LCi} - \Delta P_{LCi}, \quad i = 1, \dots, N \tag{12.12}$$

Note that during sudden increase in generation, the quantity ΔP_{LC} obtained from (12.10) will be negative, and hence, (12.10) will indicate an increase in load.

It is, however, to be noted that the frequency in a power system can fluctuate all the time. Therefore, the use of (12.10) to (12.12) can lead to unnecessary load fluctuations. To avoid this, a deadband in frequency can be used before invoking the PI controller. A demand dispatch example for smart EV charging is also discussed in [10], where the vehicle charger has been switched on and off by the aggregator. Even though smart charging is slower than continuous charging, it provides grid services by curtailing power during peaks.

12.2.1 Controlling Household Appliances

For successful implementation of DR, individual customers must participate in reducing loads during peak hours and increasing consumption when there is an increase in generation. Conversely, the utilities must also consider consumer preferences about their appliances. An example of user preferences for high-power-consuming controllable loads is given in [11], where the highest priority is given to the hot water system, followed by air conditioning such that temperatures in either of them are maintained within certain limits. The priority for clothes dryers comes next, followed by EV charging. These can be specified based on time, for example, EV gets charged by say 8 AM in the morning for the daily commute. It is assumed that each home has a home energy management (HEM) system that switches the appliances on or off depending on these priorities. The thematic diagram of a smart home with a HEM system is shown in Figure 12.8.

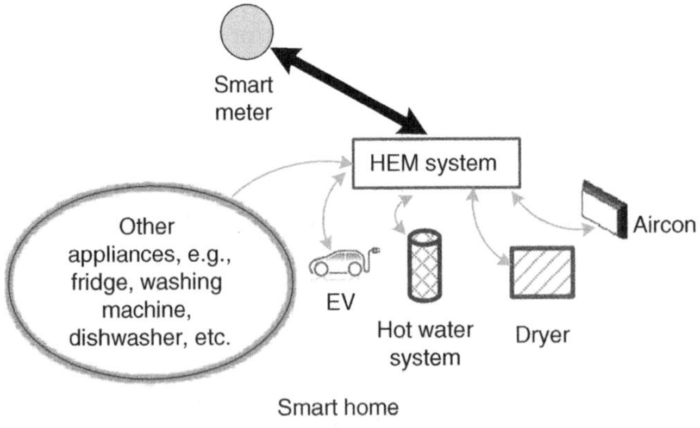

Figure 12.8 Smart home with HEM system.

Some of the domestic controllable appliances can consume more power than others. Some of them can have a mixed power consumption. For example, a clothes dryer has two cycles: low-power-consuming tumbling cycle and high-power-consuming heating cycle. In [11], it has been mentioned that the maximum power-consuming appliances are EVs, electric hot water systems, dryers, and air conditioner (aircon). In Figure 12.8, it has been shown that these devices are controlled by the HEM through bidirectional information flow. The energy consumption by other appliances, such as fridge, dishwasher, stove/oven, and washing machine, is monitored and not controlled. The HEM is in turn connected to a smart meter.

The operations of the four high-power-consuming loads are discussed in [11]. The water heater temperature must be mentioned between a maximum and a minimum limit. Since this has the highest priority, the water heater must be turned on once it falls below the minimum limit. Let the maximum and minimum levels of water heater temperatures be denoted by T_{max} and T_{min}, respectively. For a water temperature of T_{WH}, the switching status of the water heater at time n is given by

$$S(n) = \begin{cases} 0 & T_{WH} > T_{max} \\ 1 & T_{WH} < T_{min} \\ S(n-1) & T_{min} < T_{min} < T_{max} \end{cases} \tag{12.13}$$

Similarly, the aircon system temperature will be maintained within a prespecified band according to the comfort level of the householders. In general, if the room temperature goes above the comfort range, the aircon unit is switched on, provided that the energy consumption of the house does not go above the maximum limit.

The dryer operation can be cycled. For example, when energy consumption is nearing its peak, only the tumbler is allowed to operate. The heating coil can be switched on when some other appliances are switched off. However, the heating cycle must be maintained for a certain period of time to prevent heat loss from one cycle to the next. It is to be noted that a dryer need not operate every day even in colder climates, as it is unlikely that clothes will be washed every day as well. Similarly, an EV need not be charged every day as well. It is a common misconception that the EV is plugged in as soon as the householder returns home. Most auto manufacturers recommend maintaining the EV state of charge (SoC) between 30% and 80% to prolong the battery life. Thus, the EV can be charged when the electricity price is low, and the charging process can be monitored by the HEM.

The loads, in general, are defined by their energy consumption rather than their electrical model [12]. This, however, is not a problem because their dynamics will not have a major impact on the DR algorithms. There are several DR algorithms that have been proposed in the literature. However, there is no standard procedure that can be universally adopted. In [11], an intelligent HEM is proposed. A multiobjective decision-making-based DR has been proposed in [13] based on low-bandwidth two-way communication. In this method, a coordinating controller is placed at each 11 kV/440 V distribution transformer. If this controller detects the occurrence of an overload, it sends signals to the loads that can be delayed or adjusted. Once this action has taken place, the home smart meter communicates this to the coordinating controller. If the overloading condition is mitigated, no further action needs to be taken. However, the overload still persists, another set of loads must be adjusted, and this is relayed to the smart meters. However, the user priority and satisfaction factors are taken into consideration during the load adjustment. A comprehensive study of demand dispatch with different types of smart loads and PVs is presented in [14].

12.3 Cybersecurity

A power system must be guarded from cyber-physical attacks, which means both cyberattacks and physically destroying power infrastructure. While physical attacks can cause disruptions in power delivery in a localized region, cyberattacks can cause widespread disruptions in power system operations. The smart grid is heavily reliant on information and communication technology to manage electricity flow efficiently, integrate renewable energy sources, and enable real-time monitoring and control. However, this increased connectivity also opens up new avenues for cyber threats. In this regard, cybersecurity becomes an important aspect of the smart grid to protect digital communication networks and power flow control programs from attacks. Cyberattacks are typically aimed at accessing, changing, or destroying sensitive information, extorting money from users, or interrupting normal business processes. The following operation risks are emerging due to the widespread use of ICT [15]:

- *Taking Down a Server*: An attacker can access the SCADA server through the network path to take it down through either denial of service or by deleting system files. Through the SCADA server, the attacker can access other servers.
- *Gaining Control Over the System*: This is the most serious security threat in which an attacker can plant a Trojan or gain a backdoor entry into the network. Through this, the control actions can be manipulated, or false alarms can be generated, causing a widespread power supply disruption.

There are other forms of risks such as corporate or customer data being stolen and billing information being tampered with.

Some of the key aspects of cybersecurity in smart grids include:

- *System Vulnerability Management*: Identifying and addressing vulnerabilities in smart grid components, such as smart meters, sensors, and communication networks.
- *Data Integrity and Confidentiality*: Ensuring that the data transmitted and stored within the smart grid is accurate and secure from unauthorized access or tampering.
- *Network Security*: Protecting the communication networks that connect various components of the smart grid from cyberattacks. This includes the use of firewalls, intrusion detection/prevention systems, and secure communication protocols.
- *Access Control*: Implementing strict access control measures to ensure that only authorized personnel can access critical components of the smart grid.
- *Incidence Response*: Developing and implementing robust incident response plans to quickly detect, respond to, and recover from cyberattacks.
- *Advanced Threat Detection*: Utilizing advanced technologies such as machine learning and artificial intelligence to detect and respond to emerging threats in real time.

The integration of robust cybersecurity measures in smart grids is of paramount importance in order to protect against potential threats that could disrupt power supply and financial systems related to power systems.

12.3.1 False Data Injection Attacks

False data injection is one of the most prominent forms of cyberattack. This can compromise the data integrity in the SCADA system affecting the state estimation process. A false data injection attack has been discussed in [16]. From (3.92) and (3.95), the L_2 norm of the measurement residuals is given by

$$\mathbf{e} = \|\mathbf{z} - \mathbf{H}\hat{\mathbf{x}}\| \tag{12.14}$$

If $\max(\mathbf{e}) < \rho$, the bad data is assumed to be present where ρ is small threshold value. Let us assume that all the state variables are mutually independent. Then, if the sensor errors follow a normal distribution, $\mathbf{e}$ in (12.14) follows a chi-squared distribution with the stipulated degree of freedom.

Let us now assume that malicious data has been injected into the measurement vector as follows:

$$\mathbf{z}_{\alpha} = \mathbf{z} + \boldsymbol{\alpha} \tag{12.15}$$

where $\boldsymbol{\alpha}$ is a vector of malicious data injected into the measurements. Then from (3.95), the erroneous estimated state vector is given by

$$\hat{\mathbf{x}}_{\mathbf{bad}} = \left(\mathbf{H}^T\mathbf{W}\mathbf{H}\right)^{-1}\mathbf{H}^T\mathbf{W}\mathbf{z}_{\mathbf{a}} = \hat{\mathbf{x}} + \left(\mathbf{H}^T\mathbf{W}\mathbf{H}\right)^{-1}\mathbf{H}^T\mathbf{W}\boldsymbol{\alpha} \tag{12.16}$$

The L_2 norm of the estimator residuals is then given by

$$\begin{aligned}\|\mathbf{z_a} - \mathbf{H}\hat{\mathbf{x}}_{\mathbf{bad}}\| &= \left\|\mathbf{z} + \boldsymbol{\alpha} - \mathbf{H}\hat{\mathbf{x}} - \mathbf{H}\left(\mathbf{H}^T\mathbf{W}\mathbf{H}\right)^{-1}\mathbf{H}^T\mathbf{W}\boldsymbol{\alpha}\right\| \\ &= \left\|\mathbf{z} - \mathbf{H}\hat{\mathbf{x}} + \left\{\boldsymbol{\alpha} - \mathbf{H}\left(\mathbf{H}^T\mathbf{W}\mathbf{H}\right)^{-1}\mathbf{H}^T\mathbf{W}\boldsymbol{\alpha}\right\}\right\|\end{aligned} \tag{12.17}$$

Now consider that the attackers also have access to the matrix **H** and choose the attack vector $\boldsymbol{\alpha} = \mathbf{Hc}$ for any arbitrary vector **c**. Equation (12.17) can then be written as follows:

$$\|\mathbf{z_a} - \mathbf{H}\hat{\mathbf{x}}_{\mathbf{bad}}\| = \left\|\mathbf{z} - \mathbf{H}\hat{\mathbf{x}} + \left\{\mathbf{Hc} - \mathbf{H}\left(\mathbf{H}^T\mathbf{W}\mathbf{H}\right)^{-1}\mathbf{H}^T\mathbf{W}\mathbf{Hc}\right\}\right\| = \|\mathbf{z} - \mathbf{H}\hat{\mathbf{x}}\| \tag{12.18}$$

This implies that the attackers can inject false data in such a way that the bad data detection can be bypassed. This way arbitrary errors can be imposed on the state estimation, which can have a severe impact on the stable operation of the system.

From the attackers' point of view, there are two possible attack scenarios: (a) a random false data injection attack that can result in wrong state estimates and (b) a targeted false data injection attack in which the attackers target some of the state variables to inject an error into their estimates [16]. The first approach is easier to launch but may be detected easily as well. The second attack, however, can cause more damage to the system. For a state estimation process, there are several measurements made through sensors. For example, let us assume that there are m number of sensors used for the state estimation. To have a perfect attack as given by $\boldsymbol{\alpha} = \mathbf{Hc}$, the attackers must know the matrix **H** and have access to all the m number of sensors. Therefore, the ability of the attackers is constrained under the following two scenarios [16]:

- *Scenario 1 – Limited Access to the Meters*: Assuming that some of the sensors are protected and beyond reach, the attackers can only target a limited number of sensors for false data injection.
- *Scenario 2 – Limited Resources Available to Compromise Sensors*: This means that the attackers do not have the resources to compromise all the sensors and, therefore, can only target a finite number of meters.

Both these attack models are derived and discussed in detail in [16]. The countermeasures for the detection of false data injection attacks have been presented in [17].

The security of cyber-physical power systems is a crucial element of modern grids. To develop robust cyberattack countermeasures a concerted effort will be required by multidisciplinary teams, including electrical power and communication engineers, computer scientists, and mathematicians. There are several facets of cyberattacks and their prevention strategies. A review of cyberattack modeling, detection methods, and cybersecurity measures is presented in [18]. This is an excellent resource for those who are interested in this area.

12.4 Electric Vehicle (EV)

EVs are one of the components of the smart grid of the future, not only for reducing greenhouse gas emissions but also as a source of energy storage that can be utilized as and when required. Currently, most EVs charge from the grid in a mode called grid-to-vehicle (G2V). However, with the improvement in battery technology, it is expected that they can be used as mobile storage by supplying power in vehicle-to-grid (V2G) mode. In this section, several aspects of EVs are discussed.

Since 2020, EVs have been gaining traction from customers, government policy-makers, and auto manufacturers. As of 2024, there are over 80 manufacturers worldwide that are producing EVs. These include both established automotive giants and newer companies dedicated solely to EVs such as Tesla, BYD, and Cupra Born. Even high-end automotive manufacturers such as Rolls Royce, Lamborghini, Maserati, Jaguar, Porsche, Mercedes Benz, BMW, and Audi have entered the EV market. There are several benefits of EVs. Some of these are:

- *Pollution*: The most obvious one is reduced emissions because EVs do not produce any tailpipe emissions, thereby reducing environmental pollution. Particularly, the widespread use of EVs can improve the air quality in large cities, reduce smog, and reduce respiratory health problems such as asthma and bronchitis.
- *Efficiency*: Electric motors are more efficient than internal combustion (IC) engines. Therefore, EVs have higher energy conversion rates than petrol/diesel-powered vehicles.
- *Running Costs*: Since the cost of electricity is lower than fossil fuel, EVs have lower running costs. Furthermore, EV batteries can be charged during regenerative braking, which can increase the driving range of EVs. EVs have fewer moving parts, and therefore, lower maintenance costs and will not require regular oil or filter changes.
- *Driving Performance*: Electric motors can provide instant torque. Therefore, EVs can accelerate rapidly. Moreover, EVs reduce noise pollution as they are much quieter than IC engine-based vehicles.

12.4.1 Types of Electric Vehicles

Any vehicle that uses batteries is called an EV. There are three types of EVs as listed below:

- Hybrid Electric Vehicle (HEV), which runs on petrol (gasoline) but uses an electric motor to improve fuel efficiency.
- Plug-in Hybrid Vehicle (PHEV), which can have its battery charged from electric sockets; however, it runs like an HEV. This combination offers increased driving range while reducing fuel consumption.
- Battery Electric Vehicle (BEV or EV), which is exclusively powered by batteries.

HEVs combine an IC engine with an electric motor. The electric motor can provide propulsion independently or in conjunction with the IC engine. Toyota introduced Prius around the turn of this millennium as the first mass-produced HEV. There are many HEVs on the road these days, produced by different manufacturers. The main advantage of HEV is that electric propulsion is more efficient than IC engines. Moreover, the IC engines can operate at higher efficiency due to the electric propulsion backup. In general, there are three types of HEVs. These are:

- *Series HEV*: In this, the electric motor is the primary source of propulsion. The motor is driven by a battery. The IC engine acts mainly as a generator to charge the battery.
- *Parallel HEV*: In this, both the IC engine and the electric motor can drive the wheels directly, either independently or simultaneously. The IC engine drives a generator to produce electricity to charge the battery. This can occur when the IC engine produces more power than is required for propulsion, such as on highways or going downhill.
- *Series-Parallel HEV*: This can operate in either series or parallel. At low speeds or in stop-and-start traffic, the vehicle can run on electric mode alone, where the battery drives the car. It can also operate in a series hybrid mode, where the IC engine does not directly drive the vehicle but supplies power to the battery through the generator. At higher speeds or when rapid acceleration is required, both IC engine and electric motor drive the wheel simultaneously. The combined power of both IC engine and electric motor improves performance. Toyota hybrids use this technology.

The schematic diagrams of these three types of HEVs are shown in Figure 12.9.

PHEVs have larger batteries than HEVs. The batteries can be charged externally through power sockets. Once fully charged, the typical battery range of PHEVs is between 30 km and 80 km. Thus, commuters can go to their workplace and go back home without using any fossil fuel. This type of vehicle can also cure "range anxiety" as they can switch from electric to IC engine mode for long drives. PHEVs also come in series, parallel, and series-parallel configurations. The schematic diagram of a series PHEV is shown in Figure 12.10.

A PHEV can operate in either charge depletion or charge sustaining mode. In charge depletion mode, the electric motor is the main source of power. This, however, will cause depletion of the charge of the battery. Once the battery SoC reaches the minimum level, the vehicle automatically switches over to the charge-sustaining mode. In this mode, the vehicle is driven by the IC engine. However, the IC engine may also charge the battery in favorable conditions, the same as that in an HEV.

A BEV is driven by pure electric traction. It needs a high-power battery pack and its associated battery management system (BMS). A basic diagram of BEV, containing the main components, is shown in Figure 12.11. However, there are other

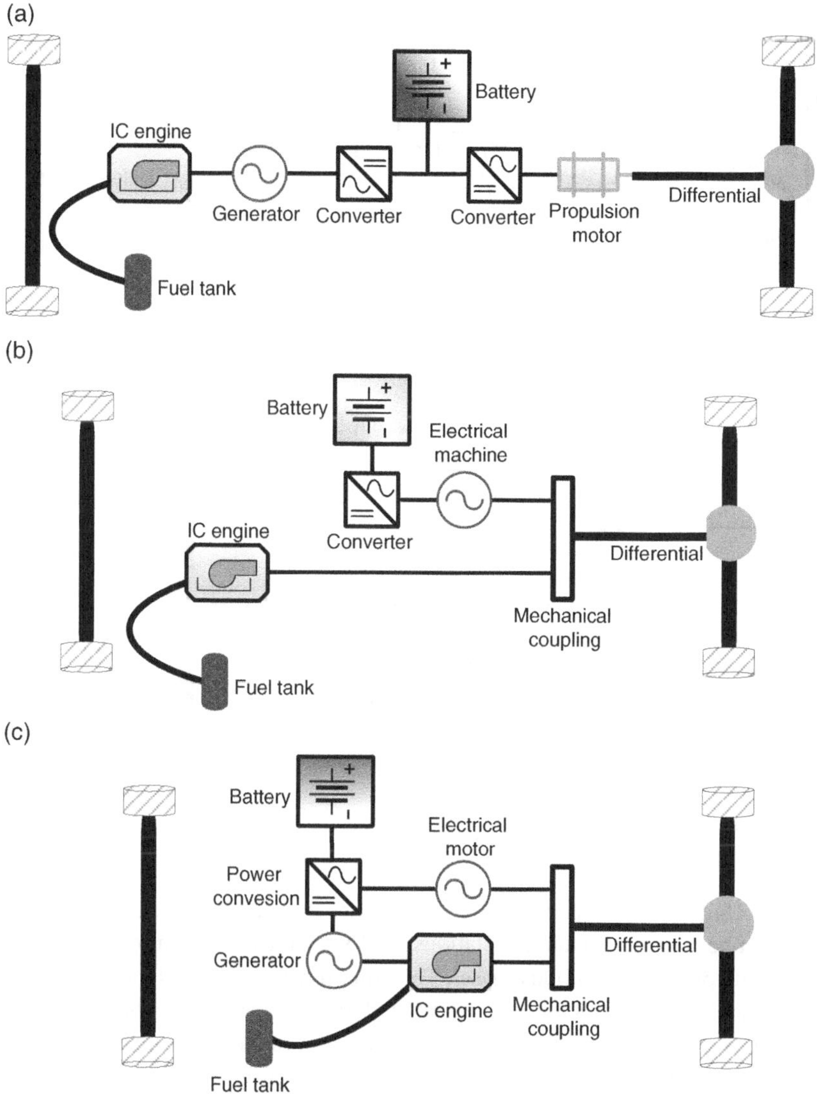

Figure 12.9 HEV configurations: (a) series, (b) parallel, and (c) series-parallel.

components, such as a small battery for ancillary purposes and a control unit, which are not shown in the figure.

Even though BEVs have some advantages over IC engine-driven vehicles, their uptakes and universal acceptance are not guaranteed. One of the reasons is the charging infrastructure, especially in a sparsely populated country such as Australia.

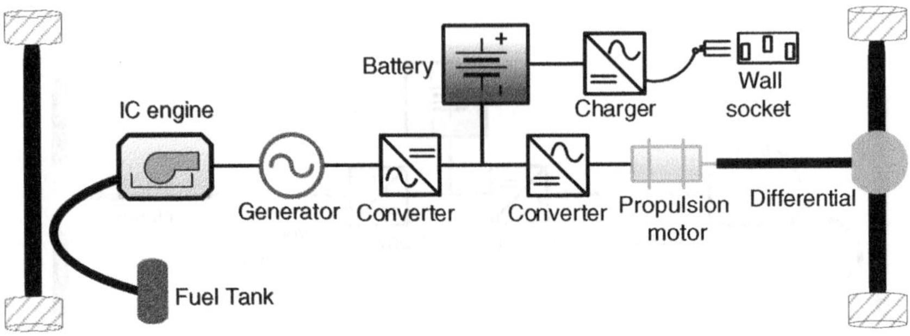

Figure 12.10 Schematic diagram of series PHEV.

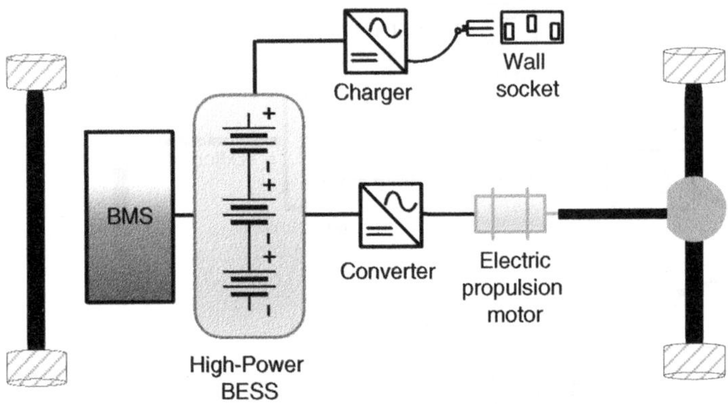

Figure 12.11 Basic schematic diagram of BEV.

This can lead to *range anxiety*. Unlike IC engines that can be filled at gas pumps, EVs must be charged at charging stations. HEVs do not suffer from range anxiety, but they do produce greenhouse gases. The other obvious reason is the higher purchase cost of EVs vis-à-vis that of conventional fossil fuel vehicles. Moreover, BEVs can also have detrimental impacts on electric utility companies. Increased penetration of BEVs can lead to higher peak demand unless customers participate in DR programs. Higher peaks can stress power grids, which may need infrastructure upgrades. Furthermore, BEVs can cause power quality problems such as voltage imbalance and harmonic injection that need to be addressed by the utilities.

12.4.2 EV Charging

There are different levels of charging for PHEVs or BEVs. These vehicles can be charged at home or at charging stations. Even though the charging voltage and current ratings can vary depending on the power supply, there are three levels of charging. These are:

- *Level 1 Charging*: Usually, these are standard household outlets of (215/230 V or 120 V). The charging process is slow and charges at a typical rate of 1.2 kW per hour. Such a charging process is suitable for PHEVs. However, it can take up to 30 hours to charge a BEV.
- *Level 2 Charging*: This is also a home charging method using 400/430 V (L-L). There are two possible options, both of which will require the installation of a wall box. The wall box can be connected to a single phase, where the BEV can charge at the rate of about 7.2 kW per hour. If the wall box is connected to three phases, the charging rate can be as high as 22 kW per hour. However, utility permission, special connection, and energy meter will be required for a three-phase connection.
- *Level 3 Charging*: This is only available at public charging stations. This usually uses DC voltage and is commonly known as DC fast charging or DC ultrafast charging. Most BEVs can be charged up to 80% of their SoC level within 20–30 minutes. There is no standard for voltage levels – they can be anywhere between 200 V and 1,000 V and the charging rate is between 50 kW and 150 kW per hour.

One of the most attractive charging features is wireless charging. If the vehicle is parked in a charging spot, it will not require cables if wireless charging is available. Moreover, it will be very desirable if the vehicle gets charged while it is driven on the road. This, however, will need infrastructure and further research before it can come to fruition. The wireless charging mechanisms are discussed below.

12.4.3 Wireless Charging

Wireless power transfer (WPT) was invented by Nicola Tesla in the late 19th century. This is known as the famous *Tesla Coil*, which produces high-voltage and low-current in an AC circuit. The schematic diagram of the Tesla coil is shown in Figure 12.12. It contains a resonant transformer, a capacitor C_1, a spark gap (SG), and a torus to suppress premature arc breakdown. The primary and secondary coils of the resonant transformer are L_1 and L_2, respectively. The primary coil has a few turns, while the secondary coil has multiple turns. The primary and secondary coils are tuned to resonate at the same frequency. The current from the supply transformer charges the capacitor. When the capacitor reaches its breakdown voltage, it discharges through SG. This will cause a spark, reducing the resistance. The current will then flow to the primary winding. This will induce a current in the secondary coil. Since the primary and secondary windings are tuned at the same frequency, the secondary voltage will be amplified significantly.

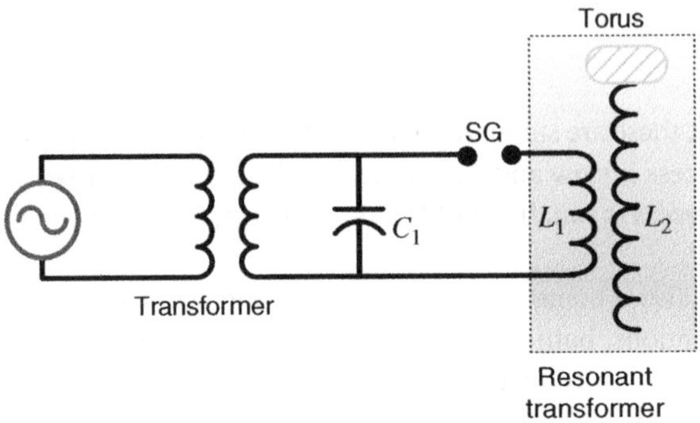

Figure 12.12 The Tesla coil.

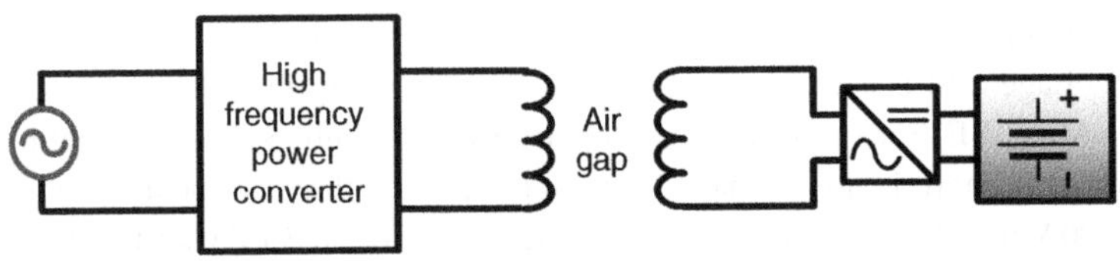

Figure 12.13 Inductive wireless power transfer.

Tesla imagined WPT using his coil, which has not been realized so far. However, it forms the basis of WPT for EV charging.

The schematic diagram of inductive WPT is shown in Figure 12.13. It contains coils that are embedded in cars and parking spots or roadways. Obviously, the air gap must be chosen such that electromagnetic induction is possible. Ferrite cores are used for flux guidance and shielding. The switching frequency is kept within 100 kHz to maintain losses low in ferrite cores. These will require very high power for operation.

In capacitive WPT, the inductor is replaced by capacitors at the two ends with an insulating material between them. The electric field created between the plates causes energy transfer from one end to the other. Capacitive WPT is used for low-power applications such as touch screens or proximity sensors. They have a short range and are not yet suitable for EV charging. Inductive WPT is a much more mature technology than capacitive WPT and has been used for charging smartphones and other portable devices. Inductive WPT can transfer a large amount of power over short or moderate distances. It is currently the preferred technology for EV charging. A review of inductive charging can be found in [19].

Dynamic WPT is a more desirable feature that can transfer power wirelessly to a moving vehicle. The first dynamic WPT system was developed by the Korea

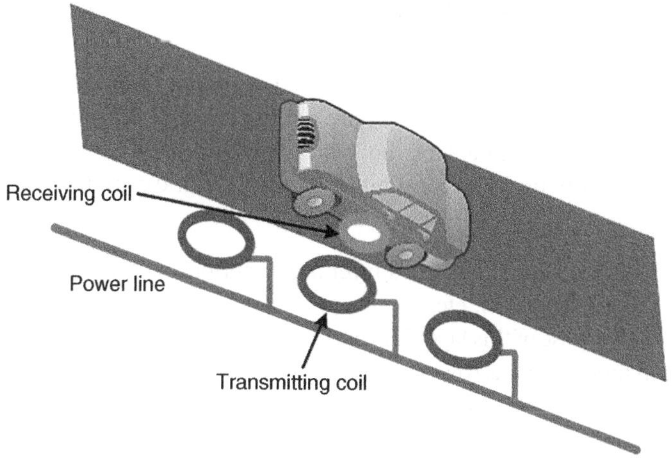

Figure 12.14 Dynamic wireless charging.

Advanced Institute of Science and Technology (KAIST) in South Korea in the year 2013. The developed technology is named the online electric vehicle (OLEV), which was tested on buses. The OLEV system was first implemented in a public transportation setting in the city of Gumi, South Korea for a 48 km route. The schematic diagram of dynamic WPT is shown in Figure 12.14. However, dynamic WPT is still in the experimental stage.

There are certain advantages of dynamic WPT. Vehicles can be charged while in motion, reducing the need for long charging stops and increasing operational efficiency. It effectively extends the range of operation of BEVs, thereby eliminating the range anxiety of the drivers. Since the vehicles can be charged when they are driven, smaller batteries will be needed, bringing the purchase cost of EVs down. Moreover, continuous charging can optimize energy use. The vehicles are charged only when needed, reducing the energy lost in idle battery storage.

There are also major drawbacks of dynamic WPT. It will require very high infrastructure costs because the coils, power lines, and transmitters must be embedded in the roadways. This will also incur high maintenance costs. Dynamic WPT is also less efficient than wired charging methods as energy losses can occur during power transfer, especially when the vehicles travel at high speeds. There are also alignment issues between the charging and receiving coils. Unless they are perfectly aligned, which itself is a tough ask, there will be energy losses, reducing efficiency further. Moreover, road safety and electromagnetic interference are critical issues that need to be addressed. Dynamic WPT technology with several applications and several projects are discussed in [20].

12.5 Smart Grid Communications

Obviously, the smart grid cannot function without effective and efficient communication systems. Communication networks are essential for the effective operation, monitoring, and management of the grid, allowing utilities to optimize energy production, distribution, and consumption. The smart grid communication architecture can have three layers. These are [21]:

- *Application Layer*: The functions of this layer are to perform DR, outage management, AMI, fraud detection, etc.
- *Power Layer*: The purpose of this layer is to ensure that generation, transmission, and distribution function cohesively together. This is a two-way communication system between the utilities and customers, including power management and control.
- *Communication Layer*: This is the heart of the communication system that provides the interconnections between all the systems and devices.

Smart grid communication is either wired or wireless. Depending on the scope of applications, there are several communication technologies available. Some of them are discussed below.

12.5.1 Smart Grid Communication Mediums

Smart grid communication mediums can either be wired or wireless. Among wired technologies, the medium of communication can either be copper- or fiber optic-based with technologies such as power line communication (PLC) and digital subscriber line (xDSL). Examples of last-mile wireless communication mediums include ZigBee, cellular communications, and Wi-Fi. Depending on the scope of applications, there are several communication technology mediums available. Some of them are discussed below:

PLC: It uses the power lines to transfer data from sender to receiver. In this technique, communication signals are injected into the power lines by modulating the data onto a carrier frequency, thereby allowing data to flow through the line with the current. It is interesting to note that the modulated signal travels through the lines, transformers, and circuit breakers. At the receiving end, the signal is demodulated to extract the data. This has the advantage that no additional wiring is required.

PLCs usually operate at lower frequencies, between 3 kHz and 500 kHz. They have low data rates, however, have longer range as they are carried through power lines. They are suitable for smart meters, remote meter readings, SCADA systems, home automation, etc. However, the signal can attenuate while transmitted over long distances and its quality can degrade due to the interference with power line noises. Moreover, because the power lines are shared, cybersecurity becomes a critical issue for this type of communication. They can be prone to unauthorized access or data theft.

Since PLC is a cost-effective solution, there are efforts to improve its performance standard. To alleviate signal attenuation and noise interference, PLC is combined with wireless systems. Furthermore, different modulation schemes are under consideration to improve transmission reliability and data transmission rate.

xDSL: Another form of wired communication medium is xDSL, which provides high-speed internet connections over traditional telephone lines, where both voice and data can be transmitted. Usually, copper lines are used in DSL but advanced DSL technologies such as very high-speed DSL (VSDL) and G.fast (fast access to subscriber terminals that delivers ultra-fast internet over copper lines) complement fiber optic networks by using the copper medium for only last-mile connectivity.

There are several advantages of using DSL. They already exist to support legacy telephone infrastructure and, therefore, will not need large-scale infrastructure investments. Usually, the DSL connections are encrypted, which is helpful to counter cyberattacks. Furthermore, DSLs have built-in redundancies, that is, if a communication path fails there are other available paths. Therefore, DSLs are good candidates for applications in DR, SCADA, dynamic load management, and self-healing operations.

However, there are a few disadvantages of DSLs as well. They might have bandwidth limitations – this can be a constraint for some advanced smart grid functions. The data rate (speed) can degrade over distance. Thus, the nodes that are further away from the central exchange to manage multiple DSL connections (also known as xDSL access multiplexer or DSLAM) can experience severe degradation. In addition, large telecommunication providers such as AT&T are phasing out this technology [22]. In this regard, fiber optic broadband ethernet connection is a much better option.

Fiber Optic Communication: Fiber optic communication networks represent another promising broadband medium with significant potential for serving as a communication backbone in smart grids. In this regard, fiber to the x or fiber in the loop (FTTx) serves as a last-mile communication medium that can carry more data and over larger distances than xDSL services. FFTx is commonly coupled with passive optical network (PON) based technology. PON provides data delivery using a point-to-multipoint design without the use of active electronic components in the distribution network. This results in simplified architecture and energy-efficient designs. Among PON technologies, Ethernet

Passive Optical Network (EPON) and Gigabit Passive Optical Network (GPON) are widely used. EPON adheres to Ethernet standards, making it cost-effective and versatile, while GPON offers higher data rates and more efficient bandwidth utilization, making it ideal for applications requiring robust performance and scalability. Together, these technologies play a critical role in enabling smart grid communication.

Overall, fiber optic communications provide high-speed, long-distance, and reliable communication networks. Given that fiber optical cables are made of glass or plastic, they serve as excellent insulators and can operate reliability in high-voltage environments. Moreover, fiber optic networks are suitable for high-density areas and are also future-proof technology. However, deploying fiber optic networks can be prohibitively expensive, particularly in low-density or remote areas where the cost of installation and maintenance is significantly higher. In such cases, wireless communication mediums present a viable alternative, offering greater flexibility and cost-effectiveness for these challenging environments.

ZigBee: It is an IEEE 802.15.4 standard-based wireless communication protocol designed for HANs in smart grids. It is inexpensive, reliable, and has lower power usage but a low data rate. Since ZigBee devices use very low power, they can be powered by batteries that can last for a long time. ZigBee has a data rate of up to 250 kbps and a range between 10 m and 100 m. Several ZigBee devices can also be meshed to form a network with lower power consumption. In a ZigBee mesh, a device can be connected to several other devices (typically ZigBee device has 16 channels). In this way, the coverage through ZigBee can be extended. Moreover, this also enables the mesh to self-heal – for instance, a router shifts connection to an alternate path in the mesh when the original path fails. ZigBee also has advanced security features such as encryption and authentication to prevent data breaches.

ZigBee can also be used in home automation to control smart devices lights and thermostats. It can facilitate communication between smart meters and appliances through energy management systems. Therefore, ZigBee is suitable for DR, energy monitoring, and energy usage management. However, ZigBee operates in a 2.4 GHz band. In this frequency range, it can face interference from other wireless devices such as Wi-Fi or Bluetooth. Furthermore, the low data rate makes ZigBee unsuitable for high-speed applications.

Cellular Communication: This is advantageous for the smart grid due to its ubiquitous usage – the technology exists all over the world. Technologies such as 3G, 4G, or 5G provide reliable wide-area communication between the various components of the smart grid. This technology can be used for real-time monitoring and control of the energy delivery system. Cellular networks have the potential to provide the backbone of several smart grid applications, such as AMI, distribution automation, DR, renewable energy integration, and monitoring the health of assets such as transformers and transmission and distribution lines.

Cellular networks have several advantages for smart grid applications, such as:

- *Wide Coverage*: Such networks are already available in urban, semi-urban, and rural areas. In general, cellular coverage is about 95%, which can be very useful for several smart grid applications.
- *Reliability*: Usually, the cellular networks are fairly reliable and are backed up with cybersecurity measures such as encryption and authentication. However, Rogers Telecom had a service outage in July 2022 that impacted 12 million users. Similarly, Optus Telecom had a security problem in September 2022 that affected about 10 million users. The latter catastrophe happened due to a code error that had not been noticed for several years.
- *Scalability*: The number of cell phone users is continuously growing. Cellular companies can scale up their systems to cater to increased data traffic.
- *Cost-effectiveness*: Cellular networks already exist. Therefore, utility companies do not need to set up the infrastructure required.

There are, however, some of the disadvantages of cellular networks. 3G and 4G networks have low latency. Furthermore, smart grid requires mission-critical applications that require ultra-reliable low latency. These include the prevention of cyber-physical attacks on smart grids. Such requirements also cannot be supported even by 5G networks [23]. Other challenges include coverage gaps, especially in the rural areas. Moreover, cellular networks can become congested in densely populated urban areas or during natural disasters, such as floods, earthquakes, or bushfires. This makes them unsuitable for mission-critical applications. However, it is expected that with the advancement in 6G networks, these requirements can be fulfilled, thereby making cellular networks viable for smart grids.

Wi-Fi: Another viable communication medium is the use of Wi-Fi, which is particularly effective for short-to-medium range devices. It operates on unlicensed frequency bands of 2.4 GHz and 5 GHz and offers several advantages, including simple and cost-effective deployments. This practicality makes it suitable for monitoring and energy management applications.

However, Wi-Fi also has several limitations, including a limited range. Moreover, depending upon the environmental factors, Wi-Fi can be susceptible to environmental challenges such as physical obstruction and the use of other household wireless transmitting devices such as microwaves or Bluetooth devices. Further, a large number of connected devices can congest the network and degrade its performance. Additionally, Wi-Fi is prone to security challenges and requires robust encryption to prevent unauthorized access.

Despite this, due to its practicality and versatility, Wi-Fi remains a viable medium of connectivity for smart grid applications.

12.5.2 Communication Requirements

Smart grid communication requirements depend on factors such as speed, data rate, and reliability. Some of the applications will require low latency and high data rates, while some other functions can be performed satisfactorily with low data rates. Latency is defined as the time delay in transmitting data from the source to the destination. Data rate is the amount of data that can be transmitted per second and is given in bits per second (bps). Even with a high data rate, latency can be significantly high due to congestion in the transmission path. Some of the smart grid applications will require low latency, and high data transmission rate at the same time.

Real-time monitoring of a large power system, PMUs, RTUs, and IEDs will be used in EMS and SCADA systems. Secured data transmission with minimum latency will be required for this purpose. Traditionally, SCADA systems data are transmitted using PLCs, which have a low data rate. However, when PMUs are used, the latency must be low. Fiber optic or cellular technologies are best suited for this purpose.

The HEM system is shown in Figure 12.8. HEM coordinates the household appliances and exchanges data with smart meters. These require low data rate transmission over short distances. Thus, technologies such as ZigBee or DSL with Wi-Fi can be used for HEM applications. DR, on the other hand, coordinates between supply and demand. There are several aspects of a DR program. The communication technology will depend on the specific DR target application. However, low-speed communication and low latency will be sufficient for most applications.

The application-specific communication requirements are discussed in [21]. Different smart grid communication technologies and their standards are discussed in [24].

12.6 Smart Grid Standards

Smart grid is an evolving field. One of the most important aspects of the smart grid is that there must be uniform standards irrespective of the geographical locations. Several organizations, such as IEEE, IEC, the American National Standards Institute (ANSI), and the International Organization of Standardization (ISO), are working on the formulation of different standards and protocols. Some of these standards are as follows:

- *IEC 61850*: For substation automation (discussed in Chapter 6).
- *IEEE 1547 is the IEEE Standard for Interconnection and Interoperability of DERs with Associated Electric Power Interfaces*: It provides requirements for operations

of DERs such as operation, testing, maintenance, and safety, as well as abnormal conditions, power quality, and islanding issues. This standard is specifically targeted toward solar PVs, wind turbines, BESSs, MGs, etc.

- *NISTIR 7628 is the NIST Guidelines for Smart Grid Cybersecurity*: It is a three-volume report that presents the development of effective cybersecurity strategies for smart grids. Volume 1 provides an overview of the objectives of cybersecurity and architectural framework. Volume 2 focuses on privacy issues such as consumer data protection. Volume 3 presents technical guidance and data analysis.
- *NIST Framework and Roadmap of Smart Grid Interoperability Standards, Release 4.0*: The purpose of this standard is to provide a comprehensive strategy for interoperability of various resources in a smart grid in a secured fashion.
- *IEC 61970-301 is Called Common Information Model Base*: This is for the operations of all major objects for electric utilities. It aims at laying the foundation of a common language that can seamlessly be integrated into applications, irrespective of how data has been generated. It can enable different utility applications in SCADA and EMS such that they can communicate using a standard model.
- *IEEE 2030 is a Standard for Smart Grid Interoperability*: This standard focuses on the integration of energy technology and information technology for the operation of electric power systems, end-user applications, and loads. The purpose of this standard is to provide a guideline for a smart grid interoperability reference model for power systems using ICT.
- *IEEE 1901 is the Standard for Broadband over Power Line Networks*: Medium Access Control and Physical Layer Specifications. This is set for high-speed communication over power lines. This is for the advancement of PLC such that high-speed data can be transmitted over existing electrical wiring for a variety of applications, such as Internet of Things (IoT), HAN and DR, and smart grid communications.
- The purpose of OpenADR standard is to facilitate DR through automated communications between the utilities and customers to manage electricity consumption in response to grid conditions, market prices, and overloading issues.
- MultiSpeak is a standardized data exchange framework that is designed for the power sector. This standard is developed by the National Rural Electric Cooperative Association (NRECA) for small and medium-sized organizations. Through this standard, the different entities can communicate using different software applications.

There are other smart grid standards that are not widely used as comprehensive as the ones presented above. They cater to some of the niche areas of smart grid applications. Some of these standards are listed below:

- *IEEE C37.118*: Synchrophasor data communication protocol for real-time monitoring and analysis of power systems.
- *ANSI C12.22*: Protocol for AMI.

- *IEC 62325*: Standards for deregulated energy market communications.
- *IEEE 2030.5 (SEP 2.0)*: Smart energy profile protocol for integrating DERs and DR systems into the grid.
- *IEC 61850-7-420*: Extensions of IEC 61850 for modeling DERs such as solar PV, wind turbines, and storage systems.
- *IEC 62116*: Standard for testing the anti-islanding functionality of grid-connected inverters.

12.7 Smart Distribution Grids

Smart grid is an all-encompassing purpose for all facets of the power delivery system. It covers power generation, including renewables. Power transmission systems will include FACTS devices, power flow control, cybersecure state estimation, and PMU. In this chapter so far, we have discussed distribution system applications such as DR. However, many renewable energy sources are being integrated into distribution systems. Therefore, there are a couple of issues that need to be addressed for the successful application of smart grid. One of them is the power quality problem that occurs due to the use of power converters and uneven single-phase PV injection. Ironically, the power quality can be improved by power converters as well (see [25, 26]). The other issue relates to the network structure and behavior is when many renewable energy sources get connected to the system. Since these sources are distributed throughout the network, they are usually referred to as distributed generators (DGs). Along with DGs, energy storage devices are increasingly being used in distribution systems. Together DGs and energy storages are termed as DERs. There are a couple of smart distribution grid strategies that will be discussed in this section.

12.7.1 Virtual Power Plant (VPP)

In essence, a virtual power plant is an algorithm for controlling and integrating decentralized DERs that can be spread over a large area such that they work together like a single power plant. The main aim of a VPP is to optimize the balance of supply and demand by optimizing energy production through diverse energy sources such as rooftop PV, battery storage systems, and EVs. Therefore, a VPP can be viewed as an energy management system for distributed DERs.

A review of energy management through VPP is given in [27], which defines the following properties of VPP:

- It is a cloud-based platform that aggregates DERs and loads.
- It uses a bidirectional communication system.
- It monitors the system in real time.

- It can trade energy among the customers and prosumers (this term is a blend of producer and customers, such as owners of rooftop PVs), as well as with utility.
- Its aim is to establish a decentralized power plant through efficient energy management through aggregation of DERs.

The main idea behind a VPP is that through the aggregation of several sources, a VPP can quickly respond to the demand and supply that can vary due to the presence of intermittent renewable generation. Thereby, a VPP can defer infrastructure upgrades and can also provide an effective spinning reserve. This is achieved by several software applications, with inputs from sensors that supply data from each connected resource and their status and load consumptions. VPPs will also have to participate in power markets such that their operations are cost-effective and profitable. A review of the market opportunities of VPP is presented in [28].

While VPP offers numerous advantages, it has several drawbacks. It had to coordinate with numerous DERs and loads in an interconnected network using a complex bidirectional communication system. Therefore, it can be prone to a single point of failure. The entire operation can be disrupted if one link goes down, as the VPP will not be able to balance demand and supply under this condition. Furthermore, VPPs may not be supported in energy markets, which will limit their revenue stream and adoption. Moreover, VPPs can be prone to cyberattacks because they involve several nodes and ICT infrastructure. Breaching a single node, an attacker can disrupt the operation of the entire system, taking down the power supply. Several customers and prosumers must participate to set up a single VPP. Usually, customers are very conservative to trial out a hitherto untried technology such as VPP. Therefore, VPP installers must overcome this barrier. Finally, setting up a VPP requires suitable coordination software, protection of each node from cyberattacks, and regular maintenance of physical and digital assets. The cost thus can be enormous and may not be beneficial for the installers and customers.

There have been some VPP trials and deployments recently. Some of these are as follows.

- *Sunrun VPP in California*: In 2023, this VPP aggregated more than 16,200 customers coordinating over 32 MW during peak hours to support the California electricity grid.
- *Tesla VPP in South Australia*: This VPP aims to cater to 50,000 customers using 7,000 Tesla Powerwall batteries with an installed capacity of 35 MW.
- *AGL VPP in South Australia*: In this trial, solar PVs and battery storage systems are connected to supply 1,000 domestic and business customers in Adelaide, South Australia supporting 5 MW of power. The project is partly funded by Australian Renewable Energy Agency (ARENA).
- *Kansai Electric Power Company (KEPCO) in Osaka, Japan*: KEPCO trial is to optimally use residential PVs and batteries in HEMS that will be connected to the *Shizen Connect* VPP platform.

Several other VPP trials are being considered. However, the outcomes of these trials will be known in the future.

12.7.2 Microgrid (MG)

MG, as the name suggests, is a small power system that mimics a large power system. It will have its own generators (mostly renewable) and loads. Figure 12.15 shows the schematic diagram of an MG that is connected to a substation. The junction between the utility substation and MG is called the point of common coupling (PCC) and it is controlled by a switch. When the switch is closed, the MG operates in the grid-connected mode, while it operates in islanded or autonomous mode when the switch is open. An MG can have PVs, wind turbines, hydrogen fuel cells, microturbines (MTs), diesel generators (DGens) (let us assume using biodiesel), and battery storage units. For a utility, an MG is a controllable load that can change its power demand depending on the prevailing conditions.

An MG can have different types of generators, some of which can be inertial, and some others can be non-inertial connected through converters. Again, some of the DERs can be dispatchable where generation can be increased or decreased on demand such as DGens, MTs, fuel cells, and batteries. On the other hand, solar and wind generators usually work in maximum power point tracking mode and only generate power depending on the prevailing weather conditions. An MG must meet its load demand in the islanded mode of operation. Thus, DR and smart load-shedding strategies must be included in the MG operation. Therefore, MG installations require meticulous planning and organization.

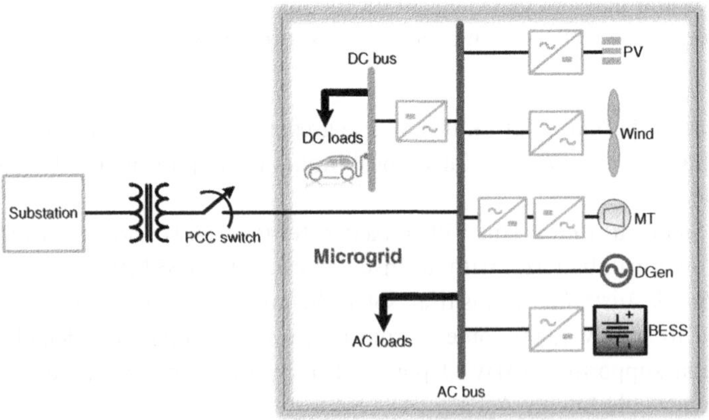

Figure 12.15 Schematic diagram of a utility-connected microgrid.

Interestingly, MGs are planned and installed for different purposes depending on the requirements of customers or installers. Some of the different types of MG installations are given below:

- *Community MGs*: They are installed to serve communities that are near the fringe of grids without having a reliable utility power supply.
- *Commercial MGs*: These serve large businesses, factories, and maybe hospitals so that power supply continuity is maintained for critical operations. For example, remote mines in Australia need their captive power supply as the grid connection is not reliable in some places.
- *Utility MGs*: These are installed by utility companies to enhance grid reliability and manage peak demand.
- *Campus MGs*: Mostly installed at university campuses to support research and development and to improve sustainability.
- *Military MGs*: The purpose of these is to protect command and control centers in the face of external threats. These will require the highest possible cybersecurity and can go off the grid in case of an eventuality.
- *Remote Area or Island MGs*: Many remote areas and small islands do not have any utility connections. Most of them rely on DGens for electricity supply, which is expensive to transport and damaging to the environment. Thus, these MGs are powered by renewable sources with possible diesel backup.

12.7.3 Microgrid Control

MG control is a complex process. Usually, it is divided into three layers: primary, secondary, and tertiary. The primary control is responsible for the control of the DERs in an MG and hence requires a fast response time. Usually, the so-called droop control, as has been discussed in Chapter 4, is employed as the primary controller. The main aim of droop control is to distribute power among the DERs in proportion to their ratings. There are several approaches to this. For high inductive, low resistance lines, P–f and Q–V droop lines are used, as shown in Figure 12.16 (a). However, distribution lines are highly resistive, therefore Q–f and P–V droop control regimes are preferred. This is shown in Figure 12.16 (b). However, resistive lines can be forced to behave as inductive by introducing a fictitious virtual inductance, as discussed in [29]. The droop gain selection is discussed in detail in [26].

The secondary control, which basically is the energy management system, is slower than the primary controller. It coordinates the primary controllers by sending commands to them to compensate for voltage and frequency deviations. Additionally, the secondary level controller can consider load and weather forecasting data to determine the best operating points of the DERs. The secondary controller does not interfere with the function of the primary controller, only improves its

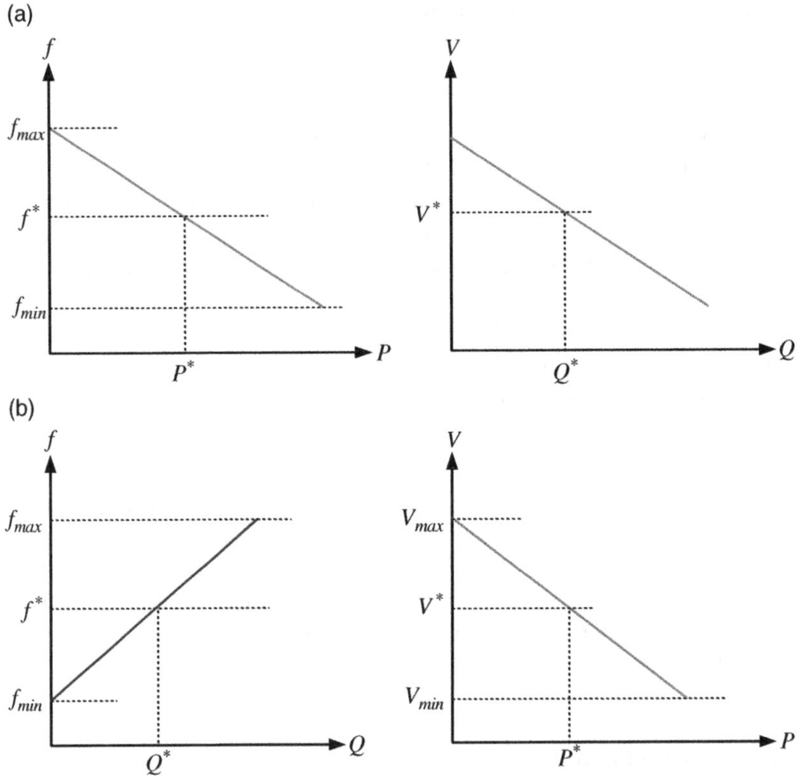

Figure 12.16 Microgrid droop control: (a) P–f and Q–V and (b) Q–f and P–V.

performance. Therefore, the data transfer rate is slow, and hence low bandwidth communication is sufficient for this level.

The tertiary level sits behind the secondary level and is mainly involved with grid operation. This level overseas the interactions between MGs, as well as between the MGs and the utility. It specifically deals with the MGs that are connected to utility systems so that the overall reliability and efficiency of the power grid are improved. Because this level is related to grid operations, it really is not an MG controller [30]. The graphical representation of the three-level control structure is shown in Figure 12.17.

There are several other topics on MGs, such as DC MGs [31] and networked MGs [32]. Some of these topics are covered in [26]. A white paper for MG for disaster management has been presented in [33]. Consider the distribution network shown in Figure 12.18, where ISS denotes an interconnecting static switch. In general, MGs are controlled by their primary controllers. Suppose a disaster has

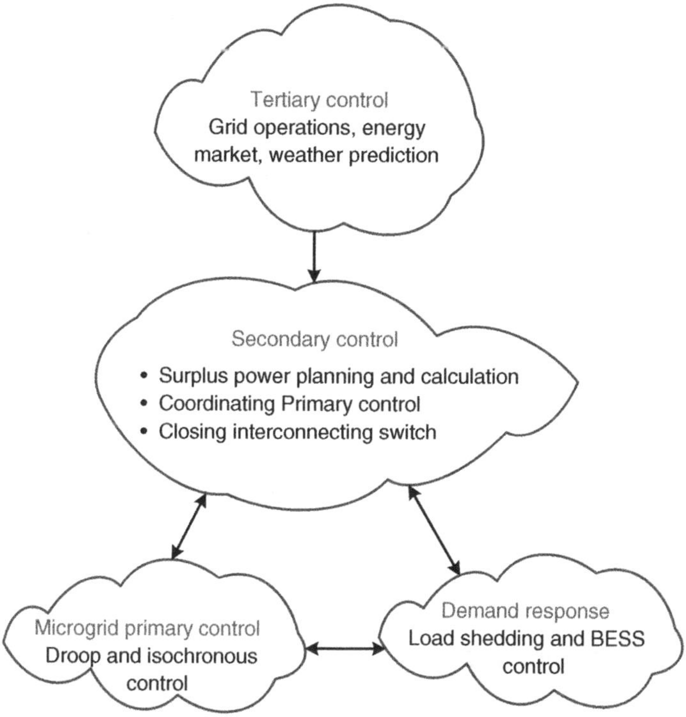

Figure 12.17 Three-level microgrid control.

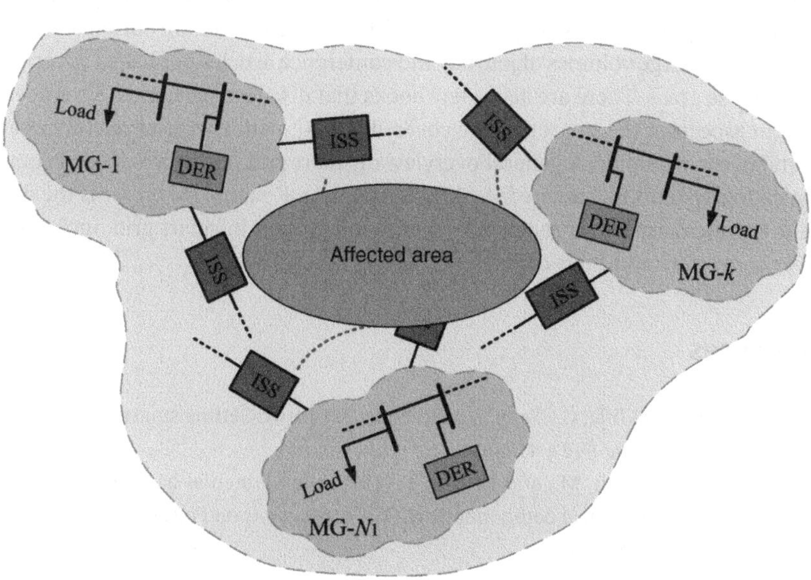

Figure 12.18 A distribution system affected by a disaster.

occurred affecting a certain area that is shown in the figure. Then a sequence of functions will be required for disaster mitigation. These are listed below:

- Those MGs that are not affected by the disaster can open the ISS to isolate from the grid.
- Subsequently, the tertiary controller first determines the area that is affected by the disaster.
- With the knowledge of the network topology, the tertiary controller determines which of the MGs can be reconnected to support a major number of loads. A graph-theoretic model or suitable machine-learning methods can be employed for this stage.
- The interconnecting switches of the chosen MGs can then be closed to restore power in different segments.

In this way, MGs can be used for self-healing the distribution network following a major outage. Note that even a single MG can be used to black start a power system following an outage, where larger generators can get synchronized to the power system using the reference set by the MG.

12.8 Concluding Remarks

Smart grid is an evolving area of research and development. It not only involves finding better utilization of power resources but also includes several other areas such as communication, cybersecurity, optimization, machine learning, and AI tools, including IoT. In this chapter, some of the basic concepts are outlined. However, there are large volumes of journal and conference articles published covering each of these areas. There are also a few books that discuss smart grids. Seventeen different aspects of the smart grid are covered in [34], with a large of references at the end of each chapter. A general overview of smart grid in the presence of communication systems is presented in [35]. The book by Kehani [36] presents the different technical aspects of renewable energy integration in smart grid, including power converters and their control.

References

1 Santacana, E., Rackliffe, G., Tang, L., and Feng, X. (2010). Getting smart. *IEEE Power and Energy Magazine* 8 (2): 41–48.

2 Phadke, A.G., Ibrahim, M., and Hlibka, T. (1977). Fundamental basis for distance relaying with symmetrical components. *IEEE Transactions on Power Apparatus and Systems* PAS-96: 635–646.

3 A. G. Phadke, T. Hlibka, M. Ibrahim and M. G. Adamiak, "A microcomputer based symmetrical component distance relay," *IEEE Conference Proceedings Power Industry Computer Applications Conference (PICA-79)*, Clevland, Ohio, 1979.
4 Phadke, A.G. (1993). Synchronized phasor measurements. *IEEE Computer Applications in Power Magazine* 6: 10–15.
5 Ghosh, A. and Ledwich, G. (2002). *Power Quality Enhancement using Custom Power Devices*. Norwell, MA: Kluwer Academic Publishers.
6 Begovic, M., Novosel, D., Karlsson, D. et al. (2005). Wide-area protection and emergency control. *Proceedings of the IEEE* 93 (5): 876–891.
7 Ledwich, G. and Vahidnia, A. (2021). *Phasors for Measurement and Control*. Springer.
8 National Energy Technology Laboratory, *Advanced Metering Infrastructure*, White Paper, Available https://www.smart-energy.com/wp-content/uploads/i/AMI%2520White%2520paper%2520final%2520021108%2520%25282%2529%2520APPROVED_2008_02_12.pdf
9 Sui, H., Wang, H., Lu, M.S., and Lee, W.J. (2009). An AMI system for the deregulated electricity markets. *IEEE Transactions on Industry Applications* 45 (4): 2104–2108.
10 Brooks, A., Lu, E., Reicher, D. et al. (2010). Demand dispatch: using real-time control of demand to help balance generation and load. *IEEE Power and Energy Magazine* 8: 21–29.
11 Pipattanasomporn, M., Kuzlu, M., and Rahman, S. (2012). An algorithm for intelligent home energy management and demand response analysis. *IEEE Transactions on Smart Grid* 3 (4): 2166–2173.
12 Shao, S., Pipattanasomporn, M., and Rahman, S. (2013). Development of physical-based demand response-enabled residential load models. *IEEE Transactions on Power Systems* 28 (2): 607–614.
13 Shahnia, F., Wishart, M.T., Ghosh, A. et al. (2012). Smart demand side management of low-voltage distribution networks using multi-objective decision making. *IET Generation, Transmission and Distribution* 6 (10): 986–1000.
14 A. Ashourpouri, *Demand Dispatch Control for Balancing Load with Generation*, Ph. D. Thesis, Curtin University, 2018.
15 Liu, J., Xiao, Y., Li, S. et al. (2012). Cyber security and privacy issues in smart grids. *IEEE Communications Surveys & Tutorials* 14 (4) Fourth Quarter: 981–997.
16 Liu, Y., Ning, P., and Reiter, M.K. (2011). False data injection attacks against state estimation in electric power grids. *ACM Transactions on Information and System Security* 14 (1) Article 13: 13.1–13.33.
17 R. B. Bobba, K. M. Rogers and Q. Wang, et al., "Detecting false data injection attacks on dc state estimation," *First Workshop on Secure Control System (SCS 2010)*, Stockholm, Sweden, 2010.

18 Yohanandhan, R.V., Elavarasan, R.M., Manoharan, P., and Mihet-Popa, L. (2020). Cyber-physical power system (CPPS): a review on modeling, simulation, and analysis with cyber security applications. *IEEE Access* 8: 151019–151064.

19 Mahesh, A., Chokkalingam, B., and Mihet-Popa, L. (2021). Inductive wireless power transfer charging for electric vehicles – a review. *IEEE Access* 9: 137667–137713.

20 Mi, C.C., Buja, G., Choi, S.Y., and Rim, C.T. (2016). Modern advances in wireless power transfer systems for roadway powered electric vehicles. *IEEE Transactions on Industrial Electronics* 63 (10): 6533–6545.

21 Gungor, V.C., Sahin, D., Kocak, T. et al. (2013). A survey on smart grid potential applications and communication requirements. *IEEE Transactions on Industrial Informatics* 9 (1): 28–42.

22 Suhaimy, N., Radzi, N.A.M., Ahmad, W.S.H.M.W. et al. (2022). Current and future communication solutions for smart grids: a review. *IEEE Access* 10: 43639–43668.

23 Tariq, M., Ali, M., Naeem, F., and Poor, H.V. (2021). Vulnerability assessment of 6G-enabled smart grid cyber–physical systems. *IEEE Internet of Things Journal* 8 (7): 5468–5475.

24 Gungor, V.C., Sahin, D., Kocak, T. et al. (2011). Smart grid technologies: communication technologies and standards. *IEEE Transactions on Industrial Informatics* 7 (4): 529–539.

25 Ghosh, A. and Ledwich, G. (2002). *Power Quality Enhancement using Custom Power Devices*. New York, MA: Springer Science+Business Media.

26 Ghosh, A. and Zare, F. (2023). *Control of Power Electronic Converters with Microgrid Applications*. Hoboken, New Jersey: IEEE Press-Wiley.

27 Rouzbahani, H.M., Karimipour, H., and Lei, L. (2021). A review on virtual power plant for energy management. *Sustainable Energy Technologies and Assessments* 47: https://doi.org/10.1016/j.seta.2021.101370.

28 Naval, N. and Yusta, J.M. (2021). Virtual power plant models and electricity markets – a review. *Renewable and Sustainable Energy Reviews* 149: https://doi.org/10.1016/j.rser.2021.111393.

29 Guerrero, J.M., de Vicuña, L.G., Castilla, M., and Miret, J. (2004). A wireless controller to enhance dynamic performance of parallel inverters in distributed generation systems. *IEEE Transactions on Power Electronics* 19 (5): 1205–1213.

30 Olivares, D.E., Mehrizi-Sani, A., Etemadi, A.H. et al. (2014). Trends in microgrid control. *IEEE-PES Task Force on Microgrid Control, IEEE Transactions on Smart Grid* 5 (4): 1905–1919.

31 Kumar, D., Zare, F., and Ghosh, A. (2017). DC microgrid technology: system architectures, ac grid interfaces, grounding schemes, power quality, communication networks, applications, and standardizations aspects. *IEEE Access* 5: 12230–12256.

32 Alam, M.N., Chakrabarti, S., and Ghosh, A. (2019). Networked microgrids: state-of-the-art and future perspectives. *IEEE Transactions on Industrial Informatics* 15 (3): 1238–1250.

33 International Electrotechnical Committee (IEC), "Microgrids for disaster preparedness and recovery with electricity continuity plans and systems," White Paper, 2013. Available http://www.iec.ch/whitepaper/microgrids/?ref=extfooter

34 Refaat, S.S., Ellabban, O., Bayhan, S. et al. (2021). *Smart Grid and Enabling Technologies*. Hoboken, New Jersey: IEEE Press-Wiley.

35 Bush, S.P. (2014). *Smart Grid Communication-Enabled Intelligence for the Electric Power Grid*. Hoboken, New Jersey: IEEE Press-Wiley.

36 Kehani, A. (2019). *Design of Smart Power Grid Renewable Energy Systems*, 3rde. Hoboken, New Jersey: John Wiley & Sons.

Index

Modern Power System, First Edition. Arindam Ghosh.

Published 2026 by John Wiley & Sons, Inc.
Companion website: www.wiley.com/go/ghoshpowersystem

IEEE Press Series on Power and Energy Systems

Series Editor: Ganesh Kumar Venayagamoorthy, Clemson University, Clemson, South Carolina, USA.

The mission of the IEEE Press Series on Power and Energy Systems is to publish leading-edge books that cover a broad spectrum of current and forward-looking technologies in the fast-moving area of power and energy systems including smart grid, renewable energy systems, electric vehicles and related areas. Our target audi- ence includes power and energy systems professionals from academia, industry and government who are interested in enhancing their knowledge and perspectives in their areas of interest.

1. *Electric Power Systems: Design and Analysis, Revised Printing*
   Mohamed E. El-Hawary
2. *Power System Stability*
   Edward W. Kimbark
3. *Analysis of Faulted Power Systems*
   Paul M. Anderson
4. *Inspection of Large Synchronous Machines: Checklists, Failure Identification, and Troubleshooting*
   Isidor Kerszenbaum
5. *Electric Power Applications of Fuzzy Systems*
   Mohamed E. El-Hawary
6. *Power System Protection*
   Paul M. Anderson
7. *Subsynchronous Resonance in Power Systems*
   Paul M. Anderson, B.L. Agrawal, and J.E. Van Ness
8. *Understanding Power Quality Problems: Voltage Sags and Interruptions*
   Math H. Bollen
9. *Analysis of Electric Machinery*
   Paul C. Krause, Oleg Wasynczuk, and S.D. Sudhoff
10. *Power System Control and Stability, Revised Printing*
    Paul M. Anderson and A.A. Fouad
11. *Principles of Electric Machines with Power Electronic Applications, Second Edition*
    Mohamed E. El-Hawary
12. *Pulse Width Modulation for Power Converters: Principles and Practice*
    D. Grahame Holmes and Thomas Lipo

13. *Analysis of Electric Machinery and Drive Systems, Second Edition*
Paul C. Krause, Oleg Wasynczuk, and S.D. Sudhoff
14. *Risk Assessment for Power Systems: Models, Methods, and Applications*
Wenyuan Li
15. *Optimization Principles: Practical Applications to the Operations of Markets of the Electric Power Industry*
Narayan S. Rau
16. *Electric Economics: Regulation and Deregulation*
Geoffrey Rothwell and Tomas Gomez
17. *Electric Power Systems: Analysis and Control*
Fabio Saccomanno
18. *Electrical Insulation for Rotating Machines: Design, Evaluation, Aging, Testing, and Repair*
Greg C. Stone, Edward A. Boulter, Ian Culbert, and Hussein Dhirani
19. *Signal Processing of Power Quality Disturbances*
Math H.J. Bollen and Irene Y.H. Gu
20. *Instantaneous Power Theory and Applications to Power Conditioning*
Hirofumi Akagi, Edson H. Watanabe, and Mauricio Aredes
21. *Maintaining Mission Critical Systems in a 24/7 Environment*
Peter M. Curtis
22. *Elements of Tidal-Electric Engineering*
Robert H. Clark
23. *Handbook of Large Turbo-Generator Operation and Maintenance, Second Edition*
Geoff Klempner and Isidor Kerszenbaum
24. *Introduction to Electrical Power Systems*
Mohamed E. El-Hawary
25. *Modeling and Control of Fuel Cells: Distributed Generation Applications*
M. Hashem Nehrir and Caisheng Wang
26. *Power Distribution System Reliability: Practical Methods and Applications*
Ali A. Chowdhury and Don O. Koval
27. *Introduction to FACTS Controllers: Theory, Modeling, and Applications*
Kalyan K. Sen and Mey Ling Sen
28. *Economic Market Design and Planning for Electric Power Systems*
James Momoh and Lamine Mili
29. *Operation and Control of Electric Energy Processing Systems*
James Momoh and Lamine Mili
30. *Restructured Electric Power Systems: Analysis of Electricity Markets with Equilibrium Models*
Xiao-Ping Zhang
31. *An Introduction to Wavelet Modulated Inverters*
S.A. Saleh and M.A. Rahman

32. *Control of Electric Machine Drive Systems*
    Seung-Ki Sul

33. *Probabilistic Transmission System Planning*
    Wenyuan Li

34. *Electricity Power Generation: The Changing Dimensions*
    Digambar M. Tagare

35. *Electric Distribution Systems*
    Abdelhay A. Sallam and Om P. Malik

36. *Practical Lighting Design with LEDs*
    Ron Lenk and Carol Lenk

37. *High Voltage and Electrical Insulation Engineering*
    Ravindra Arora and Wolfgang Mosch

38. *Maintaining Mission Critical Systems in a 24/7 Environment, Second Edition*
    Peter Curtis

39. *Power Conversion and Control of Wind Energy Systems*
    Bin Wu, Yongqiang Lang, Navid Zargari, and Samir Kouro

40. *Integration of Distributed Generation in the Power System*
    Math H. Bollen and Fainan Hassan

41. *Doubly Fed Induction Machine: Modeling and Control for Wind Energy Generation Applications*
    Gonzalo Abad, Jesús López, Miguel Rodrigues, Luis Marroyo, and Grzegorz Iwanski

42. *High Voltage Protection for Telecommunications*
    Steven W. Blume

43. *Smart Grid: Fundamentals of Design and Analysis*
    James Momoh

44. *Electromechanical Motion Devices, Second Edition*
    Paul Krause Oleg, Wasynczuk, and Steven Pekarek

45. *Electrical Energy Conversion and Transport: An Interactive Computer-Based Approach, Second Edition*
    George G. Karady and Keith E. Holbert

46. *ARC Flash Hazard and Analysis and Mitigation*
    J.C. Das

47. *Handbook of Electrical Power System Dynamics: Modeling, Stability, and Control*
    Mircea Eremia and Mohammad Shahidehpour

48. *Analysis of Electric Machinery and Drive Systems, Third Edition*
    Paul C. Krause, Oleg Wasynczuk, S.D. Sudhoff, and Steven D. Pekarek

49. *Extruded Cables for High-Voltage Direct-Current Transmission: Advances in Research and Development*
    Giovanni Mazzanti and Massimo Marzinotto

50. *Power Magnetic Devices: A Multi-Objective Design Approach*
    S.D. Sudhoff

51. *Risk Assessment of Power Systems: Models, Methods, and Applications, Second Edition*
    Wenyuan Li
52. *Practical Power System Operation*
    Ebrahim Vaahedi
53. *The Selection Process of Biomass Materials for the Production of Bio-Fuels and Co-Firing*
    Najib Altawell
54. *Electrical Insulation for Rotating Machines: Design, Evaluation, Aging, Testing, and Repair, Second Edition*
    Greg C. Stone, Ian Culbert, Edward A. Boulter, and Hussein Dhirani
55. *Principles of Electrical Safety*
    Peter E. Sutherland
56. *Advanced Power Electronics Converters: PWM Converters Processing AC Voltages*
    Euzeli Cipriano dos Santos Jr. and Edison Roberto Cabral da Silva
57. *Optimization of Power System Operation, Second Edition*
    Jizhong Zhu
58. *Power System Harmonics and Passive Filter Designs*
    J.C. Das
59. *Digital Control of High-Frequency Switched-Mode Power Converters*
    Luca Corradini, Dragan Maksimoviæ, Paolo Mattavelli, and Regan Zane
60. *Industrial Power Distribution, Second Edition*
    Ralph E. Fehr, III
61. *HVDC Grids: For Offshore and Supergrid of the Future*
    Dirk Van Hertem, Oriol Gomis-Bellmunt, and Jun Liang
62. *Advanced Solutions in Power Systems: HVDC, FACTS, and Artificial Intelligence*
    Mircea Eremia, Chen-Ching Liu, and Abdel-Aty Edris
63. *Operation and Maintenance of Large Turbo-Generators*
    Geoff Klempner and Isidor Kerszenbaum
64. *Electrical Energy Conversion and Transport: An Interactive Computer-Based Approach*
    George G. Karady and Keith E. Holbert
65. *Modeling and High-Performance Control of Electric Machines*
    John Chiasson
66. *Rating of Electric Power Cables in Unfavorable Thermal Environment*
    George J. Anders
67. *Electric Power System Basics for the Nonelectrical Professional*
    Steven W. Blume
68. *Modern Heuristic Optimization Techniques: Theory and Applications to Power Systems*
    Kwang Y. Lee and Mohamed A. El-Sharkawi
69. *Real-Time Stability Assessment in Modern Power System Control Centers*
    Savu C. Savulescu

70. *Optimization of Power System Operation*
Jizhong Zhu

71. *Insulators for Icing and Polluted Environments*
Masoud Farzaneh and William A. Chisholm

72. *PID and Predictive Control of Electric Devices and Power Converters Using MATLAB®/Simulink®*
Liuping Wang, Shan Chai, Dae Yoo, Lu Gan, and Ki Ng

73. *Power Grid Operation in a Market Environment: Economic Efficiency and Risk Mitigation*
Hong Chen

74. *Electric Power System Basics for Nonelectrical Professional, Second Edition*
Steven W. Blume

75. *Energy Production Systems Engineering Thomas*
Howard Blair

76. *Model Predictive Control of Wind Energy Conversion Systems*
Venkata Yaramasu and Bin Wu

77. *Understanding Symmetrical Components for Power System Modeling*
J.C. Das

78. *High-Power Converters and AC Drives, Second Edition*
Bin Wu and Mehdi Narimani

79. *Current Signature Analysis for Condition Monitoring of Cage Induction Motors: Industrial Application and Case Histories*
William T. Thomson and Ian Culbert

80. *Introduction to Electric Power and Drive Systems*
Paul Krause, Oleg Wasynczuk, Timothy O'Connell, and Maher Hasan

81. *Instantaneous Power Theory and Applications to Power Conditioning, Second Edition*
Hirofumi, Edson Hirokazu Watanabe, and Mauricio Aredes

82. *Practical Lighting Design with LEDs, Second Edition*
Ron Lenk and Carol Lenk

83. *Introduction to AC Machine Design*
Thomas A. Lipo

84. *Advances in Electric Power and Energy Systems: Load and Price Forecasting*
Mohamed E. El-Hawary

85. *Electricity Markets: Theories and Applications*
Jeremy Lin and Jernando H. Magnago

86. *Multiphysics Simulation by Design for Electrical Machines, Power Electronics and Drives*
Marius Rosu, Ping Zhou, Dingsheng Lin, Dan M. Ionel, Mircea Popescu, Frede Blaabjerg, Vandana Rallabandi, and David Staton

87. *Modular Multilevel Converters: Analysis, Control, and Applications*
Sixing Du, Apparao Dekka, Bin Wu, and Navid Zargari

88. *Electrical Railway Transportation Systems*
Morris Brenna, Federica Foiadelli, and Dario Zaninelli

89. *Energy Processing and Smart Grid*
James A. Momoh

90. *Handbook of Large Turbo-Generator Operation and Maintenance, 3rd Edition*
Geoff Klempner and Isidor Kerszenbaum

91. *Advanced Control of Doubly Fed Induction Generator for Wind Power Systems*
Dehong Xu, Dr. Frede Blaabjerg, Wenjie Chen, and Nan Zhu

92. *Electric Distribution Systems, 2nd Edition*
Abdelhay A. Sallam and Om P. Malik

93. *Power Electronics in Renewable Energy Systems and Smart Grid: Technology and Applications*
Bimal K. Bose

94. *Distributed Fiber Optic Sensing and Dynamic Rating of Power Cables*
Sudhakar Cherukupalli and George J. Anders

95. *Power System and Control and Stability, Third Edition*
Vijay Vittal, James D. McCalley, Paul M. Anderson, and A.A. Fouad

96. *Electromechanical Motion Devices: Rotating Magnetic Field-Based Analysis and Online Animations, Third Edition*
Paul Krause, Oleg Wasynczuk, Steven D. Pekarek, and Timothy O'Connell

97. *Applications of Modern Heuristic Optimization Methods in Power and Energy Systems*
Kwang Y. Lee and Zita A. Vale

98. *Handbook of Large Hydro Generators: Operation and Maintenance*
Glenn Mottershead, Stefano Bomben, Isidor Kerszenbaum, and Geoff Klempner

99. *Advances in Electric Power and Energy: Static State Estimation*
Mohamed E. El-hawary

100. *Arc Flash Hazard Analysis and Mitigation, Second Edition*
J.C. Das

101. *Maintaining Mission Critical Systems in a 24/7 Environment, Third Edition*
Peter M. Curtis

102. *Real-Time Electromagnetic Transient Simulation of AC-DC Networks*
Venkata Dinavahi and Ning Lin

103. *Probabilistic Power System Expansion Planning with Renewable Energy Resources and Energy Storage Systems*
Jaeseok Choi and Kwang Y. Lee

104. *Power Magnetic Devices: A Multi-Objective Design Approach, Second Edition*
Scott D. Sudhoff

105. *Optimal Coordination of Power Protective Devices with Illustrative Examples*
Ali R. Al-Roomi

106. *Resilient Control Architectures and Power Systems*
Craig Rieger, Ronald Boring, Brian Johnson, and Timothy McJunkin

107. *Alternative Liquid Dielectrics for High Voltage Transformer Insulation Systems: Performance Analysis and Applications*
Edited by U. Mohan Rao, I. Fofana, and R. Sarathi

108. *Introduction to the Analysis of Electromechanical Systems*
Paul C. Krause, Oleg Wasynczuk, and Timothy O'Connell

109. *Power Flow Control Solutions for a Modern Grid using SMART Power Flow Controllers*
Kalyan K. Sen and Mey Ling Sen

110. *Power System Protection: Fundamentals and Applications*
John Ciufo and Aaron Cooperberg

111. *Soft-Switching Technology for Three-phase Power Electronics Converters*
Dehong Xu, Rui Li, Ning He, Jinyi Deng, and Yuying Wu

112. *Power System Protection, Second Edition*
Paul M. Anderson, Charles Henville, Rasheek Rifaat, Brian Johnson, and Sakis Meliopoulos

113. *High Voltage and Electrical Insulation Engineering, Second Edition*
Ravindra Arora and Wolfgang Mosch

114. *Modeling and Control of Modern Electrical Energy Systems*
Masoud Karimi-Ghartemani

115. *Control of Power Electronic Converters with Microgrid Applications*
Armdam Ghosh and Firuz Zare

116. *Coordinated Operation and Planning of Modern Heat and Electricity Incorporated Networks*
Mohammadreza Daneshvar, Behnam Mohammadi-Ivatloo, and Kazem Zare

117. *Smart Energy for Transportation and Health in a Smart City*
Chun Sing Lai, Loi Lei Lai, and Qi Hong Lai

118. *Wireless Power Transfer: Principles and Applications*
Zhen Zhang and Hongliang Pang

119. *Intelligent Data Mining and Analysis in Power and Energy Systems: Models and Applications for Smarter Efficient Power Systems*
Zita Vale, Tiago Pinto, Michael Negnevitsky, and Ganesh Kumar Venayagamoorthy

120. *Introduction to Modern Analysis of Electric Machines and Drives*
Paul C. Krause and Thomas C. Krause

121. *Electromagnetic Analysis and Condition Monitoring of Synchronous Generators*
Hossein Ehya and Jawad Faiz

122. *Transportation Electrification: Breakthroughs in Electrified Vehicles, Aircraft, Rolling Stock, and Watercraft*
Ahmed A. Mohamed, Ahmad Arshan Khan, Ahmed T. Elsayed, and Mohamed A. Elshaer

123. *Modular Multilevel Converters: Control, Fault Detection, and Protection*
Fuji Deng, Chengkai Liu, and Zhe Chen

124. *Stability-Constrained Optimization for Modern Power System Operation and Planning*
Yan Xu, Yuan Chi, and Heling Yuan

125. *Interval Methods for Uncertain Power System Analysis*
Alfredo Vaccaro

126. *Practical Partial Discharge Measurement on Electrical Equipment*
Greg Stone, Andrea Cavallini, Glenn Behrmann, and Claudio Angelo Serafino

127. *Graph Database and Graph Computing for Power System Analysis*
Renchang Dai and Guangyi Liu

128. *The Power of Artificial Intelligence for the Next-Generation Oil and Gas Industry: Envisaging AI-inspired Intelligent Energy Systems and Environments*
Pethuru Raj Chelliah, Venkatraman Jayasankar, Mats Agerstam, B. Sundaravadivazhagan, and Robin Cyriac

129. *Microgrids: Theory and Practice*
Peng Zhang

130. *Electric Machinery and Drives: An Electromagnetics Perspective*
Nabeel A. O. Demerdash, JiangBiao He, and Hao Chen

131. *Smart Cyber-Physical Power Systems, Volume 1: Fundamental Concepts, Challenges, and Solutions*
Ali Parizad, Hamid Reza Baghaee, and Saifur Rahman

132. *Smart Cyber-Physical Power Systems, Volume 2: Solutions from Emerging Technologies*
Ali Parizad, Hamid Reza Baghaee, and Saifur Rahman

133. *Analysis of Electric Machinery and Drive Systems, Fourth Edition*
Paul C. Krause, Oleg Wasynczuk, Scott D. Sudhoff, and Steven D. Pekarek

134. *Understanding Electromagnetic Transients in Power System*
Luiz Cera Zanetta Jr.

135. *Communities for Clean Energy Justice and Equity in Grid Modernization*
Mohammadreza Daneshvar, Behnam Mohammadi-Ivatloo, and Amjad Anvari-Moghaddam

136. *Modern Power System*
Arindam Ghosh

Printed and bound by CPI Group (UK) Ltd, Croydon, CR0 4YY
17/09/2026
14964333-0006